World Sustainability Series

Series Editor

Walter Leal Filho, European School of Sustainability Science
and Research, Research and Transfer Centre "Sustainable Development
and Climate Change Management", Hamburg University of Applied Sciences,
Hamburg, Germany

Due to its scope and nature, sustainable development is a matter which is very interdisciplinary, and draws from knowledge and inputs from the social sciences and environmental sciences on the one hand, but also from physical sciences and arts on the other. As such, there is a perceived need to foster integrative approaches, whereby the combination of inputs from various fields may contribute to a better understanding of what sustainability is, and means to people. But despite the need for and the relevance of integrative approaches towards sustainable development, there is a paucity of literature which address matters related to sustainability in an integrated way.

Notes on the quality assurance and peer review of this publication

Prior to publication, the works published in this book are initially assessed and reviewed by an in-house editor. If suitable for publication, manuscripts are sent for further review, which includes a combined effort by the editorial board and appointed subject experts, who provide independent peer-review. The feedback obtained in this way was communicated to authors, and with manuscripts checked upon return before finally accepted. The peer-reviewed nature of the books in the "World Sustainability Series" means that contributions to them have, over many years, been officially accepted for tenure and promotion purposes.

Walter Leal Filho · Laís Viera Trevisan ·
Paul J. Pace · Mark Mifsud
Editors

Education for Sustainable Development: The Contribution of Universities

Springer

Editors
Walter Leal Filho
Faculty of Life Sciences
Hamburg University of Applied Sciences
Hamburg, Germany

Laís Viera Trevisan
Federal University of Rio Grande do Sul
Porto Alegre, Rio Grande do Sul, Brazil

Paul J. Pace
University of Malta
Msida, Malta

Mark Mifsud
University of Malta
Msida, Malta

ISSN 2199-7373 ISSN 2199-7381 (electronic)
World Sustainability Series
ISBN 978-3-031-86987-7 ISBN 978-3-031-86985-3 (eBook)
https://doi.org/10.1007/978-3-031-86985-3

This Springer imprint is published by the registered company Springer Nature Switzerland AG
The registered company address is: Gewerbestrasse 11, 6330 Cham, Switzerland

Contents

Climate Change Education and Gender Equity: A Challenge for Higher Education Institutions in Latin America

Juliana Restrepo Jaramillo, Ana Elena Builes-Vélez, Danny Jean Paul Mejía Holguín, Juan Diego Martínez Marín, and Luz Patricia Rave Herrera

Abstract Climate and environmental literacy are crucial for the upcoming environmental and technical decisions students will take as future professionals, such as architects, designers, and educators. Assume students do not understand the relationship between climate change and gender equity; in that case, there is no simple method to ensure they will take adequate actions against critical environmental and climate issues in their everyday lives. Therefore, this chapter aims to evidence the interconnectedness of climate change, education, and gender equity. The chapter is structured into three parts to achieve its goal. The first one explores education on climate change in Latin American Higher Education Institutions (HEIs). The second section discusses the theories of gender equity and equality and their—links with climate change education (CCE). The study's findings underscore the relevance of these issues, which align with SDGs 4, 5, and 13, while also addressing the challenges of incorporating climate change and gender equity conversations and initiatives within the university setting.

Keywords Climate change · Education · Latin America · Gender equity · SDGs

J. R. Jaramillo (✉) · A. E. Builes-Vélez · L. P. R. Herrera
School of Architecture and Design, Universidad Pontificia Bolivariana, Medellín, Colombia
e-mail: juliana.restrepo@upb.edu.co

A. E. Builes-Vélez
e-mail: ana.builes@upb.edu.co

L. P. R. Herrera
e-mail: patricia.rave@upb.edu.co

D. J. P. M. Holguín · J. D. M. Marín
School of Education and Pedagogy, Universidad Pontificia Bolivariana, Medellín, Colombia
e-mail: jean.mejia@upb.edu.co

J. D. M. Marín
e-mail: juandi.martinez@upb.edu.co

W. Leal Filho et al. (eds.), *Education for Sustainable Development: The Contribution of Universities*, World Sustainability Series,
https://doi.org/10.1007/978-3-031-86985-3_1

1 Introduction

Climate change (CC), gender equity, and equality are three of the issues of our time. Studies have shown that CC impacts are not distributed equally, and women experience more severe effects compared to men, particularly in terms of overall health and well-being (Rao et al. 2019a; b; Kwauk and Wyss 2022). Additionally, studies indicate that women tend to express greater concern about CC and are more likely to protest governmental inaction (Bush and Clayton 2023). Authors recognized that:

> Due to gender differences in work and consumption, men are more likely than women to perceive material costs from mitigation, both to themselves and to society as a whole. Moreover, climate change and decarbonization evoke considerable uncertainty about the future. (p. 591)

In this context, we can say that CC and gender are correlated regarding impacts, perception, and education. This means that women and men experience the effects of CC, interpret these effects, and—engage in education to mitigate and adapt in distinct ways. Gender disparities in climate concern and climate education correlate with socioeconomic development.

According to the Intergovernmental Panel on Climate Change (IPCC), Climate Change Education (CCE) "is about learning in the face of risk, uncertainty and rapid change" (2014). The IPCC 2022 report, which focuses on "impacts, adaptation and vulnerability" based on a significant number of studies, draws the importance of understanding the interconnection between social and physical location with climate, social, and gender justice. This chapter addresses the "adaptation gaps" between existing responses and those required to effectively tackle climate impacts. It emphasizes the importance of "inclusive, equitable, and just adaptation pathways" and highlights the need to closely consider the Sustainable Development Goals (SDGs), gender issues, and Indigenous and local knowledge and practices to bridge these gaps (IPCC 2022, 84).

Studies have highlighted the difficult interplay between education and climate change from a gender perspective, specifically examining the implications for girls' education, which serves as both a catalyst for and a response to climate change mitigation and adaptation (UNESCO 2020; UNDP 2012). Growing empirical evidence indicates increased attention to the connections between climate, gender, and education; however, more robust empirical studies, drawing closer and more nuanced links between them, remain limited (Lagi et al. 2023).

The Girls 'Education Challenge Report (2022), states that interventions that focus on girls considered to be more marginalized within education (including those with disabilities, who may be displaced by conflict, or who belong to Indigenous or ethnic minorities) are promoted as a means for social transformation. In this sense, scholars such as Kwauk and Braga (2017), Kwauk and Wyss (2022) and Winthrop and Kharas (2016), agree that other common topic in international policy discussions is the emphasis on girls' education as a vital component in addressing the climate crisis, with the potential to promote climate leadership (Winthrop and Kharas 2016) and develop and strengthen "their skills for a 'green' economy" (Kwauk and Braga 2017, 32).

The frequent emphasis on gender as a key factor in addressing climate change has sometimes reinforced a narrow view of girls' education, focusing mainly on its instrumental benefits as human capital. This perspective can place a disproportionate responsibility—and even burden—on girls (Hans et al. 2021; Eastin 2018), potentially limiting deeper, intersectional approaches to empower through climate action (Kaijser and Kronsell 2014).

To better understand these connections, some authors have proposed conceptual frameworks. For instance, Pankhurst (2022) has mapped the "multiple layered pathways between girls' education and climate change" (p. 17) illustrating how these pathways are shaped by both adverse impacts, such as forced displacement or infrastructure damage, and positive contributions, like greener behaviors and women's roles in disaster risk reduction. These pathways span across individual, community, school, and system levels, showing how education can intersect with climate mitigation and adaptation efforts.

The OCE (Office for Climate Education) promotes quality climate change education to induce behavioral changes in society necessary for the ecological transition. For Latin America, the OCE has launched a five-year pilot project to promote climate change education in the region (2021). The ALEC project (Latin America for Climate Education) intends to promote climate change literacy and education by adapting pedagogical resources and strategies to the LA context (OCE 2021). However, these efforts are not enough. According to the UN and ECLAC (2021), Latin America "continues to be the most unequal region in the world" (Aguilar 2022, 9). Inequality has grown because of the COVID-19 pandemic and has deepened poverty and equality, resulting in gender gaps and "gender-differentiated impacts of climate change" (Aguilar 2022, 9). It has shown that the climate, social, and economic crisis has affected men and women differently depending on the region in which they live, and the impact has been more significant in the countries of Latin America and the Caribbean.

In Latin American and Caribbean countries, there are challenges related to gender inequality and climate change. These challenges link the unbalanced socio-economic development of the region, the lack of climate literacy, and the lack of access to education in some of the region's populations, especially the most vulnerable ones. In addition, climate change mitigation and adaptation solutions in the region have focused on environmental and economic matters without considering human rights, girls and women's situations in vulnerable communities, and gender inequality, including the LGBT community's rights. Public policies and stakeholder discussions leave gender out of the analysis without considering the gender-differentiated effects and impacts of climate change on individuals and populations.

Although the outlook seems bleak, there is evidence of important work being done in the region by higher education institutions about climate literacy and gender equity. These efforts have been made through strategic alliances and the integration of pedagogical strategies that allow the integration of sustainability and climate change competencies, as will be seen in the following sections.

2 Methods and Instruments

This study employed a mixed-methods research approach, combining qualitative and quantitative data to comprehensively address the research goal which is to evidence the interconnectedness of climate change, education, and gender equity. The data, on these issues, were collected through a structured questionnaire featuring open-ended questions, and it was administered at Universidad Pontificia Bolivariana (UPB) in Medellín, Colombia, a higher education institution distinguished by its significant representation of women in research, teaching, and leadership roles. Therefore, the questionnaire was directed at women in leadership roles across various departments and units within the institution, providing insights into their experiences and contributions in these capacities.

Additionally, to ensure a comprehensive analysis, data triangulation was employed considering scientific publications authored by women, research projects led by women at UPB, over the past five years, and sustainability subjects also taught by women at the same institution. Thus, incorporating multiple data sources and collection methods enrich the exploration of the phenomena from varied perspectives, which enhances the study's reliability, reducing potential biases, and strengthens the validity of the findings on the constructs. Finally, the findings were examined using descriptive statistical methods and subsequently triangulated with secondary data sources, allowing an interpretation that supports a comprehensive understanding of the issues under study.

3 Education on Climate Change in Latin American Higher Education Institutions (HEIs)

As we are experiencing, the Latin America region is vulnerable to the effects of climate change, which can be seen through an increasing number of droughts, floods, and extreme weather events that affect not only millions of people in different ways but also flora and fauna (IPCC 2022; UNFCCC 2021). This vulnerability highlights climate education's importance in preparing future generations to mitigate and adapt to these challenges. Here, universities and other educational institutions play a crucial role in training teacher-researchers and professionals capable of developing groundbreaking and sustainable solutions.

Although in the LA region, there are around 3.800 universities (WHED 2023) just a small number of them have been working directly on education on climate change in the last decade; however, the interest in this topic has become relevant, scoping the idea that addressing the climate crisis is a must (Cruz Castaño and Páramo Bernal 2020). Although it is impossible to mention all of them, an analysis of 18 regional universities was conducted to describe what they are doing in terms of pedagogical approaches, challenges, and opportunities. The selected universities were Universidad de Buenos Aires (2023), Católica de Bolivia (2023), Costa Rica (2023),

Federal do Rio de Janeiro (2023), Instituto Tecnológico y de Estudios Superiores de Monterrey (2023), La Habana (2023), la República (2023), Los Andes (2023), Nacional Autónoma de México (2023), Nacional de Colombia (2023), Nacional de Córdoba (2023), Nacional Mayor de San Marcos (2023), Panamá (2023), Pontificia Bolivariana (2023), Pontificia Católica de Chile (2023), São Francisco (2023), and São Paulo (2023).

Regarding pedagogical approaches, these universities have integrated climate change into their curricula, not only in science, engineering, and economic areas but also in humanistic and social ones, to have a broader impact. Therefore, there is an interdisciplinary perspective where everybody can have a holistic view of the problem and work together to mitigate it (e.g., UNAM). They are also implementing Project-based learning, an approach that allows them to focus on contextual situations, theorize practice, and turn practice into theory to work with local communities and develop climate change adaptation solutions (e.g., USP). It is worth mentioning that they are also incorporating sustainability principles into their academic programs, which means they are teaching about climate change and promoting sustainable practices within their campuses, projecting that as a model for their communities (e.g., Uniandes and UPB).

Although these institutions look at climate change, they face different significant challenges that affect their proposals: they have limited resources, educational inequality, institutional and cultural resistance, and integration of local and traditional knowledge. It means that they operate with restricted budgets that reduce their possibilities to develop specialized programs, work with affected communities, and have access to advanced technology. These constraints affect both teacher-researches and the implementation of robust educational programs (Cruz Castaño and Páramo Bernal 2020). As for educational inequality, it could be said that there is a disparity between urban and rural universities and students which creates a climate education gap that must be addressed to ensure inclusive and equitable education (Munck et al. 2023). Universities in LA also confront institutional and cultural resistance rooted in culture, politics, and economic factors that prioritize other issues over the climate crisis to run their pedagogical proposals (UN—CEPAL 2023). The last constraint experienced is associated with the integration of local and traditional knowledge. They found that some communities have knowledge that can contribute to the adaptation and mitigation of climate change, but it is not integrated into formal curricula (McNamara and Buggy 2017).

Despite these challenges, and thanks to the efforts of different universities, obstacles are turning into opportunities to strengthen climate education. Nowadays, partnerships among institutions in other regions and with international agencies, like the Sustainable Development Solutions Network (SDSN), provide resources, knowledge, and experience to encourage better practices and initiatives. Local communities are engaged in climate change research and educational projects, which improve learning and strengthen the community's resilience. All these actions are guiding governments to design and implement public policies that promote climate change education as an essential factor for our survival, not only in higher education but all levels and types of education (Herrera-Franco et al. 2023).

Based on the Climate Change education in Latin American Universities is critical to addressing the region's environmental, social, and economic challenges. Despite the difficulties faced, LA universities are working on different strategies to train leaders and professionals who may mitigate climate change through innovative and sustainable solutions; however, to maximize that impact, universities need international collaboration, digital technology, community engagement, and supportive public policies to play the role of agents of change that may help to build a resilient and sustainable future. Thus, the convergence between academic, governmental, and community theories and practices around climate education prepares professionals and global citizens who may foster the capacities of communities to cope with and adapt to environmental changes.

4 Theory on Gender Equity and the Relation with Climate Change Education (CCE)

Globally, climate change affects women more than men (Senay 2016), especially in rural communities where life depends directly on working the land, which usually does not belong to them. This situation is not different in Colombia, where 63.7% of single-owner rural properties belong to men and only 36.3% to women (DANE 2022).

In addition to women having less access to land, their properties are smaller and used as family homes. In Colombia, the most gender-unequal regions regarding the distribution and exploitation of land are in the Amazon transition ring; this could relate to the use and exploitation of these territories, home to 40% of Colombia's forests. Many of these territories are also shared properties, some of them belonging to indigenous reservations, but even though there is a shared ownership, this does not guarantee the participation of women in decision-making about these assets, the administration of the money that is produced with them or the public representation of the family in business (DANE 2022). The deforestation of these areas is a historical problem sharpened by the pressure exerted by large landowners and ranchers and has increased after the peace agreements (Tamayo et al. 2020).

Additionally, 51.2% of women owners in Colombia declare that their main activity is domestic work (DANE 2022). In their role as custodians, their access to clean water, the availability of sanitation facilities, sewage systems, and alternatives to cooking with firewood are linked to their overall well-being. These circumstances are critical variables that increase climate change impacts, especially regarding how land is cropped and used. Most of these lands are used for livestock and planting of timber forests, which contribute to increased climate change impacts and significantly affect the environment and water sources in the territories.

Actions to adapt and mitigate climate change with a gender approach are primarily a state responsibility, providing economic resources for change and new legislations that allow women to improve their quality of life and empowerment in the use and access to land. While this happens, higher education institutions may have a crucial role in supporting women in their communities to gather information about how their livelihoods can be transformed by climate change. Likewise, education should motivate and prepare them to lead their communities. It will increase their opportunities to manage the ecosystem resources on which they depend and participate in the development of mitigation strategies and adaptation to change. This kind of analysis and decision-making will only be possible where extreme poverty is overcome. Therefore, entrepreneurship is also a key axis that can provide women with innovative alternatives to face their basic needs.

Some governments in Latin America and the Caribbean have made significant progress in developing strategies to include the gender perspective in actions against climate change. Perú is one of these countries. Their methods include the development of new national policies and developing plans that consider a gender perspective from their formulation. Training public officials in planning tools and proposing projects to face climate change with a gender perspective; promoting access to productive, financial, scientific-technological, and credit resources to foster the agricultural production and titling of land for women in poverty (ENCC 2015). In Costa Rica, they have implemented deliberative spaces between members of the government and vulnerable communities, especially women, to agree on actions against climate change, formulating policies that highlight women's efforts for social adaptation to new environmental conditions, systematizing and giving public access to information about the registration, monitoring, and forecasting of factors associated with climate change and its correlation with gender variables (MINAE 2018). In Mexico, they've included criteria for differentiated attention according to gender in their climate change policy design proposals. In Colombia, a proposal has also been designed, suggesting programs for adaptation to climate change with emphasis on a differentiated gender approach that takes into consideration the technical complexity, the resources necessary for its implementation and traceability, as well as the co-benefits that may arise from it (DNP 2016).

Likewise, a series of documents and presentations have been developed in Colombia to link the gender approach with actions against Climate Change. These documents include analyses of successful cases of adaptation and mitigation in other countries. There are also guides to understanding gender and how to incorporate this perspective in projects (MinAmbiente 2021).

Despite these governmental efforts, the public perception in a survey published in 2016 in Colombia reveals that they are not at all informed about Climate Change (75.11%). The survey indicates that people know about Climate Change mainly through television. Other sources mentioned are radio, newspapers, or the internet, but no reference is made to formal education at basic or advanced levels. TV news is usually the most recognized means of information about this phenomenon. For some of the participants in the survey, Universities develop research projects on Climate Change that frequently lead to scientific publications. Still, the possibility of their impact on non-expert populations is almost zero (IDEAM et al. 2016).

This lack of information about climate change and climate literacy cannot be addressed exclusively as a professorial transmission of knowledge; it is necessary to develop specific situated competencies and avoid abstraction since, although the majority agree on the need to reduce the impact of climate change, there may be a complete disagreement on how these challenges are specifically addressed, or how they should be prioritized (Leach 2015). In this sense, higher education institutions play a fundamental role in transforming society and are being called to start implementing pedagogical strategies that allow the training of men and women in climate change competencies and climate literacy.

5 Inclusion of the CCE and Gender Equity for Sustainable Development in Graduate and Undergraduate Programs at Universidad Pontificia Bolivariana

UPB has four chapters in the 2023 Sustainability Report (UPB 2024): Corporative Government, Collective Interests, Resilient Regeneration, and Conscious Leadership. Each chapter describes the institution's interests in achieving the SDGs, its contribution to them, and the different strategies and projects the institution has developed to implement SDGs at all levels and with all its stakeholders. To understand this report and its implications for this study, it is essential to analyze and provide all the relevant information and then put it in the context of a specific case study in the institution. The Corporative Government of UPB aims to form all educational agents regarding social responsibility, the importance of innovation, and institutional development to achieve its humanist approach. The Collective Interest chapter stimulates the necessary synergies with different agents to contribute to personal and collective social development. Resilient Regeneration introduces the report into the recovery capacity and adaptation of the institution in the challenges that appear in the new living styles and social perspectives in today's contexts. Conscious Leadership refers to the decision-making, the formation of all agents, the resources chain, and the construction of long-term strategies to achieve the SDGs.

According to the analysis made, the institution has eight focuses on the SDGs 4, 17, 3, 16, 6, 7, 13, 9. Still, the initiatives supporting SDG 5 are not referred to directly in the Report, even though the university has different strategies to contribute to it. Still, sometimes, these efforts are insufficient, or there are no means to determine their impact. These initiatives are very diverse; for example, the UPB hosted the SDG Woman Football World Cup, where girls' integration, organization, and empowerment around the SDG were promoted. The report also addresses the research efforts made by teachers and students to contribute to SDGs, mostly referring to health care in specific communities, 34% of the total new knowledge products.

However, further information on SDGs that are not prioritized is not always systematized. A previous study conducted by the authors shows that the courses in which SDGs or Sustainability theory is taught are few and, in most cases, depend on the teacher's voluntary efforts since they do not correspond to formal academic policies but to the teacher's individual criteria or interests (Builes-Vélez et al. 2024).

Regarding climate change, UPB has several initiatives, some of them related to campus administration, such as zero waste and carbon neutrality. Managers also pay special attention to campus biodiversity, which is one of the biggest green spaces in the city's center with more than 1400 trees and 20 different species of birds, among other types of animals.

According to the 2023 Sustainability Report, a survey applied to 19.765 people from different groups of interest in the academic community evaluated SDG 13 Climate Action Goal as the priority focus for community action. Information and actions to achieve this represent 80% of the last Sustainability Report. This reflects the effort made by the institution to mitigate climate change reality in the territories where the UPB has a presence. In this sense, the institution adheres to the definition of climate change, which states that:

> Climate change is caused by human activities and threatens life on Earth as we know it. With rising greenhouse gas emissions, climate change is occurring at rates much faster than anticipated. Its impacts can be devastating and include extreme and changing weather patterns and rising sea levels. (UN 2015)

According to this, the institution declares that actions to mitigate and adapt to climate change have become their main goal. To achieve this goal, the directions from the Research Vice-chancellor are to include these goals in every research proposal presented. This is linked to the institution´s prioritization of SDGs. The institution prioritizes SDG 7, 13, 4, and 6; they are listed in order of importance and percentage of development achieved in the institution (UPB 2024).

At the UPB, teachers, researchers, and administrative staff believe that the ways to accomplish these goals are related to research and projects developed in class, under the initiative of students, teachers, and administrative parties, who have personal interests in these matters related to climate change and gender equality, as the case of the in-class project developed by the Graphic Design Faculty at the UPB, which we present as follows.

5.1 A Case Study in Education for Climate Change with Gender Perspective

According to the UN, the competencies to counteract climate change must be designed "to improve education, awareness-raising, and human and institutional capacity on climate change mitigation, adaptation, impact reduction, and early warning" (UN 2015). Therefore, as Leach (2015) points out, it is necessary that the actions be situated, and education to reduce our ecological impact must also consider the situation of those who intervene in it. How can we ensure that this type of education can be informative and uses the creative potential of students and teachers to understand the problem and contribute to its solution?

In 2019 the Cooperativa Financiera de Antioquia (CFA) contacted UPB Faculty of Design. This cooperative banking organization has lines of credit and entrepreneurship training for women financed by the United States Agency for International Development (USAID). The purpose of this project was to train and support peasant producers through strategic design to shorten the chain of intermediaries—in the case of Colombia, there can be up to seven—before their products reach the final consumer to improve their income.

The students of the Image Design and Corporate Strategy course, led by Professor Luz Patricia Rave Herrera, were invited to a rural municipality in the region of Antioquia, where they met in public schools and communal houses with groups of farmers convened by the CFA. The response of the students was positive, as they wanted to support the work of the farmers using their expressive possibilities. Many of them were sensitive to the unequal conditions in the countryside that were evident through the peace and reintegration processes of the different armed actors, starting with the 2016 plebiscite in Colombia. What began as a brand image exercise soon became a powerful strategy to raise awareness amongst students regarding the sustainability and vulnerability of farmers, their vulnerability to climate change, and the unequal conditions of women in rural areas.

During the design process, farmers explained to the students the processes of growing and transforming crops into products. This exercise was touching. Although, in theory, they understood that fruits and vegetables are harvested, they questioned themselves about the challenges and difficulties involved in maintaining a crop and exploiting it. Students who have grown up in urban areas take for granted that what they eat can be found easily in the supermarket, ignoring the effort, time, and resources invested in having this availability. The use of fertilizers, the impact of weather conditions, the diversity in seeds, transportation, the selection of products according to standards of regularity and beauty perceived by distributors, and the waste of unsold food, are surprising for the majority, and once they are sensitized, they see differently, including what is damaged in their refrigerators or the high value they pay for a poorly processed product like a portion of French fries versus what a farmer received for a container of potatoes.

Between 2023 and 2024, the teachers carried out the same exercise with 34 organizations of associated producers and independent farmers, and they expanded the project's coverage to 24 more from the Rural Development Agency (ADR), an entity of the Colombian government in charge of structuring, co-financing, and executing plans of agricultural development.

The graphic material implies, in each case, a direct conversation between the students and the farmers, who teach the farmer about their trade. In many cases, the selection of photographs, the visual description of the processes, or the representation of the products are adjusted to match with reality and not with the stereotypes and imaginaries of the participants since many of them do not distinguish, for example, between a form harmful collection of one that does not damage the plants, a diseased species of a healthy one, one type of crop or plant variety of another. Exposing students to farmers implies they now know they pay 3 euros for a cup of French Fries in their favorite restaurant but that local farmers receive this same 3 euros for a whole bundle of potatoes. Now they understand the effort that was put into harvesting these potatoes. Beyond that, they understand the problem when supermarkets return vegetables because they are damaged, ugly, or when they get rotten at home. The following are some of the projects done by our students with associations.

Arrulla brand is designed for a collective of women who harvest rice in a region near the Atlantic Ocean. The name evokes a type of singing that enables children to sleep and resembles water. The Arrullas symbol shows the plant, the landscape, and a happy community with a smile. Different levels of packaging were designed, from a basic stamp to a sticker to a final real printed bag like their competitors (Fig. 1).

Cormuaris, is a cooperative project of woman that offer food, tourist guiding, cultural activities, caring for the environment, among other services to their community and tourists. A brochure was made by students to present these services. The cooperative also includes trans women as key members, this is why their rainbow is important in the branding symbol. Their slogan reads: to be, to do, to have, revealing their values as women (Fig. 2).

Asoagroemprender—Ñaam is a women's association that takes advantage of Ñame, a widespread root in the region that was exceedingly abundant and wasted because it was not transformed properly for its commercialization; today, some families earn a living through the products they make with this basis. Students developed packaging and a social media campaign to promote the use of this root, which is mainly as a useful flour (Fig. 3).

Fig. 1 Arrullas brand by Valentina Arango Palacio, Lisann Hoefer, Camila Botero Vásquez

Fig. 2 Cormuaris brand by Catalina Henao Acosta, Jorge Enrique Vuelvas Jaramillo

Fig. 3 Ñaam brand by: Susana Farley Gómez, Juan Sebastián Blanco Acevedo, Nathalia Castro Burgos

Fig. 4 Asomucampi brand by Daniela Galeano Toro, Isabel Hincapié Herrera, María Martelo Vásquez

Asomucampi is also a women's association that produces blueberries. Students made packaging designs and a social media campaign and associated different products with the logo that could be bought by consumers to foster the brand (Fig. 4).

These projects show us how education for climate change carried out from universities can transcend the traditional academic model towards more active forms of learning that result not only in knowledge of the problems of climate change and how these especially affect women but also allow a small contribution to the communities from the professional knowledge that is intended to be formed.

6 Conclusions

The challenges of Climate Change for women in Colombia are like those addressed globally. Although efforts have been made in Latin American and Caribbean governments to raise awareness of the importance of sustainable development and link gender to Climate Change action, there is still little public information on how citizens can collaborate beyond public policies to adapt or mitigate this phenomenon. Education for a positive ecological change should be implemented in cities and rural areas, based on the development of situated competencies, but it also must empower women so they can demonstrate the positive impact they can have in leading their communities. Climate Change education should be addressed not only at postgraduate levels, specific curricula, or as an optional subject but as a key objective to accomplish in all levels of learning and through every discipline, considering all of them have something valuable to provide and that we are all influenced by environmental change.

Challenges in addressing SDGs in HEIs in Latin America may be overwhelming and difficult to resolve. We consider it to be part of the change aiming to build networks for collaboration regarding SDGs and CC between universities and education institutions, not only for research. This may result in inter-institutional and transdisciplinary in-class projects that will help improve climate and gender literacy and will build capabilities and competences in former professionals in any discipline.

Empirical evidence and theory show that this relationship between gender and climate change is negative and that it can be transformed through education using pedagogical and institutional tools and strategies at different levels, allowing students and professors to have access to fundamental information to start acting and changing behaviors. Classroom-based educational projects must allow young people not only to understand the phenomena at play but also to think about solutions for the adaptation and mitigation of climate change and lead to concrete application, through actions in the classroom, school, town, or community.

The different strategies for implementing the SDGs in higher education institutions ought to be recognized by all stakeholders. They must be named and declared within institutional policies and guidelines to be appropriated by the community. Higher education institutions in Colombia can transform the future through education. They must declare and act on their declaration to positively impact future professionals in any discipline.

In this research, the findings highlight an urgent call to incorporate climate change and gender education in Colombian and Latin American institutions throughout all levels of schooling. This kind of integration however could create a climate crisis generation that is more gender sensitive and has the chance to build better resilience towards it, while increasing women empowerment in these aspects with enhanced leadership opportunities. Broader initiatives within higher education and academia that prioritize Sustainable Development Goal (SDG) 5 in institutional policies can also support transformative shifts toward bridging gender gaps in climate leadership. Therefore, implementing specific documentation guidelines can improve safety, equity, and empowerment in actual academic environments as well as the wider community.

These findings also imply that the strategies used in education could be adjusted to fulfill the specific requirements of rural and urban communities within the Colombian setting. First, if students are engaged in rural areas and learning/work are situated to local, community environmental issues, then the outcomes may be more effective than traditional approaches. But in cities, education may focus on engaging in policy and technological solutions enabling students to meet the unique climate challenges that those areas face.

Finally, in term of limitations, the first one of this study is related to the university being a recent player in the gender equity initiatives. The institution had until this year not focused attention on Sustainable Development Goal (SDG) 5 and had only recently started to set up committees tackling areas such as closing gender gaps, addressing gender equity in learning and safer campus environments for women. Over the years, this delayed focus perhaps reduced the breadth and depth of gendered climate change education initiatives within the institution, possibly affecting all students and staff movements towards a greater climate- and gender-sensitive awareness.

In addition, despite the existence of regional initiatives for a better understanding of the gender–climate change nexus, public information on how individuals, especially women, can help to adapt to and mitigate climate change in Colombia is scarce. The absence of clear information limits community involvement outside formal educational programs and policy initiatives. Another significant limitation is the lack of an integrated curriculum at all level of education that specifically addresses climate change, gender equity, and sustainability competencies. For filling this gap is essential to ensuring a systematic empowerment of students, women in particular in the nature is this research, so they can effectively and equitably understand climate issues.

7 Future Research Prospects

Research implications for the future emphasize the need for education about climate change and about gender at all levels and in every disciplinary location in Colombia and Latin America. The findings suggest that future action may include integrating sustainable development goals (SDGs), especially SDG 5 on gender equity, into policies at universities. In this way, it would enable higher education to come together in a coordinated response to climate and gender crisis. Inclusion of urban and rural needs in the curriculum, prioritizing situated learning around local issues, would lead to a more flexible and context-specific environment for future professionals working on competencies essential for effective, equitable climate leadership.

Future research should investigate the cross-sectional nature of gender and climate, including across space (regions), socioeconomic status, and urban–rural divides. These studies could provide more insight into the unique obstacles women encounter in adapting and becoming resilient to climate change. Beyond this, research on gender-sensitive climate curriculum design at other levels of education (Lopez & Andal 2024), could inform the scale-up of these and other practices for a more integrated approach in schools going forward as well (Coffman 2010), especially if they reflect an interdisciplinary curriculum which spans both natural and social sciences while integrating values-based approaches that demonstrate the multi-faceted impacts of climate change across race, class, space and time.

Other promising research could focus on how various institutional policies that specifically address gender and climate change are effective-for example, policy addressing SDG 5. Longitudinal research on the skills and career development of students associated with the implementation of such policies can be worth obtaining information on how such initiatives have an impact on reality. Finally, the design and provision of training courses for faculty on climate and gender issues would enhance the capacities of educators in providing climate literacy informed by a gender perspective. Such training would be provided through workshops, resources, and collaborative forums to ensure best practices in climate-gender education.

References

Aguilar Revelo L (2022) Women's autonomy and gender equality at the center of climate action in Latin America and the Caribbean

Builes-Vélez A, Restrepo J, Martínez JD (2024) The significance of sustainability in higher education: a view to the curricular proposal at a Colombian University. Int J Sustain High Educ 25(5):943–961. https://doi.org/10.1108/IJSHE-09-2023-0423

Bush SS, Clayton A (2023) Facing change: gender and climate change attitudes worldwide. Am Polit Sci Rev 117(2):591–608. https://doi.org/10.1017/S0003055422000752

Coffman J (2010) Broadening the perspective on scale. Eval Exch 15(1):2–3

Cruz Castaño N, Páramo Bernal P (2020) Educación para la mitigación y adaptación al cambio climático en América Latina. Educ Educadores 23(3):469–489. https://doi.org/10.5294/edu.2020.23.3.6

DANE—Departamento Administrativo Nacional de Estadística and ONU Mujeres (2022) Propiedad Rural en Colombia. Nota estadística [PDF]. Retrieved from: https://www.dane.gov.co/files/investigaciones/notas-estadisticas/jul_2022_nota_estadistica_propiedad_rural.pdf

DNP—Departamento Nacional de Planeación (2016) Plan Nacional de Adaptación al Cambio Climático - Líneas de Acción Prioritarias. República de Colombia. Retrieved from: https://colaboracion.dnp.gov.co/CDT/Ambiente/PNACC%202016%20linea%20accion%20prioritarias.pdf

Eastin J (2018) Climate change and gender equality in developing states. World Dev 107:289–305. https://doi.org/10.1016/j.worlddev.2018.02.02

ECLAC (2021) Fiscal Panorama of Latin America and the Caribbean, 2021 (LC/PUB.2021/5-P), Santiago

ENCC—Estrategia Nacional ante el Cambio Climático (2015) Ministerio del Ambiente, República del Perú [PDF]. Retrieved from: http://www.minam.gob.pe/wp-content/uploads/2015/09/ENCC-FINAL-250915-web.pdf

Girls' Education Challenge (2022) Girls' education: reaching the most marginalized girls. https://girlseducationchallenge.org/media/o45d2zj4/girls-education_reaching-the-most-marginalised-final.pdf

Hans A, Rao N, Prakash A, Patel A (eds) (2021) Engendering climate change: learnings from South Asia. Routledge India, London. https://doi.org/10.4324/9781003142409

Herrera-Franco G, Mora-Frank C, Carrión-Mero P (2023) Sustainable development in Latin American higher education institutions. In: Leal Filho W, Frankenberger F, Tortato U (eds) Sustainability in practice. World Sustainability Series. Springer, Cham. https://doi.org/10.1007/978-3-031-34436-7_7

IDEAM—Instituto de Hidrología, Meteorología y Estudios Ambientales, PNUD—Programa de las Naciones Unidas para el Desarrollo, MinAmbiente, Cancillería Colombiana and Observatorio Nacional de Ciencia y Tecnología (2016) Tercera Comunicación Nacional de Cambio Climático, ¿Qué piensan los colombianos sobre el cambio climático? Retrieved from: https://archivo.minambiente.gov.co/images/AsuntosMarinosCosterosyRecursosAcuatico/que_piensan_los_colombianos_sobre_el_cambio_clim%C3%A1tico.pdf

Instituto Tecnológico y de Estudios Superiores de Monterrey (ITESM) (2023) Project-based learning approach. Retrieved from https://www.itesm.mx

Intergovernmental Panel on Climate Change (IPCC) (2014) Climate change 2014: synthesis report. Contribution of working groups I, II and III to the fifth assessment report of the Intergovernmental Panel on Climate Change. IPCC, Geneva. https://www.ipcc.ch/report/ar5/syr/

Intergovernmental Panel on Climate Change (IPCC) (2022) Climate change 2022: impacts, adaptation and vulnerability. Cambridge University Press

Kaijser A, Kronsell A (2014) Climate change through the lens of intersectionality. Environ Polit 23(3):417–433. https://doi.org/10.1080/09644016.2013.835203

Kwauk C, Wyss N (2022) Gender equality and climate justice programming for youth in low- and middle-income countries: an analysis of gaps and opportunities. Environ Educ Res, 1–23. https://doi.org/10.1080/13504622.2022.2123894

Kwauk C, Braga A (2017) Three platforms for girls' education in climate strategies. Brookings, Washington, DC

Lagi R, Waqailiti L, Raisele K, Tyson LS, Nussey C (2023) 'Curui': weaving climate justice and gender equality into Fijian educational policies and practices. Comp Educ 59(2):305–324. https://doi.org/10.1080/03050068.2023.2188370

Leach M (2015) Gender equality and sustainable development. Routledge. London. 228 p. https://doi.org/10.4324/9781315686455

Lopez MS, Andal EZ (2024) Gender-Responsive Pedagogy and Attitude toward Sensitivity in Basic Education. TWIST 19(3):583–588

McNamara KE, Buggy L (2017) Community-based climate change adaptation: a review of academic literature. Local Environ 22(4):443–460. https://doi.org/10.1080/13549839.2016.1216954

MINAE—Ministerio de Ambiente y Energía (2018) República de Costa Rica. Política Nacional de Adaptación al Cambio Climático de Costa Rica [PDF]. Retrieved from: https://cambioclimatico.minae.go.cr/wp-content/uploads/2019/01/Politica-Nacional-de-Adaptacion-al-Cambio-Climatico-Costa-Rica-2018-2030.pdf

MinAmbiente—Ministerio del Medio Ambiente y Desarrollo Sostenible (2021) Herramientas de género y cambio climático. República de Colombia. Retrieved from: https://archivo.minambiente.gov.co/index.php/cambio-climatico/herramientas-de-genero-y-cambio-climatico

Munck R, Pinilla Y, Hodges R, Bartch C (2023) Higher education in Latin America and the Caribbean: civic engagement and the democratic mission. Machdohnil

Office for Climate Education OCE (2021) lanzamiento internacional del proyecto ALEC—América Latina para la Educación Climática. Retrieved from: https://www.oce.global/es/news/lanzamiento-internacional-del-proyecto-america-latina-para-la-educacion-climatica-alec

Pankhurst C (2022) Girls' education and climate change: a critical review of the literature. Accountability for gender equality in education (AGEE) working paper. Retrieved from: https://www.chrome-extension://efaidnbmnnnibpcajpcglclefindmkaj/https://staging.genderedddata.org/wpcontent/uploads/2022/07/Girls-Education-and-Climate-Change-Critical-Review-of-the-Literature_FINAL-2.pdf

Pontificia Universidad Católica de Chile (PUC) (2023) Curriculum in environmental engineering and geography. Retrieved from https://www.puc.cl

Rao N, Lawson ET, Raditloaneng WN, Solomon D, Angula MN (2019) Gendered vulnerabilities to climate change: insights from the semi-arid regions of Africa and Asia. Climate Dev 11(1):14–26. https://doi.org/10.1080/17565529.2017.1372266

Rao N, Mishra A, Prakash A, Singh C, Qaisrani A, Poonacha P, Vincent K, Bedelian C (2019) A qualitative comparative analysis of women's agency and adaptive capacity in climate change hotspots in Asia and Africa. Nat Clim Chang 9(12):964–971. https://doi.org/10.1038/s41558-019-0638-y

Universidad de la República (Udelar) (2023) Faculty of science programs. Retrieved from https://www.udelar.edu.uy

Senay H (2016) Gender and climate change: overview of linkages between gender and climate change. United Nations Development Programme (UNDP). https://www.un.org/en/climatechange/gender-and-climate-change.pdf

Tamayo E, Gómez-Valencia B, Morales P, Ochoa J (2020) Contención de la deforestación en resguardos indígenas. Reporte Bio: Factores de transformación y pérdida de biodiversidad. Instituto de Investigación de Recursos Biológicos Alexander von Humboldt. Retrieved from: http://reporte.humboldt.org.co/biodiversidad/2020/cap2/203/

UN (2015) Sustainable development goals. https://www.un.org/sustainabledevelopment/

UN—CEPAL (2023) The economics of climate change in Latin America and the Caribbean, 2023: financing needs and policy tools for the transition to low-carbon and climate-resilient economies. United Nations publication

UNDP (2012) Education first: unleashing potential (draft). United Nations, New York

UNESCO (2020) América Latina y el Caribe - Inclusión y educación: Todos y todas sin excepción. UNESCO

United Nations Framework Convention on Climate Change (UNFCCC) (2021) The regional impacts of climate change in Latin America and the Caribbean. UNFCC

Universidad Católica de Bolivia (UCB) (2023) Environmental engineering programs. Retrieved from https://www.ucb.edu.bo

Universidad de Buenos Aires (UBA) (2023) Specialized courses in climate change. Retrieved from https://www.uba.ar

Universidad de Costa Rica (UCR) (2023) Sustainability and environmental management programs. Retrieved from https://www.ucr.ac.cr

Universidad de La Habana (UH) (2023) Environmental science programs. Retrieved from https://www.uh.cu

Universidad de los Andes (Uniandes) (2023) Education for sustainability in Colombia. Retrieved from https://www.uniandes.edu.co

Universidad de Panamá (UP) (2023) Research on climate change. Retrieved from https://www.up.ac.pa

Universidad Nacional Autónoma de México (UNAM) (2023) Environmental studies programs. Retrieved from https://www.unam.mx

Universidad Nacional de Colombia (UNAL) (2023) Research on climate change impacts. Retrieved from https://www.unal.edu.co

Universidad Nacional de Córdoba (UNC) (2023) Scientific research and education programs on climate change. Retrieved from https://www.unc.edu.ar

Universidad Nacional Mayor de San Marcos (UNMSM) (2023) Courses and programs on sustainability and climate change. Retrieved from https://www.unmsm.edu.pe

Universidad Pontificia Bolivariana (UPB) (2023) Carbono neutro multicampus. Retrieved from https://www.upb.edu.co/

Universidad Pontificia Bolivariana (UPB) (2024) Reporte de Sostenibilidad de la Universidad Pontificia Bolivariana bajo estándares Global Reporting Initiative. Retrieved from https://repository.upb.edu.co/bitstream/handle/20.500.11912/11620/Reporte_de_Sostenibilidad_UPB_2023_compressed.pdf?sequence=4&isAllowed=y

Universidade de São Francisco (USF) (2023) Interdisciplinary research on climate change. Retrieved from https://www.usf.edu.br

Universidade de São Paulo (USP) (2023) Sustainability and climate change projects. Retrieved from https://www.usp.br

Universidade Federal do Rio de Janeiro (UFRJ) (2023) Education on climate change impacts. Retrieved from https://www.ufrj.br

Winthrop R, Kharas H (2016) Want to save the planet? Invest in girls' education. Brookings (Blog). https://www.brookings.edu/opinions/want-to-save-the-planet-invest-in-girls-education/

World Higher Education Database (WHED) (2023) Higher education institutions in Latin America

Critical Analysis of Sustainable Production Board Games

Sandro Gabriel Libretti Prestes, João Victor Encide Salla, Juliana Veiga Mendes, and Diogo Aparecido Lopes Silva

Abstract Sustainable Production is a crucial topic that necessitates a comprehensive and effective educational approach to engage both current and future generations. This work addresses the gap in Sustainable Production education within higher education, focusing on challenges in engaging students through board games. This chapter critically reviews internationally recognized and commercially available board games on Sustainable Production through a SWOT analysis. Thus far, fourteen board games have been thoroughly analyzed. The methodology involved a comprehensive analysis of user feedback and online reviews for these games. Key strengths of the reviewed games encompass elements like group cooperation, competition, interaction, result integration, social learning stimulation, and critical thinking. Weaknesses may include traits that do not provide value due to limitations in fostering cognitive and tacit learning. Opportunities include low production costs and potential for digitalization. Among the main threats is the potential lack of stimulation for learning if the games fail to adapt to digital formats for the next generations of students.

Keywords Production engineering · Gamification · Educational learning · Eco-friendly production

1 Introduction

The concept of gamification involves using game design elements and principles in non-game contexts to enhance engagement and motivation through game mechanics (Deterding et al. 2011). This approach seeks to enhance learning by fostering interactivity and engagement through tools like scoring, rewards, and challenges. The

---

S. G. L. Prestes (✉) · J. V. E. Salla · J. V. Mendes · D. A. L. Silva
Research Group On Sustainability Engineering (EngS Group), Department of Production Engineering, School of Management & Technology, Federal University of São Carlos, Sorocaba, SP, Brazil
e-mail: sandro.prestes@estudante.ufscar.br

W. Leal Filho et al. (eds.), *Education for Sustainable Development: The Contribution of Universities*, World Sustainability Series,
https://doi.org/10.1007/978-3-031-86985-3_2

goal is to improve the educational experience, promote the active participation of those involved, and facilitate learning more effectively (Bower and Thomas 2023).

Gamification stems from engagement and motivational strategies rooted in behavioral psychology and early game design concepts can be used in the most diverse areas of knowledge, including Engineering, and others, such as biology (Deterding et al. 2011), physical education (Arufe-Giráldez et al. 2022) and even for teaching organic chemistry to high school students (Lopes et al. 2024).

Education for Sustainable Production (ESP) aims to train individuals to work in production systems that minimize environmental and social impacts throughout the entire life cycle of products. According to the Ministry of the Environment (Ministério do Meio Ambiente 2020), sustainable production is characterized by the incorporation of better alternatives to minimize environmental and social impacts throughout the life cycle of goods and services, while proving to be economically viable.

This type of education is essential to ensure that current production practices do not compromise the needs of future generations (Martins and Maria 2024). In this regard, Education in the form of a Gamification tool can promote knowledge focused on environmental preservation, significantly contributing to actions and improvements to contain and reverse the negative impacts of environmental degradation (Connolly et al. 2012).

ESP began to gain strength in the mid-1950s (Skinner 1957), however, it is known that the use of gamification for ESP still lacks tools that allow for increased effectiveness in this teaching–learning process (Tilvawala et al. 2023), while also in the promotion of green jobs, i.e., those professional activities that contribute to preserving or restoring environmental quality.

The importance of ESP lies in its ability to promote more responsible and sustainable practices and jobs by fostering a paradigm shift in the way production and consumption are approached. ESP is a theme that the United Nations Environment Programme (United Nations Environment Programme 2024) considers essential for promoting, among other things, the 2030 Agenda, particularly in achieving targets related to SDG 12—Sustainable Development Goal, focused on responsible production and consumption. This educational approach is crucial for transitioning to a more sustainable world. It involves active and participatory learning methods, that are considered more effective in promoting sustainable behaviors and solving complex societal problems (Franco et al. 2024).

In the current context, gamification has gained prominence as a strategy to improve user interaction with products and services and, in this way, has become a global trend, being adopted in a variety of sectors, including education, health, marketing, and human resources. Companies have realized the potential of gamification to increase customer engagement, improve employee productivity, and promote healthy behaviors (Tilvawala et al. 2023).

This recent development has sparked the interest of researchers as a potential for creating learning environments (Reeves and Read 2009) and facilitating mass collaboration (Mcgonigal 2011). Krause and Smeddinck (2011), Shneiderman (2004), and von Ahn and Dabbish (2008) began to study the effectiveness of gamification in different contexts and its psychological impacts since the mid-2000s, when gamification evolved beyond simple reward systems, incorporating more advanced game design concepts, such as narrative, level progression, and sophisticated game mechanics, for example, based on RPG ("role-playing game").

However, the application of games or gamification for ESP is scarce, according to Tsai et al. (2021). For Chappin et al. (2017), playing is an appropriate activity for the current generations in the context of learning to promote sustainable development. Teaching through games can be an effective method for environmental education and for disseminating the concept of sustainable development.

"Serious games" related to sustainable development focus primarily on understanding sustainable development issues broadly and holistically. According to De Vries and Knol (2011), the attempt to use games as a tool to stimulate people's awareness and attitudes towards sustainability with a focus on energy saving/conservation obtained results that showed how much participants' awareness and attitudes were significantly affected. Recently, there has been an increase in research on the application of games for ESP and the topics mainly cover specific scientific fields related to sustainable development, including energy efficiency, climate change, and natural resource management (Prestes et al. 2023).

Sustainability-focused games are mainly used in higher education settings, while the most popular application is related to improving skills through simulation games, according to Hallinger et al. (2020). These games can mimic real-world problems and provide a safe environment for students to explore, experiment, and learn through play.

Gugerell et al. (2017), explain that people's awareness and understanding of sustainable development issues encourage players to adopt sustainable behaviors, and Rossano et al. (2018), highlight that game-based learning is largely considered one of the best ways to achieve learning for children and adults. Game-based learning focuses on students. The interesting learning process encourages them to acquire interrelated knowledge and skills.

Katsaliaki and Mustafee (2015), analyzed surveys of games related to ESP, and Madani et al. (2017) analyzed games related to environmental management and found that methods known as GBL, "game-based learning" are useful with special attention to the use of board games for that purpose.

GBL addresses an active methodology that is viable for the teaching–learning process from the perspective of different didactic approaches, including the simulation of those related to production operations, as they enable the creation of safe and responsive environments, which enable and provide the systemic experience of the greatest possible number of human and organizational skills. Furthermore, GBL are like cooperative games, functioning as instruments for articulating and promoting the educational process and being capable of bringing motivation and a feeling of inclusion (Salibi et al. 2021).

Despite its rapid growth in recent years, gamification faces challenges, including concerns about privacy, behavioral manipulation, and superficiality, without significant intervention in behaviors. Despite this, gamification continues to evolve and adapt, with companies, organizations, and academic spaces seeking innovative ways to integrate game elements such as new types of board games into their engagement and teaching–learning strategies (Chen and Ho 2022).

Games enable and promote interdisciplinary and systemic thinking, as well as the development of life skills, such as communication and teamwork collaboration, providing an opportunity to integrate technical and human dimensions in decision-making (Oliveira and Ferreira 2011). However, there are still no studies in the literature focused on mapping and analyzing the ESP games currently available on the market, making a critical analysis of them, which motivates this research. The objective of this work is to fill the current gap on sustainability-based gamification board games.

This chapter performed a critical analysis of current commercially sold and internationally recognized board games about ESP. To this end, the games were evaluated based on their main strengths (S), weaknesses (W), development opportunities (O), and threats (T), and ranked according to their scores (Sect. 3.3).

2 Methodology

2.1 Procedures and Document Review

The documentary review was the main research strategy to achieve the research goal, and it follows the structure shown in Fig. 1. The data search results provided a preliminary screening to understand the key characteristics of board games (BGs) for ESP. Subsequently, each BG was characterized using a SWOT matrix.

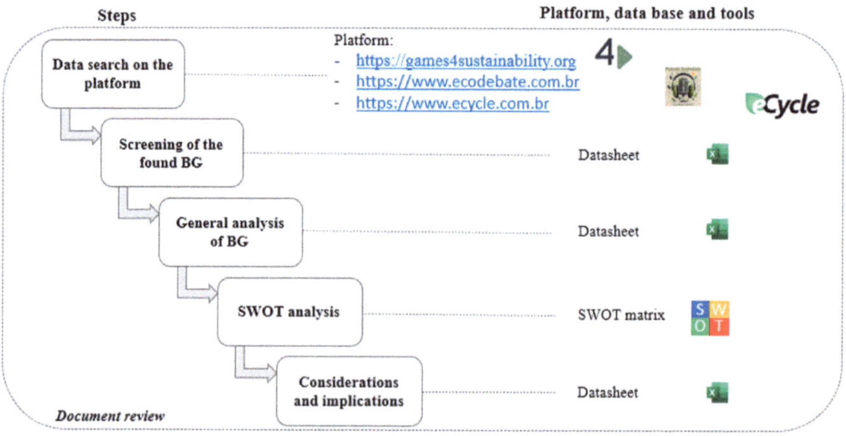

Fig. 1 Methodological flow. *Source* Own authorship

This research utilized references from platforms like 'Games4Sustainability,' which specializes in sustainability-focused games, available at: https://games4sustainability.org; on the 'Eco Debate' platform, a platform for information, articles, and news on socio-environmental issues, available at: https://www.ecodebate.com.br/; and on the 'E-cycle' platform, a platform that originated from the interest in consumer relations developed between individuals and companies and their effects on society and the environment, available at: https://www.ecycle.com.br/tabuleiro-sustentavel/.

The analysis identified 14 BGs of interest (BG-1 to BG-14). To understand the origin and purpose of each game, followed by an in-depth SWOT analysis. Data were organized using datasheets to classify the features of BGs.

The screening analysis in Fig. 1, allowed mapping the companies that manufacture the games and their respective nationalities, to identify the most influential countries in the application of BGs.

After the initial screening, each board game was analyzed in greater detail to obtain more robust information about its objectives and characteristics (general analysis of BG). Data were collected on aspects such as the number of players required for gameplay, the objectives proposed in each game, and a specific feature of its characteristics.

The SWOT analysis, from the acronym in which S is used for "Strengths", W for "Weaknesses", O for "Opportunities" and T for "Threats" was performed next. Inspired by the work of Batalha and Rachid (2008), a SWOT matrix was used to evaluate the main characteristics of the board games (BGs), generate relevant insights for future research, and support the development of new gamification themes and opportunities for ESP.

The strengths and weaknesses listed in the analysis are related to the internal environment, while the opportunities and threats enable verifications related to the external environment of an organization or analysis objective (Harrisson 2010), in the case in question, the BGs.

The use of SWOT analysis is justified by the fact that it supports the proposition of action plans for the application of clear strategies, ranging from offensive, confrontation, and reinforcement strategies to those focused on defense strategies, according to Firmino et al. (2020).

Finally, the research progressed to the stage of considerations and implications. In this phase, it was possible to rank the games based on their strengths and weaknesses (internal and external/market). A synthesis of the knowledge gained was also made, identifying the limitations of the study and suggesting topics for future research.

3 Results and Discussions

The topic in question was divided into three topics to allow the exemplification of the results of the qualitative analysis, the results of the SWOT analysis, and finally, the consolidation of the main recommendations obtained.

3.1 Qualitative Analysis of BGs for ESP

The qualitative analysis of BGs for ESP evaluated the games' origins, objectives, characteristics, and the manufacturing companies commercializing these games. Figure 2 evaluated the geographical scope of each game. In the qualitative analysis, presented in Table 1, the main objectives/characteristics and the respective game providers were established.

The analysis reveals a higher concentration of BGs for ESP in Brazil, Poland, and Germany, followed by Belgium, Switzerland, and the USA. An additional observation highlights the predominance of games developed by the Centre for Systems Solutions (CRS), a non-profit organization linking theory and practice through systemic method games.

In Brazil, it is worth highlighting an existence of companies producing games, such as Copag, Estrela, Toyster, Bigstar, and even Fundação Espaço Eco. All of this highlights the existence of an ecosystem for the economic exploration of games on such themes in the country.

The documentary research reveals a variety of BGs, ranging from knowledge-testing games on environmental issues, like New World (BG-3), to games exploring human impacts on the environment, such as Laudato Si (BG-10), and even games addressing ethics and corruption, like Do We Have a Deal? (BG-13).

In general, it is worth highlighting that the mapped BGs have different objectives, game modes, number of players, and target audiences, as pointed out in Table 1. In this sense, it is worth highlighting that the following SWOT analysis was carried out for each BG.

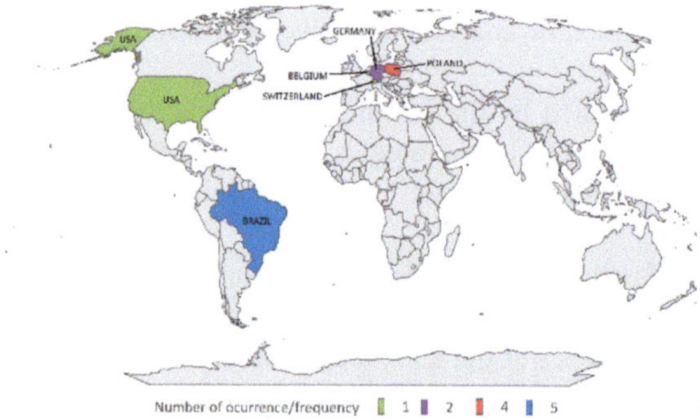

Code	Name	Country
BG-1	Eco fluxx	Brazil
BG-2	Cidades sustentáveis	Brazil
BG-3	Novo mundo	Brazil
BG-4	Coleta correta	Brazil
BG-5	Mercado sustentável	Brazil
BG-6	Trade off	Poland
BG-7	Go goals	Belgium
BG-8	Climate challenge	USA
BG-9	Sustainability in schools	Poland
BG-10	Laudato si	Poland
BG-11	Catan oil springs	Germany
BG-12	Flood resilience game	Switzerland
BG-13	Do we have a deal?	Poland
BG-14	Settlers of the green future	Germany

Fig. 2 Geographical analysis of commercial games with a focus on gamification for ESP. *Source* Own authorship

Table 1 Qualitative analysis of commercial games focusing on gamification for ESP

Code	Name	Country	Game provider	Objectives and characteristics
BG-1	Eco fluxx	Brazil	COPAG	*Objective*: To make people who are playing learn more about biology, the environment, and ecological preservation *Number of players*: Between two and six players *Fun facts*: Classified as a family-friendly and festive game that doesn't require much practice to understand the concepts
BG-2	Sustainable cities	Brazil	ECO FOUNDATION	*Objective*: To present sustainability and "socio-efficiency" in a practical way to help in decision-making *Number of players*: Up to six players *Interesting facts*: Restoring abandoned cities so that they become habitable again and in an ecologically sufficient way
BG-3	New World	Brazil	STAR	*Objective*: Answer questions about the environment to gain resources and clean up the most compromised regions of the planet *Number of players*: Between two and six players *Fun fact*: All the game material is made from recycled plastic

(continued)

Table 1 (continued)

Code	Name	Country	Game provider	Objectives and characteristics
BG-4	Correct collection	Brazil	TOYSTER	*Objective*: To teach and raise awareness among players about the correct colors for bins for selective collection *Number of players*: Between two and five players *Curiosities*: The game more aimed at children
BG-5	Sustainable market	Brazil	BIGSTAR	*Objective*: To acquire the largest number of companies and certify them, obtaining the source of resources from competitors *Number of players*: Between three and six players *Fun facts*: It's a game about business and management integrated with sustainability
BG-6	Trade-off	Poland	CRS	*Objective*: To introduce players to the trade-offs and synergies between traditional development and natural capital values *Number of players*: Multiple teams of eight to ten players (more than 3 moderators) *Fun facts*: Each version is played in two rounds

(continued)

Table 1 (continued)

Code	Name	Country	Game provider	Objectives and characteristics
BG-7	Go goals	Belgium	UNRIC	*Objective*: To help children understand the Sustainable Development Goals (SDGs), how the global goals impact, and what needs to be done daily to achieve the SDGs by 2030 *Number of players*: Between two and ten players (one moderator) *Fun fact*: It can also be used as a tool to learn a foreign language
BG-8	Climate challenge	USA	CAMBRIDGE	*Objective*: To have the highest revenue at the end of round ten. After the game, learnings are captured through a strong "debriefing", in which players reflect and learn from their results *Number of players*: More than ten players (one moderator) *Curiosities*: The game is played over 10 rounds with six teams. Each team starts at one of the six ends of the board and represents a (made-up) nation

(continued)

Table 1 (continued)

Code	Name	Country	Game provider	Objectives and characteristics
BG-9	Sustainability in schools	Poland	CRS	*Objective*: To help educators: identify projects that make business sense, organize those projects, and then integrate them into lesson plans *Number of players*: More than ten players (one moderator) *Curiosities*: The game can be played at various challenge levels, including creating a 5-year plan for a fully sustainable, zero-emissions school
BG-10	Laudato si	Poland	CRS	*Objective*: Experience the interdependence and impact that humanity has on nature, as well as see for themselves the importance of sensible management of the common good given by God *Number of players*: Two to ten players (one moderator) *Curiosities*: Players decide what they want to do—how many trees they want to cut down. One person's gain might result in another person's loss

(continued)

Table 1 (continued)

Code	Name	Country	Game provider	Objectives and characteristics
BG-11	Catan Oil Springs	Germany	CRS	*Goal*: Develop cities and roads without using oil, thus avoiding consequences for Catan's environment. Or, they can choose to use oil, resulting in irreversible damage to the environment *Number of players*: Two to ten players (no moderator) *Curiosities*: Players may experience a situation where excessive rivalry can lead to major economic and environmental losses
BG-12	Flood resilience game	Switzerland	CRS	*Objective*: Allows players to explore vulnerabilities and capabilities, leading to an advanced understanding of interdependencies and the potential for working together *Number of players*: Two to ten players (one moderator) *Curiosities*: The game is set in a community living in an area exposed to flooding, so they take on the roles of members of different groups of citizens to make decisions

(continued)

Table 1 (continued)

Code	Name	Country	Game provider	Objectives and characteristics
BG-13	Do we have a deal?	Poland	CRS	*Objective*: To learn about the dynamics of corruption, negotiations, and injustice in business, about the reporting mechanism, and how it impacts trust between different parties, among others *Number of players*: More than ten players (one moderator) *Curiosities*: The game is divided into seven stages
BG-14	Settlers of the green future	Germany	CRS	*Objective*: It aims to stimulate the building of "eco-cooperatives" and provides training in decision-making under uncertainty, environmental policy formulation, and negotiations. It promotes an understanding of the options available to achieve a green future *Number of players*: Two to ten players (one moderator) *Curiosities*: The game was created ahead of the Rio + 20 Earth Summit in June 2012, using real policy proposals and actions for sustainable development

Source Own authorship

3.2 Statistical Analysis of Responses

From a SWOT perspective, analyses were carried out covering the main strengths, weaknesses, opportunities, and threats relating to each of the games explored.

A measurement strategy was developed based on insights from documentary reviews and visits to gamification websites for ESP, assigning scores on a 1-to-10 scale for each dimension evaluated.

For the 'Strengths' and 'Opportunities' dimensions, a score closer to 10 indicates higher potential for promoting the teaching–learning process.

On the other hand, for the "Weaknesses" and "Threats" dimensions, the closer to 1, the better the score for the game, therefore, if a game scores 1 for "Threats" it means that it has very few market threats.

To allow such analysis, Table 2 was proposed, and its results are displayed in Fig. 3.

The analysis reveals that BG-2 scored the lowest among all analyzed games, presenting a low score in opportunities and strengths with an average equal to 5, therefore, below the general average of BGs (equal to 8); and showed the highest score in weaknesses and threats with an average equal to 6, while the average of the other games was less than 2.4.

On the other hand, the BG-12 game presented the best overall score, with the maximum average for opportunities and strengths and the lowest possible average for weaknesses and threats, indicating that the game has a lot of potential and can help in the teaching–learning process for ESP since there are few eminent weaknesses found per se.

From a broader point of view, Fig. 3 elucidates the predominance of higher bars for forces and opportunities and more open radars, identifying that the BGs analyzed tend to have positive scores there and opportunities yet to be explored. And as for the opportunities, considerations to explore and expand their actions for improvements, are noted in the evaluations of BG-1, BG-2, BG-3, BG-4, BG-5, BG-6, BG-10, and BG-13.

Table 2 Scores for the SWOT analysis focusing on gamification for ESP

Code	STRENGTHS	OPPORTUNITIES	MEAN (positive)	WEAKNESSES	THREATS	MEAN (negative)
BG-1	10	5	7.5	1	4	2.5
BG-2	6	4	5	6	6	6
BG-3	8	7	7.5	3	3	3
BG-4	8	7	7.5	3	3	3
BG-5	8	7	7.5	3	3	3
BG-6	10	7	8.5	2	2	2
BG-7	10	8	9	1	1	1
BG-8	10	9	9.5	1	1	1
BG-9	8	8	8	1	1	1
BG-10	7	7	7	3	4	3.5
BG-11	10	9	9.5	1	1	1
BG-12	10	10	10	1	1	1
BG-13	7	6	6.5	4	4	4
BG-14	9	9	9	1	1	1
MEAN (Strengths and Opportunities)			**8**	MEAN (Weaknesses and Threats)		**2.35**

Source Own authorship

Fig. 3 Comparative SWOT analysis (strengths, weaknesses, opportunities, and threats) of the board games evaluated in the documentary review (bars and radar). *Source* Own authorship

Similarly, an opposite behavior is noted for the dimensions related to threats and weaknesses, which have smaller bars when compared to the others and more compact radar graphs, which indicate that the games have a certain resistance to threats. Considerations on the need for interventions, regarding weaknesses for BG-2 (in particular), BG-10, and BG-13, and threats to BG-2, BG10, and BG-13.

3.3 Considerations from the Analysis of the Main ESP Games

Considering the results in Table 2 and the graphs in Fig. 3, it becomes possible to rank the BGs based on an assessment of the characteristics that promote success in games from the perspective of the SWOT score.

Considering the criteria for the characteristics of strengths and opportunities, according to which the higher the value on a scale of 1–10, the better, they are therefore considered as positive aspects. On the other hand, for the characteristics related to weaknesses and threats, the lower this value, the better for the game, these being considered as negative aspects.

The games were ranked based on the research evaluations. The SWOT matrix criteria were considered and scored in Fig. 3, considering firstly the highest average of positive points and, as a tiebreaker, the negative criteria (Table 3).

Table 3 Ranking of commercial games with a focus on gamification for ESP

Ranking	Code	Strengths	Opportunities	MEAN (positive)	Weaknesses	Threats	MEAN (negative)
1 st	BG-12	10	10	10	1	1	1
2nd	BG-8	10	9	9.5	1	1	1
2nd	BG-11	10	9	9.5	1	1	1
3rd	BG-14	9	9	9	1	1	1
3rd	BG-7	10	8	9	1	1	1
4th	BG-6	10	7	8.5	2	2	2
5th	BG-9	8	8	8	1	1	1
6th	BG-1	10	5	7.5	1	4	2.5
7th	BG-3	8	7	7.5	3	3	3
7th	BG-4	8	7	7.5	3	3	3
7th	BG-5	8	7	7.5	3	3	3
8th	BG-10	7	7	7	3	4	3.5
9th	BG-13	7	6	6.5	4	4	4
10th	BG-2	6	4	5	6	6	6

Source Own authorship

Thus, when analyzing the profile of the 14 games found so far, it is noted that there are games with different proposals, which does not invalidate the use of the SWOT tool, and therefore, it allows the establishment of a relationship of complementary information concerning what was found in the literature. Therefore, it is observed that despite this difference in scores, there is heterogeneity in the ESP games, with the SWOT categories, without the presence of a specific game standing out as the main one in the area.

BG-2 (Sustainable Cities) scores high on strengths and opportunities, and low on weaknesses and threats. This is due to the tendency to oversimplify the complex processes involved in urban planning.

The reality of smart cities involves a myriad of interconnected variables, from transportation to energy, livability, economics, and public policy. Oversimplification can lead to a distorted and superficial view of these issues, and a lack of realism can fail to capture the nuances and unpredictability that characterize the real urban environment, as the reality of cities is often chaotic, with unforeseen changes, conflicting interests, and external factors that can influence planning decisions. Other limitations in data and technology representation and a lack of historical and social context can lead this board game to fail to adequately incorporate such elements and thus leave out important social and historical implications that influence urban planning.

Other limitations may influence interactivity and responsiveness, since in board games, decisions are made statically and sequentially and, as a counterpoint, in the reality of this context, urban planning is a dynamic and interactive process, involving multiple stakeholders and responses to sudden changes.

The game might struggle to capture interactivity and responsiveness effectively, requiring thoughtful design to motivate players without making them feel as though they are merely studying.

In short, although games like "Smart Cities" provide an interesting introduction to the topic of urban planning and smart cities, it is important to recognize their limitations and not consider them as a complete or accurate representation of urban reality. On the other hand, positively, according to Table 3, the games "Flood Resilience Game" (BG-12), "Cattan Oil Springs" (BG-11), and "Climate Challenge" (BG-8) can be highlighted sequentially, which become educational and engaging and stand out for presenting essential characteristics for gamification.

The game "Flood Resilience Game", for example, in its educational aspect, addresses the existence of duality between collaboration and cooperation, variability of scenarios, and "replicability", the ability of a game to be played countless times without loss of motivation and interest and which allows permeating and reinforcing learning from the same situations, with different ways of proposing solutions to the problems presented.

The games "Cattan Oil Springs" and "Climate Challenge" also stand out, as they present other important characteristics for gamification, such as a unique and specific theme, social interaction, excellence in the quality of the product components, development of cognitive and tacit skills and accessibility to a wide audience of all ages, strengthening social interaction with engagement and fun.

On the other hand, if the focus is only on the playful part without paying attention to the motivational and content part, a product may be created that does not serve as an awareness tool and that does not fulfill its pedagogical objective, as games need to balance fun with educational content to ensure that the player remains motivated.

These aspects are important to enable greater applicability of gamified systems in teaching sustainability, exploring the finite nature of natural resources through immersion in games, and demonstrating the importance of collective action at community, national, and global levels.

In this way, games must be developed in a format that is as fun as possible and as a motivational tool, which addresses the teaching topic without the player feeling that he is "just studying".

4 Conclusions

The 14 games analyzed in this documentary review have in common the objective of promoting learning about sustainability through engaging and interactive mechanics, which apply the use of practical themes related to sustainable production and the environment, allowing players to interact with real or simulated scenarios that reinforce knowledge on the topic.

In general, these games seek to engage players in solving environmental problems, using different approaches such as cooperation and competition. Overall, the results obtained from the SWOT analysis show that games have great potential to contribute to Education for Sustainable Production (ESP), expanding the reach of traditional teaching methods.

The game that performed best for collaboration, and cooperation, among other characteristics, was the "Flood Resilience Game" (BG-12), which stood out both in its strengths and in its opportunities, due to its collaboration and cooperation characteristics. Another relevant feature was the duality between competition and cooperation, in which the game teaches about the dilemmas between personal and collective interests, simulating the interdependence necessary to deal with problems such as environmental disasters. These characteristics consolidated the "Flood Resilience Game" (BG-12) by promoting its pedagogical effectiveness, ensuring that it is engaging in ESP.

Using these games enhances the educational process by making it engaging and fostering the development of critical and reflective skills necessary for understanding sustainable practices.

By identifying the strengths and weaknesses of these games, the study provided a solid basis for future developments in the field of educational games on sustainability, as well as highlighting the importance of digitizing some of them to increase engagement among new generations.

Finally, some limitations inherent to the study can be highlighted. One of the limitations is that the study was based on a documentary analysis, which may not fully reflect the experience of using games in different educational contexts. Other perceived limitations were that some games have complex rules or require moderation, which may limit their use in classrooms or informal settings. Furthermore, the interactivity and response of the games may be limited in board games, as decisions are made in a static and sequential manner, which may affect the dynamism of the learning process.

In general, it is possible to highlight that the opportunities for future research in the field of educational games on sustainability are diverse, especially in the area of sustainable production education (ESP). In this sense, it is important for future research to explore how games can be improved to more effectively integrate the social and economic context in order to complete the sustainability triad (environmental, social, and economic). The research also suggests that gamification can be a powerful tool for teaching sustainability, opening up possibilities for new innovations in the field, such as the development of more interactive and dynamic games that better simulate the dynamism of the teaching–learning process.

In conclusion, the mapping of board games and their results contribute to advancing gamification in sustainability education, highlighting their diverse benefits and potential integration into the teaching–learning process.

Acknowledgements The authors are grateful to the "Conselho Nacional de Desenvolvimento Científico e Tecnológico" (CNPq, grant number 303343/2022-2) and to the "Fundação de Amparo à Pesquisa do Estado de São Paulo" (FAPESP), process number 2022/15134-1.

References

Arufe-Giráldez V et al (2022) Gamification in physical education: a systematic review. Educ Sci 12(8):540

Batalha MO, Rachid A (2008) Strategy and organizations. In: Batalha MO, Rachid A (eds) National confederation of industry—CNI. (2016a)

Bower M, Thomas L (2023) Gamification in higher education: a review of the literature and new directions for research. Educ Tech Res Dev 71(2):273–297

Chappin EJL, Bijvoet X, Oei A (2017) Teaching sustainability to a broad audience through an entertainment game—the effect of Catan: Oil Springs. J Clean Prod 156:556–568

Chen F-H, Ho S-J (2022) Designing a board game about the United Nations' sustainable development goals. Sustainability 14(18):11197. https://doi.org/10.3390/su141811197

Connolly TM et al (2012) A systematic literature review of empirical evidence on computer games and serious games. Comput Educ 59(2):661–686

De Vries PW, Knol E (2011) Serious gaming as a means to change adolescents' attitudes towards saving energy; preliminary results from the EnerCities case. In: Proceedings of the EDEN annual conference, Dublin, Ireland, 19–22 June 2011

Deterding S, Dixon D, Khaled R, Nacke L (2011) From game design elements to gamefulness: defining "gamification". In: Proceedings of the 15th international academic MindTrek conference: envisioning future media environments, pp 9–15

Firmino AS et al (2020) Towards Industry 4.0: a SWOT-based analysis for companies located in the Sorocaba Metropolitan Region (São Paulo State, Brazil). TT—Towards Industry 4.0: a SWOT-based analysis for companies located in the Sorocaba Metropolitan Region (SP, Brazil). Manag Prod 27(3):1–21

Franco S et al (2024) Sustainable logistics management plan and its indicators: minimum content for disclosure, awareness, and training in Brazilian Federal Universities Revista Verde, vol 1(5), pp 66–69
Gugerell K, Zuidema C (2017) Gaming for the energy transition. Experimenting and learning in co-designing a serious game prototype. J Clean Prod 169:105–116
Hallinger P et al (2020) A bibliometric review of research on simulations and serious games used in educating for sustainability, 1997–2019. J Clean Prod 256:120358
Harrisson JP (2010) Strategic planning and SWOT analysis. In: Harrison JP (ed) Fundamentals of strategic planning in health care. Health Administration Press, Chicago
Katsaliaki K, Mustafee N (2015) Edutainment for sustainable development: a survey of games in the field. Simul Gaming 46:647–672
Krause M, Smeddinck JD (2011) Game-based learning in higher education: the impact of gamification on learning environments. In: Proceedings of the 9th international conference on advances in computer entertainment technology, pp 1–5
Lopes CAP, Becker-Ritt AB, Azambuja LL (2024) Gamificação na educação de jovens e adultos (EJA): ensino de química através de sequência didática sobre necrochorume. REAMEC-Rede Amazônica De Educação Em Ciências e Matemática 12:e24034–e24034
Madani K, Pierce TW, Mirchi A (2017) Serious games on environmental management. Sustain Cities Soc 29:1–11
Martins B, Maria B (2024) Environmental education and its contributions to a more sustainable planet. Rev Verde 1(5):137–148
Mcgonigal J (2011) Reality is broken: why games make us better and how they can change the world. Penguin Press
Ministério Do Meio Ambiente (MMA) Inovar ambiental: Saiba o que é o conceito de produção sustentável e quais seus benefícios, c2020. Página inicial. Disponível em: https://inovarambiental.com.br/2020/05/04/producao-sustentavel-beneficios. Acesso em: 30 de jan. de 2023
Oliveira MA-Y, Ferreira JJP (2011) Book review: business model generation: a handbook for visionaries, game changers and challengers by Alexander Osterwalder and Yves Pigneur. Afr J Bus Manage 5(7):1459–1460. http://www.academicjournals.org/AJBM
Prestes SGL, Mergulhão RC, Silva DAL, Mendes JV (2023) Gamificação de tabuleiro para educação na produção sustentável: uma análise crítica da literatura. XXX Simpósio de Engenharia de Produção: Transformação Digital e Gestão De Operações: Desafios e Tendências, Bauru, SP, Brasil
Reeves B, Read JL (2009) Total engagement: using games and virtual worlds to change the way people work and businesses compete. Harvard Business Press
Rossano V, Roselli T, Calvano G (2018) A serious game to promote environmental attitude. In: Uskov V, Howlett R, Jain L (eds), Smart education and e-learning 2017, vol 75. Springer, pp 48–55
Salibi JGdeOR, Machado BA, Rodrigues JdeS, Souza FB, Campos Rde (2021) The use of games in a graduation course in an engineering faculty. Int Symp Proj Approaches Eng Educ 11:283–291. https://doi.org/10.5281/zenodo.5096805
Shneiderman B (2004) Designing for fun: how can we design user interfaces to be more fun? Interactions 11(5):48–50
Skinner BF (1957) Verbal behavior. Copley Publishing Group, Acton
Tilvawala K et al (2023) Values-based transformative games: from the physical to the digital. In: Proceedings of the European conference on games-based learning, vol 2023-Octob, n. section 3, pp 683–692
Tsai JC et al (2021) Using a board game to teach about sustainable development. Sustainability (Switzerland) 13(9):P1-19

United Nations Environment Programme. Global Resources Outlook 2024: bend the trend—pathways to a liveable planet as resource use spikes. International Resource Panel. Nairobi, 2024. Página inicial. Disponível em: https://wedocs.unep.org/20.500.11822/44/. Acesso em: 14 de mar de 2024

Von Ahn L, Dabbish L (2008) Designing games with a purpose. Commun ACM 51(8):58–67

Student-Centered Teaching for Sustainability Education in an Introductory Biology Course at Winona State University: A Case-Study

Bruno Borsari

Abstract Sustainability education has emerged as a common denominator among higher education institutions (HEIs) in the twenty-first century to prepare all citizens adapting to the vagaries of climate change and uncertainties brought by environmental and social shifts. The purpose of this chapter consisted in showcasing selected student-centered pedagogies in a freshman biology course that in the author's experience were effective for stimulating an education in sustainability. These aimed at enhancing awareness about sustainable development among undergraduate students at Winona State University. Between 2005 and 2019, the author served as coordinator of introductory biology courses that enrolled about 7000 students as non-science majors, or undecided. Framed by constructivism theory and the SDGs of UN Agenda 2030, four student-centered teaching methods were employed consistently, while connecting across disciplines to make learning relevant to most students, through the years. Data collected from students' assignments were substantiated by the number of scholarly publications produced by the author within this 14-year timeframe with colleagues from the Biology Department at WSU and others. This effort demonstrates an achievement of quality education for sustainable development, while fulfilling SDG 4 and its targets. A distinctive model is presented in conclusion to this work to highlight benefits and opportunities that could be gained in future efforts to implement sustainability teaching practices in the biological sciences and more curricula.

Keywords Case-study · Sustainability education · Biology · Teaching · Student-centered pedagogies · Sustainable development

---

B. Borsari (✉)
Winona State University, Winona, MN, USA
e-mail: bborsari@winona.edu

W. Leal Filho et al. (eds.), *Education for Sustainable Development: The Contribution of Universities*, World Sustainability Series,
https://doi.org/10.1007/978-3-031-86985-3_3

1 Introduction

Student-centered teaching approaches have become well established for every instruction level to make learning pertinent (Burrowes 2003; Byrne 2016), while veering education into transformative processes, applicable to resolve present and foreseeable challenges (Creighton-Hammond 1998; Sterling 2001; Leal Filho 2022). Similar approaches along this trajectory become inviolable when the teaching/learning endeavor is emphasizing sustainable development (SD), because this aims at preparing society to embrace holistic values and take actions in shifting life styles from extremely consumptive and extractive, toward frugal and adaptive in achievement of more sustainable living habits. A vast body of literature exists already in support of these claims (Borsari and Kunnas 2023; Filho et al. 2023; Leal Filho et al. 2022a), whereas an overall consensus among stakeholders indicates that schools and higher education institutions (HEIs) have potential to assist with the necessary changes in modern society for pursuing a sustainable world (Filho et al. 2023; Ashida 2023; Leal Filho et al. 2021). In the last 50-year history of this quest, as documented by the Club of Rome in the early 1970s, it was realized that inevitable constraints impede an unlimited growth of economies, due to the finite nature of planet Earth and its resources (Meadows et al. 1972). More recently, Onwueme and Borsari (2007) added that despite efforts to achieve sustainability, every individual, enterprise, or organization carry with them an inevitable sustainability deficit, making it impossible to achieve 100% sustainability. Thus, under these circumstances, the ultimate challenge for attaining SD may consist in seeking an improved balance between resource use and regeneration of the same (assuming that these are renewable). The purpose of this chapter is to present selected student-centered instruction methods that in the author's experience functioned effectively, for spurring life science education in sustainability.

2 Pedagogies for Sustainability Education

The world continues to change rapidly, posing new challenges that demand for unequivocal flexibility, innovation and adaptation to incessant environmental shifts. Hence, a sustainability education has become a learning priority for students at every level of instruction (Borsari 2012), and this should be pursued as Leal Filho (2022) pointed out, in a multidisciplinary fashion. The United Nations Agenda 2030 serves as template to achieve sustainability among world nations. Amid its 17 goals, SDG# 4 (Quality Education), plays a pivotal role in guiding the preparedness of present and future society in its quest for sustainable development. The intersectionality between multidisciplinary education framed within constructivism theory, and SDGs constitute the theoretical underpinning of this chapter, where biology becomes foundational among disciplines to lead students in understanding the functioning of ecosystems

Fig. 1 Theoretical framework for sustainability education

processes, and impacts against these caused by human activities. Within these contextual boundaries, Lord (1998) conceded that ample time should be given to students for exploring, reflecting and evaluating what is being taught, while reducing lecture time (Fig. 1).

In addition, Borsari (2012), added that the frame for sustainability education and study programs in sustainability should be eco-centric! However, ecocentrism requires embracing values like simplicity of life styles, humility and a recognition that we humans are parts of nature and not in control of it, as reductionist, anthropocentric worldviews want to make us believe (Kimmerer 2013; Borsari and Kunnas 2023). Against this background, this chapter presents selected didactic methodologies for teaching introductory biology courses, that were refined by the author during his 14-year tenure at Winona State University. This work aims at inspiring instructors to engage students more actively, while enthusing them in learning biology and its implications for sustainable development, through multifaceted approaches that connect across academic disciplines in the most inclusive manner. There is a compelling need to reconnect humanity with nature in spite of advancing computer technologies and virtual realities (Louv 2011). Now more than ever, a cultural shift towards stewardship and conservation for biodiversity is sorely needed (Wilson 2016). Yet, for realizing this vision through education, necessary reforms are becoming vital as predicated by Borsari and Mora (2020), to veer learning and teaching more emphatically, towards biophilic, eco-centric approaches.

3 Methodology

Case-study research is a distinctive mode of inquiry which can be valuable to tackle specific problems (Yin 2017). Thus, this case presents selected instruction approaches employed by the author, while teaching a freshmen biology course (BIO 118), for non-science majors between 2005 and 2019. Distinctive examples of innovative pedagogies in sustainability studies are available (Tsai et al 2022; Byrne 2016).

Initially, students' presentations and/or poster sessions were used to assess students' learning and achievement beyond the typical multiple-choice exams, or quizzes. In 2007 personal response systems devices ('clickers'), were introduced in large classes like biology at Winona State University (WSU), allowing for testing the viability of this technology when combined with case-studies, which according to Herreid (2011), engages students a step further resolving misconceptions and increasing critical thinking when learning science. In fact, Lundeberg and team (2011) discovered that combining clickers technology with case-study teaching boosted students' learning and achievement, whereas Wolter and collaborators (2011), revealed that this approach improves also students' perceptions of their understanding of science in large introductory biology classes.

The two case studies employed in this work were produced and presented by the author to his student classes in alternation (one semester the cases were taught, whereas the following semester the same topics were presented via lectures). Both teaching methods considered a sample of five multiple-choice questions that were presented to each audience before the case (pre-test), or lecture and again at the end of class (post-test), as approaches to assess each teaching method.

Another strategy consisted in employing art works making in the biology classroom. This method was inspired by a project implemented by the author in collaboration with colleagues from the Biology and Geoscience Departments at WSU, who engaged art majors and their professors in producing art works about Earth during the Precambrian Era (Borsari et al. 2008). From this project the author developed an articulated lesson plan that expanded students' artistic expression also towards music, poetry, dance and visual arts.

Lab instruction is a relevant component of BIO 118 and weighs 25% of students' final letter grade. When the weather conditions allowed (during the first half of the fall semester and during the last 3–4 weeks of each spring semester), lab instruction took place outside instead of operating from within the walls of the laboratory building. Lab exercises were experiential and considered works in both terrestrial and aquatic habitats to give students a thorough view and understanding of biodiversity and legitimization for conserving it, as a sustainability priority need. A schematic view of these teaching strategies in freshmen biology at WSU is reported together with complementary, learning objectives (Table 1).

Table 1 Student-centered teaching methods in the introductory biology course (BIO 118)

Method	Title/Activity	Learning objectives	Reference/s
Case-study teaching with clicker technology	Prairie garden of troubles	a. Learn the basic ecological characteristics of a prairie ecosystem, including prairie ecological services (soil humus formation, water conservation, pollination, biodiversity) b. Develop an appreciation for prairie habitat restoration and conservation c. Study plant biodiversity d. Be introduced to concepts of co-evolution and symbiosis among plant species and pollinators	Borsari (2011)
	Ecological footprint dilemma	a. Learn the concept of "ecological footprint" b. Understand the difference between renewable and non-renewable resources c. Measure each individual's footprint d. Propose methods to develop a sustainable community after an analysis of the footprints generated by the class	Borsari (2009)

(continued)

Table 1 (continued)

Method	Title/Activity	Learning objectives	Reference/s
Making art in the classroom	Several titles	a. Express feelings about sustainability through creative, collaborative art-work, without being inhibited by perceived inabilities to produce "good art" b. Communicate to classmates how artwork can convey a sustainability message c. Evaluate sustainability messages conveyed in artworks and transfer the acquired skills to all arts	Borsari (2016)
Students' presentations	Poster session & short talks (several titles)	a. Present with a companion, or a small group (3–4 students) a biological topic of your choice b. Engage every student in the class with the evaluation of the presenters c. Demonstrate the efficacy of this method in accord with Knowles' theory of andragogy	Borsari (1999)
Outdoor labs	Experiential activities	a. Offer opportunities to students to learn by doing in restoration projects that were accomplished on campus, or in its vicinity b. Develop an appreciation for studying both terrestrial, aquatic habitats and conservation	Borsari et al. (2014), Borsari and Vidrine (2022, 2023)

4 Results and Discussion

Here presented are the results (in chronological order), of the students' products derived from four distinctive teaching methods that were employed by the author in large freshmen classes of BIO 118, which is the introduction to biology (4-credits), with a weekly Lab component for non-science majors, at WSU.

Students activities and products consisted in:

- Small groups presentations on a topic of choice to be offered at the end of the term, or small group posters to be exhibited by the end of the semester
- Art works consisted in: drawings, poems, songs, 2-min ballet performances
- Lab reports of outdoor laboratory exercises, which included connections to sustainability
- Case-study teachings via personal response systems devices ('clickers') technology.

Data derived from the first three methodologies are reported (Table 2).

Students' oral presentations are a keystone feature of this study because these occurred every semester, whereas other activities (including the use of clickers in the classroom), were introduced at different times. This is to point out that multiple teaching methods overlapped with the intention of diversifying the instructional process and with the expectation of accommodating students' learning styles and boosting enthusiasm for learning biology.

Table 2 Students' products for BIO 118 between 2005 and 2019

Academic year	Presentations	Posters	Art works	Outdoor labs
2005–2006	14	0	0	0
2006–2007	23	34	0	0
2007–2008	28	36	0	5
2008–2009	25	32	0	9
2009–2010	33	28	0	12
2010–2011	25	42	0	0
2011–2012	26	38	0	12
2012–2013	24	37	0	3
2013–2014	21	0	0	12
2014–2015	22	0	0	0
2015–2016	13	0	182	6
2016–2017	18	0	166	9
2017–2018	9	0	173	8
2018–2019	12	0	177	7
Total	293	247	698	83

4.1 Students' Presentations

Students' oral presentations during 2005–2018 period engaged 1668 students who voluntarily accepted this challenge with 1–3 companions to replace an unsatisfactory exam score. This method applied Knowles' theory of andragogy and tested learners' ability to contribute to their own learning when motivated to prepare a 15-min talk about a biology topic of choice (Borsari 1999). Presentation scores derived from the instructor evaluation (50% of the grade), which was weighted against the mean score from the audience's scores at the end of each talk (Fig. 2).

Within the presenters' sample, 292 students (22%) excelled in this task earning a score that was at 0.90 or higher, whereas 446 students (34%), did well earning a score that was between 0.80 and 0.89. Two hundred ninety-seven students (22.6%), earned an average grade that was within 0.79 and 0.70, whereas 277 students (21.4%) scored below 70%. This experience improved students' grade point average (GPA) in the course for a majority of students (89%), whereas it did not have an effect for others (11%).

Overall, this exercise fostered learner's confidence in communicating biology concepts to a large audience, while sharpening communication skills for debating and answering questions in a public forum. Preparing for the talk was a challenging

Fig. 2 Students' presentation about discovery and use of antibiotics, while students in the audience are taking notes from the presenters' slides on their computers

Fig. 3 Students working independently to prepare their talk about the diversity of deciduous tree species at Garvin Heights park in Winona, Minnesota

task, which always required independent and consistent work. Also, most students met regularly with their instructor who made himself available for advice, resolving logistic problems, and for reviewing the presentation slides prior to students' presentation time (Fig. 3).

As alternative to the oral presentation 356 students, or (21%) of the initial population, selected to prepare and exhibit a poster about one Lab exercise of choice, that had been taught during the semester to demonstrate their understanding of the scientific process. As for the short talks in class, the posters were evaluated (on a specific evaluation form prepared by the instructor) by all students enrolled in their Lab section with the exception of the poster author/s (Fig. 4). Evaluation criteria were the same as those employed for assessing the oral presentations however, this assignment carried 15% less points than the score of oral presentations, as the instructor explained to all students on the first day of class, due to less originality in the choice of the topic and no authors' exposure to questions as their peers who had chosen to prepare the talk.

Poster sessions occurred on the last Lab meeting of the week prior to the last week of classes. Within the presenters' sample (n = 356), ninety-six (27%) excelled in this task earning a score that was at 0.90 or higher, whereas 112 students (30%) did well earning a score that was between 0.80 and 0.89, whereas 156 students (44%) scored

Fig. 4 Students' poster exhibit held in the biology lab at the end of the semester

between 0.79 and 0.70. This experience improved students' GPA in the course for all students involved in presenting a poster.

4.2 *Case Studies Analysis*

For eight years between 2006 and 2015 (except during the academic years 2010–11 because the instructor was on sabbatical leave and in 2014–15 because he had been awarded a Fulbright grant), 2264 students were taught prairie restoration and the ecological footprint/sustainability topics through lecture presentations, whereas 2002 students were taught the same topics via case studies. The two instruction strategies were alternated by the author over the course of 20 semesters. The mean scores and error bars for the pre-test and post test scores for participants in the prairie garden case (blue column) and prairie lecture (red column), are reported (Fig. 5).

The mean scores and error bars for the pre-test and post test scores for participants in the ecological footprint dilemma case (blue column) and lecture (red column), are reported (Fig. 6).

The interpretation of these results is descriptive only and should be qualified by the fact that participants were not assigned randomly to conditions and that, various differences might have characterized the student classes from semester to semester. The variability of students' samples size from one semester to the next term, was another constraint, which did not allow the use of a parametric test to determine

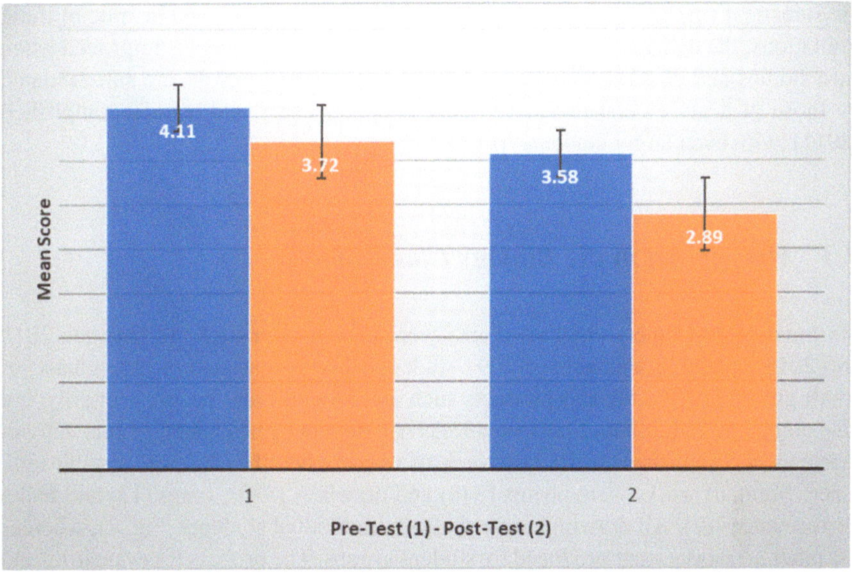

Fig. 5 Pre-test and post-test data for the prairie garden case (blue) and lecture classes (red)

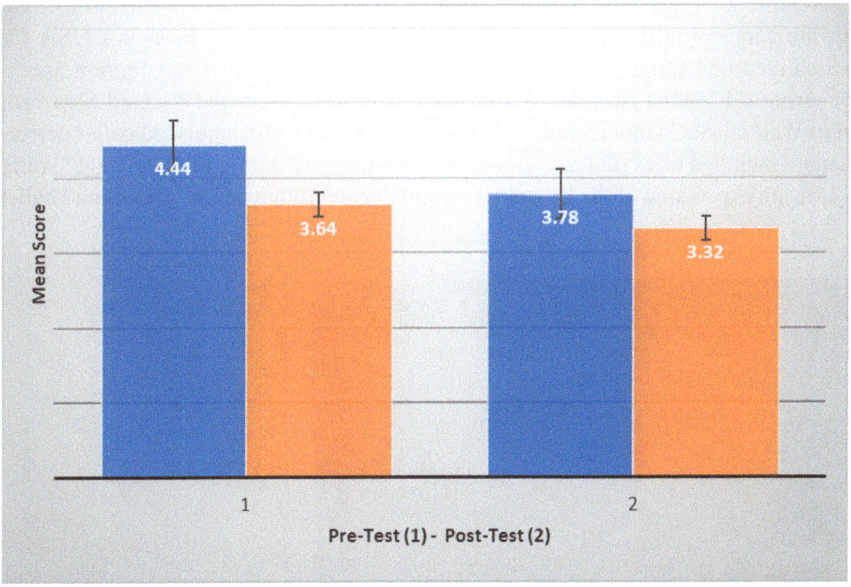

Fig. 6 Pretest and posttest data for the Ecological Footprint case (blue) and lecture classes (red)

the statistical significance of the case-study method. However, and in spite of these limitations, the data suggest that learning in large biology classes is improved when case studies and clickers technology are employed. These results are substantiated by those of a more comprehensive study that Lundeberg and team accomplished (2011), with even larger students' populations.

4.3 Making Art in the Biology Classroom

Six hundred and ninety-eight art works were produced in BIO 118 between 2015 and 2018, engaging a sample of 2094 students. These consisted of individuals' or small groups' (1–4 students) products such as: drawings and poems primarily, but also songs, or 2-min ballet performances. All students participated in this activity that was accomplished in the last week of class, when the "bio-art" exhibit took place. Main art works were poems (348) and drawings (330), songs (13) and ballet performances (7). All drawings were made by individual students (Fig. 7), whereas the other art works were produced by student groups. The process for evaluating and rewarding students' efforts in this activity followed a specific protocol proposed by Borsari (2016). A sample of students' poems is reported (Fig. 8).

Notable song titles that were produced and performed in class spanned from: "We are the genes" (on the music of Lionel Richie's and Michael Jackson's USA for Africa recorded song: "We Are the World"), to "Our non-human, common ancestor" from ("I Wanna Be Like You" song written by Robert and Richard Sherman, from Walt Disney's movie picture "The Jungle Book"). Distinguished ballet performances included titles such as: "Dancing Chromosomes during Mitosis", and "Water is Life: an expression of the hydrologic water cycle through dance". Songs and ballet

a) b)

Fig. 7 Students' drawings **a** Volcanism in the Precambrian era, **b** prairie smoke (*Geum triflorum*), iconic plant species of Midwest prairies

Bruno, Bruno, this semester was grand
I am so glad I took BIO instead of taking band
We learned about cells and all about Phylums
I was up so long studying, they could've put me in asylum

I had so much fun and learned so much stuff
Even though exam three was kind of rough
You better be there when I graduate, just to say "Cheers"
But don't you worry, Bruno

I'll bring the beers!

BIO 118 was a challenge
but BIO 118 was fun
I learned a lot of science
and now I am a little less dumb

from atoms to ecosphere
I learned it all
Now that it's over
I hope my grade won't fall

Biology is cool, Biology is fun
As I sit here watching the clock clicking down
I know there are germs all over my desk and floating around

I am ready for this semester to end

But I am glad I have learned about the bacteria
My body can withstand

Thank You for teaching me all that you know
I can now distinguish a vacuole from a mole!

The energy from the sun
Is absorbed by the plants for fun
Then, it is converted into food

then consumed to brighten the mood
of all living beings
These plants grow, grow and stand tall
Then will senesce, desiccate and fall

Giving the fungus a turn
of feasting on dead biomass
so that the carbon cycle
will never end

Fig. 8 Sample of students' poetry

performances were all group works, whereas poems were written by some individuals (32%), student pairs (48%), or small groups (15% three-student groups and 5% four-student groups). All students had been informed about the date of the bio-art exhibit on the first day of instruction, so that they had the entire term to prepare and develop their art works that were shared in class, at the end of the semester. In accord with Aghasafari and colleagues (2021), an adoption of a culturally sustaining pedagogy, inclusive of opportunities for unleashing students' artistic expressions can be supportive for all students learning biology and its implications for sustainability.

4.4 Outdoor Lab Instruction

The significance of lab, or class instruction taking place outdoor has always been the author's most passionate focus for spurring an education in sustainable development (Borsari 2012; Borsari et al. 2014; Borsari and Vidrine 2022, 2023), because this aims at reconnecting individuals with nature and biodiversity, which remain pivotal themes for biology and sustainability studies. Between 2007 and 2018, eighty-three lab exercises were taught/held in the outdoors, while taking advantage of the variety of ecological habitats that characterize the beauty and biological richness of Winona and its distinctive landscape. The WSU campus itself was recognized as "Arboretum

USA" in 2013 and provides remarkable opportunities to learn about native plant communities and their associated biota, from different terrestrial habitats (deciduous forest, prairie savannahs, restored prairie gardens), to aquatic habitats and adjacent ecotones of the riverine ecoystem like the Mississippi river and the many creeks of its watershed, that shape the landscape of this bioregion (Fremling 2005). A selection of photographic images (Figs. 9, 10 and 11), serve as qualitative data that within the timeframe of this study engaged about 3500 students.

Getting into the water to measure the habitat traits (e.g.: creek width, water depth, turbidity, velocity, pH, and nitrates), or applying the mark-recapture technique to estimate trout's population size as proposed by Brower and team (1997), were hands-on activities that enthused a broad majority of students. The distinctive approach of these labs consisted in collecting real data that allowed all learners to test hypotheses and draw conclusions about the quality of an aquatic habitat and its inhabitants like

a) b)

Fig. 9 Outdoor labs on Gilmore creek, Winona **a** assessment of the habitat physicochemical characteristics, **b** estimating the native, trout's population size through mark-recapture method

a) b)

Fig. 10 **a** Students trampling on the stubbles of grasses and forbs at WSU prairie garden (April 2014), in preparation of a prescribed burn, **b** partial view of the same prairie garden in the summer of the same year, following the burn

Fig. 11 Students assessing the diversity of soil macro-invertebrates at the author's home garden

fish, or macroinvertebrate species (Fig. 9a, b), making them realize how much water quality plays for the livelihood and well-being of all organisms, including humans.

Winona State University campus offers multiple thematic gardens that similar to outdoor labs as proposed by Kingsolver (2006) were employed directly, or adapted to the conditions and resources available at the time. Therefore, even preparing a site for a prescribed burn, or for restoration with native plants (Borsari and Vidrine 2023) became valuable learning experiences, at the onset of each new growing season (Fig. 10a, b).

The health of plant communities depends upon the health of biota inhabiting soils thus, making the soil ecosystem a wonderful environment to consider for studying biology (Weil and Brady 2017). The biological focus on soil was for investigating soil life as these decomposer species are the micro-engines of the carbon cycle, whose functioning is key for achieving regenerative and healthy ecosystems (Gliessman 2000; Vidrine et al. 2024) (Fig. 11).

In common with the living labs presented by Leal Filho and colleagues (2022a), the outdoor labs described here aim at achieving SDG 4 (Quality Education) yet, whereas SDG 11 is the second objective of living labs, the outdoor labs in biology courses can embrace a larger group of SDGs.

5 Conclusion and Final Comments

The experience of teaching and learning in a freshmen biology course as reported here, can play a pivotal role for an amplification of sustainability education on this campus and beyond. This case-study is a descriptive work that highlighted didactic approaches in a specific research context. Hence, a possible application of its findings elsewhere constitutes a major limitation unless necessary adaptations of the pedagogies presented here are enacted. In addition to this, it remains important to overcome existing constraints that still persist within HEIs and not only for those regarding living labs (Leal Filho et al. 2022a), governance (Filho et al. 2023), social values (Leal Filho et al. 2022c), curricula (Borsari and Kunnas 2023; Leal Filho et al. 2021), or even more philosophical themes like spirituality (Leal Filho et al. 2022b).

The benefits of constructivism theory when learning biology were demonstrated through unremitting efforts of engaging students in the education process and this can truly happen, when they become more responsible for their own learning (Lord 1998; Borsari 1999). Another benefit of this pedagogy is that when students are actively engaged, they optimize their education and this enables them to apply knowledge beyond the classroom setting. From here connecting knowledge to sustainability in search of effective solutions to resolve current and foreseeable problems in conserving planetary health (Borsari and Vidrine 2022), can be just a short step away! The core of transformative learning embodied by an education in sustainability is that most students will remain passionate life-long learners, enhancing the probabilities that their knowledge may trigger the auspicious, needed changes for developing a more sustainable world.

The core of the matter consists in reforming education, while investing more resources in the preparation of teachers. The latter could become most relevant especially if these are faculty members in the biological sciences because their decision-making is extremely important when adopting, or adapting eco-centric learning approaches for an education in sustainability (Louv 2011; Borsari 2012; Leal Filho 2022). Nonetheless, an education for sustainability should be eco-centric and place-based, where learning occurs in environments abundant with experiential activities. These take place in nature mostly, to appreciate the biodiversity of ecosystems and to comprehend the many reasons for preserving this resource for future generations. Consequently, an understanding of conservation biology across curricula is of paramount importance to create awareness in this field because it is through a handful of its basic principles like coevolution, extinction, populations' genetics, pandemics, biogeochemical cycles and energy flow, that an education for sustainability can truly succeed in veering humanity toward a sustainable development. However, Berry and her team (2018) discovered that among stakeholder groups for example, there is always an assortment of views about biodiversity, explaining that species appraisals may not justify always an obliteration of nature thus, amplifying variances among these groups and their support for either an eco-centric, or an anthropocentric worldview. Therefore, it is imperative to boost education in sustainability with an emphasis on conservation, rehabilitation and sustainable management of ecosystems, while engaging all students in understanding the need to restore landscapes to an improved status of ecological functioning. This awareness can be achieved through outreach efforts similar to those provided for farmers, gardeners, ranchers, homeowners by the extension personnel of the U.S. colleges of agriculture. Borsari and Vidrine (2023) envision curriculum modules with an abundance of hands-on activities that could become transformative for educating all learners in sustainability. Shifting the emphasis of curricula (from pre-k to college-level), towards ecocentrism is the necessary reform in education, which is needed to develop a culture for stewardship and sustainability (Borsari and Mora 2020). Thus, reconnecting people with nature while restoring human well-being and planetary health is the trajectory professed also by the novel 'One-health' paradigm that aims to sustainably balance and optimize the health of people, animals and ecosystems (Borsari and Vidrine 2022).

Herein, a unifying model for developing study programs in sustainability education is proposed. It was inspired by the Permaculture Tree model created by Mollison and Holmgren (1978), which exemplifies the evolutionary process of realizing a valuable product from an initial idea. These authors developed and employed their paradigm to organize the elements of permaculture design where the preliminary idea (the starting construct of the whole process), stands within the tree trunk. In a similar fashion, sustainability education conceived as a well-structured plan of studies, or curriculum is the keystone idea within the tree trunk, at the center of this symmetrical model, which solidly, supports the roots that are the system's sources, with the canopy and its fruits, or products (Fig. 12).

The roots underground represent the broadest array of human disciplines and these are the sources of knowledge, necessary to convert ideas into more products. The

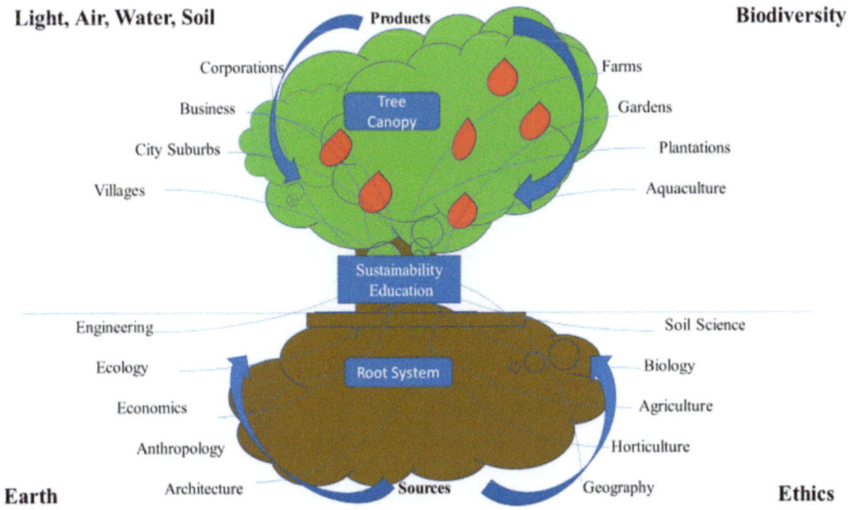

Fig. 12 Adaptation of the permaculture tree model for a wider implementation of an education for sustainable development

latter appear as the red fruits hanging from the tree canopy, representing the tangible outcomes of knowledge 'put to work' to resolve human problems with sustainability as a priority. Similarly, sustainability education is located at the convergence of the network that connects every source with the products, highlighting the inclusiveness and complexity of the model. At its corners are the Earth, the only home capable of supporting life as we know it, together with abiotic resources (light, air, water, soil), biodiversity and ethics, which represent concept features of a sustainability education. The ethical principles of present civilization embrace every source and product, and are essential for maintaining the balance within the tree model, where sustainability education remains its fulcrum.

Aldo Leopold's writings from his Land Ethic essay expanded the "community" concept to include all biotic and abiotic components of the Earth (soil, water, plants, and animals)—"the land". Hence, a land ethic is the moral protocol that emerges from these interconnected relationships to increase conservation strategies in biology. In a similar fashion, an ethical framework for land conservation should be revered as priority for establishing and advancing an education in sustainability. Auspiciously, the model emerged from this case-study research will enhance further reforms in support of creative pedagogies for sustainability education.

References

Aghasafari S, Bivins K, Nordgren B (2021) Arts integration and culturally sustaining pedagogy: supporting bi/multilingual high school learners in biology. J Interdisc Stud Educ 10(1):59–81. https://ojed.org/index.php/jise/article/view/3172

Ashida A (2023) The role of higher education in achieving the sustainable development goals. In: Urata S, Kuroda K, Tonegawa Y (eds) Sustainable development disciplines for humanity. Sustainable Development Goals Series. Springer, Singapore. https://doi.org/10.1007/978-981-19-4859-6_5

Berry PM et al (2018) Why conserve biodiversity? A multi-national exploration of stakeholders' views on the arguments for biodiversity conservation. Biodivers Conserv 27:1741–1762. https://doi.org/10.1007/s10531-016-1173-z

Borsari B (2012) Curriculum framework for sustainability education. Acad Exch Q 16(1):74–78

Borsari B, Richardson R, Scott Plummer A et al (2008) Earth's Precambrian era as a common evolutionary theme in two art courses at Winona State University. Evo Edu Outreach 1:172–178. https://doi.org/10.1007/s12052-008-0028-9

Borsari B, Kunnas J (2023) Historical memory and eco-centric education: looking at the past to move forward with the 2030 agenda for sustainable development. In: Leal Filho W, Azul AM, Doni F, Salvia AL (eds) Handbook of sustainability science in the future. Springer, Cham. https://doi.org/10.1007/978-3-031-04560-8_40

Borsari B, Mora C (2020) Education reform through a systems approach for sustainable development. In: Leal Filho W, Azul A, Brandli L, Özuyar P, Wall T (eds) Quality education. Encyclopedia of the UN Sustainable Development Goals. Springer, Cham

Borsari B, Vidrine MF (2022) Planetary health begins on campus: enhancing students' well-being and health through prairie habitat restoration. In: Leal Filho W (ed) Handbook of human and planetary health. Climate change management. Springer, Cham. https://doi.org/10.1007/978-3-031-09879-6_14

Borsari B, Vidrine MF (2023) Envisioning prairies as curricular focus for climate change education. In: Leal Filho W, Sima M, Lange Salvia A, Kovaleva M, Manolas E (eds) University initiatives on climate change education and research. Springer, Cham. https://doi.org/10.1007/978-3-031-25960-9_56-1

Borsari B, O'Connor K, Mundahl N (2014) Design process and reconstruction of a prairie garden at Winona State University: a case study. In: Glaser A (ed) Proceedings of the 2nd Biennial conference on the conservation of America's grasslands. 12–14 Aug 2013. National Wildlife Federation and Kansas State University, Manhattan, pp 72–74

Borsari B (1999) Practical applications of the andragogical theory assumptions in introductory biology courses. adopting an educational paradigm that targets the adult learner. J Coll Sc Teach 28(5):311–314

Borsari B (2009) The ecological footprint dilemma. Available at: https://www.nsta.org/ncss-case-study/ecological-footprint-dilemma

Borsari B (2011) Prairie garden of troubles. Available at: https://www.nsta.org/ncss-case-study/prairie-garden-troubles

Borsari B (2016) Making and assessing art in the sustainability classroom. In: Byrne L (eds) Learner-centered teaching activities for environmental and sustainability studies. Springer, Cham. https://doi.org/10.1007/978-3-319-28543-6_43

Brower JE, Zar JH, von Ende CN (1997) Field and laboratory methods for general ecology, 4th edn. WCB McGraw-Hill. Boston, MA. ISBN 0-697-24358-3

Burrowes PA (2003) A student-centered approach to teaching general biology that really works: Lord's constructivist model put to test. The Am Biol Teacher 65(7):491–502

Byrne LB (2016) Learner-centered teaching activities for environmental and sustainability studies. Springer, Cham. https://doi.org/10.1007/978-3-319-28543-6

Creighton-Hammond S (1998) Greening the ivory tower. Improving the environmental track record of universities, colleges, and other institutions. MIT Press, Cambridge, MA. https://doi.org/10.7551/mitpress/3367.001.0001

Filho WL, Abubakar IR, Mifsud MC et al (2023) Governance in the implementation of the UN sustainable development goals in higher education: global trends. Environ Dev Sustain. https://doi.org/10.1007/s10668-023-03278-x

Fremling CR (2005) Immortal River. The Upper Mississippi in ancient and modern times. The University of Wisconsin Press. Madison, Wisconsin

Gliessman SR (2000) Field and laboratory investigations in agroecology. CRC Press. Boca Raton, Florida. ISBN 1-55670-445-6

Herreid CF (2011) Case study teaching. New Dir Teach Learn 128:31–40. https://doi.org/10.1002/tl.466

Kimmerer RW (2013) Braiding sweetgrass: indigenous wisdom, scientific knowledge, and the teachings of plants. Milkweed Editions

Kingsolver RW (2006) Ecology on campus. Lab manual. Pearson Education, Inc., San Francisco, California. ISBN 0-8053-8214-3

Leal Filho W, Frankenberger F, Salvia AL, Azeiteiro U, Alves F, Castro P, Will M, Platje J, Lovren VO, Brandli L, Price E (2021) A framework for the implementation of the sustainable development goals in university programmes. J Clean Prod 25(299):126915

Leal Filho W, Ozuyar PG, Dinis MAP et al (2022a) Living labs in the context of the UN sustainable development goals: state of the art. Sustain Sci 18:1163. https://doi.org/10.1007/s11625-022-01240-w

Leal Filho W, Lange Salvia A, Ulluwishewa R, Rimi Abubakare I, Mifsfud M, LeVasseur TJ, Correia V, Consorte-McCrea A, Paço A et al (2022b) Linking sustainability and spirituality: a preliminary assessment in pursuit of a sustainable and ethically correct world. J Clean Prod 380:135091. https://doi.org/10.1016/j.jclepro.2022.135091

Leal Filho W, Levesque V, Sivapalan S et al (2022c) Social values and sustainable development: community experiences. Environ Sci Eur 34:67. https://doi.org/10.1186/s12302-022-00641-z

Leal Filho W (2022) Pursuing sustainable development across disciplines. In: Leal Filho W, Portela de Vasconcelos CR (eds) Handbook of best practices in sustainable development at university level. World Sustainability Series. Springer, Cham. https://doi.org/10.1007/978-3-031-04764-0_27

Lord T (1998) Cooperative learning that really works in biology teaching: using constructivist-based activities to challenge student teams. The Am Biol Teacher 60(8):580–588

Louv R (2011) The nature principle: reconnecting with life in a virtual age. Algonguin Books, Chapel Hill

Lundeberg MA, Kang H, Wolter B et al (2011) Context matters: increasing understanding with interactive Clicker Case studies. Educ Tech Res Dev 59:645–671. https://doi.org/10.1007/s11423-010-9182-1

Meadows DH, Meadows DL, Randers J, Behrens WW III (1972) The limits to growth. A report for the Club of Rome's project on the Predicament of Mankind. Universe Books. New York

Mollison B, Holmgren D (1978) Permaculture One: a perennial agriculture system for human settlements. Corgi Publishing, Australia & New Zealand. ISBN 10: 0552980609/ISBN 13: 9780552980609

Onwueme I, Borsari B (2007) The sustainability asymptogram: a new philosophical framework for policy, outreach and education in sustainability. Int J Sustain Higher Ed 8(1):44–52. https://doi.org/10.1108/14676370710717580

Sterling S (2001) Sustainable education. Re-visioning learning and change. Green Books Ltd., Devon, UK

Tsai N, Kang H, Chang J, Cassese K (2022) Adapting existing curriculum for equitable learning experiences. Science Scope 45(5):44–51. https://doi.org/10.1080/08872376.2022.12291478

Vidrine MF, Borsari B, Pastorek M, Allen CM (2024) Rewilding the Cajun Prairie: efforts in land rehabilitation ensure a sustainable future in Southwestern Louisiana and beyond. In: Leal Filho W, Salvia AL, Portela de Vasconcelos CR (eds) An agenda for sustainable development research. World Sustainability Series. Springer, Cham. https://doi.org/10.1007/978-3-031-65909-6_6

Weil RR, Brady NC (2017) The nature and properties of soils, 15th edn. Prentice Hall. Upper Saddle River, NJ. ISBN-13:978-0-13-325448-8

Wilson EO (2016) Half-Earth: our planet's fight for life. Liveright, New York

Wolter BHK, Lundeberg M, Kang H, Herreid CF et al (2011) Students' perceptions of using personal response systems ('Clickers') with cases in science. J Coll Sc Teach 40(4):14–19

Yin RK (2017) Case study research and applications, design and methods, 6th edn. Sage Publications, Inc. https://us.sagepub.com/en-us/nam/case-study-research-and-applications/book250150

Can Climate Anxiety and Awareness Drive Academic Behaviour in Higher Education?

Enzo Ferrari, Lorraine Whitmarsh, Nancy MacCann-Alfaro, Claudia Cáceres-González, Elvia Melara, Eva Hernadez-Gallego, and Camilo Ruiz

Abstract Climate change is one of the most pressing global challenges that requires urgent action from individuals, organizations, and institutions. In the context of higher education, students, professors, and researchers play a crucial role in advancing climate action through their learning, teaching, and research endeavours. However, we know little about the factors that influence their engagement and commitment to climate and sustainability action. This study aims to fill this gap by identifying and analysing the specific factors that influence climate action among students and staff in higher education, with the objective of developing effective strategies to enhance their involvement. The study draws on the results of an online survey in which 9121

E. Ferrari (✉) · C. Ruiz
EMC3 Research Group and Department of Math and Science Education, Universidad de Salamanca, Salamanca, Spain
e-mail: enzoferrari@usal.es

C. Ruiz
e-mail: camilo@usal.es

L. Whitmarsh
Department of Psychology, University of Bath, Bath, UK
e-mail: lw2253@bath.ac.uk

N. MacCann-Alfaro
Department of Social Work, Universidad de Atacama, Copiapo, Chile
e-mail: nancy.maccann@uda.cl

C. Cáceres-González
Escuela de Posgrado, Universidad Nacional de Itapua, Encarnación, Paraguay
e-mail: ccaceres@uni.edu.py

E. Melara
Department of Mathematics and Sciences, Universidad Tecnologica Centroamericana, Tegucigalpa, Honduras
e-mail: teslap53@unitec.edu

E. Hernadez-Gallego
Office of Environmental Quality and Sustainability, Universidad de Valladolid, Valladolid, Spain
e-mail: oficina.calida.ambitental@uva.es

W. Leal Filho et al. (eds.), *Education for Sustainable Development: The Contribution of Universities*, World Sustainability Series,
https://doi.org/10.1007/978-3-031-86985-3_4

students and academics from six universities in Europe and Latin America participated. The results show that climate education and research demonstrated associations with early and mid-career researchers, duration of work/study, and academic discipline, with engineering staff and students showing the highest level of involvement. Moreover, we found that climate anxiety and awareness of climate research groups significantly explain academic behaviour among students and academics. These findings inform targeted approaches to mobilize action within higher education institutions by focusing on the identified influencing factors. The outcomes emphasize the criticality of enhancing climate literacy and sustainability initiatives within higher education through tailored strategies that address these key factors. Higher education institutions are responsible for incorporating climate education, research, and sustainable practices to address the urgent issues of sustainability and climate change.

Keywords Climate action · Higher education · Climate change · University research · Climate anxiety

1 Introduction

Climate change remains one of the most significant global challenges, necessitating immediate and comprehensive action across all sectors of society. Effective climate action cannot be achieved solely through technological change. An essential component is the adoption of a comprehensive approach that encompasses the promotion of climate literacy, and fostering behaviour change (IPCC 2018).

Universities, recognised as bastions of knowledge and ingenuity, carry a singular responsibility to guide climate action by incorporating sustainability into their curricula, research, and campus operations (Timmerman & Metcalfe, 2009). Through their commitment to advancing climate education, promoting sustainability practices, and supporting research on environmental issues, universities can play a crucial role in global climate change mitigation and serve as a source of inspiration for future generations to take meaningful action (Hess and Collins 2018).

Despite the recognised importance of higher education institutions in promoting sustainability, there is limited understanding of the factors that influence engagement and commitment to climate action among students and staff (Lozano et al. 2019). Recent studies have emphasized the need to explore these factors to enhance the effectiveness of sustainability initiatives within universities (Han et al. 2023; Leal Filho et al. 2018a, b, c; Lozano et al. 2019).

Combating climate change demands significant changes in practices across a range of environments, including academic institutions (Nielsen et al. 2021). However, much of the current research on climate-related behaviour has centred on actions conducted within domestic settings, thereby frequently overlooking the influence of behaviour exhibited in professional and educational contexts. Nevertheless, the active participation of students and staff is essential for universities to accomplish decarbonization. Despite the prevalent climate awareness within academic institutions, significant obstacles hinder behavioural change (Latter et al. 2024). Conversely, recent research suggests that psychological factors like climate anxiety may serve as a catalyst for individual climate action, particularly among younger individuals, thus warranting significant attention within university communities (Whitmarsh et al. 2022).

This study aims to answer the research question: What factors predict climate action in higher education, including professional behaviour? There is a limited understanding of the factors that motivate or hinder climate engagement among key stakeholders, despite the acknowledged importance of this role. Therefore, the study aims to fill a gap in the existing literature by examining the factors that influence climate action in professional contexts within higher education. By doing so, it hopes to provide valuable insights that can improve the effectiveness of climate initiatives in academic institutions and contribute to sustainability efforts on a larger scale.

2 Literature Review

Understanding factors that impact academic conduct in climate education and research is vital to cultivate a more engaged and proactive academic community. Key predictors of academic behaviour in relation to climate change include institutional factors (e.g. career stage, discipline), social factors (e.g. presence of climate action teams) and psychological factors (e.g. climate anxiety).

2.1 Career Stage and Duration of Work/Study

The engagement of academics in climate education and research activities is significantly influenced by their career stage. Recent studies show younger and mid-career academics display heightened participation in climate-related academic behaviour when contrasted with their senior peers (Adlit and Adlit 2022; Chowdhury et al. 2021). Similarly, students with a longer duration of university education have higher levels of engagement in climate-related activities (Leal Filho et al. 2018a, b, c).

2.2 Academic Discipline

Academic field also plays a crucial role in shaping climate-related academic behaviour. According to Kim et al. (2016); also Shealy et al. 2021), among the academic community, both engineering faculty and students demonstrate the most significant levels of involvement in climate education and research. Conversely,

Latter et al. (2024) argues that health professions exhibit the lowest levels of academic climate action.

2.3 *Awareness of Climate Actors (Teams and Research Groups)*

Beyond structural factors, the presence of climate research groups in academic institutions can greatly impact academic behaviour. When individuals have knowledge of active research groups or climate action teams that concentrate on climate issues, their propensity to participate in related initiatives and contribute to the wider academic discussion on climate change increases (Leal Filho et al., 2021).

2.4 *Psychological Factors: Climate Anxiety*

Psychological factors are also highly influential in forecasting academic behaviour associated with climate change. Climate anxiety, characterized by distress over environmental future, has been found to motivate individuals to become more engaged in climate-related pursuits (Whitmarsh et al. 2022). The presence of anxiety can serve as a catalyst, compelling both students and academics to actively participate in climate education and research as a means of addressing their apprehensions (Atta et al. 2024; Dablander et al. 2023).

2.5 *Present Study*

The purpose of this study is to explore the determinants of climate action in higher education institutions, thereby closing the existing knowledge gap. Ultimately, this study highlights the essential role that higher education institutions must fulfil in tackling climate change. Through the integration of comprehensive climate education, the promotion of sustainable practices, and the support of research initiatives centred around climate action, these institutions can empower their communities to effectively tackle urgent sustainability challenges. The study's results offer valuable insights into the methods that can be employed to strengthen participation and drive significant climate action among students and academics in higher education.

3 Methods

3.1 Research Design

This study employed a cross-sectional survey design to investigate the factors influencing climate action among students and staff in higher education institutions. The goal was to identify key predictors of engagement in climate-related activities to inform strategies for enhancing involvement. A survey was carried out online using a Qualtrics questionnaire, targeting academics and students of eight universities from the United Kingdom, Spain and Latin-America. An email was used to distribute the instrument to the entire university community, which comprised a total of 111,502 individuals. The number of usable responses received was 8,111, indicating a response rate of 7%. Data collection took place from November 9th to November 30th, 2023. The study received ethical approval from the psychology ethics committee at the lead author's university.

3.2 Participants

In terms of universities, the sample was Salamanca 17.6%, Valladolid 8.5%, Nacional de Itapua 5.6%, Atacama, 6.7%, Tecnológica Centroamericana 13.9% and Bath 47.8%. With respect to gender and role, the sample was a strong reflection of the university community, with 42.4% being male, 53.8% female; and 3.8% other; 68% were undergraduate students, 11% Master's students, 6% Doctoral students, 5% Early Career Academics, and 10% Senior Academics; details of the sample are shown in Table 1.

In terms of the duration of employment at the university, 30% had been employed for a period ranging from 1 to 5 years. In addition, 18% of the employees had served for 6 to 10 years, while 9% had been dedicated for 11 to 15 years. Finally, a significant 44% of the academics had a remarkable tenure of over 16 years at the university. In relation to the duration of students' studies at the university, 43% had studied for 1 year, 22% for 2 years, 15% for 3 years, and 20% for 4 years.

On the other hand, levels of climate anxiety were very similar for students and academics (60% mild, 39% moderate and 1.5% severe).

3.3 Measures

Academic behaviour or professional climate action was assessed using an item that asked participants *if their studies or work included research or learning about climate change or sustainability*. We asked participants to rate their level of involvement in

Table 1 Participants' characteristics from each population

	Students			Academics	
	N	%		*N*	%
School or Faculty			School or Faculty		
Art and Humanities	910	13.2	Art and Humanities	231	18.48
Experimental Sciences	1262	18.3	Experimental Sciences	268	22.37
Health Sciences	669	9.7	Health Sciences	183	14.51
Social Sciences	2096	30.4	Social Sciences	316	26.50
Engineering and Architecture	1951	28.3	Engineering and Architecture	219	17.79
Climate anxiety			Climate anxiety		
Mild	4123	59.8	Mild	731	60.1
Moderate	266	38.7	Moderate	469	38.5
Severe	103	1.5	Severe	17	1.4

the topic using a three-point scale: 3 for a major part of their course or work, 2 for a minor part, and 1 for no involvement.

Awareness of climate action actors (teams or research groups) was measured with the item 'Before today, were you aware of the University has a Climate Action Team (CAT) or a specific research group on climate change? The response options for this question were categorized as either Aware (2) or Unaware (1).

Climate anxiety was measured using a condensed version of Clayton and Karazsia's (2020) Climate Change Anxiety Scale (CCAS), which consisted of three items: 'I think, why can't I handle climate change better?', 'I have problems balancing my concerns about climate change with the needs of my family' and 'Thinking about climate change makes it difficult for me to concentrate'. This measure used a five-point frequency response scale from 'never' (1) to 'almost always' (5). The scale demonstrated an adequate level of reliability ($\alpha = 0.72$).

3.4 Data Analysis

For all analyses, we made use of the complete dataset obtained in 2023. In Sect. 4.1, we explain the Kruskal–Wallis test we performed to identify the groups that had the highest levels of learning, teaching, and research in the fields of climate change and sustainability. The test was useful in analyzing predictors among various demographic groups. In Sect. 4.2, we employ correlational analyses to examine the associations between all the predictors. Section 4.3 employs logistic regression to explore the predictors of engagement in climate change and sustainability studies or work. The analysis further presented odds ratios (ORs) and 95% confidence intervals (CIs).

The variable of interest was the degree of participation in academic or professional endeavours related to climate change and sustainability. The software program employed for statistical analysis was Jamovi (version 2.3.18.0).

4 Results

4.1 *Participant Demographics and Involvement of Studies or Work with Climate Change and Sustainability*

Participant demographic data and median involvement in learning/teaching, as well as research on climate change and sustainability levels, are presented in Table 2. The median score exhibited a significant variation among participants with diverse demographic characteristics ($p < 0.01$). The results of the study demonstrate that Master degree students prioritize studies and research related to climate change and sustainability (H (2) = 16.51, $p < 0.001$), compared with Doctoral students ($p < 0.001$). Comparing schools and faculties, health sciences students typically exhibit lower scores on studies and research related to climate change (H (4) = 375.48, $p < 0.001$). In contrast, Engineering and Design were the students with higher levels on behaviour academic ($p < 0.001$). Similarly, Engineering and Design academics tend to have higher levels of involvement in teaching and research on climate change and sustainability (H (4) = 107.03, $p < 0.001$). In contrast, Health academics had the lowest levels of behaviour compared to other schools and faculties ($p < 0.001$), with the exception of academics in the arts and humanities ($p = 0.87$). In addition, the study found that academics who have accumulated over 16 years of work experience tend to have significantly less engagement in climate change and sustainability research compared to their younger peers (H (3) = 37.18, $p < 0.001$). Likewise, students with one year of study showed less academic behaviour compared to all students who had been studying for longer (H (3) = 40.58, $p < 0.001$). On the other hand, the results show notable disparities in the awareness of climate action actors, with the aware group showing the highest level of academic behaviour among both students (U = 30,356, $p <$; 0.001) and academics (U = 30,856, $p < 0.001$). This is clear in the variable score for academic behaviour, where students in the aware group had a median score of 2.0 compared to 1.0 for the unaware group, and academics in the aware group had a median score of 2.0 compared to 1.0 for the unaware group.

4.2 *Correlational Analyses*

Several variables were found to be linked to levels of academic behaviour through correlation analyses (see Table 3). Regarding students, there was evidence of a positive association between climate anxiety and academic engagement with climate/

Table 2 Involvement in professional climate action levels stratified by members of the university′ characteristics

	Students						Academics				
	Frequency (%)						Frequency (%)				
	Med	IQR	Major	Minor	No		Med	IQR	Major	Minor	No
Type of student						Type of student					
Undergraduate*	1.0	1.0	15.2	34.2	50.6	Early Career	1.0	1.0	22.0	28.6	49.3
Master	1.0	1.0	24.8	24.7	50.5	Senior Academic	1.0	1.0	23.4	26.1	50.5
Doctoral	1.0	1.0	20.0	17.4	62.6	School or Faculty***					
School or Faculty***											
Art and Humanities	1.0	1.0	9.1	31.1	59.8	Art and Humanities	1.0	1.0	13.6	16.8	69.6
Experimental Sciences	1.0	1.0	12.4	24.4	63.2	Experimental Sciences	2.0	1.0	21.5	32.2	46.4
Health Sciences	1.0	1.0	5.0	23.5	71.5	Health Sciences	1.0	1.0	8.5	16.9	74.6
Social Sciences	1.0	1.0	12.7	38.5	48.8	Social Sciences	1.0	1.0	21.5	27.9	50.6
Engineering and Architecture	2.0	2.0	31.7	32.3	36	Engineering and Architecture	2.0	2.0	40.0	30.5	29.5
Studying time***						Working time					
1 year	1.0	1.0	15.2	29.9	54.9	1 to 5 years	1.0	1.0	17.3	14.3	68.4
2 year	1.0	1.0	17.1	34.5	48.4	6 to 10 years	1.0	1.0	16.4	21.5	62.1
3 year	1.0	1.0	17.6	32.7	49.7	11 to 15 years	1.0	1.0	20.6	18.6	60.8
4 year	1.0	1.0	17.2	35.6	47.2	16 years or more	1.0	1.0	9	15.6	75.4
Awareness of climate action actors						Awareness of climate action actors					
Yes***	1.0	2.0	18.6	39.6	41.8	Yes***	1.0	2.0	25.8	32.2	42.0
No	1.0	1.0	12.6	87.4	53.1	No	1.0	1.0	19.1	27.9	52.9

Note. * $p < 0.05$, ** $p < 0.01$, *** $p < 0.001$

sustainability ($r_s = 0.11, p < 0.001$). Similarly, increased academic engagement was associated with awareness of climate action teams or groups ($r_s = 0.20$, $p < 0.001$). In terms of academics, we obtained similar outcomes; higher climate anxiety ($r_s = 0.08$, $p < 0.001$), was associated with increased involvement of climate change and sustainability in their teaching and research. In a similar vein, there was a significant relationship between awareness of climate action actors ($r_s = 0.08$, $p < 0.001$).

Table 3 Correlations between different variables

	Students		Academics	
Variable	1	2	1	2
1.Academic behaviour	-		-	
2. Climate anxiety	0.11***		0.08***	
3. Awareness of climate actors	0.20***	0.02	0.08***	0.02

Note. * $p < 0.05$, ** $p < 0.01$, *** $p < 0.001$

4.3 Regression Analysis: Predicting Involvement of Studies or Work and Research with Climate Change and Sustainability

Since multiple groups of students and academics demonstrated increased engagement in studying or working on climate change and sustainability, and since there were several variables associated with this involvement, we performed a logistic regression analysis to explore the individual effects of each predictor on studying, teaching, and researching climate change and sustainability.

The student model included five different variables (the type of student, the school or faculty they belonged to, their level of climate anxiety, awareness of climate action actors and years studying at the university). The full model, which included all predictors, was found to be statistically significant ($\chi 2\ (11) = 689.35, p < 0.001$). This indicates that the model was able to accurately distinguish between students who were involved in climate change and sustainability studies and research. The overall model accounted for 12% (Nagelkerke R^2) of the variation in academic behaviour related to climate change and sustainability studies. Based on the data presented in Fig. 1, it can be concluded that all independent variables made a statistically significant and distinct contribution to the model. Master's degree students ($OR = 1.65$, 95% CI 1.27–2.16, $p < 0.001$) had the strongest climate change and sustainability involvement in their learning and research (65% more likely) compared to doctoral students. Similarly, students who showed greater academic behaviour were found to be more prevalent in the fields of engineering and design ($OR = 8.58$, 95% CI 10.94–18.47) compared with health sciences students. On the other hand, increased climate anxiety ($OR = 1.25$ (25% more likely), 95% CI 1.16–1.35, $p < 0.001$) and awareness of climate actions actors ($OR = 2.13$ (2 times more likely), 95% CI 1.87–2.43, $p < 0.001$), predicted increased the involvement of learning and research on climate change and sustainability. Similarly, academic behaviour increased with 2, 3 and 4 years of study at university (OR = 1.20 (20% more likely), 95% CI 1.87–2.43, $p < 0.05$; OR = 1.27 (27% more likely), 95% CI 1.87–2.43, $p < 0.01$; OR = 1.44 (44% more likely), 95% CI 1.87–2.43, $p < 0.001$ respectively).

Likewise, the academic model is composed of five distinct variables (academic type, school or faculty, climate anxiety, awareness of climate action actors, and work duration). The overall model, incorporating all predictors, demonstrated statistical

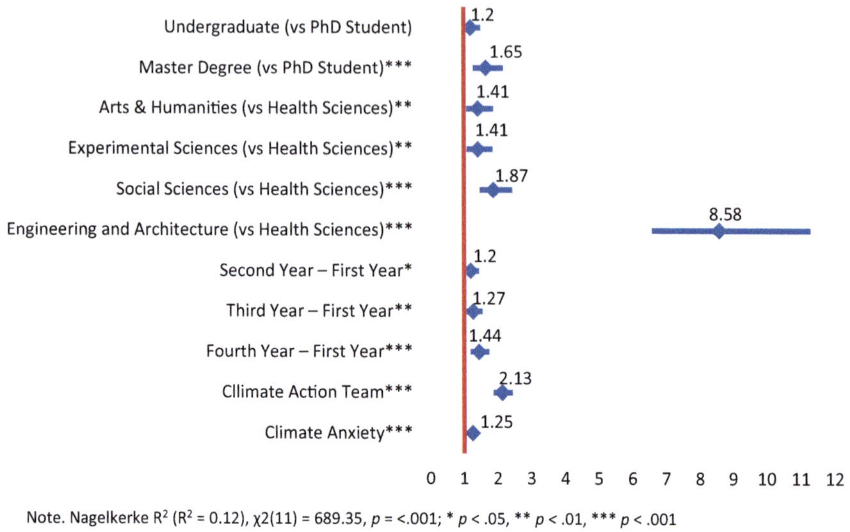

Fig. 1 Logistic regression analysis on students of academic behaviour (Involvement of studies with climate change and sustainability)

significance ($\chi 2(11) = 87.32$, $p < 0.001$). The model showed its ability to differentiate between academics engaged in climate change and sustainability work (teaching or researching), with a significance level of less than 0.001. In terms of academic behaviour, the model showed an explanation of 9% (Nagelkerke R^2) of the variance. Among the independent variables, only three made a statistically significant contribution to the model, as depicted in Fig. 2. Academics in the fields of engineering and design were significantly more inclined towards teaching and researching climate change and sustainability (*OR* = 5.59, 95% *CI* 3.03–10.61) compared with health sciences academics. Likewise, academics with 11 to 15 years work experience are twice as likely to engage in academic climate behaviour compared to their counterparts with more than 16 years working at the university (*OR* = 2.26, 95% *CI* 1.39–3.70). Similarly, academics with awareness of climate actors (OR = 2.18 (2 times more likely), 95% *CI* 1.50–3.10, $p < 0.001$), were more likely to be involved with learning and research on climate change and sustainability. Surprisingly, academics with increased climate anxiety ($p = 0.07$) and seniority of academics ($p = 0.90$), did not predict academic behaviour.

5 Discussion

The aim of this study was to investigate primary predictors of academic behaviour in climate education and research within higher education. It emphasised career stage, duration of work or study, academic discipline, awareness of climate actions

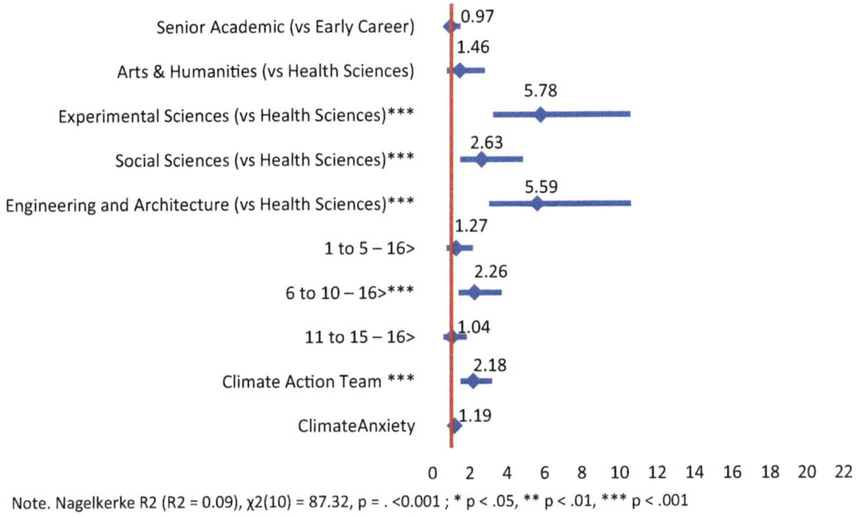

Fig. 2 Logistic regression analysis on academics of academic behaviour (Involvement of work with climate change and sustainability)

actors and psychological factors like climate anxiety. The results of our study demonstrate clear patterns when comparing students and academics, explaining the specific influence of these factors on the participation of each group in climate-related activities.

The time students spend at the university has a substantial influence on their participation in climate-related initiatives. According to our study, students with longer periods of enrolment, exhibit a higher propensity for involvement in climate education and research. This is under the concept that extended exposure to academic settings nurtures a deeper comprehension and commitment to matters concerning sustainability. Students who have dedicated more time to their university studies are presented with enhanced opportunities to engage in climate-related projects and coursework, consequently fostering heightened participation (Leal Filho et al. 2018a, b, c; Wodika and Middleton 2020). **Academics at the early and mid-career stages** exhibited higher levels of engagement in climate-related education and research when compared to their senior counterparts. The findings align with the research conducted by Chowdhury et al. (2021), which indicates that early-career scholars tend to prioritize emerging global challenges such as climate change, influenced in part by the demands of establishing a successful research trajectory. Mallon et al. (2023) suggests that younger generations of learners and faculty are increasingly inclined to seek congruence between their personal values and institutional missions in relation to climate action. This phenomenon could be attributed to the increasing tendency of younger generations of students and educators to expect a convergence between their own ethical beliefs and the environmental objectives of academic institutions (Mallon et al., 2023).

The influence of **academic discipline** on climate-related behaviour is particularly evident among students and academics. Our research reveals that the field of engineering exhibits the highest levels of involvement in activities related to climate change. This outcome aligns with the conclusions drawn by Kim et al. (2016) and Shealy et al. (2021). This is likely a result of the direct relevance of their studies to practical climate solutions. Engineering students frequently participate in practical projects aimed at tackling environmental concerns, thereby deepening their comprehension of climate science and fostering active involvement in sustainability endeavours (Axelithioti et al., 2023). On the other hand, academics in the health sciences felt less comfortable and less responsible in teaching and researching about climate change (Beck et al., 2013). The reasons for the disparity in perception among medicine and health sciences researchers, who do not view their field as equally relevant in tackling climate change compared to those in other disciplines, are not yet clear (Latter et al., 2024).

About the **type of student**, our study shows that Master's degree students exhibit significantly greater engagement in teaching and learning activities focused on climate change, in comparison to both doctoral and undergraduate students. The growing urgency to address climate change and the demand for capable professionals to take the lead is reflected in the increased emphasis on climate change education and research at the Master's level (Leal Filho et al., 2021).

Awareness of climate teams within academic institutions significantly influences the engagement of both students and academics in climate-related activities. Being aware of these teams is of utmost importance for students to cultivate engagement. The results align with previous findings (Apostol & Delos Santos, 2023), emphasising the significance of institutional visibility in fostering student engagement in sustainability initiatives. In a similar vein, a study revealed that students who receive information about research groups' presence and activities display a higher inclination towards involvement in climate-related initiatives (Villavicencio Calzadilla et al., 2018).

For academics, awareness of climate research groups significantly influenced their academic climate behaviour. Our findings are consistent with those of Latter et al., (2024) suggesting that institutions with active sustainability research groups exhibit increased faculty engagement in climate-related research and teaching. Likewise, a particular research study revealed that academics who possess awareness of and connections to active research groups that concentrate on climate issues have a higher propensity to engage in collaborative projects and make significant contributions to the broader academic discourse surrounding climate change (Leal Filho et al., 2023).

Climate anxiety played a major role in shaping the academic behaviour of both students and academics in the field of climate education and research. Students, influenced by climate anxiety, actively sought opportunities to engage in climate-related learning and activities in order to address their environmental concerns. Maran and Begotti (2021) found similar results, showing that students with higher climate anxiety could be more inclined to engage in environmental advocacy and sustainability initiatives. Likewise, the presence of climate anxiety has been recognised as a key factor in motivating academics to take part in climate-related research and education. These results align with the conclusions drawn by Atta et al. (2024), showing that scholars who experience climate anxiety are more inclined to incorporate climate-related matters into their research and teaching. This shows that emotional responses to climate change may play a role in fostering involvement in climate education and research. Although anxiety can serve as a driver of action, institutions should exercise caution when relying on it as a primary motivator and should help individuals experiencing distress caused by climate concerns.

6 Implications

This study highlights the importance of higher education institutions in addressing climate change and provides recommendations for enhancing their role. The findings suggest universities should focus on incorporating more **climate-related content into their curricula** (Fahey, 2012), particularly in undergraduate and Master's programmes outside of engineering. Concerningly, health researchers and students engage in climate studies far less than other disciplines, despite the profound health implications of climate change. Reframing climate change and sustainability as not only technological challenges, but also social, economic, health and cultural ones, will expand green career options for young people and foster more interdisciplinary solutions.

Additionally, providing **support for Master's programs**, such as resources and research opportunities, can further engage students in climate education and research. **Faculty development** and leadership training should also empower early and mid-career academics to drive climate-related initiatives. The study also emphasises the need for universities to address climate anxiety by providing **mental health support** and resources. Furthermore, increasing the visibility of climate research groups within the academic community and encouraging participation can promote collaboration and engagement (Molthan-Hill et al., 2019). Overall, these strategies can help universities foster greater involvement among students and academics in climate change efforts.

7 Conclusions

This study has contributed important knowledge on the primary factors that influence engagement in climate education and research within the context of higher education. Our research findings demonstrate that career stage, study or work duration, academic discipline, student type, climate anxiety, and awareness of climate research groups serve as significant indicators for participation in climate-related activities.

A notable characteristic of Master's students is their elevated engagement levels, which are fostered by focused and applied learning experiences. Engineering disciplines show significant involvement perhaps due to the dominant technological framing of climate solutions. Furthermore, climate anxiety, despite its difficulties, can act as a compelling catalyst for both students and academics to engage in proactive measures.

These insights indicate that higher education institutions can have a vital role in advancing global sustainability efforts through curriculum design enhancement, graduate programme support, faculty empowerment, mental health concern addressing, and research group visibility promotion.

However, the study's cross-sectional design limits causal inferences, and cultural differences may affect the generalizability of the findings. Future research should employ longitudinal studies across a broader range of institutions to explore causal relationships and assess the impact of targeted interventions to promote climate action in higher education.

8 Limitations

Although this study offers valuable insights into the factors that impact climate education and research engagement in higher education, it is important to acknowledge several limitations. First, the study depends on self-reported data, which could be susceptible to bias, given that participants may overstate or understate their engagement in climate-related activities. Second, the study is limited to a specific set of institutions, which may not fully represent the diversity of higher education environments globally. Third, the cross-sectional design of the research hinders the ability to establish definitive causal relationships between the predictors and engagement outcomes. Finally, the study primarily centres on quantitative data, potentially neglecting qualitative factors such as personal motivation or institutional culture that could contribute to a better understanding of variations in climate engagement.

9 Future Research

Subsequent investigations should endeavour to expand upon the discoveries of this study by rectifying its constraints and examining supplementary variables that impact involvement in climate education and research within higher education.

Longitudinal studies have the potential to offer a more comprehensive understanding of the evolution of engagement over time, shedding light on causal relationships between predictors and outcomes. To enhance the generalizability of the findings and account for the diversity of academic environments, it is advisable to broaden the scope of the research to include a more extensive range of global institutions. Furthermore, the integration of qualitative approaches, such as interviews and case studies, has the potential to unveil the intricate motivations and institutional cultures that propel climate action. Conducting research on the influence of policy changes, interdisciplinary approaches, and emerging technologies on climate education and research would be beneficial for future investigations.

References

Adlit M, Adlit MF (2022) Teaching climate change: a systematic review from 2019–2021. Int J Adv Res Publ (IJARP) 5(5):12–19. www.ijarp.org

Apostol RL, Delos Santos LSP (2023) The mediating effect of school climate on the relationship between academic self-concept and student engagement. Eur J Educ Stud 10(11):336–358. https://doi.org/10.46827/ejes.v10i11.5084

Atta MHR, Zoromba MA, El-Gazar HE, Loutfy A, Elsheikh MA, El-ayari OSM, Sehsah I, Elzohairy NW (2024) Climate anxiety, environmental attitude, and job engagement among nursing university colleagues: a multicenter descriptive study. BMC Nurs 23(1):1–14. https://doi.org/10.1186/s12912-024-01788-1

Axelithioti P, Fisher RS, Ferranti EJS, Foss HJ, Quinn AD (2023) What are we teaching engineers about climate change? Presenting the MACC evaluation of climate change education. Educ Sci, 13(2). https://doi.org/10.3390/educsci13020153

Beck A, Sinatra GM, Lombardi D (2013) Leveraging higher-education instructors in the climate literacy effort: factors related to university faculty's propensity to teach climate change. Int J Clim Chang: Impacts Responses 4(4). https://doi.org/10.18848/1835-7156/CGP/v04i04/37181

Chowdhury MTA, Ahmed KJ, Ahmed MNQ, Haq SMA (2021) How do teachers' perceptions of climate change vary in terms of importance, causes, impacts and mitigation? A comparative study in Bangladesh. In: SN social sciences, Vol 1(7). Springer International Publishing. https://doi.org/10.1007/s43545-021-00194-7

Clayton S, Karazsia BT (2020) Development and validation of a measure of climate change anxiety. J Environ Psychol 69:1–11. https://doi.org/10.1016/j.jenvp.2020.101434

Dablander F, Sachisthal M, Cologna V, Strahm N, Bosshard A, Grüning N, Green A, Brick C, Aron AR, Haslbeck J (2023) Climate change engagement of scientists. Nat Clim Chang 14:1–80. https://doi.org/10.1038/s41558-024-02091-2

Fahey SJ (2012) Curriculum change and climate change: inside outside pressures in higher education. J Curric Stud 44(5):703–722. https://doi.org/10.1080/00220272.2012.679011

Han S, Zha Q, Xie J (2023) Understanding Chinese universities' strategic responses to the 'Double first-class' initiative and the potential outcome on higher education diversity: an exploratory study. Stud High Educ 48(11):1624–1638. https://doi.org/10.1080/03075079.2023.2206426

Hess DJ, Collins BM (2018) Climate change and higher education: assessing factors that affect curriculum requirements. J Clean Prod 170:1451–1458. https://doi.org/10.1016/j.jclepro.2017.09.215

IPCC (2018) Framing and Context. In: Global Warming of 1.5°C. An IPCC Special Report on the impacts of global warming of 1.5°C above pre-industrial levels and related global greenhouse gas emission pathways, in the context of strengthening the global response to the. In: Elgizouli I, Fischlin A, Gao X (eds) Global warming of 1.5°C. An IPCC Special Report on the impacts of global warming of 1.5°C above pre-industrial levels and related global greenhouse gas emission pathways, in the context of strengthening the global response to the threat of climate change, pp 47–92. https://www.ipcc.ch/site/assets/uploads/sites/2/2019/02/SR15_Chapter1_Low_Res.pdf

Kim J, Schmöcker JD, Fujii S (2016) Exploring the relationship between undergraduate education and sustainable transport attitudes. Int J Sustain Transp 10(4):385–392. https://doi.org/10.1080/15568318.2014.961108

Latter B, Demski C, Capstick S (2024) Wanting to be part of change but feeling overworked and disempowered: researchers' perceptions of climate action in UK universities. PLOS Climate 3(1):29. https://doi.org/10.1371/journal.pclm.0000322

Leal Filho W, Aina YA, Dinis MAP, Purcell W, Nagy GJ (2023) Climate change: why higher education matters? Sci Total Environ 892:164819. https://doi.org/10.1016/j.scitotenv.2023.164819

Leal Filho W, Azeiteiro U, Alves F, Pace P, Mifsud M, Brandli L, Caeiro SS, Disterheft A (2018) Reinvigorating the sustainable development research agenda: the role of the sustainable development goals (SDG). Int J Sust Dev World 25(2):131–142. https://doi.org/10.1080/13504509.2017.1342103

Leal Filho W, Brandli LL, Becker D, Skanavis C, Kounani A, Sardi C, Papaioannidou D, Paço A, Azeiteiro U, de Sousa LO, Raath S, Pretorius RW, Shiel C, Vargas V, Trencher G, Marans RW (2018) Sustainable development policies as indicators and pre-conditions for sustainability efforts at universities: fact or fiction? Int J Sustain High Educ 19(1):85–113. https://doi.org/10.1108/IJSHE-01-2017-0002

Leal Filho W, Morgan EA, Godoy ES, Azeiteiro UM, Bacelar-Nicolau P, Veiga Ávila L, Mac-Lean C, Hugé J (2018) Implementing climate change research at universities: barriers, potential and actions. J Clean Prod 170:269–277. https://doi.org/10.1016/j.jclepro.2017.09.105

Leal Filho W, Sima M, Sharifi A, Luetz JM, Salvia AL, Mifsud M, Olooto FM, Djekic I, Anholon R, Rampasso I, Kwabena Donkor F, Dinis MAP, Klavins M, Finnveden G, Chari MM, Molthan-Hill P, Mifsud A, Sen SK, Lokupitiya E (2021) Handling climate change education at universities: an overview. Environ Sci Eur 33(1). https://doi.org/10.1186/s12302-021-00552-5

Lozano R, Barreiro-Gen M, Lozano FJ, Sammalisto K (2019) Teaching sustainability in European higher education institutions: assessing the connections between competences and pedagogical approaches. Sustainability 11(6):1–17. https://doi.org/10.3390/su11061602

Mallon WT, Deas D, Good ML (2023) Reasons for optimism about academic medicine's actions against climate change. Acad Med 98(11):1243–1246. https://doi.org/10.1097/ACM.0000000000005331

Maran DA, Begotti T (2021) Media exposure to climate change, anxiety, and efficacy beliefs in a sample of Italian University students. Int J Environ Res Public Health 18(17). https://doi.org/10.3390/ijerph18179358

Molthan-Hill P, Worsfold N, Nagy GJ, Leal Filho W, Mifsud M (2019) Climate change education for universities: a conceptual framework from an international study. J Clean Prod 226:1092–1101. https://doi.org/10.1016/j.jclepro.2019.04.053

Nielsen KS, Nicholas KA, Creutzig F, Dietz T, Stern PC (2021) The role of high-socioeconomic-status people in locking in or rapidly reducing energy-driven greenhouse gas emissions. Nat Energy 6(11):1011–1016. https://doi.org/10.1038/s41560-021-00900-y

Shealy T, Katz A, Godwin A (2021) Predicting engineering students' desire to address climate change in their careers: an exploratory study using responses from a U.S. National survey. Environ Educ Res 27(7):1054–1079. https://doi.org/10.1080/13504622.2021.1921112

Timmerman N, Metcalfe AS (2009) From policy to pedagogy: the implications of sustainability policy for sustainability pedagogy in higher education. Can J High Educ 39(1):45–60

Villavicencio Calzadilla P, Mauger R, Du Plessis A (2018) Climate change communication in higher education Institutions: the case of the North-West University in South Africa. In: Handbook of climate change communication: Vol. 3: case studies in climate change communication, Vol 3, pp 241–255. https://doi.org/10.1007/978-3-319-70479-1_15

Whitmarsh L, Player L, Jiongco A, James M, Williams M, Marks E, Kennedy-Williams P (2022) Climate anxiety: what predicts it and how is it related to climate action?. J Environ Psychol 83:101866. https://doi.org/10.1016/j.jenvp.2022.101866

Wodika AB, Middleton WK (2020) Climate change advocacy: exploring links between student empowerment and civic engagement. Int J Sustain High Educ 21(6):1209–1231. https://doi.org/10.1108/IJSHE-03-2020-0091

Citizen Science: Unlocking Skills and Experience for University Students in Hungary

Zoltán Péter Alföldi

Abstract Understanding biodiversity-both in our immediate surroundings and globally-is particularly critical in the era the global biodiversity loss. Citizen science presents a valuable opportunity to engage with this challenge, and university students can make significant contributions to such programs. However, as revealed by the findings of this study conducted in one of the largest university in the field of natural sciences in Hungary, the majority of students lack essential knowledge about citizen projects of great relevance and how to effectively participate in them. This indicates the need for integrations of the use of important online community databases and participations in citizen science programs in higher education.

Keywords Citizen science · University students · Biodiversity platforms · iNaturalist

1 Introduction

Citizen science—also referred to as *community science* (Carr 2004)—is defined as" a method of integrating public outreach and scientific data collection locally, regionally, and across large geographical scales" by Cooper et al. (2007). In essence, citizen science involves volunteer citizens contributing to scientific programs (Kruger and Shannon 2000). A program qualifies as citizen science if more than 50 percent of the data are collected by volunteers (Chandler et al. 2016). The history of citizen science is intertwined with traditional scientific endeavors, marked by numerous notable contributions (Miller-Rushing et al. 2012). The field of citizen science has seen exponential growth in recent years (McKinley et al. 2017), with increasing applications in environmental research and recognition of citizen-generated data for scientific purposes (Pocock et al. 2017). Robust regulations and conservation

Z. P. Alföldi (✉)
Institute of Wildlife Management and Nature Conservation, Hungarian University of Agriculture and Life Sciences, Keszthely, Hungary
e-mail: alfoldi.zoltan.peter@uni-mate.hu

W. Leal Filho et al. (eds.), *Education for Sustainable Development: The Contribution of Universities*, World Sustainability Series,
https://doi.org/10.1007/978-3-031-86985-3_5

strategies require substantial, accurate species data (Heberling et al. 2021). Thus, citizen science, with its diverse types (Dillon et al. 2016) and models (Cooper et al. 2007) emerges as a strategic tool to enhance environmental policies, supported by social and economic factors to achieving the Sustainable Development Goals (SDGs), as indicated by the Bio Innovation Service (2018) and Turbé et al. (2019). When carefuly designed, executed, and evaluated, citizen science projects produce high-quality data that significantly contribute to scientific research and education. They also raise environmental awareness (Wals et al. 2014) and address critical societal challenges (Kobori et al. 2016; McKinley et al. 2017).

The SDGs, established by the United Nations (UN) General Assembly, represent a globally recognized multi-sector policy framework (Griggs et al. 2014). Numerous citizen science projects can effectively contribute to nearly all 17 SDGs (Turbé et al. 2019; Fraisl et al. 2020). For example, the Citizen Science Global Partnership, initiated in collaboration with the United Nations Environment Programme (UNEP), supports global coordination of citizen science activities aligned with the SDGs (Bowser et al. 2020). Similarly, the European Joint Research Centre (JRC) has developed an inventory of citizen science projects relevant to environmental policy and the SDGs (EC-JRC 2018).

Citizen science can generate large datasets that inform resource management and conservation strategies (McKinley et al. 2017). These projects enable monitoring of ecosystems at various scales (Conrad and Hilchey 2011) and facilitate access to local knowledge, aiding the implementation of conservation initiatives (Kobori et al. 2016). Engaging local communities in data collection enhances their commitments to environmental issues and encourages their participations in specific conservation actions (Devictor et al. 2010).

Scientists and non-government organizations (NGOs) are increasingly rely on citizen volunteers to enhance their capacities to monitor and manage natural resources, track different species, and protect natural ecosystems. Citizen science projects are pivotal in identifying declines in threatened species (Pocock et al. 2017; Stephenson and Stengel 2020) and in the early detection of invasive species (Gallo and Waitt 2011). These activities, across various scales and fields yield substantial social, economic, and ecological benefits, fostering a deeper understanding of ecosystem functions and structures (Dickinson et al. 2012).

Beyond environmental contributions, citizen science provides personal benefits to participants, including opportunities to spend time outdoors, strengthen their connections to nature, and engage with with like-minded individuals, thereby enhancing overall well-being (Bell et al. 2008; Toomey and Domroese 2013). This engagement builds social capital and fosters collaborations among societal partners (Greenwood 2007; Silvertown 2009). For university students, citizen science offers a platform to meet educational requirements, engage in scientific research, and gain practical experiences in natural and urban environments (Mitchell et al. 2017; Roche et al. 2020; Stevenson et al. 2021; Potsikas et al. 2023).

Volunteers not only contribute data collections, but also serve as higly educated participants in various projects of significant scientific value (Lakshminarayanan 2007). The credibility of data collected by amateur scientists remains a fundamental concern (Freitag et al. 2016; Bowser et al. 2020). To ensure reliability, expert verification of collected data is essential for citizen science (Crall et al. 2011; Kosmala et al. 2016). The Ten Principles of Citizen Science was defined by the European Citizen Science Association (Robinson et al. 2018). Several professional citizen science organizations, such as the European Citizen Science Association, have emerged to support this field, and since 2016, the journal of Citizen Science: Theory and Practice, was launched. National and international organizations, government agencies, and universities integrated citizen science into their activities (Ellwood et al. 2017).

With advancements in information technology, citizen scientists can easily participate in diverse research projects using popular smartphone apps and online platforms (Andrachuk et al. 2019). One of the most popular platforms is iNaturalist (www.inaturalist.org, Boone and Basille 2019), a multitaxa platform and joint initiative of the California Academy of Sciences and the National Geographic Society founded in 2008 at the University of California, Berkeley (Nugent 2018). This online platform provides data validation, online discussions, and training, fostering lifelong learning and public engagement in biodiversity conservation (Freitag et al. 2016; Callaghan et al. 2022). It became a globally used online platform for the observations and identifications of fungi, and plant and animal species (Aristeidou et al. 2021), and helping citizens to record and learn biodiversity at different scales. By uploading observations–images with GPS coordinates–volunteers contribute to a global database, promoting biodiversity research and education for both professional and amateur naturalists (Bell et al. 2008). High quality "research grade" data are published to the international research network of Global Biodiversity Information Facility (GBIF), from which professionals can use data for biodiversity analyses globally (GBIF, 2024). The avian biodiversity platform eBird (EOD) is a major source of this type of data, being one of the most popular site (Sullivan et al. 2014).

2 Materials and Methods

The Hungarian University of Agriculture and Life Sciences (MATE), as the largest institution in the fields of agriculture and related disciplines, offers considerable potential for engaging its approximately 15,000 students in citizen science projects. This study aimed to assess the knowledge and attitudes of randomly selected MSc students with their different specializations across three MATE campuses (Gödöllő, Kaposvár and Keszthely) towards citizen science. The main goal of this study was to assess the knowledge on and attitude towards citizen science in general, for Hungarian university students with various natural science majors. University students, due to their interests and education, could potentionally be one of the main target groups of participants for citizen science programs, as they could contribute very effectively to the successes of these projects. On the other hand, online citizen science platforms can also be usefully integrated into university education programs.

Based on the findings of Hitchcock et al. (2021), that the global online biodiversity platform iNaturalist was proved to be effective at increasing students' engagement in biodiscovery and interest in improving bioliteracy, the observation data uploaded from Hungary into this database were recorded between 2008 and 2023 here. Also, based on the literature, EOD was considered here as an important international biodiversity platform for using across many disciplines (Sullivan et al. 2014).

A questionnaire comprising ten open-ended and semi-structured questions was distributed in March 2024 to gauge students' knowledge, opinions, experiences and attitudes regarding citizen science. Out of one hundred online questionnaires sent out, 89 responses were received and evaluated. The gender distribution of responders was nearly equal (53.9% and 46.1% for women and men, respectively). In terms of age, 44.9% of the responders were between 19 and 24 years, whereas 16.9% were between 25 and 30 years, and 38.2% were older than 30 years. All returned questionnaires were suitable for analysis.

3 Results

First, the evolution of species observation data uploaded to major international online biodiversity platforms most popular in Hungary were analized over the past 16 years. These records support the findings of Chandler et al. (2016), that the great majority of citizen science records available through GBIF comes from the eBird Observation Dataset (EOD). Data recorded at the GBIF and EOD platforms have shown significant growth over this time period, whereas research-grade observations submitted to iNaturalist have only recently started to rise (Fig. 1). These trends align with similar patterns observed in other European countries and globally, providing opportunities for local and international naturalists to better understand the rich biodiversity of Hungary. Data recordings of Hungarian naturalists at Observation.org, which is another global biodiversity platform for citizen science and monitoring established in 2004, did not show continuous growth.

When asked about their understanding of citizen science, nearly three-quarters of students (73%) reported either lacking adequate knowledge (29.2%) or having no information at all (43.8%), as shown in Fig. 2b. An even higher proportion of them (94%) indicated no prior involvement in citizen science activities (Fig. 2b). These results highlight an urgent need for professional scientists in universities, research institutions, NGOs and civil associations to introduce courses and campaigns that raise awareness about citizen science.

Despite limited knowledge and experience, 92% of students recognized the societal value of citizen science (Fig. 3a). Over half of them viewed it as important for education in general, while more than 40% considered citizen science necessary across all social sectors. Only less than 8% of the students believed that citizen science should be limited to professionals or deemed it unnecessary to have more citizen science programs. Regarding the future of citizen science, nearly two-thirds of the students (63.6%) anticipated an expanded role in all societal domains (Fig. 3b). Additionally, almost one-quarters (23.4%) foresaw its applications in specialized areas, while only 13% expressed skepticism about its growing importance.

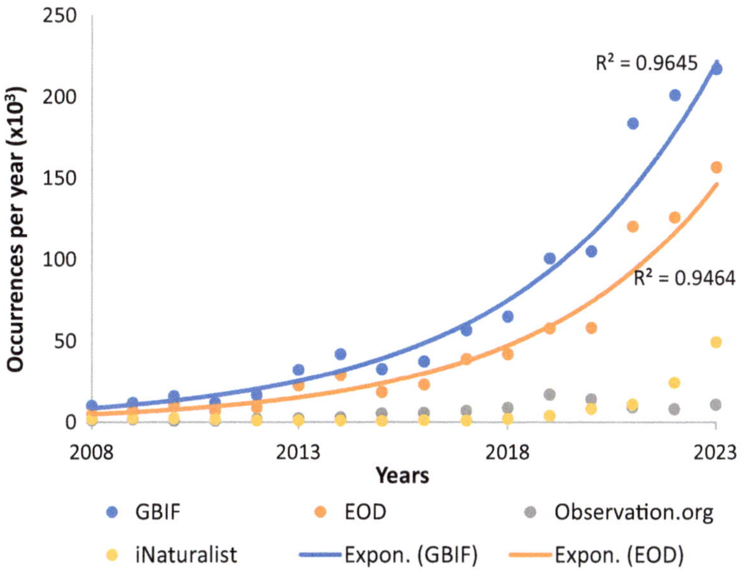

Fig. 1 The dynamics of Hungarian records uploaded to major international online biodiversity platforms (GBIF: Global Biodiversity Information Facility, EOD: eBird Observation Dataset, iNaturalist: Research-grade observations) and fitted trend curves with regression coefficients

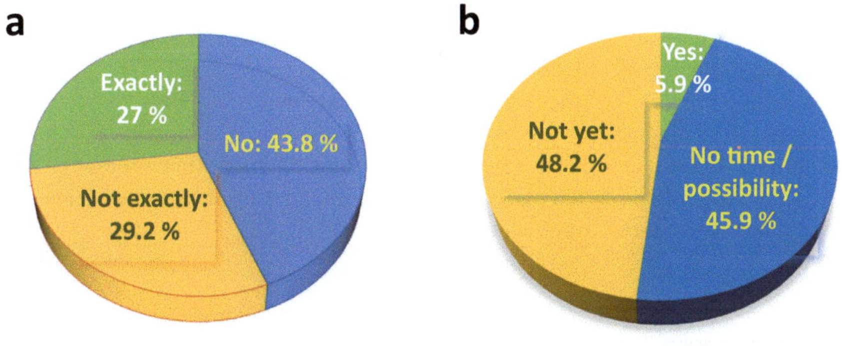

Fig. 2 Distribution of the responses of Hungarian university students about their knowledge on the meaning of citizen science **a** and their participations in citizen science projects **b**

When asked about the utilizations of citizen science, nearly two-thirds (65.1%) of the respondents could not define any social area of use (data not shown), which reflect the lack of knowledge for the students. This indicates the need for a significant role for university education in this field.

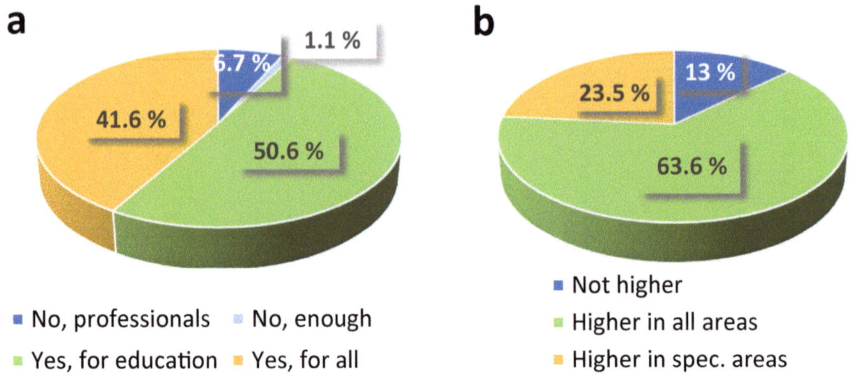

Fig. 3 Distribution of the responses of Hungarian university students about the necessity of citizen science **a** and their perspectives on its future roles **b**

4 Discussion

These results revealed that 73% of the students lacked adequate knowledge about citizen science, and 94% of them had never participated in such programs. On the other hand, over 90% of them recognized its importance for society, with 65% being uncertain about its practical applications. However, some limitations could influence the broad-scale applicability of these preliminary results, especially the lower number of university students involved in this study. More students from more disciplines need to be involved in future research in order to achieve more comprehensive results for wider applications. But these findings emphasize the responsibility of educators and professionals to promote awareness of the benefits and opportunities associated with citizen science. Incorporating online citizen science platforms, such as iNaturalist, into university curricula can open new scientific perspectives for students and provide them with useful experiences.

5 Conclusions

International online biodiversity platforms are increasingly used by Hungarian professional scientists as well as by amateur naturalists. However, university students often lack the proper knowledge about these databases, as indicated here. According to them, there is a great need for community participations in science. Encouraging the use of online citizen science tools in classrooms could bridge the gap between theoretical knowledge and practical applications. By leveraging these platforms, students can explore biodiversity in their local rural and urban environments, gaining a deeper understanding of the natural world while contributing to global datasets. Furthermore, the integration of citizen science into education aligns with SDGs, offering a pathway to cultivate future generations of environmentally conscious citizens.

References

Andrachuk M, Marschke M, Hings C, Armitage D (2019) Smartphone technologies supporting community-based environmental monitoring and implementation: a systematic scoping review. Biol Conserv 237:430–442. https://doi.org/10.1016/j.biocon.2019.07.026

Aristeidou M, Herodotou C, Ballard HL, Young AN, Miller AE, Higgins L, Johnson, RF (2021) Exploring the participation of young citizen scientists in scientific research: the case of iNaturalist. PLoS One 16(1):e0245682. https://doi.org/10.1371/journal.pone.0245682

Bell S, Marzano M, Cent J, Kobierska H, Podjed D, Vandzinskaite D, Reinert H, Armaitiane A, Grodzinska-Jurczak M, Mursic R (2008) What counts? Volunteers and their organisations in the recording and monitoring of biodiversity. Biodivers Conserv 17(4):3443–3454. https://doi.org/10.1007/s10531-008-9357-9

Bio Innovation Service (2018) Citizen science for environmental policy: development of an EU-wide inventory and analysis of selected practices. Final report for the European Commission, DG Environment under the contract 070203/2017/768879/ETU/ENV.A.3

Boone ME, Basille M (2019) Using iNaturalist to contribute your nature observations to science: WEC413/UW458, 6/2019. EDIS (4). https://doi.org/10.32473/edis-uw458-2019

Bowser A, Cooper C, de Sherbinin A, Wiggins A, Brenton P, Chuang T-R, Faustman E, Haklay M, Meloche M (2020) Still in need of norms: the state of the data in citizen science. Citiz Sci: Theory Pract 5(1):18. https://doi.org/10.5334/cstp.303

Callaghan CT, Mesaglio T, Ascher JS, Brooks TM, Cabras AA, Chandler M, Cornwell WK, Ríos-Málaver IC, Dankowicz E, Dhiyáulhaq NU, Fuller RA, Galindo-Leal C, Grattarola F, Hewitt S, Higgins L, Hitchcock C, Hung K-LJ, Iwane T, Kahumbu P, Kendrick R, Kieschnick SR, Kunz G, Lee CC, Lin C-T, Loarie S, Medina MN, McGrouther MA, Miles L, Modi S, Nowak K, Oktaviani R, Waswala Olewe BM, Page J, Petrovan S, Saari C, Seltzer CE, Seregin AP, Sullivan JJ, Sumanapala AP, Takoukam A, Widness J, Willmott K, Wüster W, Young AN (2022) The benefits of contributing to the citizen science platform iNaturalist as an identifier. PLoS Biol 20(11):e3001843. https://doi.org/10.1371/journal.pbio.3001843

Carr AJL (2004) Why do we all need community science. Soc Nat Res 17:841–849

Chandler M, See L, Copas K, Bonde AMZ, Claramunt López B, Danielsen F, Legind JK, Masinde S, Miller-Rushing AJ, Newman G, Rosemartin A, Turak E (2016) Contribution of citizen science towards international biodiversity monitoring. Biol Conserv. https://doi.org/10.1016/j.biocon.2016.09.004

Conrad CC, Hilchey KG (2011) A review of citizen science and community-based environmental monitoring: issues and opportunities. Environ Monit Assess 176:273–291. https://doi.org/10.1007/s10661-010-1582-5

Cooper CB, Dickinson J, Phillips T, Bonney R (2007) Citizen science as a tool for conservation in residential ecosystems. Ecol Soc 12(2):11. http://www.ecologyandsociety.org/vol12/iss2/art11/

Crall AW, Newman GJ, Stohlgren TJ, Holfelder KA, Graham J, Waller DM (2011) Assessing citizen science data quality: an invasive species case study. Conserv Lett 4:433–442. https://doi.org/10.1111/j.1755-263X.2011.00196.x

Devictor V, Whittaker RJ, Beltrame C (2010) Beyond scarcity: citizen science programmes as useful tools for conservation biogeography. Divers Distrib 16:354–362. https://doi.org/10.1111/j.1472-4642.2009.00615.x

Dickinson JL, Shirk J, Bonter D, Bonney R, Crain RL, Martin J, Phillips T, Purcell K (2012) The current state of citizen science as a tool for ecological research and public engagement. Front Ecol Environ 10(6):291–297. https://doi.org/10.1890/110236

Dillon, J, Stevenson RB, Wals AEJ (2016) Introduction to the special section moving from citizen to civic science to address wicked conservation problems. Conserv Biol 30(3):450–455. https://doi.org/10.1111/cobi.12689

Ellwood ER, Crimmins TM, Miller-Rushing AJ (2017) Citizen science and conservation: Recommendations for a rapidly moving field. Biol Conserv 208:1–4. https://doi.org/10.1016/j.biocon.2016.10.014

EC-JRC (2018) An inventory of citizen science activities for environmental policies. European Commission (EC), Joint Research Centre (JRC), http://data.europa.eu/89h/jrc-citsci-10004

Fraisl D, Campbell J, See L, Wehn U, Wardlaw J, Gold M, Moorthy I, Fritz S (2020) Mapping citizen science contributions to the UN sustainable development goals. Sustain Sci 15(6):1735–1751. https://doi.org/10.1007/s11625-020-00833-7

Freitag A, Meyer R, Whiteman L (2016) Strategies employed by citizen science programs to increase the credibility of their data. Citiz Sci Theory Pract 1(1):2. https://doi.org/10.5334/cstp.6

Gallo T, Waitt D (2011) Creating a successful citizen science model to detect and report invasive species. BioSci 61(6):459–465. https://doi.org/10.1525/bio.2011.61.6.8

GBIF (2024) Global biodiversity information facility. https://www.gbif.org/en/

Griggs, D, Stafford Smith M, Rockström J, Öhman MC, Gaffney O, Glaser G, Kanie N, Noble I, Steffen W, Shyamsundar P (2014) An integrated framework for sustainable development goals. Ecol Soc 19(4):49. https://doi.org/10.5751/ES-07082-190449

Greenwood JJD (2007) Citizens, science and bird conservation. J Ornithol 148(Suppl 1):S77–S124. https://doi.org/10.1007/s10336-007-0239-9

Heberling JM, Miller JT, Noesgaard D, Weingart SB, Schigel D (2021) Data integration enables global biodiversity synthesis. PNAS USA 118:e2018093118. https://doi.org/10.1073/pnas.2018093118 PMID: 33526679

Hitchcock C, Vance-Chalcraft H, Aristeidou M (2021) Citizen science in higher education. Citiz Sci Theory Pract 6(1):22:1–4. https://doi.org/10.5334/cstp.467

Kobori, H, Dickinson JL, Washitani I, Sakurai R, Amano T, Komatsu N, Kitamura W, Takagawa S, Koyama K, Ogawara T, Miller-Rushing AJ (2016) Citizen science: a new approach to advance ecology, education, and conservation. Ecol Res 31:1–19. https://doi.org/10.1007/s11284-015-1314-y

Kosmala M, Wiggins A, Swanson A, Simmons B (2016) Assessing data quality in citizen science. Front Ecol Environ 14:551–560. https://doi.org/10.1002/fee.1436

Kruger LE, Shannon MA (2000) Getting to know ourselves and our places through participation in civic social assessment. Soc Nat Res, 13:461–478. https://doi.org/10.1080/089419200403866

Lakshminarayanan S (2007) Using citizens to do science versus citizens as scientists. Ecol Soc 12(2):r2. http://www.ecologyandsociety.org/vol12/iss2/resp2/

McKinley DC, Miller-Rushing AJ, Ballard HL, Bonney R, Browna H, Cook-Patton SC, Evans DM, French RA, Parrish JK, Phillips TB, Ryan SF, Shanley LA, Shirk JL, Stepenuck KF, Weltzin JF, Wiggins A, Boyle OD, Briggs RD, Chapin SF III, Hewitt DA, Preuss PW, Soukup MA (2017) Citizen science can improve conservation science, natural resource management, and environmental protection. Biol Conserv 208:15–28. https://doi.org/10.1016/j.biocon.2016.05.015

Miller-Rushing A, Primack R, Bonney R (2012) The history of public participation in ecological research. Front Ecol Environ. 10:285–290. https://doi.org/10.1890/110278

Mitchell N, Triska M, Liberatore A, Ashcroft L, Weatherill R, Longnecker N (2017) Benefits and challenges of incorporating citizen science into university education. PLoS ONE 12(11):e0186285. https://doi.org/10.1371/journal.pone.0186285

Nugent J (2018) INaturalist: citizen science for 21st-century naturalists. Sci Scope 41(7):12

Pocock MJO, Tweddle JC, Savage J, Robinson LD, Roy HE (2017) The diversity and evolution of ecological and environmental citizen science. PLoS one 12(4):e0172579. https://doi.org/10.1371/journal.pone.0172579

Potsikas M, Prouska K, Efthimiou G, Plakitsi K, Kornelaki A-C (2023) Citizen science practice around lake Pamvotis and the Ioannina castle: using iNaturalist to foster connectedness to nature in citizens and university students. Int J Geoherit Parks 11(3):450–463. https://doi.org/10.1016/j.ijgeop.2023.07.002

Roche J, Bell L, Galvão C, Golumbic YN, Kloetzer L, Knoben N, Laakso M, Lorke J, Mannion G, Massetti L, Mauchline A, Pata K, Ruck A, Taraba P, Winter S (2020) Citizen science, education, and learning: challenges and opportunities. Front Sociol 5:613814. https://doi.org/10.3389/fsoc.2020.613814

Robinson LD, Cawthray-Syms JL, West SE, Bonn A, Ansine J (2018) Ten principles of citizen science. In: Hecker S, Hacklay M, Bowser A, Makuch Z, Vogel J, Bonn A (eds) Innovation in open science, society and policy. UCL Press, London, UK

Silvertown J (2009) A new dawn for citizen science. Trends Ecol Evol 24:467–471. https://doi.org/10.1016/j.tree.2009.03.017

Stephenson PJ, Stengel C (2020) An inventory of biodiversity data sources for conservation monitoring. PLoS ONE 10:e0242923. https://doi.org/10.1371/journal.pone.0242923

Stevenson R, Merrill C, Burn P (2021) Useful biodiversity data were obtained by novice observers using iNaturalist during college orientation retreats. Citiz Sci Theory Pract 6(1):1–12. https://doi.org/10.5334/cstp.407

Sullivan BL, Aycrigg JL, Barry JH, Bonney RE, Bruns N, Cooper CB, Damoulas T, Dhondt AA, Dietterich T, Farnsworth A, Fink D, Fitzpatrick JW, Fredericks T, Gerbracht J, Gomes C, Hochachka WM, Iliff MJ, Lagoze C, La Sorte FA, Merrifield M, Morris W, Phillips TB, Reynolds M, Rodewald AD, Rosenberg KV, Trautmann NM, Wiggins A, Winkler DW, Wong W-K, Wood CL, Yu J, Kelling S (2014) The eBird enterprise: an integrated approach to development and application of citizen science. Biol Conserv 169:31–40. https://doi.org/10.1016/j.biocon.2013.11.003

Toomey AH, Domroese MC (2013) Can citizen science lead to positive conservation attitudes and behaviors? Human Ecol Rev 20(1):50–62

Turbé A, Barba J, Pelacho† M, Mugdal S, Robinson LD, Serrano-Sanz F, Sanz F, Tsinaraki C, Rubio JM, Schade S (2019) Understanding the citizen science landscape for european environmental policy: an assessment and recommendations. Citiz Sci Theory Pract 4(1):34. https://doi.org/10.5334/cstp.239

Wals AEJ, Brody M, Dillon J, Stevenson RB (2014) Convergence between science and environmental education. Science 344:583–584. https://www.science.org/doi/10.1126/science.1250515

Communicating Sustainability Competencies: A Case Study of Sustainability and Learning Design

Paul Astles and Catriona Matthews

Abstract Embedding sustainability competencies offers non-sustainability-focused courses an opportunity to meaningfully connect with sustainability. However, there is no universally applicable way to frame and communicate sustainability competencies, skills, activities and learning outcomes (LOs). Recognising this challenge, the Open University Learning Design (OULD) team created a series of sustainability skills cards (SSCs). These cards connect sustainability competencies with the Open University's existing institutional frameworks and use of language and provide colleagues with a model to use in course design. This chapter explores the importance of context when integrating sustainability into curriculum design. It also demonstrates how student voices can provide valuable insights into how to communicate sustainability competencies, using a case study focused on the design and comprehension of the SSCs. Student engagement during the SSCs design resulted in valuable changes to the contextualisation and presentation of the cards: for example, adapting the SSCs' tone to be more 'personal' and, therefore, engaging. Key findings show context matters for sustainability skills as students interpret sustainability in a multitude of ways, from recycling to applying skills in future careers. Findings suggest students want sustainability to be meaningful in their learning. This may involve guidance to connect with and reflect on sustainability in course-specific, individual, local, or societal contexts. This chapter concludes that resources such as the OU's SSCs, that clearly connect sustainability competencies to the institutional context, can help to guide colleagues in connecting sustainability to LOs and/or activity design, kick-start discussions about embedding sustainability into a student learning journey, and empower the integration of relevant context-specific sustainability skills within course design.

Keywords Learning design · Sustainability competencies · Student engagement · Empowering colleagues

---

P. Astles (✉) · C. Matthews
The Open University, Milton Keynes, UK
e-mail: paul.astles@open.ac.uk

C. Matthews
e-mail: catriona.matthews@open.ac.uk

W. Leal Filho et al. (eds.), *Education for Sustainable Development: The Contribution of Universities*, World Sustainability Series,
https://doi.org/10.1007/978-3-031-86985-3_6

1 Introduction

Sustainability competencies can empower individuals to 'reflect on their own actions' and provide cross-cutting skills necessary for both addressing sustainability challenges and working towards the UN's 17 Sustainability Development Goals (SDGs) (UNESCO 2017). A point of significant debate in the literature is around which combination of competencies are key for supporting individuals to 'transform their own lifestyles and contribute to societal transformation towards sustainability' (Rieckmann, 2012; Wiek et al. 2011; Bianchi 2020; Redman and Wiek 2021). This chapter does not contribute to this ongoing debate but rather builds upon the work of authors such as Lwin (2024) and Wilhem et al. (2019) who have argued that competencies are 'fundamentally context bound' and that our focus should shift to ensuring competencies are successfully integrated and implemented (Wiek and Redman, 2022). Integrating sustainability competencies within courses can help to create authentic opportunities for learners to build the 'knowledge, skills, attitudes and values that enable effective, embodied action in the world with respect to real world sustainability problems, challenges, and opportunities, according to the context' (Bianchi 2020).

The pivotal role of higher education institutions (HEIs) in supporting sustainable development has been emphasised throughout the literature. Authors such as Mokski et al. (2023) and Chankseliani and McCowan (2020) highlight the critical positioning of HEIs in creating knowledgeable citizens with sustainable attitudes and values. However, realising this role by integrating sustainability competencies into educational institutions involves pedagogical transformations that are often at odds with traditional educational systems (Jones et al. 2010). Furthermore, the successful transformation of competency frameworks into tangible developmental experiences for students requires not only successful integration into the design of course material but, in many cases, the upskilling and convincing of academic and course design staff, particularly where academic subjects may not historically or easily connect with sustainability. Hesitation, resistance, or a lack of capacity on the part of individuals and institutions can therefore inhibit the successful integration of competencies (Jones et al. 2010).

Institutions can begin to address these challenges through implementing learning design methodologies which place student experience at the centre of course design. This action would support the evidence-based inclusion of active and transformative pedagogies (Lozano et al. 2017; Cotton et al. 2010; UNESCO 2017). A complimentary approach is to involve students in the design and evaluation of education. Work by Lowe and El Hakim (2020) demonstrates the ways students can support educators to develop accessible, inclusive, and engaging learning opportunities. In the context of sustainability, engaging with students supports the framing of sustainability competencies such that other students are more likely to perceive them as relevant. The case study described in this chapter provides an example of how student voices can inform the prioritisation of sustainability skills and the development of supportive training to enable their integration into the curriculum and its design.

This chapter details how the Open University Learning Design (OULD) team took a step towards extending the evidence base around what student involvement in the process of supporting Education for Sustainable Development (ESD) to be embedded in curricula can look like. The approach outlined in this chapter could also be adopted by institutions looking to use or adapt the OULD sustainability skills cards (SSCs) to their context.

1.1 The Role of Higher Education Institutions in Sustainability

HEIs play a pivotal role in supporting the SDGs and progressing sustainable development more broadly through their role in 'human formation, knowledge production and innovation' (Chankseliani and McCowan, 2020). ESD sees institutions preparing learners with the 'appropriate awareness, attitude, behaviours, and mindset for addressing the world's issues in the name of ESD competencies' (Rieckmann 2012) and developing professionals who may directly inform key policy decisions and societal institutions (Mokski et al. 2023).

There is increasing evidence that students want HEIs to demonstrate stronger connections to sustainability in their courses (Carr-Shand 2022; Kaplan 2021; Unite Group 2021). The latest in a series of yearly surveys run by Students Organising for Sustainability (SOS) in the UK found 89% of the approximately 9000 respondents felt their place of study should actively incorporate and promote sustainable development, while 82% would like to see sustainable development actively incorporated and promoted in all courses (SOS 2023). If, as Sterling (2001) posited, 'the key to a more sustainable and peaceable world is learning' then the key to connecting learning with sustainability is through its design.

Lwin (2024) identified three main approaches to integrating ESD into the curriculum.

1. Integrating ESD-related content into existing models or by re-designing the existing content of a programme or course.
2. Developing standalone, tailored degrees or diplomas on sustainable development or a university-wide compulsory course.
3. Using a non-explicit or hidden curriculum approach to transmit ethos and values.

The highest standard is the first approach where ESD and, by extension, sustainability competencies become 'transversal and intrinsic in education' (Bianchi 2020). However, this can only be achieved when institutions move away from knowledge-centric to competency-centred approaches through an overhaul of curriculum design and teaching methodologies (Wilhelm et al. 2019; Toro-Troconis et al. 2023). Competency-based education focuses on 'the ability of students to develop important knowledge, values, aptitudes, and attitudes necessary to address complex issues they will encounter in their future personal lives and professional careers' (Lozano et al. 2017). The difficulty of navigating transformative changes within the boundaries of a largely traditional educational system has been raised within the literature (Jones et al. 2010; Wiek and Redman, 2022).

Even where competencies have been successfully integrated into curriculum design, there are further obstacles to be navigated in the form of teacher confidence and competencies which require teaching staff to go 'beyond being instructors of specific content…' to become '…dynamic members of the classroom…' (Leal Filho et al. 2021). Although this chapter does not cover the challenge of developing staff sustainability competencies and knowledge, it is important to note that doing so provides the foundations of successful curriculum design. If instructors are unable to clearly articulate competencies, then students likewise will struggle (Brundiers et al. 2020). The resources discussed in this chapter were part of a wider upskilling and confidence-building approach led by OULD. The SSCs were positioned to support academics and course design colleagues who already have an awareness of sustainability to apply their knowledge by identifying overlaps and potential points of synergy between their courses and related sustainability competencies.

2 Sustainability Competencies

There has been much debate about the definition and use of 'competencies' and 'key competencies' for sustainability. Arguably the most influential was Wiek et al. (2011) framework which provided the foundation for numerous academic sustainability competency frameworks (Rieckmann 2012; Lozano et al. 2017; Brundiers et al. 2020), as well as influential institutional frameworks such as UNESCO's (2017). There has been significant debate around the terminology and structure of competency frameworks. Reviews undertaken by Bianchi (2020) and Redman and Wiek (2021) indicate that despite the 'terminological differences' there is 'substantial convergence' regarding what 'change agents need to be capable of to advance social transformations to sustainability.' Indeed, Wiek and Redman (2022) have recently called for an end to 'reinventing competencies in sustainability' and suggest that the future focus should be on how to successfully support institutions to holistically integrate, implement, and build staff and student confidence around sustainability competencies.

Given the plethora of frameworks available, narrowing down which to use can be a challenge. Wilhem et al. (2019) have posited that competencies are 'fundamentally context bound' and that 'generalised models make little sense.' This argument is expanded by Lwin (2024) in their PHD thesis which explores ESD in the context of Myanmar. Lwin suggests that the successful integration and implantation of sustainability competencies requires contextualisation to the 'social, cultural, environmental, political and philosophical factors of a country.' However, a balance needs to be found between the adaptation of competencies to specific contexts and the use of consistent language to describe competencies. This helps to make them measurable and transferable between education institutions and into employment.

The OULD resources (Calder et al. 2023b) use the Advance HE and QAA (2021) adaptation of the UNESCO (2017) framework rather than blending multiple frameworks. However, as will be discussed later in this chapter, our approach looked to integrate the framework into our context by connecting with institutional Employability (The Open University, n.d.), Digital and Information Literacy (DIL) (The Open University, n.d.), and Inclusive Curriculum frameworks (Veuger et al. 2023). It is worth acknowledging some of the limitations of the UNESCO framework put forward by Lwin (2024). The framework is western-centric as, despite being reviewed by scholars and experts from different countries, the main developing authors were from Europe. Lwin argues that the framework lends itself to the creation of 'laundry lists' of competencies which could contribute to the full cohort of interlinked competencies not being integrated. Furthermore, it has been critiqued for entangling what are termed 'general' or 'academic' competencies with sustainability competencies (Wiek and Redman 2022). This, Wiek and Redman note, makes it challenging for students to distinguish 'general' or 'academic' competencies from key sustainability competencies and impacts their 'operationalisation.' However, in our context we have found the visibility of academic competencies, such as critical thinking, to be helpful. They helped to bridge the conceptual gap for colleagues creating courses that don't typically or easily connect with sustainability, making it easier to identify where connections can be made in a non-tokenistic manner.

3 Learning Design and Sustainability Competencies

Designing courses that develop sustainability competencies requires a move away from traditional pedagogies to those which engage learners with processes of 'reflection, action and experience' (Bianchi 2020; Lozano et al. 2017). The centrality of transformative pedagogies to ESD and sustainability competencies has been readily discussed in the literature with authors outlining the importance of certain pedagogical approaches (Lozano et al. 2017; Cotton et al. 2010; UNESCO 2017), case studies, and bridging the gap between theory and practice through the 'operationalisation of competencies' (Sterling 2012; Advance HE and QAA 2021; Wilhem et al. 2019; Lozano et al. 2017). The multiplicity of elements involved in integrating sustainability into curriculum design has led to CoDesigns, a collaborative group of universities and professional bodies, creating a framework which aims to break down 'the intricate process of curriculum design into smaller, manageable tasks' supported by an eight-week bootcamp (Toro-Troconis 2023).

Our approach to embedding sustainability in the curriculum is informed by the adaptive approach to learning design taken at the Open University (OU). Supporting good learning design, much like supporting sustainability, is a long journey and, along the way, you meet individuals at different stages of knowledge, confidence, and cooperation. In the design and production of courses at the OU, the OULD team use evidence-based approaches such as the ICEBERG framework (Reedy 2021; Van Ameijde et al. 2016) as the foundation of our practice. We adopt supportive

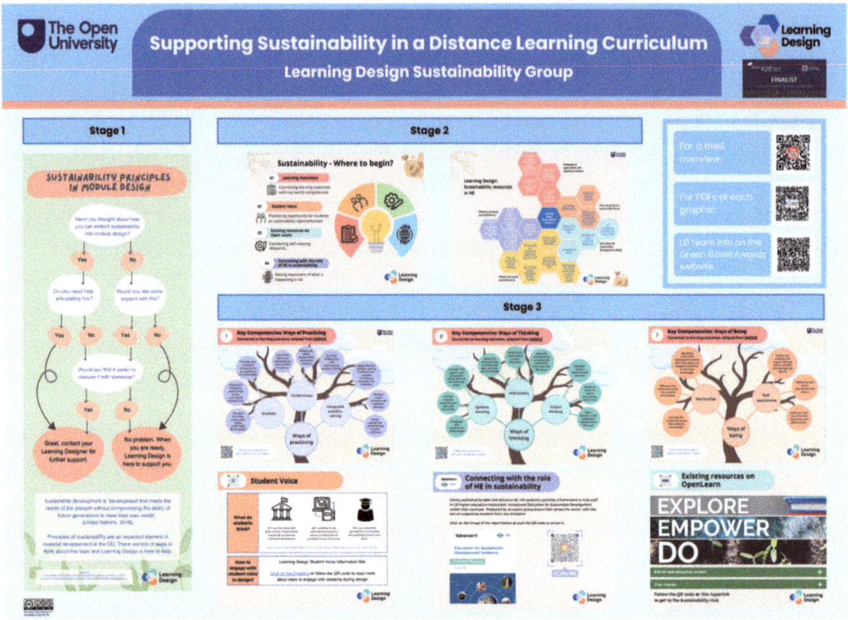

Fig. 1 Poster created to showcase the OULD Sustainability group's resources

approaches depending on the need, knowledge base, and engagement level of authors of course materials while keeping students' experiences as a central focus.

Integrating sustainability into existing practices involves navigating competing requirements within module design processes such as external accreditation, professional requirements, and institutional Employability, DIL, and Inclusive Curriculum frameworks. The OULD team initially created a set of resources which summarised knowledge and learnings from the sustainability literature to begin conversations with teams involved in module design and production (Calder et al. 2024, see Fig. 1). The aim was to start an ongoing dialogue with the ambition of meaningfully integrating sustainability into the curriculum. Taking a resource-led approach removed barriers for colleagues who were time poor or uncertain where to begin amid the range of existing resources. The next step on our journey was the creation of a series of SSCs which then helped to translate theory relating to sustainability competencies into practice and move towards practical support for colleagues.

4 The Sustainability Skills Cards

The OULD team created of a set of SSCs (Calder et al. 2023b, see Fig. 2) focused on supporting colleagues to not only identify points of connection between their courses and the UNESCO (2017) sustainability competencies, but to begin to turn those connections into actionable and measurable learning opportunities. The functionality of the cards is based on a belief that, for both staff and students, opportunities to engage with sustainability should be meaningful and non-tokenistic (Astles et al. 2023). This is in part informed by student responses to an internal OU 'Question of the Month' survey regarding sustainability and higher education which was shared in 2021 with a panel of approximately 2000 students. The 34 responses (response rate 1.7%) largely aligned with wider institutional surveys by indicating a desire for HEIs to take the initiative and model sustainability practice. They also suggested that where subjects don't traditionally or easily connect with sustainability, it should not be shoehorned in, but only included where it could be meaningfully integrated in a non-tokenistic manner. The SSCs therefore form part of a design approach that engages colleagues in reflection about how sustainability can meaningfully connect to their context and how this authenticity can be translated into learning opportunities for students.

The SSCs focus on the eight UNESCO (2017) sustainability competencies and split them into the overarching themes of 'ways of being, thinking and practising' provided by the QAA and Advance HE (2021) adaptation of the framework. The SSCs are action-focused and designed to help colleagues translate sustainability competencies into LOs and/or activities. The overlap of LOs and activities in the SSCs support the development of constructive alignment which has been found by Stamov Roßnagel et al. (2021) to support student perceptions of courses as useful,

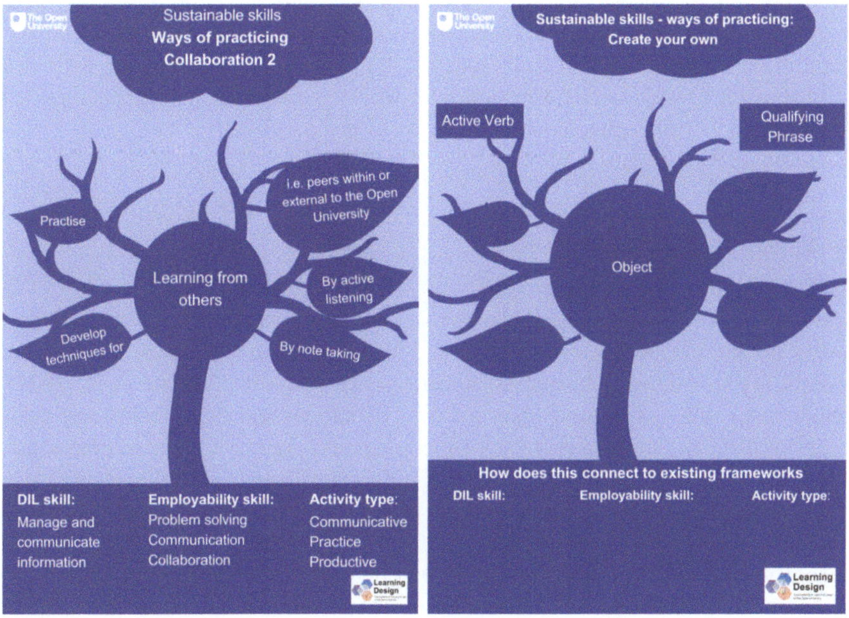

Fig. 2 Example of SSC content after feedback from students

and to create higher efforts in learning and increased enjoyment (Biggs and Tang, 2011).

The SSCs use the Mann (2004) model, a way of breaking down the competencies into LOs. Moon (2004) and Mann explain that LOs are statements about what a student will be able to do by the end of a given timeframe. Therefore, LOs are intended to be student-focused statements that describe the outcome of what the learning journey should be. The UNESCO (2017) competencies were reviewed by the OULD team to identify measurable language. This shifted the focus of the SSCs from describing possible sustainability competencies to outlining measurable outcomes. The 'Engaging student voices' section of this chapter details how students helped to further develop the SSCs.

The SSCs do not provide a comprehensive list of 'objects' or 'outcomes' related to each competency, as one of their main focuses is to encourage users to reflect on which element(s) of each competency connect with the context of their subject or module. Likewise, the SSCs do not visualise or explicitly encourage reflection about the interconnections between sustainability competencies; instead, they make connections with OU frameworks. Although Weik et al. argues that this encourages the use of competencies as 'laundry lists', we suggest it provides an accessible way to engage with competencies which allows them to be adjusted to suit institutional and module-specific contexts (Wiek et al. 2011). The resources also aim to avoid the trap of 'greenwashing' (Netto et al., 2020) by emphasising the need to contextualise sustainability and providing a model which supports active and transformative pedagogical approaches. The SSCs provide a way into sustainability through the language of LOs. They begin to create points of commonality that encourage educators to reflect on the needs of their students, the future tasks that students will need to fulfil as they progress to further study or work, and how these can be supported through meaningful engagement.

5 Engaging Student Voices

Approaches that aim to engage student voices are based on the premise that students have a unique perspective on learning and teaching and should therefore have an opportunity to influence how these practices are realised (Tuhkala et al. 2021). This chapter will not expand upon the vast debate around the benefits and challenges of different approaches to student engagement; rather it looks to support the belief that providing opportunities for students to contribute to educational processes makes education 'more accessible, its practices more inclusive and the learning more engaging' (Lowe and El Hakim 2020). Student engagement is also an important feature of ESD. It supports meaningful opportunities to develop the UNESCO sustainability competency of 'collaboration' and contributes towards the goal of empowering learners who will one day enable social transformation.

The SSCs were developed in consultation with students at the OU using an institutional student engagement mechanism established in 2016 called the

Curriculum Design Student Panel (CDSP). The CDSP provides students with voluntary consulting opportunities to contribute to curriculum design through surveys, testing curriculum content, interactive media and tools, providing feedback on prototypes, and sharing past study experiences. Since 2016, over 150 activities have taken place and over 3000 students from various levels of study and a range of faculties have contributed. The benefits of student engagement mechanisms such as the CDSP include helping students and staff to understand (or at least reflect on) each other's perspectives, supporting the transformation of thinking about teaching practices, and providing staff with a 'deeper understanding of student needs' (Cook-Sather et al. 2014).

6 Method

Student feedback for the purpose of developing the SSCs was captured using a survey circulated to the CDSP student volunteers. A survey format was chosen as it allowed students flexibility around their study schedules whilst still allowing rich data to be gathered. The aims of the survey were to gather feedback on the vision, format, and function of the SSCs and to gain an insight into how students define sustainability and perceived sustainability competencies. Ahead of completing the survey students were provided with a short description of the UNESCO competencies, a prototype version of the SSCs and a vision statement for a possible use case of the SSCs to help contextualise them. The survey contained a mixture of question types. Four questions were quantitative in nature, asking students to respond to a Likert scale or multiple-choice question. Five questions were qualitative, asking students to expand on why they had chosen an answer or provide further comment. The qualitative questions asked students to reflect on how the content of the SSCs related to their understanding of sustainability, share whether they found the cards difficult or easy to understand and why, articulate which skills they would most like to see embedded in courses, and provide general feedback on the format, vision, and functionality of the SSCs. In addition to the survey, an online student forum was available for students to post questions or comments for staff or other students to respond to during the period of the survey. The survey was made available to the whole CDSP which, at the time, consisted of 1723 students. The survey was open for two weeks and 33 students responded (a 2% response rate). The qualitative data, including the forum responses, were analysed using the six-phase process of thematic analysis defined by Braun and Clarke (2006) with the aim of identifying common themes relating to the utility of the SSCs and students' comprehension of sustainability. The quantitative data outputs were represented as a frequency, showing the total number of responses to the relevant questions.

7 Survey Responses

Student feedback about the design of the SSCs was overwhelmingly positive. In total, 76% of respondents answered that the cards linked 'well', 'very well' or 'extremely well' with their understanding of sustainability. Most of the feedback about the context of the SSCs was also positive. Students indicated that the SSCs were 'easy to understand', 'clearly defined', and presented in a 'logical and systematic format'. Constructive feedback highlighted the need for a clearer explanation about the function of the cards and provided advice about the formatting and language used. In response to this feedback, we created a 'setting the scene' page for each SSC deck which covered the aims of the SSCs, as well as key insights from students about sustainability drawn from the CDSP activity.

One respondent, despite acknowledging the importance of building the skills defined in the SSCs, questioned how the content of the cards related to sustainability. This comment aligns with a challenge identified by Brundiers et al. (2021) about the inclusion of base academic competencies in the UNESCO framework which they argue compliment sustainability competencies but should not be included in the framework because they are 'developed in every quality academic programme'. As previously mentioned, we have found that 'academic'-focused competencies can help academics and students identify connections with sustainability and bring it into their context. Therefore, the main focus of the CDSP activity was to understand how students define and conceptualise sustainability. These insights would then inform how we frame sustainability competencies in our work with academics and module design colleagues and how they likewise communicate and design sustainability competencies meaningfully into their modules.

As argued by Angus-Cole et al. (2020), if students' perceptions do not align with the context and aims of sustainability elements within the curriculum, then the perceived relevance of sustainability's inclusion will be 'radically' affected. If ESD is perceived as 'superficial' and disconnected from students' fields of study, for example where sustainability is conflated with scientific solutions to climate change and a student is studying English Literature, then 'students may not be motivated to adopt knowledge and willingness to act' (Hyytinen et al. 2023).

Most responses to the question, 'As a student what does sustainability mean to you?' were related to either 'reserving resources' (n = 12), 'environment/climate change' (n = 9) or sat across both these categories (n = 2). Several responses paraphrased or quoted the famous definition from the 1987 UN Brundtland report of 'meeting the needs of the present without compromising the ability of future generations to meet their own needs' (UNWCED 1987). The 2022–2023 annual survey run by SOS (2023) similarly found that most students understood sustainable development in line with the 1987 Brundtland definition. A small number of responses took a broader view of sustainability connecting it to 'personal development' (n = 4), 'skills' (n = 3) or 'thinking globally' (n = 1). A smaller number still were 'unsure' (n = 1), felt it didn't connect to their study subject (n = 1) or was 'vague' (n = 1) and stating, 'it means a lot'.

These findings suggest that students may not immediately recognise or be interested in the notion of sustainability competencies if they do not connect with their own/current understanding of sustainability. One way to address this challenge is through framing. Framing is described by Sandri (2021) as a 'tool to help educators connect sustainability to the disciplinary context and future professional practice of learners.' Sandri argues that 'good practice in teaching and learning in sustainability education…necessitates that sustainability is framed in a way in which everyone sees how they fit in necessary societal transformation.' As previously mentioned, the creation of the 'setting the scene' page on the SSCs aimed to begin to frame the resources for colleagues. To further support colleagues to both frame sustainability in their own and students' contexts we developed a workshop focused around using the SSCs. The workshop aimed to not only introduce colleagues to sustainability competencies and provide a space for them to explore and debate how they connect to their context, but also to further structure the development of activities to advance sustainability competencies. This process, supported by the SSCs, helps colleagues to consider what sustainability might look like, how they might meaningfully frame sustainability in the context of their subject, and communicate this with students, in a way that encourages them to expand or adjust their perception rather than rejecting sustainability outright.

The CDSP activity also asked students to share what they felt the top three skills from the QAA and Advanced HE (2021) adaptation of the UNESCO (2017) competencies are from their perspective, asking them to consider what they would most like to see embedded in their courses. Awareness of students' perceived interest relating to specific skills not only gives further insight into student definitions of sustainability, but also highlights where more thoughtful framing and evidence may be needed to scaffold student understanding of the applicability of skills that they may not see as important as others. 'Problem solving' ($n = 11$), 'critical thinking' ($n = 9$) and 'collaboration' ($n = 6$) were the skills with the highest number of student responses. Given these skills fall largely into the category of 'general academic skills', it could be interpreted that they were chosen due to their familiarity amongst students and their perceived applicability across academic and sustainability contexts. One student did, however, link their choice to sustainability stating that 'achieving sustainability will depend largely on our ability to collaborate and solve problems together.' The skills least chosen by students largely sat within the 'normative' and 'anticipatory' competencies. This may, in part, have been due to uncertainty about what the 'skill' is asking students to do. One student indicated confusion as to what 'negotiating sustainable development values, principles, goals, and targets in a context of uncertain knowledge' was referring to. The 'flowery' language of the competencies was also highlighted by another student as making the description of the skills 'unnecessarily convoluted.' This response encapsulates the challenges of using a sustainability framework which was created to sit across multiple contexts.

These results somewhat align with the findings of Hyytinen et al. (2023) who noted a positive response from students about the value of critical thinking in their survey of almost 800 students in a Finnish University. However, collaboration was less favourably viewed by their respondents, although it is worth noting that their questionnaire covered a mix of 39 academic and sustainability competencies. Hyytinen et al. (2023) findings also provide initial evidence that the area of study can impact the interest that a student has in the context of the different competencies. For example, they note that those studying humanities subjects tended to view collaboration less favourably than those in other subject areas. Due to the small number of responses to our survey, we were unable to analyse for subject-based trends in the data but recognise this type of evidence would be useful for future work supporting the integration of sustainability competencies.

8 Limitations and Future Work

The low survey response rate is a limitation that impacts the generalisability of our results, particularly in the context of promoting a focus on certain skills. The self-selection bias inherent in the use of the CDSP, a panel made up of voluntary, unpaid students, makes it more likely that the data is not representative. For example, students who felt strongly or had an existing interest in sustainability or sustainability competencies may have been more likely to respond. A further limitation is the context of the predominantly online and largely asynchronous institution in which this survey took place. Applicability to traditional higher education settings may vary depending on the nature of their course design. In addition, the results explored in this chapter have been largely taken from participants in the global north with western-focused values. To address this, future work would benefit from including a cross section of insights from online and in-person institutions from across the globe.

The results explored in this chapter used student feedback collected via a survey. Whilst this is an effective means of gathering feedback, it does not allow students to engage with the process as co-creators or partners. These approaches would provide students with greater opportunities to input into decision-making (Lowe and El Hakim 2020). Whilst students as partners is often held by education professionals as the most meaningful and impactful approach to student engagement, the significant time investment required to plan and successfully implement this approach can be a barrier (Lowe and El Hakim 2020). Future work should explore how to engage with students as partners from the outset.

9 The Impact of the Sustainability Skills Cards

The primary focus of this chapter was to explore the process of creating the SSCs. Evaluation of their impact and reception amongst colleagues is ongoing. Preliminary feedback is positive with colleagues commending how the SSCs focus their conversations on practical approaches to connect with sustainability in their course design.

Future work relating to the SSCs may focus on their reception by module design colleagues or their impact on pedagogical approaches to course design.

The SSCs began as an ambition of the OULD sustainability group. However, since their creation in 2023, they have become increasingly integrated into the OU's module design and production processes. At the time of writing this chapter, the SSCs, alongside further sustainability resources created by the OULD team (Calder et al., 2023b), have supported the embedding of ESD in seven modules, impacting approximately 2000 students. During 2023 the SSCs supported development activities for approximately 135 colleagues across the OU, including reflection on how sustainability connects with LOs and existing DIL and Employability frameworks. The creativity, innovation, and collaboration involved in the development and subsequent use of the SSCs has been recognised externally by the EAUC who awarded the OULD team's sustainability endeavours 'Highly Commended' in the 2023 Green Gown Awards (Calder et al., 2023a). These awards recognise exceptional sustainability initiatives being undertaken by tertiary education institutions.

10 Conclusion

This chapter has demonstrated the importance of considering how sustainability competencies are framed and communicated in institutional contexts using a UK-based, open, distance University as an example. It is argued that creation of resources, such as the SSCs, that clearly connect sustainability competencies to the institutional context, can help to guide colleagues in connecting sustainability to LOs and/or activity design, kick-start discussions about embedding sustainability into a student learning journey, and empower the integration of relevant context-specific sustainability skills within course design. The SSC case study furthermore illustrates how engaging with student voices can meaningfully inform the generation of resources and approaches that empower students and staff to make connections with their own disciplinary contexts.

Our work has also shown how learning design can provide a spark for wider institutional change. The SSCs demonstrated how sustainability competencies could sit within the institutional context and across existing frameworks and provided a common language to help break down barriers to their implementation, particularly for courses that may not traditionally connect with sustainability. This chapter suggests that utilising resources such as the SSCs provides a foundation of support which can be adapted depending on where colleagues are in their sustainability journeys. Embedding this process in course design and supporting colleagues to draw on existing knowledge and connect the competencies to their own context, rather than starting from scratch or adding competencies retrospectively, increases the likelihood of meaningful framing of sustainability competencies.

Furthermore, the outcomes from the SSC case study have shown how student voice can also provide insights into how sustainability may need to be framed to help students see meaning in the competencies that courses are seeking them to develop as they may not immediately recognise or be interested in the notion of sustainability competencies if they do not connect with their own/current understanding of sustainability.

The three main learnings taken from our work to create and implement the use of the SSCs are:

1. Focus on identifying non-tokenistic actions that can meaningfully support change, knowledge building or skills development within your context.
2. Create opportunities for student input into resource design—their ideas, comments and critiques are invaluable.
3. Recognise that everyone's journey towards taking meaningful non-tokenistic actions to embed sustainability will be different.

If you want to develop SSCs of your own, use, or adapt the ones created by the OULD, you can follow the model and approach outlined in this chapter.

References

Advance HE, QAA (2021) Education for sustainable development guidance. https://www.advance-he.ac.uk/knowledge-hub/education-sustainable-development-guidance

Angus-Cole KL, Eaton R, Dawes M (2020) The design and delivery of a workshop to support curriculum development, education for sustainability and students as partners: sustainability in your curriculum—identify, improve, inspire! Emerald Open Res 1(9):1–18. https://doi.org/10.1108/eor-09-2023-0003

Astles P, Calder K, Jones H, Openshaw J (2023) What is non-tokenistic sustainability in a distance learning curriculum? [Presentation]. In: EDEN conference 2023, Dublin, pp 122–124. https://www.eden-europe.eu/wp-content/uploads/2023/07/EDEN-2023-Annual-Conference-Proceedings.pdf

Bianchi G (2020) Sustainability competences: a systematic literature review. Jt Res Cent (Eur Comm). https://doi.org/10.2760/200956

Biggs J, Tang C (2011) Teaching for quality learning at university, 4th ed. McGraw Hill Education & Open University Press

Braun V, Clarke V (2006) Using Thematic Analysis in Psychology. Qual Res Psychol 3(2):77–101. https://doi.org/10.1191/1478088706qp063oa

Brundiers K, Barth M, Cebrián G, Cohen M, Diaz L, Doucette-Remington S, Dripps W, Habron G, Harré N, Jarchow M, Losch K, Michel J, Mochizuki Y, Rieckmann M, Parnell R, Walker P, Zint M (2020) Key competencies in sustainability in higher education—toward an agreed-upon reference framework. Sustain Sci 16(1):13–29. https://doi.org/10.1007/s11625-020-00838-2

Calder K, Astles P, Johns H, Matthews C, Openshaw J (2023) Green gown awards UK & Ireland. https://www.greengownawards.org/the-open-university-learning-design-sustainabil.

Calder K, Astles P, Johns H, Matthews C, Openshaw J (2023). Sustainability resources, OU learning design team blog. https://www.open.ac.uk/blogs/learning-design/?page_id=1555.

Calder K, Astles P, Johns H, Matthews C, Openshaw J (2024) Going for green: learning design sustainability champions. In: Hayes-Budgen S (ed) OU learning design team blog. https://www.open.ac.uk/blogs/learning-design/?p=1884

Carr-Shand S (2022) Climate change crisis should be embedded in education, say students and staff. Climate Action. https://blogs.bath.ac.uk/climate-action/2022/04/29/climate-change-crisis-should-be-embedded-in-education-say-students-and-staff/

Chankseliani M, McCowan T (2020) Higher education and the sustainable development goals. High Educ 81(1):1–8. https://doi.org/10.1007/s10734-020-00652-w
Cook-Sather A, Bovill C, Felten P (2014) Engaging students as partners in learning and teaching: a guide to faculty. John Wiley & Sons, Incorporated
Cotton D, Winter J (2010) It's not just bits of paper and light bulbs: a review of sustainability pedagogies and their potential for use in higher education. In: Sterling S (eds) sustainability education: perspectives and practice across higher education. Routledge, pp 39–54. https://doi.org/10.4324/9781849776516
Hyytinen H, Laakso S, Pietikäinen J, Ratvio R, Ruippo L, Tuononen T, Vainio A (2023) Perceived interest in learning sustainability competencies among higher education students. Int J Sustain High Educ 24(9):118–137. https://doi.org/10.1108/ijshe-06-2022-0198
Jones P, Selby D, Sterling S (2010) 'Introduction'. In: Jones P, Selby D, Sterling S (eds) Sustainability education: perspectives and practice across higher education. 1st ed. Taylor and Francis Group, pp 1–16
Kaplan A (2021) Prepare for student sustainability demands to go through the roof. THE Campus. https://www.timeshighereducation.com/campus/prepare-student-sustainability-demands-go-through-roof
Leal Filho W, Levesque VR, Salvia AL, Paço A, Fritzen B, Frankenberger F, Damke LI, Brandli LL, Ávila LV, Mifsud M, Will M, Pace P, Azeiteiro UM, Lovren VO (2021) University teaching staff and sustainable development: an assessment of competences. Sustain Sci 16:101–116. https://doi.org/10.1007/s11625-020-00868-w.
Lowe T, El Hakim Y (2020) An introduction to student engagement in higher education. In: Lowe T, El Hakim Y (eds) A handbook for student engagement in higher education: theory into practice. Routledge, pp 3–26. https://doi-org.libezproxy.open.ac.uk/https://doi.org/10.4324/9780429023033
Lozano R, Merrill M, Sammalisto K, Ceulemans K, Lozano F (2017) Connecting competences and pedagogical approaches for sustainable development in higher education: a literature review and framework proposal'. Sustainability 9:1889
Lwin B (2024) Exploring education for sustainable development in Myanmar: concepts and models to influence the integration in, and the transformation of, higher education Institutions [Doctoral dissertation, The Open University]. Open Research Online. https://oro.open.ac.uk/98089/
Mann SJ (2004) Guidelines for writing aims and intended learning outcomes at the programme and course level. Glasgow University
Mokski E, Leal Filho W, Sehnem S, Andrade Guerra JBSOD (2023) Education for sustainable development in higher education institutions: an approach for effective interdisciplinarity. Int J Sustain High Educ 24(1):96–117. https://doi-org.libezproxy.open.ac.uk/https://doi.org/10.1108/IJSHE-07-2021-0306
Moon J (2004) Linking levels, learning outcomes and assessment criteria. https://www.ehea.info/media.ehea.info/file/Learning_Outcomes_Edinburgh_2004/77/4/040701-02Linking_Levels_plus_ass_crit-Moon_577774.pdf
Netto SV, Sobral MFF, Ribeiro ARB, Soares GRDL (2020) Concepts and Forms of greenwashing: a systematic review. Environ Sci Eur 32(1):1–12. https://doi.org/10.1186/s12302-020-0300-3
Redman A, Wiek A (2021) Competencies for advancing transformations towards sustainability. Front Educ 6. https://doi.org/10.3389/feduc.2021.785163
Reedy K (2021) Learning from practice: refreshing the OU activity types framework. OU Learning Design team blog. https://www.open.ac.uk/blogs/learning-design/?p=1280
Rieckmann M (2012) Future-oriented higher education: Which key competencies should be fostered through university teaching and learning? Futures 44(2):127–135. https://doi.org/10.1016/j.futures.2011.09.005
Sandri O (2021) Providing a 'point of entry': approaches to framing sustainability in curriculum design in Higher Education. Aust J Environ Educ 37(1):56–68. https://doi.org/10.1017/aee.2020.19

Stamov Roßnagel C, Fitzallen N, Lo Baido K (2021) Constructive alignment and the learning experience: relationships with student motivation and perceived learning demands. High Educ Res Dev. 40(4):838–851. https://doi.org/10.1080/07294360.2020.1787956
Sterling S (2001) Sustainable education: re-visioning learning and change. Schumacher Society
Sterling S (2012) The future fit framework: an introductory guide to teaching and learning for sustainability in HE. Higher Education Academy Publication. https://www.plymouth.ac.uk/students-and-family/sustainability/sustainability-education/resources
Students Organising for Sustainability (2023) Sustainability skills survey 2022–23. https://www.sos-uk.org/research/sustainability-skills-survey
The Open University (n.d.) DIL framework. https://www5.open.ac.uk/library-skills-framework/DIL-framework
The Open University (n.d.) Employability and the Open University. https://www.open.ac.uk/courses/careers/employability
Toro-Troconis M, Inzolia Y, Ahmad N (2023) Exploring attitudes towards embedding education for sustainable development in curriculum design. Int J High Educ 12(4):42–54. https://doi.org/10.5430/ijhe.v12n4p42
Tuhkala A, Ekonoja A, Hämäläinen R (2021) Tensions of student voice in higher education: involving students in degree programme curricula design. Innov Educ Teach Int 58(4):451–461. https://doi.org/10.1080/14703297.2020.1763189
UNESCO (2017) Education for sustainable development goals: learning objectives. UNESCO, Paris
Unite Group (2021) Survey: Students call for strong action on climate change. https://www.unitegroup.com/articles/survey-students-call-for-strong-action-on-climate-change
UNWCED (1987) Report of the world commission on environment and development—Our common future. https://www.un.org/ga/search/view_doc.asp?symbol=A/42/427&Lang=E.
van Ameijde J, Weller M, Cross S (2016) Designing for student retention: the ICEBERG model and key design tips. The Open University. https://www.open.ac.uk/blogs/learning-design/wp-content/uploads/2020/09/ICEBERG-booklet-compressed.pdf
Veuger SJ, Butler D, Wood P, Potter A (2023) Inclusive frameworks in online STEM teaching and learning. In: Keengwe J (ed) Handbook of research on innovative frameworks and inclusive models for online learning. IGI Global, pp 28–51. https://doi.org/10.4018/978-1-6684-9072-3.ch002
Wiek A, Redman AD (2022) What do key competencies in sustainability offer and how to use them. In: Vare P, Lausselet N, Rieckmann M (eds) Competences in education for sustainable development. Sustainable development goal series. Springer, pp 27–34. https://doi.org/10.1007/978-3-030-91055-6_4
Wiek A, Withycombe L, Redman CL (2011) Key Competencies in sustainability: a reference framework for academic program development. Sustain Sci 6(2):203–218. https://doi.org/10.1007/s11625-011-0132-6
Wilhelm S, Förster R, Zimmermann AB (2019) Implementing competence orientation: towards constructively aligned education for sustainable development in university-level teaching-and-learning. Sustainability 11(7):1891

Conceptualising and Evaluating the Impact of Business Training Programmes Delivered to Marginalized Individuals

Tracey Dodd

Abstract In the context of education for sustainable development, scholars, policymakers, and quality accreditors around the globe challenge business schools and management academics to extend their impact beyond citations to the real world. Many academics have responded by becoming involved in business training programmes that seek to increase the well-being of marginalised individuals. Yet, the impact of these programmes remains under-theorised. Through the lens of the integral pheno-practice of well-being framework, this chapter addresses this gap, presenting a conceptual framework that illustrates how these programmes can both positively and negatively influence individual- and societal-level well-being. In doing so, the chapter raises many institutional questions for those involved in training programmes designed to advance the SDGs, pointing to promising avenues for future research and action.

Keywords Business schools · Education for sustainable development · Integral pheno-practice of well-being · Management education · Training programmes

1 Introduction

Increasing attention is being paid to the role of management education in advancing sustainable development and the United Nations Sustainable Development Goals (SDGs) (Dean et al. 2018). Business schools worldwide have introduced management training programs aimed at supporting the SDGs, particularly those focused on enhancing the well-being of marginalized individuals. These initiatives operate on the expectation that equipping individuals with new skills will empower them to address both personal and societal challenges. However, significant questions persist about the effectiveness, appropriateness, and long-term impact of such programs,

T. Dodd (✉)
Adelaide Business School, GAICD, MSPPM, Research Development, Governance Institute of Australia, University of Adelaide, Adelaide, Australia
e-mail: tracey.dodd@adelaide.edu.au

W. Leal Filho et al. (eds.), *Education for Sustainable Development: The Contribution of Universities*, World Sustainability Series,
https://doi.org/10.1007/978-3-031-86985-3_7

particularly regarding their alignment with sustainable development goals (Fayolle 2013; Productivity Commission 2019).

Existing literature underscores the vital role business schools can play in advancing the SDGs but also raises concerns about the effectiveness of their efforts. For example, scholar such as Fayolle (2013) and Prieto et al. (2021) highlight the transformative potential of management education (also see Alaref, Brodmann, and Premand 2020). Yet, Spicer et al. (2021) and others (see Halkic and Arnold 2019) caution against performative approaches that prioritise symbolic rather than substantive changes. At the program level, Härtel (2010) argues that while management training initiatives can foster resilience and innovation, they often lack clear links to measurable outcomes, hindering their ability to demonstrate meaningful impact (also see Nabi et al. 2017; Dodd et al. 2022).

Further investigation is essential to evaluate whether business training programs genuinely advance the SDGs or risk reinforcing existing inequalities (Anosike 2018; Morgan and Leung, 1980; Simpson 2006). Research suggests that poorly designed or implemented programs may inadvertently perpetuate rather than alleviate social issues, including inequality and exclusion (Nabi et al. 2017; Piketty 2020). Reay (2018) critiques many initiatives for reinforcing existing power dynamics, offering short-term benefits without addressing systemic barriers. These critiques underscore the need for critical, context-sensitive approaches to program design and evaluation, as well as robust impact measurement frameworks (Nabi et al. 2017; Dodd et al. 2022).

Programs explicitly designed to address the SDGs are increasingly common across universities. For example, the Consulting and Business Development Center at the University of Washington offers training to marginalized individuals, reporting efforts to address "structural barriers that disproportionately affect under-represented minority-owned businesses as well as businesses in rural, urban, and suburban communities with historically high unemployment rates" (University of Washington 2024, para 2; also see Wang 2020). Similarly, in Africa, initiatives engage marginalized youth with the dual aim of improving their stewardship within communities and fostering environmental sustainability (Hope 2012; Nafuko and Helen 2010). However, the absence of comprehensive impact measurement frameworks makes it difficult to assess whether these initiatives are achieving their intended outcomes and contributing meaningfully to the SDGs.

Addressing these challenges requires a shift in how the SDGs are conceptualised within management education. Traditionally framed as "wicked problems" due to their complexity and interconnectedness (Rittel and Webber 1973), these issues can alternatively be approached as "grand challenges" that demand collaborative, interdisciplinary solutions (Spicer et al. 2021). By adopting this perspective, business schools can position themselves as leaders in addressing global challenges through innovative and evidence-based approaches.

This chapter responds to these gaps by exploring the role of business training programs in advancing the SDGs through a critical and conceptual lens. It aims to contribute to the broader discourse by proposing a framework for evaluating the impact of these programs, with particular attention to their alignment with sustainable development and their potential to deliver both individual and societal benefits.

2 Conceptualizing Impact: The Challenge of Business Training to Advance the SDGs

Business schools are increasingly engaging more directly with real-world issues linked with the SDGs, yet rigorous and relevant methods for conceptualizing and evaluating such efforts remain underdeveloped (Friedland and Jain, 2022; MacIntosh et al. 2017). Traditionally, "scholarly impact" has been measured through publications, citation counts, and journal rankings, metrics that prioritize quantity over substance (Aguinis et al. 2014; Özbilgin 2009). These measures offer limited insight into the broader concept of impact, which organizations like the OECD (2021) define as change—whether at the system or individual level—resulting from an influential action. Furthermore, citation metrics fail to provide the space for scholars to reflect on and advance their societal role, signalling a need for more nuanced research and conceptualization around scholarly impact, especially in the context of sustainable development (Haley et al. 2020; Härtel 2010).

One area that warrants further exploration is the intersection of research and teaching (Mutch 2021). MacIntosh et al. (2017) and Boyer (1990) suggest that academic impact may be most effectively achieved through teaching. By critically reflecting on their teaching practices and applying new knowledge, academics can advance social development and learning. This approach promotes a multifaceted evaluation of impact that includes changing practice (self-reflexivity), influencing others (social outreach), advancing professionally (self-development), and fostering new ideas (social impact). By rethinking the role of scholars in education, the concept of scholarly impact can move beyond research-driven models to include "research-informed" frameworks that prioritize the needs and experiences of others (Mutch 2021). This also allows for the simultaneous consideration of both internal and external stakeholders, enriching the relationship between academia and society (Fotaki & Prasad 2014; Friedland and Jain, 2022).

The complexity of delivering business training that aims to address the SDGs highlights the necessity of this reflexive, adaptive approach. Without it, scholars risk exacerbating precarious conditions (Girei 2017; Reay 2018). For example, marginalized individuals are often the target of business schools who offer training programs to advance the SDGs. This includes people who are underrepresented in social systems, such as Indigenous people, those with disabilities, ethnic minorities, low socio-economic groups, and individuals from remote areas (Becker and Luthar 2002; Sanders et al. 2018). It also extends to veterans (Hunter-Johnson et al. 2021) and individuals identifying as LGBTQ + (Lee 2021). Business training in this context refers to non-degree programs, short courses, and micro-credentials offered by or in partnership with higher education institutions (Wang 2020).

Policymakers and industry leaders advocate for such training programs, believing they will alleviate poverty and increase well-being, aligned with the SDGs (Ataei et al. 2020; Chakravarty et al. 2019). For example, organizations like the Lumina Foundation and the OECD (2020) call for continued public funding of higher education, including for short-term programs and micro-credentials (Matthews et al. 2018; Salmi 2020). Accreditation bodies like the Association to Advance Collegiate Schools of Business AACSB (2024) also encourage institutions to demonstrate positive social impact through these initiatives.

However, understanding how training addresses the structural causes of marginalization—such as systemic racism and economic inequality—is still in its infancy (Piketty 2020; Prieto et al. 2021). Marginalized groups are not monolithic (Sitzmann and Campbell, 2021), and business frameworks may not align with diverse knowledge systems, cultural contexts, or approaches (Doucette et al. 2021; Zoogah 2021). Research shows that while business training can enhance employment, wealth, social mobility, and self-determination, all aligned with the Sustainable Development Goals (Fairlie 2013; Khavul et al. 2013), its success depends on a commitment to social justice and intergenerational equality (Hickel 2021; Levin 2017; Wadhwani and Viebig 2021).

History demonstrates that even modest goals, such as improving economic outcomes, face significant challenges (Bradley et al. 2012; Grimm 2020). Without well-considered goals and teaching strategies, education can exacerbate rather than mitigate marginalization (Piketty 2020; Reay 2018). Tensions between individual liberty and collective welfare often emerge, particularly in marginalized communities (Prieto et al. 2021). Balancing these tensions requires collective approaches grounded in principles of access and equality (Piketty 2020; Reay 2018), such as "cooperative business theory" (Prieto et al. 2021).

To capture the true impact of business training, it is essential to understand broader social implications (Bruton et al. 2013). These issues go beyond management education and relate to normative models of sustainable development (World Commission on Environment and Development 1987) and "degrowth" (Hickel 2021), which aim to reduce inequality and ensure intergenerational justice. The development of new models to conceptualize and measure the impact of management education remains a work in progress.

Thus, there is a need to critically examine the assumptions underlying business training programs and define what constitutes optimal results. While this may seem an insurmountable "wicked problem" (Rittel and Webber 1973), it can also be viewed as a "grand challenge" (Spicer et al. 2021), consistent with alternative impact frameworks like the integral pheno-practice of well-being (IPW) framework (Painter-Morland et al. 2017). As explored next, the IPW extends the analysis of impact to include social, individual, and temporal dimensions, acknowledging the complexities and paradoxes inherent in the process of well-being.

3 Methodology

This chapter employs a conceptual research approach to explore the role of business training programs in advancing sustainable development, particularly in their alignment with the SDGs. Conceptual research is well-suited for addressing complex and underexplored topics as it synthesises existing literature, identifies theoretical gaps, and proposes frameworks for further inquiry (Jabareen 2009; Gilson and Goldberg 2015). The IPW framework is used as the central theoretical lens to critically examine how these programs influence individual and societal well-being. The IPW framework provides a holistic perspective, making it particularly apt for understanding the multifaceted outcomes of initiatives designed to address sustainability and well-being challenges. Building on insights from the literature that has examining training programs delivered to the marginalized (e.g., Dodd et al. 2022), the chapter extends understanding of how the IPW framework to consider how university training programmes can both positively and negatively influence individual- and societal-level well-being. This new framework addresses the gaps identified in the literature by offering a structured approach to evaluating the effectiveness of business training programs as they relate to SDGs. It considers both intended and unintended outcomes, providing criteria for assessing program impact at both individual and societal levels. Furthermore, it proposes pathways for future research to refine and test the framework in empirical settings. This conceptual approach is particularly well-suited to the study's aims, given the limited availability of comprehensive frameworks that critically evaluate the intersection of business training, well-being, and sustainable development. By synthesising diverse perspectives and leveraging the IPW framework, this chapter contributes to academic discourse and lays the groundwork for future empirical research to build on its theoretical insights.

4 Assessing Business Training Programs for Marginalized Groups: Integrating the IPW and SDGs

As introduced above, the IPW framework provides a multidimensional lens through which the impact of initiatives aimed at marginalized groups can be understood (Painter-Morland et al. 2017). Rather than focusing solely on immediate, measurable outcomes, such as income generation or employment rates, the IPW framework integrates subjective, intersubjective, and objective dimensions of well-being. It emphasizes that well-being is not static but a dynamic process of interrelational becoming well that evolves over time and unfolds through complex social, personal, and material interactions. This broader, more nuanced understanding of well-being is particularly valuable when evaluating business training programs targeting marginalized individuals, as it allows for a more comprehensive analysis of both individual and community-level impacts.

In applying the IPW framework to the evaluation of business training programs, it becomes evident that these initiatives align closely with several of the United Nations SDGs, particularly SDG 4 (Quality Education), SDG 8 (Decent Work and Economic Growth), and SDG 10 (Reduced Inequality). For instance, SDG 4 emphasizes the need for inclusive and equitable quality education, with an aim to promote lifelong learning opportunities for all. While business training programs for marginalized groups contribute to this goal by increasing access to education, the IPW framework pushes evaluators to consider how such programs affect participants beyond knowledge acquisition. It suggests a deeper examination of how these educational opportunities impact subjective well-being, such as personal empowerment, and intersubjective dimensions, such as changes in community dynamics and social cohesion. Scholars such as Sitzmann and Campbell (2021) argue that educational interventions must be tailored to the specific cultural and contextual needs of marginalized individuals, an approach that aligns well with the IPW's focus on the subjective and intersubjective dimensions of well-being.

Similarly, SDG 8, which seeks to promote inclusive and sustainable economic growth, decent work, and full employment for all, is central to many business training programs. These programs are designed to increase employability and foster entrepreneurship among marginalized individuals. However, as Prieto et al. (2021) have shown, such initiatives can sometimes introduce neoliberal ideals that inadvertently undermine collective well-being in marginalized communities, creating tensions between individual economic progress and broader social welfare. By including both the subjective and intersubjective dimensions of well-being, the IPW framework ensures that business training programs designed to promote SDG 8 are evaluated with an eye toward the broader social justice implications. This might include not only the material benefits, such as increased income, but also how programs foster or hinder community cohesion, intergenerational equity, and the dismantling of systemic barriers like racial or gender-based discrimination, as discussed by Piketty (2020) and Hickel (2021).

Further, the IPW framework's multidimensional approach to well-being aligns with the goals of SDG 10, which focuses on reducing inequality within and between countries. Business training programs are often seen as key interventions for promoting social mobility and reducing inequality, but as Sitzmann and Campbell (2021) argue, marginalized groups are not homogeneous. Programs that fail to consider the unique needs and contexts of different communities' risk exacerbating, rather than reducing, inequalities. The IPW framework encourages evaluators to consider a broader range of outcomes, not just economic measures, but also the subjective experiences of empowerment, self-determination, and collective well-being. In doing so, it aligns closely with the broader aims of SDG 10 by ensuring that business training programs are designed to promote long-term equity and social justice, rather than reinforcing existing disparities.

Incorporating the IPW framework into the evaluation of business training programs for marginalized groups offers a more robust and comprehensive mechanism for aligning these programs with the SDGs. While traditional approaches may focus on immediate and quantifiable outcomes, such as employment rates or income growth, the IPW framework encourages evaluators to consider the broader, more intangible effects of such programs. This includes changes in community cohesion, empowerment, and intergenerational equity–outcomes that are crucial for achieving the transformative, long-term goals of the SDGs. For example, SDG 4's emphasis on quality education is expanded through the IPW framework's focus on culturally sensitive and contextually appropriate interventions that support both individual and collective well-being. Similarly, SDG 8's focus on decent work and economic growth is deepened by the framework's insistence on considering the broader social and economic conditions of marginalized communities, ensuring that these programs promote sustainable, inclusive growth. SDG 10's goal of reducing inequality is also supported by the IPW's multidimensional approach, which ensures that business training programs address the structural and systemic barriers that sustain marginalization.

By integrating the IPW framework into the evaluation of training programs, higher education institutions and policymakers can assess the full range of impacts these programs have on marginalized individuals and their communities. The framework allows for a more comprehensive understanding of how business training programs contribute to both individual and societal well-being, ensuring that these programs align with the broader goals of sustainable development. In this way, the IPW framework supports a more holistic and just approach to evaluating the impact of business training programs, ensuring that they contribute to the long-term goals of reducing inequality, promoting inclusive education, and fostering sustainable economic growth. This aligns with the transformative aspirations of the SDGs and provides a roadmap for creating more equitable and sustainable educational interventions.

5 Expanding the Framework of Impact

Broadening the conceptualization of how business training programs are evaluated to encompass the SDGs and the IPW framework opens up new possibilities for measuring program effectiveness. This approach reframes the traditional evaluation metrics used for these programs, moving beyond short-term, individual-objective outcomes to a more holistic view of impact. Despite criticism that business schools may sometimes reinforce, rather than redress, social exclusion (Fotaki and Prasad 2015; Özbilgin, 2009; Vijay and Nair 2021), a shift toward SDG-aligned training programs and the application of the IPW framework provides a path to illustrate the broader, positive impact that these programs can have. This expanded framework allows scholars to demonstrate evidence of outcomes that reach beyond the immediate, focusing on long-term, systemic changes.

Current literature in this space often focuses on specific outcomes, such as improving young people's autonomy in making employment choices as they transition out of foster care or empowering marginalized individuals by helping them overcome the constraints imposed by their past (Wang 2020). For instance, business training programs have been shown to aid in prisoners' reentry into society, bringing about profound identity shifts. One participant in Wang's study stated, "The program gave me confidence… I didn't know what my worth was. So, I learned all of that through the program: my worth, my confidence" (Wang 2020, p. 9). Themes of life satisfaction and hope for the future also emerge and other themes in existing work (e.g., see Alaref et al. 2020).

Other studies have shown that business content can not only increase employability but also enhance social capital. Taylor et al. (2004) argue that business schools are well-placed to contribute to the creation of social capital, particularly for socially disadvantaged entrepreneurs. For policymakers and scholars, this reinforces the potential of training programs to improve outcomes for marginalized individuals as they relate to the SDGs. However, not all findings are positive. Calderon, Cunha, and De Giorgi (2020) note that a business literacy program in rural Mexico improved participants' ability to measure business outcomes without changing actual outcomes themselves, highlighting the limits of training in isolation from broader systemic change. Similarly, Littenberg-Tobias and Reich (2020) critique the capacity of Massive Open Online Courses (MOOCs) to democratize access to education, revealing gaps in existing platforms, with supply-based pedagogies showing limited effectiveness in improving the skills, knowledge, and well-being of marginalized learners (also see Dodd et al. 2022).

The IPW framework pushes for a broader evaluation of business training programs, suggesting that well-being outcomes should also be considered, such as how programs influence feelings of autonomy, satisfaction, or hope. For example, while Hinson and Amidu (2006) examine the skills gained from internet adoption in Ghana, their study does not extend to the well-being outcomes these skills may influence. Similarly, other studies focus narrowly on educational engagement without considering broader impacts on participants' lives.

The IPW framework also highlights the importance of institutional impact. For example, Rodríguez-Hernández et al. (2021) found that institutions involved in these programs must be open to learning about and addressing the barriers marginalized groups face. This echoes Subotzky's (1999) call for a joint ownership model that meets the needs of both the institution and the community. However, through a systematic review of research investigating the efficacy of training programs delivered to the marginalized, Dodd et al. (2022) reveal a research-practice gap, with few integrated evaluations published in peer-reviewed journals. Littenberg-Tobias and Reich (2020) also show the limitations of current approaches in increasing access to higher education, while Prieto et al. (2021) argue for the need to integrate new knowledge systems into business schools to address social challenges.

Tensions within these programs are also apparent. While training can help marginalized individuals expand their capabilities and networks, it cannot, on its own, address the structural issues that contribute to their exclusion, such as systemic racism (Prieto et al. 2021; Vijay and Nair 2021). Furthermore, neoliberal ideas like wealth creation may conflict with the social and community values that training programs should promote (Painter-Morland et al. 2017). The literature shows that happiness and well-being are not always tied to economic success (Easterlin 1974;

Blanchflower and Oswald 2008), raising questions about how training programs should be designed to balance economic development with social and ecological well-being.

This expanded evaluation framework raises several conceptual questions for future research, particularly in relation to the SDGs and scholarly impact:

- To what extent are business schools aligning their training programs for marginalized groups with the broader goals of the SDGs, such as reducing inequalities (SDG 10) and promoting inclusive and equitable education (SDG 4)? Are these programs being designed with the specific needs of marginalized communities in mind, or are they shaped by more traditional, market-driven objectives?
- How are business schools measuring the long-term impact of these programs beyond immediate economic outcomes, and are they incorporating multidimensional well-being metrics aligned with the IPW framework? Could a broader focus on social, environmental, and psychological well-being contribute to a more comprehensive evaluation of their contributions to the SDGs?
- What systemic barriers, both within business schools and at an institutional level, limit the development of programs that directly contribute to SDGs like sustainable livelihoods and social inclusion? How can these barriers be addressed to better support scholars in designing and delivering programs with lasting, transformative impact?
- How can business schools better integrate marginalized voices into the design and evaluation of their programs? To what extent are these programs fostering true empowerment, as envisioned by SDG 16 (Peace, Justice, and Strong Institutions), rather than reinforcing existing inequalities?

These questions challenge business schools to not only assess the efficacy of their programs but to reconsider their role in contributing to the global mission of achieving the SDGs. They also illustrate the institutional pressures that might affect both the delivery and evaluation of business training programs. For instance, some business schools struggle to secure funding for such initiatives or fail to integrate them into academic workloads. Additionally, there are questions about whether scholars involved in these programs experience greater job satisfaction and whether institutions are adequately supporting this work.

Ecological sustainability is another area where tensions arise. While many training programs focus on economic growth, Nafuko and Helen (2010) show that some also consider how education can drive progress toward environmental sustainability. However, more work is needed to explore the trade-offs between economic development and ecological well-being, such as how advancements in one area might undermine progress in another (Painter-Morland et al. 2017). The IPW framework addresses these trade-offs by considering the multiple dimensions of development—economic, social, and environmental—within the context of business education.

Future studies must build on this expanded framework, examining not only the content and pedagogy of training programs but also how they can better serve marginalized individuals as related to the SDGs. Integrating IPW and SDGs into program evaluation can ensure that the voices of marginalized communities are considered in program design and outcomes (Rodríguez-Hernández et al. 2021). Programs should be developed in collaboration with communities, not imposed upon them, as O'Brien et al. (2019) emphasize.

This chapter sets the stage for future program design that integrates IPW and SDG-aligned outcomes, offering insights into how scholars can deepen their understanding of the links between pedagogy and social impact. For practitioners, the chapter highlights the need for goal alignment, where both the institution and the community benefit from training programs (Subotzky, 1999). Accreditation agencies, such as the AACSB (2024), can play a crucial role in promoting these values, emphasizing the importance of socially inclusive education. Ultimately, successful programs will require not only delivering knowledge but also fostering broader cultural and systemic change, aligning with the SDGs' emphasis on inclusive education and reduced inequalities.

6 Limitations, Future Prospects, and Constraints

This chapter explores the potential and challenges of business training programs aimed at advancing the SDGs, with a focus on their impact on marginalized individuals. However, several limitations and constraints must be acknowledged.

One of the primary limitations is the lack of comprehensive, widely accepted frameworks for measuring the impact of these training programs. While some evaluations exist, there remains a significant gap in understanding the full extent of their effects on both individual well-being and broader societal outcomes (also see Nabi et al., 2017; Dodd, Graves, & Hentzen 2022). The absence of robust measurement tools makes it difficult to definitively assess the efficacy of these programs in addressing systemic issues such as inequality and social exclusion. This limitation is compounded by the diversity of contexts in which these programs operate, from different countries to various industries, which makes generalisation of findings challenging.

Another constraint lies in the theoretical and conceptual framing used in this paper. While the IPW framework provides a useful lens for understanding the multidimensional nature of well-being, it may not fully capture all the nuances of the diverse and complex ways in which business training can impact marginalized communities. This is particularly true in regions where cultural and socio-economic factors may differ significantly from those seen in Western contexts, and where the SDGs themselves may be interpreted and applied in varying ways.

Looking to the future, there are significant prospects for advancing the understanding and effectiveness of business training programs aimed at the SDGs. The development of more rigorous impact measurement frameworks is a critical area for future research. Such frameworks would need to account for both short-term and long-term outcomes, not only in terms of economic success but also in relation to social and environmental sustainability. Further research could also focus on the intersections between different SDGs and how they might be collectively addressed through business training, potentially offering a more holistic approach to fostering sustainable development.

Moreover, future studies could explore the ways in which business schools and other educational institutions can refine their curricula and training programs to better address the root causes of social inequality, while also considering the limitations of these initiatives in broader systemic contexts. A more nuanced approach could also investigate the potential unintended consequences of these programs, such as reinforcing existing power dynamics or creating dependency, rather than empowering marginalized individuals.

7 Conclusions

In sum, this chapter explores the intersection of education for sustainable development and the IPW framework, shedding light on the impact of business training programs aimed at contributing to the SDGs. As scholars, policymakers, and accreditation bodies call for business schools to enhance their real-world relevance, this analysis highlights both the potential benefits and challenges these programs pose to individual and societal well-being. The IPW framework presented offers a nuanced understanding of how such initiatives can contribute to positive outcomes, while also recognizing the risk of negative consequences, such as inadvertently reinforcing social inequalities.

The chapter creates a case for the need to critically examining the institutional structures and practices that shape the effectiveness of these programs. By addressing gaps in the current literature and proposing pathways for future research, this work highlights the need for more comprehensive impact measurement frameworks and deeper theoretical exploration into the complex ways these programs influence marginalized communities. The chapter also calls attention to the limitations of current conceptual frameworks and the potential unintended consequences of business training programs, urging caution in their design and implementation.

Implications for business schools and other educational institutions are significant. The insights garnered here underscore the need for a collaborative approach that prioritizes the voices and needs of marginalized communities. It is essential to foster a deeper commitment to social inclusion and sustainable development, ensuring that training programs are not only aligned with the SDGs but are also effectively contributing to social justice and empowerment.

Looking forward, the chapter calls for ongoing reflection and innovation within business education, encouraging scholars and practitioners to extend their influence beyond academic citations to achieve meaningful, transformative impacts in the real world. As business schools continue to evolve, their role in advancing the SDGs must be carefully evaluated, with an emphasis on developing programs that are both effective and accountable. Future research should focus on refining measurement frameworks and exploring the broader implications of these educational initiatives, with the goal of improving their capacity to drive lasting social change.

References

Aguinis H, Shapiro D, Antonacopoulou E, Cummings T (2014) Scholarly impact: a pluralist conceptualization. Acad Manage Learn Educ 13(4):623–639

Alaref J, Brodmann S, Premand P (2020) The medium-term impact of entrepreneurship education on labor market outcomes: experimental evidence from university graduates in Tunisia. Labour Econ 62(C):1–18

Anosike P (2018) Entrepreneurship education knowledge transfer in a conflict Sub-Saharan African context. J Small Bus Enterp Dev 25(4):591–608

Association to Advance Collegiate Schools of Business (AACSB) (2024) AACSB accreditation overview. https://www.aacsb.edu/educators/accreditation/business-accreditation. Accessed 24 Sept 2024

Ataei P, Karimi H, Ghadermarzi H, Norouzi A (2020) A conceptual model of entrepreneurial competencies and their impacts on rural youth's intention to launch SMEs. J Rural Stud 75:185–195

Becker B, Luthar S (2002) Social-emotional factors affecting achievement outcomes among disadvantaged students: closing the achievement gap. Educ Psychol 37(4):197–214

Blanchflower D, Oswald A (2008) Is well-being U-shaped over the life cycle? Soc Sci Med 66:1733–1749

Boyer E (1990) Scholarship reconsidered: priorities of the professorate. Jossey-Bass, San Francisco, CA

Bradley S, McMullen J, Artz K, Simiyu E (2012) Capital is not enough: innovation in developing economies. J Manage Stud 49(4):684–717

Bruton G, Ketchen D Jr, Ireland R (2013) Entrepreneurship as a solution to poverty. J Bus Ventur 28(6):683–689

Calderon G, Cunha JM, De Giorgi G (2020) Business literacy and development: evidence from a randomized controlled trial in rural Mexico. Econ Dev Cult Change 68(2):507–540

Chakravarty S, Lundberg M, Nikolov P, Zenker J (2019) Vocational training programs and youth labor market outcomes: evidence from Nepal. J Dev Econ 136(C):71–110

Dean BA, Gibbons B, Perkiss S (2018) An experiential learning activity for integrating the United Nations sustainable development goals into business education. Soc Bus 8(4):387–409

Dodd T, Graves C, Hentzen J (2022) Impact and university business training courses delivered to the marginalized: a systematic review. Acad Manage Learn Educ 21(3):449–469

Doucette M, Gladstone J, Carter T (2021) Indigenous conversational approach to history and business education. Acad Manage Learn Educ 20(3):473–484

Easterlin RA (1974) Does economic growth improve the human lot? Some empirical evidence. In: David PA, Reder MW (eds) Nations and households in economic growth: essays in honor of Moses Abramovitz. Academic Press, New York, pp 89–125

Fairlie R (2013) Entrepreneurship, economic conditions, and the great recession. J Econ Manage Strat 22(2):207–231

Fayolle A (2013) Personal views on the future of entrepreneurship education. Entrep Reg Dev 25(7–8):692–701

Fotaki M, Prasad A (2014) Social justice interrupted? Values, pedagogy, and purpose of business school academia. Manage Learn 45(1):103–106

Fotaki M, Prasad A (2015) Questioning neoliberal capitalism and economic inequality in business schools. Acad Manage Learn Educ 14(4):556–575

Friedland J, Jain T (2022) Re-framing the purpose of business education: crowding-in a culture of moral self-awareness. J Manage Inq 31(1):15–29
Gilson LL, Goldberg CB (2015) Editors' comment: so, what is a conceptual paper? Group Org Manage 40(2):127–130
Girei E (2017) Decolonising management knowledge: a reflexive journey as practitioner and researcher in Uganda. Manage Learn 48(4):453–470
Grimm J (2020) Cognitive frames of poverty and tension handling in base-of-the-pyramid business models. Bus Soc 61(8):2070–2114
Haley U, Cooper C, Hoffman A, Pitsis T, Greenberg D (2020) Learning and education strategies for scholarly impact: influencing regulation, policy, and society through research. Academy of Management Learning & Education Special Issue Call. https://aom.org/docs/default-source/publishing-with-aom/amle_special_issue_call_strategies_for_scholarly_impact.pdf?sfvrsn=713ff196_2. Accessed 24 Sept 2024
Halkic B, Arnold P (2019) Refugees and online education: student perspectives on need and support in the context of (online) higher education. Learn Media Technol 44(3):345–364
Härtel C (2010) From the editors: International students, representation of global management issues, and educating from a paradigm of human well-being. Acad Manage Learn Educ 9(4):585–590
Hickel J (2021) What does degrowth mean? A few points of clarification. Globalizations 18(7):1105–1111
Hinson R, Amidu M (2006) Internet adoption amongst final year students in Ghana's oldest business school. Libr Rev 55(5):314–323
Hope K (2012) Engaging the youth in Kenya: Empowerment, education, and employment. Int J Adolesc Youth 17(4):221–236
Hunter-Johnson Y, Liu T, Murray K, Niu Y, Suprise M (2021) Higher education as a tool for veterans in transition: battling the challenges. J Contin High Educ 69(1):1–18
Jabareen Y (2009) Building a conceptual framework: philosophy, definitions, and procedure. Int J Qual Methods 8(4):49–62
Khavul S, Chavez H, Bruton G (2013) When institutional change outruns the change agent: the contested terrain of entrepreneurial microfinance for those in poverty. J Bus Ventur 28(1):30–50
Lee C (2021) Promoting diversity in university leadership: the argument for LGBTQ + specific leadership programmes in higher education. Perspect Policy Pract High Educ 25(3):91–99
Levin H (2017) The economic payoff to investing in educational justice. In: Siljander P, Kontio K, Pikkarainen E (eds) Schools in transition: linking past, present, and future in educational practice. Springer, New York, pp 161–188
Littenberg-Tobias J, Reich J (2020) Evaluating access, quality, and equity in online learning: a case study of a MOOC-based blended professional degree program. Internet High Educ 47:1–11
MacIntosh R, Beech N, Bartunek J, Mason K, Cooke B, Denyer D (2017) Impact and management research: exploring relationships between temporality, dialogue, reflexivity and praxis. Br J Manag 28(1):3–13
Matthews K, Garratt C, Macdonald D (2018) The higher education landscape: trends and implications. UQ Discussion Paper 2018. https://core.ac.uk/download/pdf/158604032.pdf. Accessed 24 Sept 2024
Morgan B, Leung P (1980) Effects of assertion training on acceptance of disability by physically disabled university students. J Couns Psychol 27(2):209
Mutch A (2021) Business education in the UK polytechnic tradition: uncovering alternative approaches through historical investigation. Acad Manage Learn Educ 20(3):407–422
Nabi G, Liñán F, Fayolle A, Krueger N, Walmsley A (2017) The impact of entrepreneurship education in higher education: a systematic review and research agenda. Acad Manage Learn Educ 16(2):277–299
Nafukho FM, Helen Muyia MA (2010) Entrepreneurship and socioeconomic development in Africa: a reality or myth? J Eur Ind Train 34(2):96–109

O'Brien E, Cooney T, Blenker P (2019) Expanding university entrepreneurial ecosystems to under-represented communities. J Entrep Public Policy 8(3):384–407
OECD (2020) Tackling coronavirus (COVID-19): focus on social challenges. https://www.oecd-ilibrary.org/social-issues-migration-health/oecd-policy-responses-to-coronavirus-covid-19_5b0fd8cd-en. Accessed 24 Sept 2024
OECD (2021) Social impact measurement for the social and solidarity economy. https://www.oecd-ilibrary.org/docserver/d20a57ac-en.pdf?expires=1635725289&id=id&accname=guest&checksum=1CED329D2355CC7D36BE8FB1663CA2B4. Accessed 24 Sept 2024
Özbilgin M (2009) From journal rankings to making sense of the world. Acad Manage Learn Educ 8(1):113–121
Painter-Morland M, Demuijnck G, Ornati S (2017) Sustainable development and well-being: a philosophical challenge. J Bus Ethics 146(2):295–311
Piketty T (2020) Capital and ideology. Harvard University Press, Cambridge, MA
Prieto L, Phipps S, Stott N, Giugni L (2021) Teaching (cooperative) business: the 'Bluefield experiment' and the future of Black business schools. Acad Manage Learn Educ 20(3):320–341
Productivity Commission (2019) The demand driven university system: a mixed report card. Commission Research Paper, Canberra
Reay D (2018) Miseducation: inequality, education and the working classes. University of Bristol, Policy Press
Rittel H, Webber M (1973) Dilemmas in a general theory of planning. Policy Sci 4:155–169
Rodríguez P, Izuzquiza D, Cabrera A (2021) Inclusive education at a Spanish university: the voice of students with intellectual disability. Disabil Soc 36(3):376–398
Salmi J (2020) COVID's lessons for global higher education. Lumina Foundation. https://www.luminafoundation.org/wp-content/uploads/2020/11/covids-lessons-for-global-higher-education.pdf. Accessed 24 Sept 2024
Sanders J, Munford R, Boden J (2018) Improving educational outcomes for at-risk students. Br Edu Res J 44(5):763–780
Sitzmann T, Campbell E (2021) The hidden cost of prayer: religiosity and the gender wage gap. Acad Manage J 64(4):1016–1048
Simpson R (2006) Masculinity and management education: feminizing the MBA. Acad Manage Learn Educ 5(2):182–193
Spicer A, Jaser Z, Wiertz C (2021) The future of the business school: finding hope in alternative pasts. Acad Manage Learn Educ 20(3):459–466
Subotzky G (1999) Alternatives to the entrepreneurial university: new modes of knowledge production in community service programs. High Educ 38(4):401–440
Taylor, D., Jones, O., & Boles, K. (2004). Building social capital through action learning: An insight into the entrepreneur. Educ Train 46(5):226–235
University of Washington (2024) Consulting and business development center. https://foster.uw.edu/centers/consulting-and-business-development-center/. Accessed 27 Sept 2024
Vijay D, Nair VG (2021) In the name of merit: Ethical violence and inequality at a business school. J Bus Ethics 169(3):541–554
Wadhwani D, Viebig C (2021) Social imaginaries of entrepreneurship education: The United States and Germany, 1800–2020. Acad Manag Learn Educ 20(3):342–360
Wang Q (2020) Higher education institutions and entrepreneurship in underserved communities. High Educ 2(9):1–19
World Commission on Environment and Development (1987) Our common future. Oxford University Press, Oxford
Zoogah D (2021) Historicizing management and organization education in Africa. Acad Manage Learn Educ 20(3):382–406

Education for Sustainable Urban Development in a Global Context—Study of the Academia's Contribution and Cross-Sectoral Cooperation in the UN HABITAT 11th World Urban Forum in Katowice

Karolina Thel

Abstract The aim of my paper is to critically analyze the potential of the World Urban Forum (WUF) for education for sustainable development through the prism of academic contribution and cross-sectoral cooperation. Based on the analysis of the literature on the subject and qualitative research. I address the motivation and capacity, as well as social and physical opportunities to undertake such activities. I describe the personal and material scope of education for sustainable urban development. The final conclusions focus on the lesson learned from the WUF for ESD. The analysis of the WUF covers different rationales for the key issue relates to cross-sectoral cooperation, with Political Administration and Academia as leading players. The main conclusion concerns their motivations for collaboration. When motivations relate to similar resources (like prestige), the risk of competition increases. When motivations relate to different resources—held by one party and in demand by the other (like financial resources), then the skills to negotiate, define common goals and rules for sharing resources become crucial. I argue, that WUF can be seen as a forum supporting evidence-based policy, but also practice-based education, offering a database of case studies for ESD.

Keywords Cross-sectoral cooperation · Sustainable urban development · World Urban Forum · Education for sustainable development

K. Thel (✉)
Faculty of Artistic Research and Curatorial Studies, Academy of Fine Arts in Warsaw, Warsaw, Poland
e-mail: karolina.thel@asp.waw.pl

Center for Environmental Education and Research, University of Malta, Msida, Malta

Faculty of Sociology, University of Warsaw, Warsaw, Poland

W. Leal Filho et al. (eds.), *Education for Sustainable Development: The Contribution of Universities*, World Sustainability Series,
https://doi.org/10.1007/978-3-031-86985-3_8

1 Introduction

The aim of the presented article is to analyze cross-sectoral cooperation in the field of education for sustainable development (CSC ESD) with reference to the case study of the UN-Habitat 11th Session of World Urban Forum held in Katowice, Poland, in June 2022—WUF11 (IISD 2022a). The World Urban Forum (WUF) is a non-legislative technical forum in which experts can exchange views, debate and discuss the challenges of urbanization. UN HABITAT has organized and run the WUF sessions every second year since 2002 as the world's leading gathering on urban issues (IISD 2022b). It is a mass event (over a dozen thousand participants each) with special requirements in terms of organization and security. It takes place in a formula that assumes cross-sectoral cooperation (CSC) and the involvement of numerous stakeholders, including the community of universities and research institutions (hereinafter referred to as Academia).

Declared goals of WUF relate directly to the vision of sustainable development promoted by the UN. They are defined as follows: raising awareness around sustainable urbanization; improving the collective knowledge of sustainable urbanization through inclusive open debates; exchanging best practices and good policies; increasing cooperation among stakeholders and sharing of lessons learned. Referring to the last indicated objective, my article tries to answer the question: what is the lesson learned from cross-sectoral cooperation within the WUF framework and how it can support ESD in the urban context?

In analyzing Academia's contribution to the co-creation of the WUF11, I was particularly interested in the underlying motivations and capacity associated with CSC SD. The aim was also to analyze the difficulties and barriers to introducing this type of educational activities. I also wondered to what extent the experiences with WUF support learning organizations to improve the quality of activities for SD education.

2 Literature Review

Sustainability, in its broadest sense, is the ability of systems to survive and maintain diversity and productivity over time. The perspective on sustainability shows that cities are part of a larger, global ecosystem. Urban sustainability is the practice of achieving a level of economic and social development that does not alter the balance of the ecosystem (Slavin 2011). At the same time, 'a city from the perspective of sustainability is not the sum of private property, interests and resources, but a common good' (Crutzen and Stoermer 2000). Cities are seen as centers of knowledge, sources of innovation and drivers of growth (Acs 2005; Florida 2005; Yigitcanlar et al. 2008), but this comes at a price. There is growing pressure on limited resources. Contemporary concepts of sustainable urban development encompass a range of issues, including, but not limited to: resource efficiency, sustainable mobility; spatial order, the shaping of compact urban structures, the creation of high quality public spaces, the development of the local economy and labor market, the protection of historical and cultural heritage, the protection of the environment and the promotion of biodiversity, social cohesion, quality of life, social participation, competitiveness, the use of modern technologies (UN 2016). Sustainability can be seen as an umbrella concept, encompassing various visions of urban development, including the green city, happy city, etc.

Sustainable urban development is about taking responsibility for the long-term shaping of the economy, environment and urban society in such a way as to reduce the multi-faceted costs. The demands associated with sustainability require a radical revision of values and form a huge agenda of economic, social and political issues to be re-evaluated, as well as a set of dilemmas and problems on which consensus is difficult to reach (Breheny 1992). In cities, the density of relationships is high, followed by an increased potential for contradictions and dilemmas, which affects the practical implementation of the concept of sustainable development. Thus, for more effective implementation of SD, there is a need for education and involvement of the public, including in resource management and conflict resolution (Kronenberg 2010).

Education has a significant role in enhancing inclusive and transformative urban development (Chitereka et al. 2022). Education for sustainable development—as indicated by numerous authors—is currently one of the most critical challenges of the UN and the EU (Leal Filho et al. 2021; Lorek et al. 2023; UNESCO 2007, 2009, 2020, 2021). Activities for sustainable urban development, as well as education in this area, take place at different levels—global, regional, international, national and local—where it is possible to distinguish additionally the macro-local dimension (urban functional areas, metropolitan areas), as well as the micro-local level (districts, housing estates, special zones inside the city). They are implemented by different stakeholders—Academia, NGOs, national, regional and local authorities etc. This is where the challenges of coordinating educational activities arise.

I see also a second challenge related to the education for sustainable urban development. Urban development challenges relate to different dimensions: political, economic, social, environmental, cultural, organizational, etc. Many of development problems are too complex for any single agency to effectively [handle] and interorganizational cooperation is required to find effective solutions to such problems (Brown and Tandon 1992; Gray 1989; Raišienė and Baranauskaitė 2018). Based on the assumption that different stakeholders possess distinctive assets (including, e.g. knowledge, experience, financial resources, organizational facilities) that can be combined in a productive manner to solve complex problems. 'Sustainable development and the interplay between its ecological, social, and economic dimensions can be regarded as a highly complex task. As a logical consequence, educating for sustainable development also has a complex character. Traditional unidirectional educational processes are only of very limited use when educating for sustainable development (...) besides analytical capabilities and deterministic process planning, dealing with the complex problem of sustainable development requires creativity, social competencies and specific communication skills in order to cope with the dynamic change that characterizes the developments in most facets of society and nature' (Steiner and Posch 2006).

In order to build an up-to-date, reliable and interdisciplinary educational offer in the field of sustainability—where we could traditionally expect that universities would play a leading role (Popowska and Sady 2024; Purcell et al. 2019; Renfors 2024)—it is not enough for researchers to collect and analyze data and disseminate research results alone, because 'environmental knowledge alone does not automatically lead to behavioral change' (Dür and Keller 2018). Instead, what is needed is a dynamic process of mutual learning from case studies as part of the collaboration of different sustainability stakeholders (Matu and Perez-Johnston 2024; McBain et al. 2024).

In light of the current ESD paradigm, related to the idea of global citizenship, 'an adult education course which delivers solely knowledge-based content about environmental issues or about international or sustainable development (…) must incorporate the skills, values and attitudes dimension and encourage the learner to take positive action' (The Council of Europe 2024; Welsh Government 2012). Thus, the emphasis of ESD should not so much be on competencies related to knowledge acquisition, but on fostering collaboration competency as one of the necessary key resources required to deal with the complexity of current global challenges (UNESCO 2017). At the same time, this somehow forces a change in the area of the education sector itself and points to the need to be open to different types of external cooperation, including especially cross-sectoral partnerships (UNESCO 2019).

In the broadly understood science and education for sustainable development, there is a clear undercurrent of networking between representatives of Academia, politics, business, NGOs and citizens (Barnes and Phillips 2000). I am fully aware that the term 'cooperation' has become a catchphrase for almost any kind of inter-organizational or interpersonal relations (Gajda 2004). A characteristic feature of the cross-sectoral cooperation is that the undertaken activities bring mutual benefits to partners they could not achieve through the implementation of these activities on their own (Rama Murthy et al. 2021). This convention of CSC SD is followed, for example, by urban laboratories and networks such as URBACT or Eurocities. CSC SD is sometimes stimulated by the need to raise funds for research and implementation projects, which is evident in the EU project funding policies (e.g. LIFE or Interreg program).

Apart from pragmatic considerations, building partnerships between Academia and stakeholders, creating practices, solutions, and platforms for cooperation can be perceived as ways of broadening the engagement of modern universities in social life of communities and local and regional development (Hintea et al. 2022; Profiroiu and Brișcariu 2021). In the context of Academia's activities, CSC SD is most often referred to as a so-called third mission, 'where teaching and research are taken as the first and second missions with any applied works in either sphere considered other or third' (König and Evans 2013; Trencher et al. 2014).

To conclude this section of the article, I would like to address the prerequisites of CSC SD. Number one is the involvement of leaders from different backgrounds, including Academia (Sady et al. 2019). The importance of this aspect is reflected

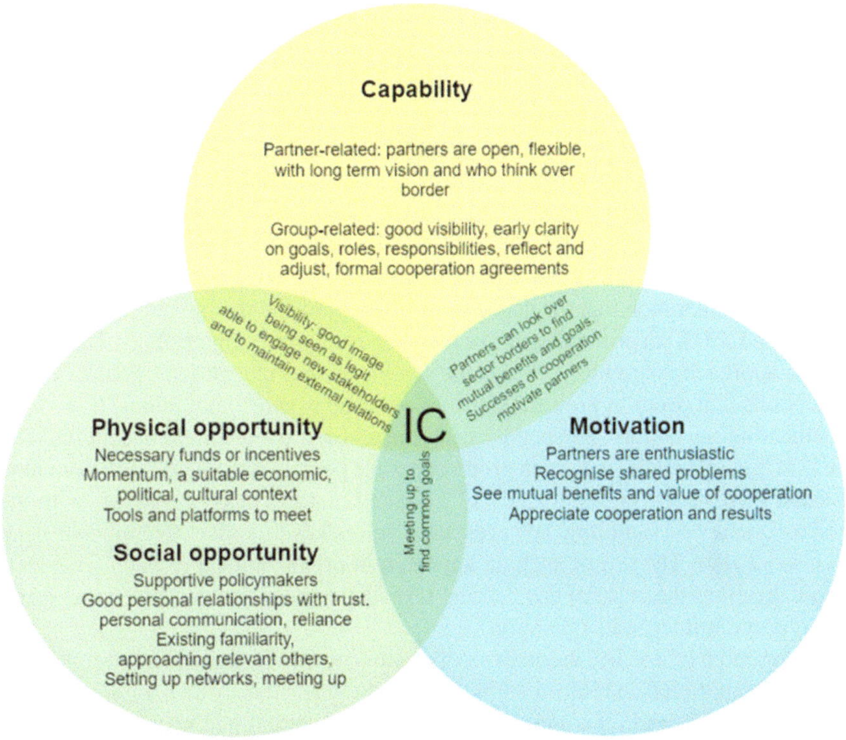

Fig. 1 Facilitators for intersectoral cooperation. *Source* van der Vliet et al. (2022)

in the ESD trend promoting education model based on improving the quality of leadership (Broman et al. 2017; Leal Filho et al. 2020; Parkin 2010; Visser and Courtice 2011). This is important because articles about cross-sectoral collaboration often contain a mental shortcut: 'collaboration between institutions'. Meanwhile, in reality, these are not 'institutions' that collaborate, but people—and not 'all people' either, but most often the leaders. This refers to the personal aspect of the issue. As for the material scope, cross-sectoral collaboration refers to both cognitive aspects and psychological aspects. For CSC SD to thrive, there needs to be an adequate level of capacity and motivation, as well as social and physical opportunity (Gazley 2016; van der Vliet et al. 2022) (Fig. 1).

3 Methodology and Theoretical Basis of the Research

The analyses for the presented article were carried out with reference to the four key prerequisites of CSC SD that, in my opinion, condition its use as a tool to support ESD: capacity, motivation, social opportunity and physical opportunity. I sought answers to the following research questions relating to the preparation of WUF11: I. What is the stakeholder motivation for CSC SD? II. What is the importance of stakeholder capacity for CSC SD? III. What are the lessons learned from the WUF for ESD?

Cross-sector partnership research is characterized by widely dispersed and multidisciplinary theoretical as is the case with its methodological approaches employing a multitude and mixture of methods, which has resulted in a toolkit that has grown large and heavy to carry (van Tulder and Keen 2018). Faced with the multiplicity of possible research paths to answer the research questions posed, I decided at the initial stage of research preparation that the data collection method and theoretical framework of analysis should be (in principle) similarly rooted in cultural anthropology. The ethnographic approach is oriented towards the study of social interactions, behaviors and perceptions that occur in groups, teams, organizations and communities. It involves documenting the culture, perspectives and practices of people in these settings. The aim is to 'get inside' the way each group of people perceives the world. Ethnographic research involves an in-depth study of a limited number of observations, with the need for qualitative sampling: key events/key actors/key processes (Hammersley and Atkinson 2019). The use of such an approach in organizational research enables a deeper dive into the cultural context and reveals the meanings, practices and social relations that influence its operation.

The research is based on the analysis of existing materials (documentation, website content, promotional and informational material), but primarily on a 4-month participant observation and ethnographic interviews (individual and group) with representatives of the main stakeholders of educational activities during WUF (academic experts from various institutions, the Ministry, the local government of the organizing city, UN Habitat, members of the WUF youth council). I sought to answer questions about the opportunities and challenges of CSC SD and the perceived importance of WUF for sustainable urban development and education in this regard. In my work, I use participatory research approach, as I have been professionally involved in the activities of various WUF11 preparatory committees. I have had the opportunity to observe different stakeholders of CSC SD, to look at the activities from the inside, behind the scenes and also to talk to them in the unstructured interview formula. Familiarity with the context and the language adapted to the type of interviewee, as well as conducting the research in places close to the interviewees and where they felt at ease, allowed for obtaining rich empirical material.

The preparation of WUF involved various actors. Often the same person represented different backgrounds and institutions. Thus, for the purpose of the study, it was necessary to make a simplification—in my proposed approach, two groups have been distinguished—'Political Administration' and 'Academia'. In the first group, I classified representatives of national, regional and local public administration, including ministries, national agencies, municipalities, as well as representatives of international organizations, especially UN. The second group included representatives of universities and research institutions. This division resulted from the nature of anthropological research and referred to the respondents' membership in a particular cultural group, their knowledge of its cultural codes, language, value system and work ethos.

4 Results

The results of my research will be presented with reference to the conceptual framework developed by van der Vliet et al. (2022), presented in the literature review part of this article. It contains the following elements: motivation, capacity, social and physical opportunity.

4.1 *Physical and Social Opportunity*

The whole concept of WUF can be understood as an educational activity with a strong emphasis on peer-to-peer education. The WUF agenda consists of different categories of events, which are summarized below (Table 1).

The involvement of universities in the main formats of WUF events can be divided into two categories—institutional involvement, e.g. in the form of hosting a specific event, and personal involvement of academics as panelists or participants. In addition to the main formats of events mentioned above, involvement of Academia in the preparation for WUF11 was carried out in the following formulas:

- opening of the Learning Zone as a part of the WUF in the City Events, that focus on the idea of sustainable urbanization spreading throughout the whole city of Katowice. It aimed to involve residents, schools and NGOs more widely.
- training and mentoring led by academic experts for more than a 100 Polish municipal governments involved in the Action Plan for Cities program (as a part of the so-called 'WUF legacy'), in which cities implemented projects within working groups focused on specific SDGs (IRMiR 2022a),
- participation of students in the Youth Council WUF11, which implemented an educational program around sustainable urban development addressed mainly to people aged 15–30.

The wider substantive involvement of the scientific community in the preparation of the WUF was visible in the activities of the Institute of Urban and Regional Development (IRMiR), which is a research and implementation unit working closely with the government and local government sectors in Poland on development-related topics. IRMiR prepared the concept of the event track during WUF (European Track) organized around the theme 'Transforming European cities: for a better Europe, for a better world' (IRMiR 2022b). It was dedicated to the key development challenges of today's European cities: the green city, equitable and inclusive city, sustainable urban mobility and affordable housing and was developed in cooperation with networking

Table 1 Characteristics of the events during WUF11 and main elements of the narrative

Type of event	Characteristics	Main elements of the narrative	Comments
Assemblies	Stage to give voice to key stakeholder groups from broader segments of society to support a decision-making process and the New Urban Agenda	Challenges, shared vision, motivation	
Dialogues	High profile events that set the agenda for policy and action throughout the forum and beyond	Challenges, objectives, activities at different levels	
Roundtables	Section-oriented sessions enable peers to exchange views on key urban issues	Challenges, evidence-based approaches to SD	One event was dedicated to Academia in relation to SDG 4, education and academic support for public policy-making
Special Sessions	A number of events on topics of importance of the host region	Mixed-depending on the subject	
Networking events	Opportunities for participants with shared interests in specific areas to exchange ideas, share knowledge and experiences and strengthen partnerships	Motivation, good practices, benefits of cooperation	Involvement of universities as hosts of some sessions
Training events	Events aimed at transferring practical skills and knowledge in specific areas	Challenges, good practices	Involvement of universities as hosts of some sessions
Voices from cities	Events created in order to produce effective action-based solutions to the SD problems	Challenges, action at different levels, good practices	

(continued)

Table 1 (continued)

Type of event	Characteristics	Main elements of the narrative	Comments
SDGs in action	Highlighting hands-on examples of the SDGs implementation	Mixed-depending on the subject	
European track transforming European cities: for a better Europe, for a better world	The block of events dedicated to the urbanization in EU, host by the Polish Ministry of Development Funds and Regional Policy, the Institute of Urban and Regional Development and other partners	Demonstration of the European context and model of sustainable urban development	

Source Author's compilation based on: UNHABITAT (2022)

institutions such as URBACT, European Urban Knowledge Network and METREX (IRMiR 2022c).

The elaborate program of events during WUF was developed through cross-setoral cooperation of various stakeholders related to urban development, including international organizations, governments, local and regional governments, non-governmental institutions, universities, think-tanks, research units, NGOs and business representatives. In my research, I specifically focused on Academia—Political Administration collaboration. The stakeholders revealed different attitudes during the preparation and implementation of WUF, in their behavioral, emotional and awareness-related components.

4.2 *Motivation*

In this study I define motivation as "the driving and pulling forces which result in persistent behavior directed toward particular goals" (Morgan et al. 1993, p. 268) and as "the factors that direct and energize the behavior of humans (Feldman 2015, p. 287). I do not distinguish between external and extrinsic motivation, but I only wanted to indicate the mentioned "factors" and "pulling forces", that direct toward the collaborative behavior.

The research revealed seven key types of stakeholder motivation for collaboration. Three of these are common to academics and politicians, so the main dynamic here is either collaboration or competition. These motivations relate to promoting one's own activities, branding and demonstration a particular political affiliation.

Table 2 Types of motivation of key WUF11 stakeholders and leading relationship dynamics

Types of motivation	Key stakeholders		Leading relationship dynamics
	Political administration	Academia	
Looking for a stage to promote one's own activities	x	x	Cooperation-competition continuum
Building the image and rank of the institution	x	x	
Revealing political preferences	x	x	
Expert certification of clerical activities	x		Cooperation based on exchange
The need to outsource some tasks (exceeding organizational capabilities or inconvenient)	x		
Maintaining good relations with opinion-forming circles in the field of public policies	x		
Establishing contacts with potential stakeholders interested in financing research activities		X	

Source Author's compilation based on qualitative research

The second type of relationship dynamic I identified is based on the access to resources, which are owned by one institution or the other and can be acquired through exchange. These motivations relate to endorsement of political decisions by experts, the need to outsource tasks, relations with opinion leaders and the possibility to obtain funding for research (Table 2).

4.3 Capacity

A large event such as WUF comes at a high organizational and financial cost in preparing the event, as well as the responsibility associated with UN branding. Stakeholders become involved in the co-creation of the event in the expectation of certain reputational effects and profits resulting from the arrival of visitors to the country and city. In this configuration, the world of Academia acts as an executive body subscribing to a specific convention, within a framework set by the main political players.

At the same time, Academia, as a provider of the elements of the agenda (proposals for specific events) had considerable freedom to take up topics of its own choice and thus promote its own research. Local universities are also a venue for attracting potential WUF participants, which helps to meet the attendance expectations of the main organizer.

Financial considerations are a major constraint to Academia's greater institutional involvement in cross-sectoral collaboration in organizing WUFof. The fact that a conference is not strictly of a scientific nature may constitute a barrier to support participation of scholars in such an event as these expenses cannot be assessed as a scientifically parameterized activity. Often, the costs are borne by academics themselves or by entering into cooperation with public institutions as experts and consultants, which affects their independence as researchers (Table 3).

5 Discussion and Conclusion

The research has shown what types of motivation and capacity shape cross-sectoral collaboration within WUF, with Political Administration and Academia as leading players. When motivations for collaboration are similar, there is a risk of competition. Conversely, when motivations relate to resources that are held by one party and in demand by the other, then the skills to negotiate, define common goals and rules for sharing resources become crucial. According to my research, the scale of WUF activities necessitates CSC SD, as only the combined capacities of different institutions are able to respond to organizational challenges. The analysis of WUF covers different rationales for the key issue relates to cross-sectoral cooperation. The exploratory nature of the article has made it possible to identify analytical leads that can be developed through further research.

The aim of my article indicated in the introduction was to show lessons learned from cross-sectoral cooperation within the WUF framework. The first conclusion in this respect is that WUF can be seen as a forum supporting evidence-based policy, but also practice-based education. 'Increased importance of scientific knowledge for political decision-making has led to stronger demands for *scientific consensus*. For scientific experts to effectively impact on and influence political decisions, their knowledge base must be generally accepted among other scientists (…) politics frames the work of experts and how expert knowledge is used, and thereby expertise is reduced to a tool in managing political conflicts' (Soneryd and Sundqvist 2023). Given this, ESD within WUF is based on scientific research validated by a body of experts, but this body of experts is selective and supportive of the policy-based narrative. The policy-based narrative, in turn, refers in many places not to how things are, but to how things should be. Additionally, much of the education during WUF is based on case studies that are not analyzed in the context of broader relationships or generalizations. It is thus a narrative that does not fit into the traditional scientific discourse, in which the dominant trend is to seek to make hypotheses plausible in as many cases as possible.

In the context of the potential of WUF to support ESD, I would like to highlight two issues. Firstly, the implementation of ESD in the WUF formula provides a valuable

Table 3 Rationale and main themes in WUF11 capacity analyse

Issue relates to cross-sectoral cooperation	Chief rationale	Main theme emerging from the statements of the respondents
Philosophy	Cooperation as a necessity Cooperation as a necessary evil	'This project is beyond the capacity of a single institution' 'Now everyone is only concerned with WUF and not the real work'
Objectives	Community (declared) ⇳ Institutional (partisan)	'Everyone must be satisfied' 'We take care of our European priorities' 'We have something to be proud of' 'This event must bring a return in some way'
Attitudes to risk	Innovation ⇳ Conventionalism	'There are places where you can afford more' 'This is not the place for controversial topics' 'we draw on previous experience'
Management	Authoritarian ⇳ Participatory	'These decisions are taken at the top and we follow them' 'Organisational requirements stand above the merits' 'There must be political will, we are waiting for approval' 'We have been discussing for three hours, and actually what has been agreed?'
Impact	Significant ⇳ Marginal	'Great opportunity and prestige' 'I wonder if anyone will be interested in this later' 'Lots of fireworks, little effect' 'Event like any other'
Relationship building	Organisation of new structures ⇳ Use of existing structures	'We are developing interesting contacts' 'You can see that people know each other here' 'Everyone wants to show off' 'Politicians always manage to reach agreement with each other'

Source Author's compilation based on qualitative research

perspective of the global view of urban development. This is a different perspective from the Eurocentric trend of ESD dominant in Europe. The Eurocentric narrative around development issues lacks complexity and excludes alternative voices (i.e., attitudes, beliefs, and practices) from a more realistic story (Jordan et al. 2024). The geographically diverse nature of the issues raised during WUF makes the issues of urban development disparities around the world more realistic and encourages thinking about urban sustainability not in the context of single urban centers, but of the entire global urban settlement network. Secondly, referring to the need to build ESD based on interdisciplinary and cross-sectoral case studies, I consider WUF a valuable source of inspiration for expanding the case study base for use in university curricula (Huckvale and Van Riper 2019).

The WUF narrative focuses on three main agendas—the agenda of SD challenges, the agenda of goals and the agenda of actions taken, which are most often discussed in relation to good practices. I understand good practice 'as anything that has been tried and shown to work in some way—whether fully or in part but with at least some evidence of effectiveness—and that may have implications for practice at any level elsewhere' (Serrat 2017). WUF can be seen as a database for obtaining case studies for ESD, as long as 'good practices' are subjected to reliable scientific analysis.

The scale and complexity of the sustainability challenge makes it apparent that there is a need for a two-pronged flow—from theory to practice, but also from practice to theory. The former approach seems to be widely understood within the university community. ESD is to be conducted in such a way as to bring about a change in social attitudes. The stated aim of UNESCO's ESD for 2030 education program is 'to bring about the personal and societal transformation that is necessary to change course' (UNESCO 2020). It seems that the opposite relationship, i.e. the incorporation of experiences from practice into education, sometimes arouses less interest, as evidenced by the asymmetry in the number of scientific publications devoted to both issues.

The reasons for this may be varied, but in this article I would like to identify what I consider to be three key ones. Firstly, it is more difficult for academics to access knowledge resources derived from practice than it is to use strictly academic resources, and this extra effort is rarely rewarded in university evaluation systems. Secondly, when interacting with the outside world, academics often act as experts, which gives rise to the need to maintain the facade of this social role and not to disclose ignorance (even when it actually occurs). This can result in a distance in dealing with external institutions. Thirdly, an important barrier to transfer from practice to knowledge can be 'stereotypes that exist among sectors and the considerable differences in modes of operation, sectoral cultures and power differentials. Factors that can seriously threaten the cooperative relationship and associated development outcomes include co-optation and inequitable distribution of costs and benefits. Additionally, future cooperation can be undermined by the bad experiences actors may have had in the past' (Kalegaonkar and Brown 2000). The determinants I have identified are exploratory and further research is needed around the determinants of dissemination of practical knowledge from intersectoral cooperation experience to ESD.

In terms of benefits for Academia associated with ESD, WUF provides an opportunity for broad dissemination of research results and promotion of the brand (personal as well as institutional). It also provides an opportunity to strengthen existing (rather than establish new) contacts with representatives of other sectors. One of the non-obvious benefits for Academia, in my opinion, is something that academics do not perceived as a benefit, but as a cost, namely the need to adapt the ESD language to different target groups as members of the WUF audience. 'As human contacts and negotiation are essential to serve the purposes of sustainable development worldwide, language as a shared means of communication is worth paying attention to' (Toppo and Rahman 2020). Effective communication is pivotal in fostering a positive shift towards sustainable practices (Idiaye 2022), and the hermetic languages of different sectors can be a serious barrier to communication and collaboration (Rice and Giles 2017).

At the same time, 'most of the platforms discussing sustainable development focus upon the technical, administrative, commercial and economic aspects related to progress. Language and communication find only a trivial mention and are referred to as if having secondary importance' (Ezeh et al. 2020; Suchitra 2023). The most common publications are on national languages and the related barriers in ESD. Few publications touch on the issue of language differences in the light of CSC SD. I see a need for further in-depth research also in this area.

References

Acs ZJ (2005) Innovation and the growth of cities. In Capello R, Nijkamp P (eds) Urban dynamics and growth: advances in urban economics (contributions to economic analysis, vol 266). Emerald Group Publishing Limited, Leeds, pp 635–658. https://doi.org/10.1108/S0573-8555(2005)0000266021

Barnes N, Phillips P (2000) Higher education partnerships: creating new value in the environment sector. Int J Sustain High Educ 1:182–190. https://doi.org/10.1108/1467630010328252

Breheny M (1992) Sustainable development and urban form. European Research in Regional Science. Pion Limited, London

Broman G et al (2017) Science in support of systematic leadership towards sustainability. J Cleaner Prod 140:1. https://doi.org/10.1016/j.jclepro.2016.09.085

Brown L, Tandon R (1992) Multiparty cooperation for development in Asia. IDR. https://www.syngs.info/files/multiparty-cooperation-for-development-in-asia.pdf

Chitereka C, Mtetwa E, Ruparanganda W, Chirisa I (2022) Education for inclusive and transformative urban development. In: Brears RC (ed) The Palgrave encyclopedia of urban and regional futures. Palgrave Macmillan, Cham. https://doi.org/10.1007/978-3-030-87745-3_70

Crutzen PJ, Stoermer E (2000) The "Anthropocene." Glob Change Newsl 41:17

Dür M, Keller L (2018) Education for sustainable development through international collaboration. A case study on concepts and conceptual change of school-students from India and Austria on gender equality and sustainable growth. Educ Sci 8:187. https://doi.org/10.3390/educsci8040187

Ezeh N, Udaba R, Obiwolu A (2020) The role of language in achieving the world's sustainable development goals (SDGS). Eur J Engl Lang Lit Stud 8(6):53–61. https://eajournals.org/ejells/wp-content/uploads/sites/23/2020/09/The-Role-of-Language-in-Achieving-the-Worlds-Sustainable-Development-Goals-SDGS.pdf

Feldman RS (2015) Essentials of understanding psychology. McGraw-Hill Education, New York

Florida R (2005) Cities and the creative class. Routledge, London. https://doi.org/10.4324/9780203997673

Gajda R (2004) Utilizing collaboration theory to evaluate strategic alliances. Am J Eval 25(1):65–77. https://doi.org/10.1016/j.ameval.2003.11.002

Gazley B (2016) The current state of interorganizational collaboration: lessons for human service research and management. Hum Serv Organ Manag, Leadersh Gov 41(1):1–5. https://doi.org/10.1080/23303131.2015.1095582

Gray B (1989) Collaborating: finding common ground for multiparty problems. Jossey-Bass, San Francisco

Hammersley M, Atkinson P (2019) Ethnography: principles in practice. Routledge, London. https://doi.org/10.4324/9781315146027

Hintea CE, Hamlin RE, Neamtu B (2022) University and community: an essential partnership for the future. Transylvanian Rev Admin Sci 2022:70–98. https://doi.org/10.24193/tras.SI2022.6

Huckvale MU, Van Riper I (2019) Using case studies in the higher education classroom: case studies in higher education—what's the big idea? In Baron A, McNeal K (eds) Case study methodology in higher education. IGI Global Scientific Publishing, pp 47–59. https://doi.org/10.4018/978-1-5225-9429-1.ch003

Idiaye L (2022) Exploring the roles of language in sustainable development goals. In: Conferences on research in management and social sciences, Christopher University. https://www.academia.edu/102770110/EXPLORING_THE_ROLES_OF_LANGUAGE_IN_SUSTAINABLE_DEVELOPMENT_GOALS?uc-sb-sw=63376510

IISD—International Institute for Sustainable Development (2022a) 11th session of the World Urban Forum (WUF11), Earth Negotiations Bulletin. https://enb.iisd.org/world-urban-forum-wuf11

IISD (2022b) Summary of the eleventh session of the World Urban Forum: 26–30 June 2022. https://enb.iisd.org/world-urban-forum-wuf11-summary

IRMiR (2022b) Urban policy observatory. https://obserwatorium.miasta.pl/wuf11/sciezka-europejska/

IRMiR (2022c) Urban policy observatory, urban track WUF11. Concept note. https://obserwatorium.miasta.pl/wp-content/uploads/2022/04/WUF11_Koncepcja-Sciezki-europejskiej-1.pdf

IRMiR—Institute for Urban and Regional Development (2022a) Action plan for cities. https://pdm.irmir.pl/

Jordan S, Baker CC, Wingenbach G, Landaverde R (2024) Land Grant University participants' eurocentric attitudes about agriculture: an ideological constraint to achieving the 2030 agenda for sustainable development. Sustainability 16(6):2410. https://doi.org/10.3390/su16062410

Kalegaonkar A, Brown D (2000) Intersectoral cooperation: lessons for practice. Inst Dev Res Rep 16(2). https://pdf.usaid.gov/pdf_docs/Pnacp367.pdf

König A, Evans J (2013) Introduction: experimenting for sustainable development? Living laboratories, social learning and the role of the university. In: König A (ed) Regenerative sustainable development of universities and cities: the role of living laboratories. Edward Elgar, Cheltenham, pp 1–26

Kronenberg J (2010) Związki gospodarki, społeczeństwa i środowiska. In: Kronenberg J, Bergier T (Esd) Wyzwania zrównoważonego rozwoju w Polsce. Fundacja Sędzimira, Warsaw, p 28

Leal Filho W, Eustachio JHPP, Caldana ACF, Will M, Lange Salvia A, Rampasso IS, Anholon R, Platje J, Kovaleva M (2020) Sustainability leadership in higher education institutions: an overview of challenges. Sustainability 12(9):3761. https://doi.org/10.3390/su12093761

Leal Filho W, Frankenberger F, Lange Salvia A, Azeiteiro U, Alves F, Castro P, Will M, Platje J, Orlovic Lovren V, Brandli L, Price E, Doni F, Mifsud M, Veiga Ávila L (2021) A framework for the implementation of the sustainable development goals in university programmes. J Cleaner Prod 299. https://doi.org/10.1016/j.jclepro.2021.126915

Lorek E, Lorek A, Koczur W (2023) Education for sustainable development in Polish institutions of higher education—present and future. Econ Environ 1(84):116–132. https://doi.org/10.34659/eis.2023.84.1.513

Matu J, Perez-Johnston A (2024) The importance of incorporating lived experience and identity in promoting cultural diversity and sustainability in community college and education: a case study of Community College of Allegheny County. Int J Sustain High Educ 25(3):470–488. https://doi.org/10.1108/IJSHE-05-2023-0166

McBain B, Phelan L, Ferguson A, Brown P, Brown V, Hay I, Horsfield R, Taplin R, Tilbury D (2024) Collaboratively crafting learning standards for tertiary education for environment and sustainability. Int J Sustain High Educ 25(2):338–354. https://doi.org/10.1108/IJSHE-11-2022-0363

Morgan CT, King RA, Weisz JR, Schopler J (1993) Introduction to psychology. Tata McGraw-Hill Publishing Company Limited, Delhi

Parkin S (2010) The positive deviant: sustainability leadership in a perverse world. Earthscan, London

Popowska M, Sady M (2024) Universities' journey towards sustainability—systematic literature review. Int J Sustain High Educ 25(3):596–615. https://doi.org/10.1108/IJSHE-08-2022-0280

Profiroiu MC, Brişcariu MR (2021) Universities as 'drivers' of local and regional development. Transylvanian Rev Admin Sci 62 E/2021:134–152. https://doi.org/10.24193/tras.62E.8

Purcell WM, Henriksen H, Spengler JD (2019) Universities as the engine of transformational sustainability toward delivering the sustainable development goals: "living labs" for sustainability. Int J Sustain High Educ 20(8):1343–1357. https://doi.org/10.1108/IJSHE-02-2019-0103

Raišienė A, Baranauskaitė L (2018) Investigating complexity of intersectoral collaboration: contextual framework for research. Contemp Res Organ Manag Admin 6(1). https://cris.mruni.eu/server/api/core/bitstreams/586a119d-5653-47ac-a415-ac3b0cfe0874/content

Rama Murthy S, Roll K, Colin-Jones A (2021) Ending business-non-profit partnerships: the spinout of social enterprises. Scand J Manag 37(1). https://doi.org/10.1016/j.scaman.2020.101136

Renfors S (2024) Education for the circular economy in higher education: an overview of the current state. Int J Sustain High Educ 25(9):111–127. https://doi.org/10.1108/IJSHE-07-2023-0270

Rice R, Giles H (2017) The contexts and dynamics of science communication and language. J Lang Soc Psychol 36(1):127–139. https://doi.org/10.1177/0261927X16663257

Sady M, Żak A, Rzepka K (2019) The role of universities in sustainability—oriented competencies development: Insights from an empirical study of Polish universities. Admin Sci 9(3):62. https://doi.org/10.3390/admsci9030062

Serrat O (2017) Knowledge solutions. Tools, methods, and approaches to drive organizational performance. Springer Nature, Singapore. https://doi.org/10.1007/978-981-10-0983-9

Slavin M (2011) Sustainability in America's cities. Creating the green metropolis. Island Press, Washington DC. https://doi.org/10.5822/978-1-61091-028-6

Soneryd L, Sundqvist G (2023) Science and democracy: a science and technology studies approach. University Press, Bristol. https://doi.org/10.46692/9781529222159

Steiner G, Posch A (2006) Higher education for sustainability by means of transdisciplinary case studies: an innovative approach for solving complex, real-world problems. J Cleaner Prod 14(9–11):877–890. https://doi.org/10.1016/j.jclepro.2005.11.054

Suchitra D (2023) Significance of language in sustainable development. Int J Arts Educ Res 12/3. https://ijaer.org/admin/uploads/paper/file1/NN8g8ssg9nnamP0PTl1eJQ==2.pdf

The Council of Europe (2024) Education for sustainable development and global citizenship. https://www.coe.int/en/web/education/4.7-education-for-sustainable-development-and-global-citizenship

Toppo N, Rahman M (2020) The role of language in sustainable development: multilingualism and language literacy in India. Problemy Ekorozwoju—Prob Sustain Dev 15(1):89–93. https://bibliotekanauki.pl/articles/371757

Trencher G, Yarime M, McCormick KB, Doll C, Kraines SB (2014) Beyond the third mission: exploring the emerging university function of co-creation for sustainability. Sci Public Policy 41(2):151–179. https://doi.org/10.1093/scipol/sct044

UN HABITAT (2022) The eleventh session of the world urban forum. Programme, https://www.gov.pl/web/wuf11/program2
UN (2016) The new urban agenda. https://habitat3.org/the-new-urban-agenda/
UNESCO (2007) The UN decade of education for sustainable development 2005–2014: the DESD at a glance. https://unesdoc.unesco.org/ark:/48223/pf0000141629
UNESCO (2009) Review of contexts and structures for education for sustainable development. https://unesdoc.unesco.org/ark:/48223/pf0000187757
UNESCO (2017) Education for SDG learning objectives. https://doi.org/10.54675/CGBA9153
UNESCO (2019) Framework for the implementation of education for sustainable development—ESD beyond 2019. https://unesdoc.unesco.org/ark:/48223/pf0000370215.locale=en
UNESCO (2020) Education for sustainable development. A roadmap. https://doi.org/10.54675/YFRE1448
UNESCO (2021) Berlin Declaration on education for sustainable development. https://unesdoc.unesco.org/ark:/48223/pf0000381228
van der Vliet N, den Broeder L, Romeo-Velilla M, Staatsen B, Kruize H, Friedrich B, Schuit AJ (2022) Facilitators and barriers of intersectoral co-operation to promote healthier and more environmentally friendly behaviour: a qualitative evaluation through focus groups for the INHERIT project. BMC Public Health 22:617. https://doi.org/10.1186/s12889-022-12974-8
van Tulder R, Keen N (2018) Capturing collaborative challenges: designing complexity-sensitive theories of change for cross-sector partnerships. J Bus Ethics 150:315–332. https://doi.org/10.1007/s10551-018-3857-7
Visser W, Courtice P (2011) Sustainability leadership: linking theory and practice. Social Science Research Network, Rochester NY, USA
Welsh Government (2012) Education for sustainable development and global citizenship. A common understanding for the adult and community learning sector. https://www.gov.wales/sites/default/files/publications/2018-02/education-for-sustainable-development-and-global-citizenship-a-common-understanding-for-the-adult-and-community-learning-sector.pdf
Yigitcanlar T, Velibeyoglu K, Martinez-Fernandez C (2008) Rising knowledge cities: the role of urban knowledge precincts. J Knowl Manag 12(5):8–20. https://doi.org/10.1108/13673270810902902

Greening the Audit: How Universities Can Shape Tomorrow's Sustainability Assurance Professionals

Sakshi Girdhar and Kristina Linke

Abstract Sustainability is a gift in the wake of expanded global environmental consciousness of all stakeholders. Organizations face enormous challenges to deal with sustainability reporting. One of the challenges is lack of sustainability auditors to provide assurance over sustainability reporting. The implementation of the Corporate Sustainability Reporting Directive (CSRD) in 2024 has significantly increased the demand for certified auditors proficient in sustainability assurance. Consequently, there is an upcoming demand to equip students with specialized expertise in sustainability reporting and assurance. Universities play a vital role in shaping the future workforce capable of tackling these challenges. This underscores the urgent need for universities to re-evaluate and enhance their educational programs to provide the necessary knowledge and skills in sustainability assurance. However, this endeavor is not without obstacles; integrating sustainability into standardized education encounters complexities due to the multifaceted nature of the subject and the stringent regulations governing the audit educational standards generally and audit certification particularly. Therefore, our chapter provides a critical review of the pivotal role universities play in navigating these challenges within audit curricula, particularly concerning sustainable development. By presenting potential solutions, we aim to highlight how universities can support the audit profession in developing human capital to effectively incorporate sustainability principles into sustainability assurance, preparing auditors for the field's evolving demands.

Keywords Sustainable development · Curriculum development · Audit · Assurance · Sustainability · CSRD · Universities · Higher education

S. Girdhar (✉) · K. Linke
University of Groningen, Groningen, The Netherlands
e-mail: s.girdhar@rug.nl

K. Linke
e-mail: k.linke@rug.nl

W. Leal Filho et al. (eds.), *Education for Sustainable Development: The Contribution of Universities*, World Sustainability Series,
https://doi.org/10.1007/978-3-031-86985-3_9

1 Introduction

Up until the recent introduction of the Corporate Sustainability Reporting Directive (CSRD)—Directive 2022/2464/EU (European Parliament & Council of the European Union 2022), the assurance of sustainability reports has long been a largely voluntary practice (Alsahali and Malagueño 2022). Corporate Social Reporting (CSR) emerged in the 2000s and the European Union (EU) introduced the Non-Financial and Diversity Information Directive (referred to as the 'Non-Financial Reporting Directive'—NFRD)—Directive 2014/95/EU (European Parliament & Council of the European Union 2014)—for corporate reports. Previous scholars have identified many shortcomings in the current approach to NFRD. These include the lack of specific standards, the combination of mandatory and voluntary items, and the failure of a clear directive to guarantee that firms are making genuine contributions to the sustainable development goals (Pizzi et al. 2021). Building on the development of the NFRD directive requirements, the EU has proposed a new directive called the CSRD, which introduces considerably more rigorous reporting obligations for companies falling within its scope. The CSRD is also intended to address the criticism regarding CSR reporting quality, an example of which is corporate greenwashing (García-Sánchez et al. 2021). The new directive will ultimately apply to more than 49,000 organizations across the EU from 2024, starting gradually with large stock listed firms and increasingly extending to small and medium sized companies in the coming years (PWC 2023). Therefore, in response, the auditing profession is positioning itself to prepare its professionals to meet sustainability assurance demands.

Traditionally, 'the objective of the auditor is to identify and assess the risks of material misstatement, whether due to fraud or error' (IFAC 2010) and provide assurance to improve capital market transparency and stakeholder satisfaction. The traditional audit relies on concrete and verifiable evidence. However, CSR regulatory reforms have pushed the audit profession towards the nexus of an evolving assurance landscape. The role of an auditor now extends beyond the conventional scope of attesting financial information; in addition they must also provide assurance regarding the non-financial information related to companies. The implicit assumption in the expanded role of auditors is that they will receive adequate education, knowledge and skills to deal with such emergent issues of sustainable development goals (Ng et al. 2017).

The audit assurance process has evolved over more than 100 years and has undergone rigorous methodology and accounting standards, independent oversight mechanisms to test compliance and lawsuits and penalties/fines for audit firms in case of breach of professional standards (Boiral and Gendron 2011). This process has resulted in a well-established educational curriculum across the globe. However, with the changing demands for skills that enable auditors to provide sustainability assurance, the profession and society place certain expectations, whether justly or not, on universities to take on the task of teaching the necessary skills and knowledge areas crucial to the auditing profession (Siriwardane et al. 2014). Even though, universities have demonstrated a growing commitment towards facilitating the promotion and implementation of sustainability based education (Aleixo et al. 2018) embedding sustainability into standardized education encounters complexities due to the multifaceted nature of the subject and the stringent regulations governing the audit curriculum. In addition, the short time frame for the CSRD implementation presents a great challenge: to achieve a strategic alignment of universities and professional associations by developing curriculum that can integrate the demands of the directive and provide an effective toolkit enabling students to handle future market demands.

This chapter focuses particularly on the Netherlands within the EU as it stands out in the European context. The EU Directive has not yet been integrated into Dutch audit law. Furthermore, the Dutch audit profession is known to be heavily regulated by independent oversight bodies (Jahn and Loy 2023) and attracts strict penalties and fines in the case of a breach in professional standards (Humphrey et al. 2018). The strict environment thus has a spillover effect on auditors' training and education by placing a higher bar to qualify as an auditor. This setting is complicated by a review of the audit sector initiated by the Dutch Ministry of Finance. In its final report, the independent review commission concludes that the entire training process for auditors needs to be thoroughly overhauled (De Vreis and Fonteijn 2023). Any initiative to develop an audit-related sustainability curriculum is hampered by the uncertainty of how far-reaching this revision will be. Thus, in this chapter, we discuss how universities must navigate a field characterized by complicated directives and the stringent audit educational regime. We try to answer: How can universities educate audit professionals in sustainability assurance? In the following sections, we will discuss the general challenges faced by universities in incorporating CSRD and impediments in the auditors' curriculum development process particularly in the Dutch setting and how to overcome them.

2 General Challenges Faced by Universities

2.1 The Availability of Resources

Imparting sustainability education to prepare audit professionals for the future requires universities to transform their curriculum. Integrating specialized education based on sustainability is by no means an easy task, as the university needs to tackle its resource barriers (financial and non-financial). In this regard, Glennan (1998), demonstrated that the curriculum design implementation is significantly influenced by resource constraints, as lack of resources can sometimes make the implementation less effective or weak. In addition, accreditors place reasonable emphasis on availability of resources for academics to ensure quality education. Therefore, funding becomes an imperative resource for universities to deliver high quality education. However, public funding of universities has become a recurring topic in Western European countries and governments are reducing funding due to uncertain economic conditions and increased public pressure to allocate public resources more efficiently, particularly in higher education (Chevaillier and Eicher 2002). Many countries are now introducing performance-based measures to allocate public funds for higher education. Developing/redesigning a sustainability-based curriculum for audit professionals could be an expensive development, as it requires staff with sustainability expertise. In addition, if the enrollments of students decline, it will further jeopardize the position of universities to acquire public funding which could depend on the numbers of diplomas awarded. Thus, a university must make informed decisions about the appropriate allocation of resources within the faculties/ departments as the new changes in the program affect performance of the university and could present financial risks.

2.2 Bureaucracy

The surge in accountability parallels the growth in bureaucratization of university education (Churcher and Talbot 2020). There are two levels of bureaucracy in universities—internal and external. The internal bureaucracy involves a complex web of bureaucratic boundaries comprised of rules and procedures needed to meet the requirements of transparency for quality assurance and accountability. For instance, Hurlimann et al. (2013) argue that conducting essential curriculum reviews involves lengthy timelines for completion—first within programs and faculties and then through broader university procedures and government reporting requirements—which, they showed, often deterred and frustrated the development process.

External bureaucracy involves regulations from government and other accreditors such as professional accounting bodies; whereas internal bureaucracy involves the university's processes for addressing these regulations and additional rules. Universities are increasingly moving from an independent organization to a co-opted organization where external pressures, exerted from several directions, are taking over the control of curriculum development, objectives and their everyday activities (Abidin 2020). In particular, disciplines such as law, accounting and medicine attract more regulations for the education sector as they impact society and public well-being in general. In the light of the new CSRD directive, there is an increased need for cooperation between the state, professional bodies and accreditors to change the curriculum and prepare professionals for the pending developments in the sustainability assurance market. However, the pace of the changes in the reporting and auditing curriculum is significantly impeded by bureaucratic procedures due to the involvement of several stakeholders in the education sector. For example, the bureaucracy in the education sector is amplified by the heightened involvement of the government/ministry of education to obtain funding, in order to achieve accreditations and re-accreditation from multiple accrediting bodies (such as AACSB, EQUIS). The accreditation process is perceived as a shift in power from educators to the managers and bureaucrats that inhibits improvement and innovation (Harvey 2004). In addition, this process also brings implications for the faculty members as they become more of a bureaucrat than an educator, scholar or facilitator of change in the area of sustainability (Greenwood et al. 2010). The content and development of reporting/auditing curriculum is particularly impacted by interference of professional bodies imposing special requirements for e.g., additional courses or practical training approved by an independent accrediting body/professional associations in order to grant permission to successful candidates to practice a profession. Alexander and Hjortsø (2019), argued that quality assurance processes and legitimacy often hinder universities from responding promptly to the contemporary needs of the labor market, thereby making curriculum development slow and less relevant if implemented too late.

Given the internal and external bureaucratic procedures, the timeframe to embed the educational requirements based on sustainability can be challenging. Moreover, the aforementioned line of reasoning gives rise to a debate within the education sector regarding the optimal means of integrating sustainability into the audit education. This debate encompasses a range of considerations including the potential benefits of offering sustainability courses as electives or mandatory requirements, the suitability of discipline-specific or cross-disciplinary approaches, and the relative importance of sustainability as a core or peripheral element within academic programs.

2.3 Auditing Education Requirements

The educational trajectory to become a certified auditor is a combination of university and professional education. It is uncontested that universities form a cornerstone in auditors' education because most worldwide accountancy bodies require applicants to hold a degree in accountancy or a comparable accounting degree. For EU-countries—Directive 2006/43/EC (European Parliament & Council of the European Union 2006)—defines the minimum requirements for access to the audit profession as well as exceptions. In order to register as an auditor, the following requirements must be fulfilled (Article 6 & 10): First, a university entrance or equivalent level of education; second, a completed course of theoretical instruction including examination; third, practical training for at least three years and fourth, a successful examination of professional competence of university final or equivalent examination level. Although Directive 2006/43/EC (European Parliament & Council of the European Union 2006, 2022)—intended to harmonize access to the audit profession in EU-countries, it leaves member states instead with a broad range of implementation choices. It is up to the individual member states to decide upon the distribution of educational tasks and oversight among universities, professional and oversight bodies. Recently, Jahn and Loy (2023) developed an auditor access requirement index and classified EU member states accordingly. They find five clusters of countries with low, lower moderate, upper moderate, high and very high access requirements and conclude that despite ongoing harmonization efforts within the EU the access requirements to the audit profession still vary considerably.

Directive 2006/43/EC (European Parliament & Council of the European Union 2006)—serves as the basis upon which CSRD builds to define requirements for corporate sustainability reporting audits. CSRD (Article 65) specifies that auditors have "the necessary level of theoretical knowledge of subjects relevant to the assurance of sustainability reporting and the ability to apply such knowledge". However, the challenge presented by the CSRD directive is that it is relatively recent and does not explicitly delineate the specific competencies, knowledge, or skills that auditors must possess in order to become certified sustainability assurance providers. In addition, the current assurance providers in the market have very limited knowledge and it remains fragmented among experts across different disciplines thereby making it hard to identify the competent assurance providers in the market (Boiral et al. 2024). Given the range of exhaustive access requirements to the profession, the role of universities in educating auditors with sustainability knowledge depends on the country setting. In particular, curriculum development is only possible when taking the requirements of professional and oversight bodies into account.

2.4 *Students*

Major audit firms in the market recognize CSR assurance service as a market opportunity as well as a threat (Fernandez-Feijoo et al. 2018). A report from KPMG finds that 59% of the world's largest 250 global companies (G250) with CSR reporting, involve external assurance; of these, two-thirds are major audit firms (KPMG 2013). However, audit firms that were used to perform services in a regulated market have to compete with other service providers such as consultants specialized in CSR (Cohen and Simnett 2015; Deegan et al. 2006). This implies that there is a growing demand for auditors educated in sustainability but the supply of educated specialists may continue to be in deficit. Worldwide there is a dramatic decrease in accounting students. Candidates for the U.S. CPA exam dropped by 33% in the period 2016–2021(Burke and Polimeni 2023). The Dutch professional body of accountants (Nederlandse Beroepsorganisatie van Accountants, hereafter, NBA) initiated an ambassador program in which professional auditors visit high schools to promote the profession (Nederlandse Beroepsorganisatie van Accountants 2024). The sustainability angle of auditing may increase the attractiveness for future students, a generation for whom sustainability and climate change is important. Horne et al. (2023), find experimental evidence showing that study participants with prosocial values were more interested in sustainability assurance job postings relative to those pertaining to financial assurance job postings. CSRD implementation offers an opportunity for universities to design a curriculum that fulfills the needs of students as well as employers. However, universities need to design audit curricula to meet requirements of government, audit firms, professional bodies and regulators. Design choices relate to an optimal mix of knowledge and professional skills. In this context, previous scholars have been discussing the role universities should occupy in relation to employers or professional education (Crawford et al. 2013; Howieson et al. 2014).

The perspective of the student has received little attention and this one-sided approach is no longer effective. The CSRD has created an opportunity to make the profession more attractive for students and universities by integrating more sustainability into the audit curriculum, reviving interest and meeting the growing demands of the audit service sector. As students become a more significant factor for designing a curriculum, audit firms need to build a team of diversified experts from different disciplines to meet the complexity of sustainability assurance standards, resulting in several questions that universities face: From which disciplines should students come? Should they have technical or audit background? What should teaching staff specialize in to meet these educational needs? In the next section, we present the specific case of a Dutch public university.

2.5 *Case—Netherlands*

The aforementioned challenges will be examined in the context of the Netherlands, with a particular focus on the empirical setting of the University of Groningen.

2.5.1 Dutch Setting

The Dutch auditing environment is distinguished by a considerable degree of external bureaucracy in comparison to other European countries. According to Jahn and Loy (2023), the Netherlands is among the three European countries (Belgium, France, Netherlands) with the highest educational entry requirements for auditors. Dutch legislators demand stricter entry requirements than those laid out in—Directive 2006/43/EC (European Parliament & Council of the European Union 2006). The auditing profession is entirely regulated by Dutch law (Overheid 2024), which defines institutions involved in auditing oversight and education. The Dutch Professional Body of Accountants defines the professional profile for auditors and a revision is expected in Summer 2025 (Nederlandse Beroepsorganisatie van Accountants 2024). Building on this profile the Dutch Accountancy Educational Standards Board (henceforth called 'CEA') determines educational objectives, called "end-terms", for different professional profiles (Article 49, number 2a), accredits which educational programs meet requirements (Article 49, number 2b) and monitors quality of practical training (Article 49, number 2c). Universities, as education providers, are bound to design auditing curricula within the framework of end-terms. The idea is that well educated auditors will lead to higher audit quality (Van Linden and Hardies 2018) which in turn leads to higher educational access standards in practice.

However, strict entry requirements come at a price. The CEA prescribes the entire curriculum of auditors. The educational objectives comprise 12 disciplines (for example statistics, law, economics, financial accounting). The CEA determines European Credit Transfer (ECTS) and Accumulation System points for each discipline. Throughout the entire theoretical training these add up to 200 ECTS (CEA 2021) which is more than a full-time bachelor program (180 ECTS). Dutch universities offer the whole trajectory from Bachelor-, Master-, and Post-Master programs in auditing to become a certified auditor. It is up to universities to decide how theoretical training is distributed between the three program phases. When educational objectives are initially addressed as part of Bachelor- and/or Master Programs, this shortens Post-Master programs. All universities make use of this option, but their choices differ, resulting in Post-Master programs with varied lengths, lasting 1.5 to 2.5 years. When substantial changes are required, as is the case with CSRD implementation, universities must uphold the CEA end-terms. Consequently, curriculum innovation initiatives are very difficult. The complexity of CSRD legislation cannot be covered by simply adding another course to an existing program; instead, a new curriculum needs to be designed. Development of a truly innovative sustainability audit curriculum without CEA approval involves financial and accountability risks for universities, making it a rather unattractive option. The CEA educational approach has elicited criticism from multiple stakeholders, among which are universities, students and audit firms. Educators at universities feel constrained by the current CEA framework, notwithstanding major changes such as CSRD. Students currently experience auditor education as inflexible and lengthy and this has resulted in declining student numbers. Students either do not choose auditing as a career path or quit the educational program without completion (NBA-CEA 2022). Consequently, audit firms suffer from an insufficient inflow of young audit professionals; CSRD assurance requirements increase the demand of auditors, thereby worsening the certified auditors' shortage. In its final report, a review commission implemented by the government advises a thorough revision of the entire auditor education, leaving the initiative up to the profession (De Vreis and Fonteijn 2023). The NBA expects to publish the new professional auditor profile in summer 2025 (Nederlandse Beroepsorganisatie van Accountants 2024a) and the CEA is currently working in anticipation of the profile pertaining to new end-terms (CEA 2021). In addition to this major educational revision, the CEA is also awaiting CSRD legislation to be approved by Dutch Parliament (overheid). As of September 2024, the issue of CSRD end-terms is still pending. In this situation, the CEA is waiting for the Dutch Parliament to pass CSRD law, meanwhile universities not only lack incentives to develop sustainability assurance programs, they must also bear additional financial risks. This cumbersome process inhibits innovation and universities, students and audit firms must tolerate the status quo.

2.5.2 University of Groningen Setting

The University of Groningen (UG) was founded in 1614 and is today a leading research and teaching university. There are eleven faculties with a broad academic spectrum that spans medical sciences, science and engineering, economics and business, and philosophy. The university's strategic plan 2021–2026, called Making Connections (University of Groningen 2020), focuses on societal themes, globalization, and sustainable business operations (University of Groningen 2022, p. 12). The strategic plan establishes interdisciplinary programs in research and teaching as top priority. In an auditing context, an important question to address is the optimal combination of auditing and subject matter expertise, needed for an audit team to provide CSR assurance (Cohen and Simnett 2015). Subject matter experts are knowledgeable experts in their area and complement audit teams (Wright and Bhattacharjee 2018). A comparable discussion around industry specialization shows that audit firms have a competitive advantage, whereby industry specialization is defined as knowledge of varying business practices, expertise developed from auditing similar clients in different locations, professional staff, and the capability to develop specific industry training and protocols (Carson 2009). CSRD goes beyond industry specialization, it requires information that can only be assessed by subject matter experts, for example ecological biologists, geologists and water experts. However, these experts lack audit skills. Audit teams in the future will therefore most likely need to have a different team composition to maneuver CSRD challenges. UG is well positioned to play a role in research and education in tackling this interdisciplinary challenge. UG has four interdisciplinary schools that embrace education, research and outreach. Each school has a specific focus and can be linked to the three pillars of CSRD: (1) Environment (Wubbo Ockels School, energy transition and climate adaptation); (2) Social (Aletta Jacobs School, healthy aging and healthy society; Jantina Tammes School, digital innovation, AI, and knowledge society) and Governance (Rudolf Agricola School, governance, politics, and sustainable society) (University of Groningen 2022, pp. 51–52). Moreover, UG promotes inclusive, widely accessible, interdisciplinary, challenge-based, and community-oriented teaching. To make this possible the university aims to bring together various disciplines, cooperating with representatives from society and the professional field (University of Groningen 2022, p. 22). Consequently, each faculty translates the UG strategy into a tailored strategic faculty plan. UG research and teaching setting is prepared to embrace the challenge of developing new CSRD audit teams; however, the slow implementation of CSRD legislation and the related response of professional and educational standard setting, cause delays in audit curriculum development. The UG cannot progress without guidance from the NBA and the CEA since curriculum development is costly and the UG is accountable to the Dutch government for efficient usage of public funds.

2.5.3 Curriculum Implications Auditing

Curriculum development can be viewed from an internal and an external perspective. The internal perspective comprises the culture, operations, people, strategy and institutional setting (Khan and Law 2015); whereas the external perspective is geared towards knowledge-intensive education that produces competent professionals for the labor market and society, and enables individuals to pursue a meaningful and successful career (Mäkinen and Annala 2010). Curriculum development in auditing is substantially restricted by external CEA requirements. The magnitude of CSRD legislation and implementation can be compared to the impact of information technology on society, education and on the auditing profession. To address the impact from the field of information technology the NBA created the professional profile of IT-auditor called NOREA. Likewise, the CSRD is extensive and requires expertise that goes far beyond standard auditor education and, as argued above, a new type of auditor and audit team composition needs to be developed. As CEA guidance is scarce, small steps only are possible, thus prohibiting the implementation of an innovative audit sustainability curriculum. This results in the UG not using the infrastructure available to design sustainability program(s) that would address labor market needs as well as student wishes for sustainability education.

From an internal perspective, the UG auditing program management is not fully able to respond to the labor market and students' needs by addressing important CSRD topics in current auditing courses. This approach remains superficial and is therefore not suited for training students adequately. An example is the new key concept of Double Materiality (impact materiality and financial materiality) which is complex and demanding. European Sustainability Reporting Standards (ESRS)—Directive 2023/2772/EU (European Parliament & Council of the European Union 2023)—define impact materiality as: "A sustainability matter is material from an impact perspective when it pertains to the undertaking's material, actual or potential, positive or negative, impacts on people or the environment over the short-, medium- or long-term" (ESRS, Article 3.4, number 43). Consequently, an auditor has to assess a firm's impact on the environment, an illustration of this would be the case of "Water & Marine Resources" (ESRS, E3) in which water consumption, —withdrawals, —discharges (lakes and rivers), —discharges in the oceans and extraction and use of marine resources must be assessed. Students can be alerted to the inherent challenges, for example whether subject matter experts have to be involved, but more than creating awareness is not possible. Interested master students who wish to engage more in sustainability related subjects can only be directed to extracurricular activities or advised to pursue a second master program related to sustainability. UG Faculty of Economics and Business (FEB), with the exception of the MSc Accountancy & Controlling, all offer focus areas (Healthy Society, Digitalization and Artificial Intelligence, Future Prosperity and Sustainability, and Energy Transition and Climate Change) as part of master programs. These program focus areas are considered extracurricular activities for MSc Accountancy & Controlling students who choose to follow them. A similar situation applies to bachelor students who specialize in Accountancy & Controlling where the entire program is compulsory. Other bachelor programs include university level minors at the end of the third year enabling students to broaden their perspectives. FEB offers a minor in Energy Transition from Multidisciplinary Perspectives but for Accountancy & Controlling bachelor students it would mean again an extra-curricular activity. A particular design choice in the FEB—to have one of the shortest Dutch post-master accountancy programs to become a certified auditor—is a reason for the rigidity of bachelor and master programs. On one hand, it is clearly a disadvantage, but on the other hand, a short post-master program has the advantage of a reduced workload when students step into their professional careers. A NBA-CEA report (2022) confirms the view that students wish to receive more freedom to pursue their interests and needs of their work environment.

To conclude, the room for program management to innovate, based on well-developed UG infrastructure is limited, instead students have to follow extra-curricular activities. As argued earlier, stringent regulation causes constraints for innovative curriculum development. However, for the flipside of auditing, that is, management accounting / controlling, the situation is different. This profession is not regulated and it is up to universities to develop adequate programs. At FEB, program management (under the umbrella of MSc Accountancy & Controlling) developed and marketed a master program sustainability management & controlling, all within one-year. This example proves that fast innovative responses are possible. Amazingly, auditing students consider pursuing this master program as a second degree.

3 Discussion and Conclusion

The implementation of the CSRD is a landmark event and requires universities to re-think current auditing and accounting curricula. Innovative professional auditing education at universities is characterized by many challenges. Universities face scarce resources that have to be optimally used which is difficult in an environment with decreasing auditing student numbers. External bureaucracy involves government regulations and other accreditors' requirements thereby limiting universities room for maneuver. Internal bureaucracy potentially limits curricula reviews further. As the EU defines educational boundaries for the trajectory to become certified auditor in very broad terms, the implementation differs significantly among EU member states. The CSRD uses the same approach of boundary definition, currently requiring member states to translate the directive again into national law. Finally, auditors have to be prepared for a role change from financial auditor to one which includes non-financial auditing expertise. These challenges were discussed in the context of the Netherlands.

External bureaucratic requirements appear to be the limiting factor of auditing curriculum development in the Netherlands. Substantial changes in the current program are not possible because the CEA requirements would be endangered and consequently, students' ability to become a certified auditor would also be impacted. Innovation needed to serve society and students remains on-hold or delayed, despite its urgency. The case of the UG shows that existing infrastructure, financially and intellectually remains unused although students voice their interest. This results in a profession that is increasingly unattractive to students due to its lengthy and tough requirements. Audit firms are already experiencing shortages on the work floor and the CSRD continues to further increase personnel demand. Although the CEA is aware of the need to restructure auditor's education and has investigated actual changes (NBA-CEA 2022), revised end-terms are still expected to be presented in one to two years' time. Slow innovations may worsen the inflow of auditing students in the Netherlands in the coming years. These strict regulatory requirements come at a price especially when compared to other European countries, with a less restrictive regulatory environment, where it may be found much easier to address the needs of the professional market and students. Given all these challenges, and the rapidity with which they must be addressed, the training and educational programs remain underdeveloped.

Recommendations to eliminate the rigidity in the Dutch educational system for certified auditors are twofold. The first is for a significant reduction of the CEA end-terms, so that they cover only the post-master education and limit the length of the study to 60 ECTS, which translates into a period of two years. This would then leave the determination of bachelor- and master program content up to universities. Since universities are subject to a strict accreditation regime by the Accreditation Organization of the Netherlands and Flanders (NVAO) that monitors educational quality, this change requires trust in universities to maintain an adequate educational program for auditors. A second recommendation as suggested by (Buijink and Mertens 2024) goes further and advocates the elimination of the CEA and introduces instead a university post-master program with a NVAO accreditation. By providing opportunities to simplify auditor education and create room for innovations, both recommendations change the current approach significantly.

References

Abidin M (2020) Reinventing higher education bureaucracy through entrepreneurial values. Univ J Educ Res 8(12):6482–6490

Aleixo AM, Azeiteiro U, Leal S (2018) The implementation of sustainability practices in Portuguese higher education institutions. Int J Sustain High Educ 19(1):146–178

Alexander IK, Hjortsø CN (2019) Sources of complexity in participatory curriculum development: an activity system and stakeholder analysis approach to the analyses of tensions and contradictions. High Educ 77:301–322

Alsahali KF, Malagueño R (2022) An empirical study of sustainability reporting assurance: current trends and new insights. J Account Organ Chang 18(5):617–642

Boiral O, Gendron Y (2011) Sustainable development and certification practices: lessons learned and prospects. Bus Strateg Environ 20(5):331–347

Boiral O, Brotherton MC, Talbot D (2024) What you see is what you get? Building confidence in ESG disclosures for sustainable finance through external assurance. Bus Ethics, Environ Responsib 33(4):617–632

Buijink W, Mertens G (2024) Wetenschappelijk onderzoek moet natuurlijk centraal staan in de universitaire accountantsopleiding. https://www.accountant.nl/discussie/opinie/2024/7/wetenschappelijk-onderzoek-moet-natuurlijk-centraal-staan-in-de-universitaire-accountantsopleiding/

Burke JA, Polimeni RS (2023) The accounting profession is in crisis. CPA J 93

Carson E (2009) Industry specialization by global audit firm networks. Account Rev 84(2):355–382

CEA (2021) EINDTERMEN Accountantsopleidingen. https://www.cea.nl/siteassets/archief/eindtermen/eindtermen-accountantsopleidingen/actuele-eindtermen/rapport-eindtermen-2016-versie-1.4-geldig-vanaf-01-09-2021.pdf

Chevaillier T, Eicher J-C (2002) Higher education funding: a decade of changes. High Educ Eur 27(1–2):89–99

Churcher M, Talbot D (2020) The corporatisation of education: bureaucracy, boredom, and transformative possibilities. New Formations 100(100–101):28–42

Cohen JR, Simnett R (2015) CSR and assurance services: a research agenda. Audit: J Pract Theory 34(1):59–74
Crawford L, Helliar C, Monk EA (2013) Generic skills in audit education. In: Audit education. Routledge, pp 5–21
De Vreis M, Fonteijn C (2023) Druk en tegendruk. Slotrapportage kwartiermakers toekomst accountancy. https://www.rijksoverheid.nl/documenten/rapporten/2023/10/30/druk-en-tegendruk-slotrapportage-kwartiermakers-toekomst-accountancy
Deegan C, Cooper BJ, Shelly M (2006) An investigation of TBL report assurance statements: UK and European evidence. Manag Audit J 21(4):329–371
European Parliament & Council of the European Union (2006) Directive 2006/43/EC of the European Parliament and of the Council of 17 May 2006 on statutory audits of annual accounts and consolidated accounts, amending Council Directives 78/660/EEC and 83/349/EEC and repealing Council Directive 84/253/EEC. https://eur-lex.europa.eu/legal-content/EN/TXT/PDF/?uri=CELEX:32006L0043&qid=1732883119532
European Parliament & Council of the European Union (2014) Directive 2014/95/EU of the European Parliament and of the Council of 22 October 2014 amending Directive 2013/34/EU as regards disclosure of non-financial and diversity information by certain large undertakings and groups Text with EEA relevance. http://eur-lex.europa.eu/legal-content/EN/TXT/PDF/?uri=CELEX:32014R0537&from=EN
European Parliament & Council of the European Union (2022) Directive (EU) 2022/2464 of the European Parliament and of the Council of 14 December 2022 amending Regulation (EU) No 537/2014, Directive 2004/109/EC, Directive 2006/ 43/EC and Directive 2013/34/EU, as Regards Corporate Sustainability Reporting. https://eur-lex.europa.eu/legal-content/EN/TXT/?uri=CELEX:32022L2464
European Parliament & Council of the European Union (2023) Commission delegated regulation (EU) 2023/2772 of 31 July 2023 supplementing Directive 2013/34/EU of the European Parliament and of the Council as regards sustainability reporting standards. https://eur-lex.europa.eu/legal-content/EN/TXT/?uri=CELEX%3A32023R2772
Fernandez-Feijoo B, Romero S, Ruiz S (2018) Financial auditor and sustainability reporting: does it matter? Corp Soc Responsib Environ Manag 25(3):209–224
García-Sánchez I-M, Hussain N, Khan S-A, Martínez-Ferrero J (2021) Do markets punish or reward corporate social responsibility decoupling? Bus Soc 60(6):1431–1467
Glennan Jr TK (1998) New American Schools after six years. Rand, Distribution Services, Santa Monica, CA
Greenwood R, Díaz AM, Li SX, Lorente JC (2010) The multiplicity of institutional logics and the heterogeneity of organizational responses. Organ Sci 21(2):521–539
Harvey L (2004) The power of accreditation: views of academics. J High Educ Policy Manag 26(2):207–223
Horne E, Loftus S, McCoy SS, Winn AM (2023) Attracting the next generation of accountants: the joint impact of sustainability emphasis and social value orientation on accounting career perceptions. Audit: J Pract Theory 43(1):191–209
Howieson B, Hancock P, Segal N, Kavanagh M, Tempone I, Kent J (2014) Who should teach what? Australian perceptions of the roles of universities and practice in the education of professional accountants. J Account Educ 32(3):259–275
Humphrey C, Canning M, O'Dwyer B (2018) Audit quality and inspection in the Netherlands: the importance of an intellectual approach to experiential learning and practice advancement1. Maandblad Voor Accountancy En Bedrijfseconomie 92(1/2):7–19
Hurlimann A, March A, Robins J (2013) University curriculum development–stuck in a process and how to break free. J High Educ Policy Manag 35(6):639–651
IFAC (2010) International Standard on Auditing (ISA) 315. Identifying and assessing the risks of material misstatements through understanding the entity and its environment. https://www.ifac.org/sites/default/files/downloads/a017-2010-iaasb-handbook-isa-315.pdf

Jahn P, Loy T (2023) Audit in Europe—a comparison of access requirements into the audit profession across the European Union. Account Europe 20(2):244–271

Khan MA, Law LS (2015) An integrative approach to curriculum development in higher education in the USA: a theoretical framework. Int Educ Stud 8(3):66–76

KPMG (2013) The KPMG survey of corporate responsibility reporting 2013. https://assets.kpmg.com/content/dam/kpmg/pdf/2013/12/corporate-responsibility-reporting-survey-2013.pdf

Mäkinen M, Annala J (2010) Meanings behind curriculum development in higher education. Prime 4(2):9–24. http://urn.fi/urn:nbn:uta-3-1004

NBA-CEA (2022) Stip aan de horizon. https://www.cea.nl/siteassets/archief/themas/stip-aan-de-horizon/rapport-project-stip-aan-de-horizon---contouren-van-een-toekomstbestendige-en-studeerbare-opleiding.pdf

Nederlandse Beroepsorganisatie van Accountants (2024a) https://www.nba.nl/themas/herijking-beroepsprofiel/

Nederlandse Beroepsorganisatie van Accountants (2024b) Onze ambassadeurs. https://www.nba.nl/opleiding/accountant-worden/onze-ambassadeurs/

Ng AW, Leung TC, Lo JM (2017) Developing sustainability competence for future professional accountants: the integrative role of an undergraduate program. Handb Theory Pract Sustain Dev High Educ 1:119–136

Overheid (2024) https://wetten.overheid.nl/

Pizzi S, Venturelli A, Caputo F (2021) The "comply-or-explain" principle in directive 95/2014/EU. A rhetorical analysis of Italian PIEs. Sustain Account, Manag Policy J 12(1):30–50

PWC (2023) PwC study 2023: the majority of organizations already collect CSRD-related KPIs. https://www.pwc.nl/nl/themas/sustainability/esg/corporate-sustainability-reporting-directive/de-corporate-sustainability-reporting-directive-csrd-heeft-nu-al-impact.html

Siriwardane HP, Kin Hoi Hu B, Low KY (2014) Skills, knowledge, and attitudes important for present-day auditors. Int J Audit 18(3):193–205

University of Groningen (2020) Strategic plan 2021–2026. https://www.rug.nl/feb/organization/mission-vision-values/strategic-plan-2021-2026.pdf

University of Groningen (2022) Annual report. https://www.rug.nl/about-ug/profile/facts-and-figures/jaarverslagen-pdf/rug-annual-report-2022.pdf

Van Linden C, Hardies K (2018) Entrance requirements to the audit profession within the EU and audit quality. Int J Audit 22(3):360–373

Wright NS, Bhattacharjee S (2018) Auditors' use of formal advice from internal firm subject matter experts: the impact of advice quality and advice awareness on auditors' judgments. Contemp Account Res 35(2):980–1003

How to Be a Change Agent at Schools? Development and Implementation of a Virtual Course for Pre-service Teachers' Sustainability Education

Nataliia Demeshkant, Tobias Hoppe, Armin Lude, Verena Raissa Bodenbender, Katarzyna Potyrała, and Kees-Jan Van Oorsouw

Abstract Environmental impact and climate change do not recognize national borders. Therefore, collaborating with international partners in this field is paramount. In the twenty-first century, universities have a crucial role in combining teaching and research not only with the transfer of knowledge but also with the renewal of the society. From this perspective, we consider pre-service teachers as agents of positive change. In this article, we present the development and implementation of a virtual and inclusive course for Pre-Service Teachers' Sustainability Education. In cooperation with five universities across Germany, the Netherlands, and Poland, an online course curriculum was developed to enhance pre-service teachers' sustainability competences, with a focus on education for sustainable development, globalization, and world citizenship. The course is structured around three key modules: (1) Introduction to the sustainable development issues, (2) World Citizenship, and (3) Globalization. The content of each module was developed and delivered by lecturers from each partner's university. The paper reports preliminary findings on the aims and expectations following the practical testing of the course with students.

N. Demeshkant (✉)
Institute of Pedagogy, University of the National Education Commission, Krakow, Poland
e-mail: nataliia.demeshkant@uken.krakow.pl

T. Hoppe · V. R. Bodenbender
Freiburg University of Education, Freiburg, Germany

A. Lude
Ludwigsburg University of Education, Ludwigsburg, Germany

K. Potyrała
Cracow University of Technology, Kraków, Poland

K.-J. Van Oorsouw
HAN University of Applied Sciences, Arnhem, The Netherlands

W. Leal Filho et al. (eds.), *Education for Sustainable Development: The Contribution of Universities*, World Sustainability Series,
https://doi.org/10.1007/978-3-031-86985-3_10

Keywords Education for sustainable development · International virtual ESD course · Future teachers' ESD competencies

1 Introduction

Sustainable development needs change agents that drive and facilitate the transformation towards a more sustainable future (Redman and Wiek 2021). Among other professions, teachers can serve as change agents for sustainability, while education has been identified as a driving force to promote sustainable development (UNESCO 2020). In addition to the necessity of facilitating sustainable development through change agents, it represents shared responsibility that must ultimately be pursued through efforts that are crossing national borders (Bianchi et al. 2022). The EU-funded "Net(t)work for the Development and Exchange of Virtual and Inclusive Sustainability Education" (NETT-DEVISE) project addresses these necessities assuming that pre-service teachers in different European countries will be able to motivate and enable young people to engage in the transformation towards eco-friendlier and more justice-oriented societies. Virtual exchange allows for collaboration and provision of international learning experiences.

In this article, we introduce a virtual and inclusive course designed for pre-service teachers' education for sustainable development (ESD virtual course). The curriculum was developed in collaboration with five universities: two in Germany, two in the Netherlands, and one in Poland. The GreenComp Framework, provided by the European Commission, served as the basis for the course's goals, content and teaching methods. To provide international experience and facilitate cross-national perspectives, pre-service teachers from the participating universities were invited to take part in the project ESD course. Virtual learning enabled broader attendance, thereby increasing the inclusiveness of the course. The course was evaluated after the first semester through a self-reported survey completed by the ESD virtual course participants. The data were analysed using qualitative content analysis. The results from the survey were discussed regarding the further development of the curriculum.

The NETT-DEVISE project represents an example of how universities involved in teacher education can cooperate on an international level with the aim to equip teachers to be change agents for sustainability. The data collected from the first round of participants provide insights into good practices and potential improvements for the curriculum and its implementation in the university course. In the following sections, we first outline how virtual exchange facilitates international experience and collaboration. We explain how the principles of virtual exchange were applied when developing an international ESD course. Then, we introduce the ESD competence framework underlying the course curriculum, followed by a description of its key aspects. Finally, we present and discuss the course evaluation.

2 Development and Implementation of the Virtual ESD Course Curriculum

2.1 Application of Inclusive Intercultural Collaboration

For some of the participating universities involved in the NETT-DEVISE project, the development of virtual exchange seminars first became an issue for their internationalization under the impact of the COVID-19 pandemic. Until then, virtual exchanges were mainly known in foreign languages, if at all, although they had been around since the 1980s (Stevens Initiative 2019). Now, this form of collaboration suddenly became the only way to maintain international contacts. Out of this necessity, however, a new understanding of virtual exchange collaboration emerged, which continued to bring its benefits to internationalization efforts even in the aftermath of the pandemic.

This type of cooperation has various advantages. Virtual cooperation has the potential to provide an intercultural experience for those students who, for a variety of reasons, do not wish to spend time studying abroad. These reasons may be related to health, financial situation or family circumstances. For these students, the internationalization opportunities at their home universities are, or have been, generally quite limited. In this respect, there is great potential for inclusive internationalization in offering virtual exchange cooperations.

There are a lot of studies on the concept of virtual exchange. One of the well-known definition was provided by Robert O'Dowd: virtual exchange is the application of online communication tools to bring together classes of learners in geographically distant locations with the aim of developing their foreign language skills, digital competence and intercultural competence through online collaborative tasks and project work (O'Dowd 2017).

Owing to the specific characteristics of virtual exchange formats which involve cooperation in internationally mixed groups, working on the project tasks and presenting them together in English allow students to expand relevant soft skills, intercultural and digital competence. Virtual exchange thus offers students low-threshold and equal opportunity access to intercultural experiences during their studies. In this way it is an outstanding tool for the internationalization at home.

Thus, virtual exchange offers also other advantages. One aspect is the climate-friendliness of this type of internationalization. Collaboration can take place across several continents without the need for air travel. Furthermore, the cost-effectiveness of this form of exchange is a positive factor. It is also important for the virtual seminars to be integrated into the curricula of participating universities, to assure that students obtain credits. The ESD virtual course presented in this study ensured the integration in the curricula of all the partner universities. This made the participation in online taught course very attractive for students.

2.2 Competencies for Teachers as Change Agents for Sustainable Development

In NETT DEVISE project, prospective teachers were the addressees of the virtual exchange. Teachers can serve as change agents and multipliers in advancing sustainable development (Otto et al. 2020). Recognizing the critical role educators play, there has been a growing focus on integrating sustainable development principles into teacher education programs within higher education institutions (Fischer et al. 2022).

To effectively prepare teachers for this role, various frameworks have been proposed to identify and develop the necessary competencies (e.g. the CSCT model (Sleurs 2008), the KOM-BiNE model (Rauch et al. 2008), ECE Model (Bertschy et al. 2013), the UNECE model (UNECE 2012) and the Green Comp (Bianchi et al. 2022).

All of these frameworks provide guidelines on what competencies teachers should acquire and how they can be effectively instilled, they are distinct in their target audiences, structure, and specific competencies and key elements they emphasize (Wiek et al. 2011; Rieckmann 2018). For our project we choose the GreenComp Framework, developed by the European Commission. It offers a structured approach to fostering sustainability competencies and provides a set of sustainability competencies for individuals, aiming to support lifelong learning for sustainability across various sectors and levels of education (Bianchi et al. 2022).

The GreenComp Framework is built on the recognition that systemic change towards sustainability is a shared responsibility and a global necessity. The authors emphasize that this framework complements existing frameworks through "[making] a contribution to help make the competences portable and promote mobility in the EU for a full participation in European society." (Bianchi et al. 2022, p. 7).

GreenComp identifies four key competence areas crucial for sustainability:

(1) Embodying Sustainability Values: this area encompasses competencies such as valuing sustainability, supporting fairness, and promoting environmental stewardship. These competencies aim to instil a deep appreciation and commitment to sustainability principles.
(2) Embracing Complexity in Sustainability: this involves understanding and navigating the intricate and interconnected nature of sustainability challenges, fostering systems thinking and holistic approaches.
(3) Envisioning Sustainable European Futures: this competence area focuses on developing the ability to generate and work towards sustainable visions for the future, fostering forward-thinking and innovative approaches.
(4) Acting for Sustainability: this includes competencies related to taking practical actions and advocating for sustainability, translating knowledge and values into concrete steps.

Each of these areas is interlinked and equally important, providing a holistic approach to sustainability education. For example, within the area of 'embodying sustainability values', the competencies of valuing sustainability, supporting fairness, and promoting nature are designed to work together to foster a comprehensive understanding and commitment to sustainability.

According to the GreenComp Framework, "a sustainability competence empowers learners to embody sustainability values, and embrace complex systems, in order to take or request action that restores and maintains ecosystem health and enhances justice, generating visions for sustainable futures" (Bianchi et al. 2022, p. 12). This definition underlines the multifaceted nature of sustainability competencies, which include not only knowledge and skills but also attitudes and values. Integrating these competencies into teacher education programs is crucial. By doing so, future teachers are equipped with the attitudes, skills, and knowledge necessary to engage in and promote sustainable practices in both their personal and professional lives. This integration helps to ensure that teachers can effectively serve as change agents, fostering a culture of sustainability within their classrooms and communities.

2.3 Curriculum Development and Implementation in the Virtual Course for Pre-service Teachers' Sustainability Education

The virtual ESD course curriculum is the result of a collaborative effort by academic and administrative staff from the partner universities involved in the NETT DEVICE project. The experience of these universities in delivering ESD courses at their own institutions served as the foundation for developing a unified virtual ESD course for future teachers across different countries. This process involved selecting the best practices from each partner, integrating the chosen content into a coherent program, and identifying engaging methods and formats to deliver the course content effectively.

It is noteworthy that the ESD course programs developed by the project partners from Germany and the Netherlands stand out for their comprehensiveness. These programs encompass not only ecological and natural aspects but also social and economic dimensions. Both the content and the pedagogical approaches employed in these courses are highly engaging, prioritizing active learning, independent problem-solving, collaborative teamwork, and the cultivation of research skills for producing reports on sustainable development issues. The exemplary practices of the partner universities were analyzed, leading to the development of an international remote ESD course tailored for future teachers.

As a result of close cooperation, a remote course program was developed consisting of three modules: (1) Basics of ESD, (2) World Citizenship, and (3) Globalization (Fig. 1).

The content of each module was prepared and delivered by lecturers from each partner university. The first module of this course provided an introduction to basic concepts related to sustainable development, including sustainable development (SD)

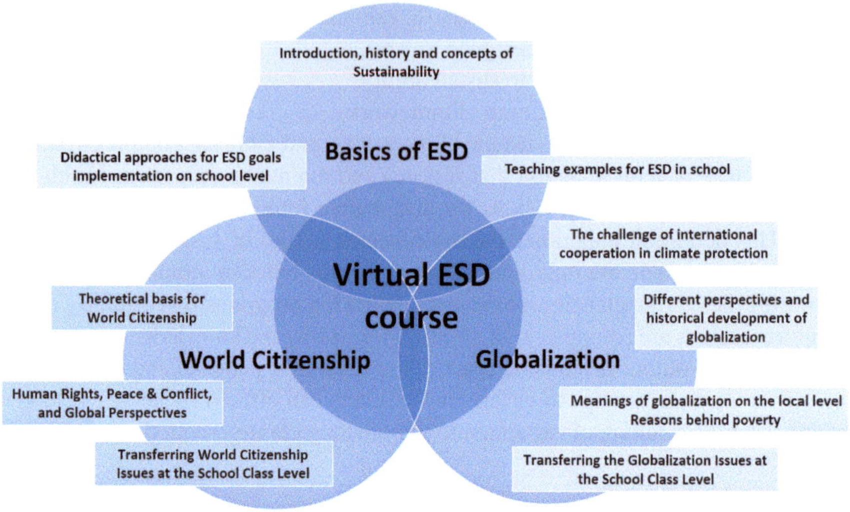

Fig. 1 Structure of the virtual course for pre-service teachers' sustainability education

goals, methods, and historical approaches to SD concepts. It also addressed how to apply various SD topics in specific educational settings. The assumption was that active participation in discussions, completion of assignments, and the utilization of provided resources would contribute to a well-rounded and practical understanding of how to integrate sustainability education into students' teaching repertoires.

The second thematic module focused on topics related to world citizenship, emphasizing the promotion of well-being not only for oneself but also in contributing to the welfare of others. It was presumed that when students who are future citizens develop skills enabling self-regulation and a disposition that promotes social contribution, they will be more likely to engage in prosocial behaviour. The module includes three main topics. The first is the theoretical basis for world citizenship, which encourages learners to explore and define global citizenship and its implementation in educational environments. The second covers human rights, peace and conflict, and global perspectives, emphasizing the importance of understanding these concepts within the context of world citizenship. Finally, the third section focuses on transforming world citizenship issues at the classroom level, where students discussed practical ways to integrate global citizenship topics into school curricula and classroom activities.

The third module of the ESD course addresses globalization issues, helping students understand sustainability in both global and local contexts. It involves such issues as measuring prosperity, sustainable livelihoods, and quality of life. It also explains concepts related to a sustainable economy and the challenges of international climate negotiations. The module was designed with the expectation that students would develop the skills to articulate the potential and challenges associated with alternative prosperity indexes and to facilitate the exchange of ideas and resources on addressing these measurement issues with school students. The module covers four main topics: the challenges of international cooperation in climate protection; an introduction to various perspectives and the historical development of globalization; the fundamental causes of poverty; and the implications of globalization at the local level, particularly in relation to ecological footprints. The final section addresses practical strategies for integrating globalization issues into school curricula and classroom activities.

A key feature of the learning structure across all the modules is the integration of independent study, group discussions, and practical weekly assignments. Students are expected to engage with a variety of resources, including academic articles, videos, and websites, to enhance their understanding of sustainable development issues. Some assignments must be completed in mixed international groups to ensure cross-country discussions during preparation.

Additionally, students began the course with an introductory module on intercultural communication before delving into the ESD content. This segment aimed to unify the international cohort of students participating in the virtual ESD course into a cohesive team. It also sought to familiarize them with potential cultural differences, strategies for overcoming these differences, the specifics of working in international groups, and the principles of collaboratively developing ESD projects within such diverse teams.

The structure and content of the ESD virtual course were designed to cultivate key competencies for sustainable development in pre-service teachers. The course aimed to stimulate their interest in the relevant issues and to enhance their ability to implement sustainable development goals both in their daily lives and in their professional roles as educators.

3 Evaluation of the ESD Virtual Course and Further Development

3.1 Methods and Materials

The study is a qualitative analysis of student feedback on the ESD virtual course designed for pre-service teachers sustainability education. It aims to investigate students' self-reported learning experiences from the pilot testing of this course. The presented results are based on feedback from German, Dutch, and Polish students who participated in the course from January to April 2024 ($n = 19$ participants). The study is scheduled to be repeated in the coming autumn semester of 2024, with the aim of increasing the study group to 48 participants. All the participants who completed the survey are studying to become teachers. Of these, 75% aspire to teach in primary schools, while 25% aim to teach in secondary schools. Additionally, 80% of the participants already have teaching experience.

The analysis was guided by the following research question: *What learning outcomes did participants experience during the four-month ESD online course?*

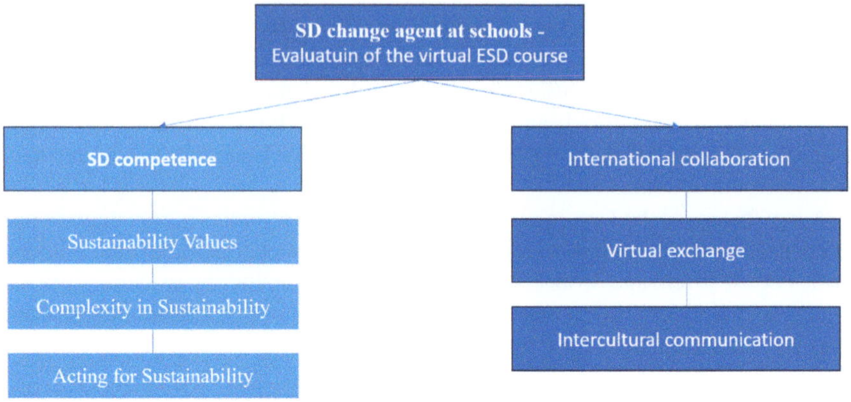

Fig. 2 Categories and related subcategories according to the theoretical concept of the study

Data for the study were collected through a self-reported survey. Student teachers were given an electronic questionnaire to reflect on their specific learning experiences after completing the course. The evaluation process provided insights into the course's effectiveness and substantive relevance. It was anticipated that the results from student feedback would help identify deficiencies or gaps in the course content, thereby enabling targeted improvements.

Thematic analysis (Braun and Clarke 2006) was employed to identify, analyse, and report themes within the data. The obtained data were reviewed by the researchers for accuracy, resulting in the development of coding categories and sub-categories aligned with the study's objectives. The categories and sub-categories were developed inductively and compared to the key descriptors for sustainable competencies in the GreenComp Framework (Bianchi et al. 2022). Figure 2 illustrates the emerging themes based on the theoretical concept of the study, which includes the GreenComp Framework, and the related sub-categories identified during the analysis.

3.2 *Results*

Following the theoretical framework of this study, the learning outcomes experienced by participants during the four-month online course were analyzed using two categories: SD competency and international collaboration. The answers to the survey questions obtained from respondents made it possible to distinguish the codes that complement the relevant subcategories. Table 1 presents a summary of the results, organized by categories, sub-categories, codes, and examples of respondents' responses.

Analysis of the obtained data shows that participation in the virtual ESD course fostered the development of key *SD competences*, as outlined in the GreenComp Framework developed by the European Commission.

Table 1 Summary of the results from the evaluation of the virtual ESD course based on students' opinions

Categories	Sub-categories	Codes	Examples of student responses
SD competence	Sustainability values	Words associations with virtual ESD course content	*Planetary Boundaries, Mutual respect and understanding, Critical thinking, Hope*
		Important for school students to learn about sustainability at school	*… They can pass on their awareness to those around them, they can take action themselves, they learn to take responsibility themselves, knowledge about the problem is the first step towards a better world*
	Complexity in sustainability	Reflections about SD issues after the virtual ESD course completing	*…. There is so much material and theory on the subject of sustainability and that it is much bigger than I always thought*
		Awareness of the importance of SD issues in everyday life	*…. Only now I 've realized the impact of our way of life has on everyone's lives.*
	Acting for sustainability	Readiness to take SD action in everyday life	*Buy locally, Green travel, Only consume when necessary, Vegan diet, Sustainable clothing* *Use a bike or public transport instead of cars* *Switch off lights and radiators, Sort out the rubbish, Avoid takeaway food, Be mindful that all of our actions have a global impact*
International collaboration	Virtual exchange	Opportunity to study with international students form home setting	*… The main advantages are the opportunity to connect with people from diverse cultural and geographical backgrounds without the need to travel. This not only reduces our carbon footprint but also makes the learning experience more inclusive and accessible*
		Flexibility of learning offers	*… I could engage with course materials and discussions at my own pace, which allowed me to balance my studies with other commitments*

(continued)

Table 1 (continued)

Categories	Sub-categories	Codes	Examples of student responses
	Intercultural communication	Exchange of experiences with international peers	*… I was able to gain different perspectives on sustainability challenges and solutions from other countries peers, which broadened my understanding of global issues*
		Communication and teamwork skills	*… Collaborating on projects with peers from diverse backgrounds also helped me develop better communication and teamwork skills*

In their feedback, students highlighted several values associated with the course related to SD issues, namely Planetary Boundaries, Mutual Respect and Understanding, Critical Thinking, and Hope. This result suggests that the course effectively fosters the development of sustainability values. Significantly, students, as future teachers, recognize the importance of teaching sustainability in schools. They emphasize that this education helps develop a sense of responsibility for the world in the younger generation.

The key competence of Complexity in Sustainability is illustrated through students' reflections after completing the course. Most participants were surprised by the extensive material and theory on sustainability, discovering it to be far broader than they had previously thought. Some noted that they only now realize the profound impact of human activities on everyone's lives. These reflections demonstrate that participation in the virtual ESD course fosters critical thinking and holistic approaches to SD issues.

An equally important competence is the willingness to take practical actions for sustainability and to apply the gained knowledge in everyday life and professional work. When asked about specific activities to make the world a more sustainable place, each student listed a number of actions they undertake in their daily lives (Table 1).

The second category of analysis was *International Collaboration.* Students highly appreciate the opportunity to collaborate on SD issues virtually with peers from other countries. They highlighted that participating in the virtual ESD course has been an incredibly rewarding experience. One of the main advantages is the opportunity to connect with people from diverse cultural and geographical backgrounds without the need to travel. An important advantage of the virtual ESD course was that students had access to course materials and discussions at their own pace, allowing them to balance course meetings with other commitments.

Undoubtedly, an important sub-category of the respondents' learning outcomes is *Intercultural Communication.* Students emphasized that one of the most valuable aspects of the virtual course was the exchange of experiences with international peers.

Engaging with students from different countries allowed them to observe how sustainability challenges are addressed in various contexts. During discussions and group tasks, students shared their local practices, success stories, and the obstacles they face in implementing SD activities in their home regions. This exchange of knowledge was incredibly eye-opening and significantly enriched their understanding of sustainability on a global scale.

The students' responses revealed a sub-category related to the improvement of *communication and teamwork skills* as a key outcome of participating in the virtual ESD course.

The concept of communication encompasses listening, understanding different perspectives, exchanging points of view, being open, and respecting others' ideas. Mutual understanding was mentioned frequently as well. Additionally, engaging in dialogue helped participants develop better communication and teamwork skills.

3.3 Discussion and Conclusions

The past twenty years have witnessed a rise in the study and application of Higher Education for Sustainable Development (Lozano et al. 2017). The main focus was on educating future generations of professionals in sustainability, which fostered the integration of SD content into university courses. Most of these initiatives were centered on the development and delivery of competencies with an emphasis on sustainability. Future teachers, as key agents of social transformation and progress, should be particularly well-prepared to integrate the concept of SD into their teaching and educational work (Pegalajar-Palomino et al. 2021). However, the situation regarding teacher training education programs, particularly in relation to the challenges of globalization and efforts toward a better future, varies across different countries (Evans et al. 2021). This was also confirmed by the project's experience, particularly in developing the joint international virtual ESD course for future teachers. Reaching consensus required identifying previous good practice and finding common points of support for the programme.

The analysis of the published examples indicates the need for international cooperation in sharing good practices and developing new, innovative didactic and substantive solutions. Therefore, the collaboration undertaken to develop a program for educating future teachers and to implement its elements was necessary. However, studies have shown that achieving SD goals, in connection with providing high-quality education for global citizenship within teacher education has received little attention in the literature (Nketsia et al. 2020). Nketsia et al. (2020) suggested using a conceptual framework based on four important themes: lifelong learning, equity, inclusion, and quality. Regarding the definition of quality in teacher education, they clarified—that it can be supported through relevant knowledge, skills, and practices that equip students with effective teaching methods to guide curriculum and instruction. Several studies have identified interconnected critical components. The obtained results aligns significantly with previous innovations while introducing a new arrangement of content within the adopted conceptual framework, along with new approaches to achieving educational goals in connection with technological advancements and progress in understanding how students learn.

The results confirm the relationship between students' knowledge, awareness, and attitudes, and highlight that the joint negotiation of the world by individuals-based on experience and social relations—is a crucial element in promoting and developing the concept of participatory pedagogy. A good starting point was the active involvement of both teachers and students in the process of acquiring and deepening knowledge, followed by knowledge sharing, building a network of engaged individuals, and collaborative activities. This approach has been promoted previously and proven to be effective (Potyrała 2017).

The results pertain to students' knowledge of the basic concepts of sustainable development, as well as their declarations of pro-environmental behaviour and reflections on the factors that motivate young people to take action for the benefit of both the local environment and the entire planet. Knowledge can serve as a starting point for awareness, but actual actions often differ significantly from the declared behaviours (Potyrała et al. 2023). That is why mutual relationships, cooperation, meetings, and the shared motivation to engage in project tasks proved to be so important, serving as a measurable outcome of the education program's implementation.

The conducted research leads us to conclude the need to restructure the education system for sustainable development at all levels (Evans et al. 2021; Potyrała et al. 2023). Current findings allow for the formulation of new didactic recommendations for Education for Sustainable Development (ESD). A one-way transmission of knowledge is insufficient as a foundation for shaping students' activities and attitudes. Instead, practices such as student collaboration in solving real-world tasks, situational learning, design thinking, and engaging in critical reflection are effective ESD strategies, as confirmed by our findings.

References

Bertschy F, Künzli C, Lehmann M (2013) Teachers' competencies for the implementation of educational offers in the field of education for sustainable development. Sustainability 5:5067–5080. https://doi.org/10.3390/su5125067

Bianchi G, Pisiotis U, Cabrera Giraldez M (2022) GreenComp the European sustainability competence framework (No. JRC128040). Joint Research Centre (Seville site)

Braun V, Clarke V (2006) Using thematic analysis in psychology. Qual Res Psychol 3(2):77–101

Evans N, Inwood H, Christie B, Ärlemalm-Hagsér E (2021) Comparing education for sustainable development in initial teacher education across four countries. Int J Sustain High Educ 22(6):1351–1372. https://doi.org/10.1108/IJSHE-07-2020-0254

Fischer D, King J, Rieckmann M, Barth M, Büssing A, Hemmer I, Lindau-Bank D (2022) Teacher education for sustainable development: a review of an emerging research field. J Teach Educ 73(5):509–524. https://doi.org/10.1177/00224871221105784

Lozano R, Merrill MY, Sammalisto K, Ceulemans K, Lozano FJ (2017) Connecting competences and pedagogical approaches for sustainable development in higher education: a literature review and framework proposal. Sustainability 9(10):1889. https://doi.org/10.3390/su9101889

Nketsia W, Opoku MP, Saloviita T, Tracey D (2020) Teacher educators' and teacher trainees' perspective on teacher training for sustainable development. J Teach Educ Sustain 22(1):49–65. https://doi.org/10.2478/jtes-2020-0005

O'Dowd R (2017) Virtual exchange and internationalizing the classroom. Training Lang Cult 1(4):8–24. https://doi.org/10.29366/2017tlc.1.4.1

Otto IM, Donges JF, Cremades R, Bhowmik AK, Hewitt RJ, Lucht W, Schellnhuber HJ (2020) Social tipping dynamics for stabilizing earth's climate by 2050. Proc Natl Acad Sci 117(5):2354–2365. https://doi.org/10.1073/pnas.1900577117

Pegalajar-Palomino MDC, Burgos-García A, Martinez-Valdivia E (2021) What does education for sustainable development offer in initial teacher training? A systematic review. J Teach Educ Sustain 23(1):99–114. https://doi.org/10.2478/jtes-2021-0008

Potyrała K, Tuszyńska L, Makiewicz M (2023) Pedagogical students as ecological education animators. Forum Pedagogiczne 13(1):409–423 (in Polish)

Potyrała K (2017) iEducation. Synergy of new media and didactics. Wydawnictwo Naukowe UP, Kraków (in Polish)

Rauch F, Streissler A, Steiner R (2008) Kompetenzen für bildung für nachhaltige entwicklung (KOM-BiNE). Konzepte und anregungen für die praxis. Bmuk

Redman A, Wiek A (2021) Competencies for advancing transformations towards sustainability. Front Educ 6:1–11. https://doi.org/10.3389/feduc.2021.785163

Rieckmann M (2018) Chapter 2—Learning to transform the world: key competencies in ESD. In: Leicht A, Heiss J, Byun WJ (eds) Issues and trends in education for sustainable development. UNESCO, Paris, pp 39–59

Sleurs W (ed) (2008) Competencies for ESD (Education for sustainable development) teachers. A framework to integrate ESD in the curriculum of teacher training institutes. Comenius 2.1 project 118277-CP-1-2004-BE-Comenius-C2.1. https://unece.org/fileadmin/DAM/env/esd/inf.meeting.docs/EGC/CSCT_Handbook_11_01_08.pdf

Stevens Initiative (2019) 2019 Virtual exchange impact and learning report. https://www.stevensinitiative.org/resource/virtual-exchange-impact-and-learning-report/

UNECE Competences for ESD (2012) Learning for the future. Competences in education for sustainable development. https://unece.org/info/Environment-Policy/Education-for-Sustainable-Development/pub/3098

UNESDOC Digital library (2020) Education for sustainable development: a roadmag. https://unesdoc.unesco.org/ark:/48223/pf0000374802

Wiek A, Withycombe L, Redman CL (2011) Key competencies in sustainability: a reference framework for academic program development. Sustain Sci 6(2):203–218

Implementation of Vocational Education and Training for Sustainable Development Through Cooperation Between Research and Practice

Robert Hantsch and Harald Hantke

Abstract The 17 Sustainable Development Goals (SDGs) provide a framework for sustainable development, yet companies encounter significant challenges in their implementation. Vocational education and training for sustainable development (VETSD) is crucial in equipping individuals and organizations to address these challenges. This paper examines the integration of VETSD within the vocational education and training system in Germany and the role of universities in this process. Based on the authors' experiences with pilot projects through the Federal Institute for Vocational Education and Training (BIBB), the paper presents these projects as a form of design-based research (DBR) that fosters collaboration among practitioners, researchers, and policymakers to develop innovative solutions. The article is structured into five sections: (1) a definition of design-oriented research in vocational education, (2) characteristics of pilot projects within DBR, (3) an overview of pilot projects focusing on "Vocational Education for Sustainable Development" from 2012 to 2023, (4) presentation of two projects from the federal programm "Sustainability in the occupation" (Nachhaltig im Beruf), and (5) reflections on the collaboration between science and practice. This examination illustrates how VETSD can be effectively integrated into vocational education and training, fostering sustainable practices in enterprises.

Keywords VETSD · ESD · Vocational education and training · Research and practice · Design based research

---

R. Hantsch (✉)
Institute of Business and Economic Education, University of Rostock, Rostock, Germany
e-mail: robert.hantsch@uni-rostock.de

H. Hantke
Department of Economic Didactics at the Institute of Social Science Education, Leuphana University Lüneburg, Lüneburg, Germany
e-mail: harald.hantke@leuphana.de

W. Leal Filho et al. (eds.), *Education for Sustainable Development: The Contribution of Universities*, World Sustainability Series,
https://doi.org/10.1007/978-3-031-86985-3_11

1 Introduction Vocational Education and Training for Sustainable Development (VETSD) in Germany

Since the Brundtland Report of 1987, the concept of sustainable development has increasingly become ingrained in public awareness. The definition emphasizes both intra- and intergenerational perspectives as well as the necessity of considering economic and social aspects alongside environmental protection (Hauff 1987, p. 46). This comprehensive approach shapes the guiding principle of sustainable development. The "Fridays for Future" movement, upcoming laws on packaging reduction, and securing social standards in supply chains are all examples of this. The role of education in sustainable development is crucial. Specifically, vocational education and training (VET) is identified as a critical area where key innovations take place during transformation processes (Melzig et al., 2021; Müller et al. 2023). Consequently, vocational education and training for sustainable development (VETSD) aims to enable people at all levels, from skilled laborers to management, to assume responsibility, operate efficiently and sustainably, and shape globalization in a fair and socially compatible manner while considering values and conflicts of interest (Dietrich et al. 2007). The modernization of the standard vocational training positions (Standardberufsbildpositionen) of the Federal Institute for Vocational Education and Training (BIBB) in Germany in 2021 demonstrates that there's no longer a question of whether sustainability should be included in VET, but how it can be implemented practically and effectively. A key factor in accomplishing this is the qualification of educators, the transformation of companies and schools into sustainable learning environments, and the strengthening of cooperation between schools and businesses. As a result, intensive research and design processes are needed, for which neither practical concepts nor scientific theories, models, or empirical findings will provide guidance at some point. To address these challenges, cooperative approaches between science and practice should be used (Hemkes et al., 2017). In reality, the concept that scientists and practitioners work together on practical problems is not part of mainstream scientific practice, nor is it part of universities' self-perception (Nowotny et al., 2001).

This article investigates the role of design-based research in integrating research, practice, and innovation to advance VETSD in Germany. An analysis of the conceptual and methodological differences between design-based and conventional research approaches is presented in the first section of the article. We establish a theoretical framework in this section in order to understand how design-based research can address the challenges of educational innovation. Following that, the article traces the evolution of pilot projects in Germany that employed design-oriented research to foster VETSD. The initiatives, which have spanned the past decade, have generated practical solutions as well as theoretical advancements, providing significant insights into the development of VETSD. Continuing, the article discusses two contemporary design-oriented research projects funded under the federal program "Sustainability in the Occupation" (Nachhaltig im Beruf). These projects are centered on the professional development of VET-educators. A comparative analysis of these projects is conducted in the article's concluding section, which highlights the shared characteristics and divergent approaches of the projects. A critical reflection on the lessons learned is included as part of a synthesis of key lessons learned, offering recommendations for future research and practice in the field of sustainable vocational education.

2 Design-Based-Research

In educational research, radical innovations rarely succeed, and evolutionary innovations that better meet educational context needs struggle to gain traction due to a lack of quick successes and support (Reinmann, 2005). While science is often viewed as an innovator, it is rarely associated with education as opposed to the natural or engineering sciences. It is unclear whether this is due to the complexity of context-dependent phenomena, limited funding, or other factors.

In the following section, we will focus on the science-practice relationship in the context of debates surrounding "Mode 2" knowledge production (Gibbons et al., 1994). This relationship has long been a subject of scientific discussion. Within conventional educational research, the roles and assignments between science and practice seem clearly defined, often characterized by a significant distance between the two. Science views practice as a research object, whereas practice relies on science for legitimacy or for new insights (Dilger & Euler, 2018). It is worth questioning whether this remains true in light of increasing skepticism toward science (Mede et al. 2020). In general, it is unclear which scientific findings are used and for what purposes, and which are rejected for many reasons. In light of current crises, this undermines the supposedly clear assignment of roles. No longer can we assume that the public will follow scientific findings inevitably. As an example, Kahan et al. (2010) demonstrated that individuals with hierarchical and individualistic worldviews are more likely to reject the scientific consensus on climate change, likely because they perceive environmental risks as a justification for restricting commerce and industry. In this regard, cognitive enlightenment campaigns fall short, as they merely present scientific information without making it accessible to different audiences. In other words, scientific skepticism is not merely the result of knowledge gaps. It is essential to portray science realistically, including its limitations, uncertainties, and logic. In this way, not only would we be able to clarify the quality of scientific findings, but we would also be able to emphasize their limitations. There is no panacea in science; it cannot resolve certain issues, such as the meaning of life, and must accept uncertainty as a constant companion (Bogner, 2021; Parthey-Müller et al., 2023).

While this article cannot provide a comprehensive discussion of how to address scientific skepticism, it should be clear that communication between science and practice is highly complicated, particularly in the context of the climate crisis. Consequently, science must find ways to integrate its findings effectively and participatory into public discourse, development, and the tensions that accompany such integration. Especially true in educational research, which is dominated by "belief mode" research, which establishes or refutes new ideas and supports or refutes claims, based on the verification and justification of knowledge. This is undoubtedly a significant aspect of scientific thinking and action, yet one might question whether the exclusive application of this mode is adequate for addressing today's societal challenges (Bereiter & Scardamalia, 2003). Alternatively, Bereiter and Scardamalia (2003) conceptualize a "design mode" that includes a wide range of activities focused on producing and improving knowledge: theorizing, inventing, designing, identifying promising ideas, and seeking improved approaches. Design modes have been discussed and conceptualized in various ways since the 1990s (Reinmann, 2005). We utilize the term "Design-Based Research" (DBR) as a synonym for various design-oriented approaches in this contribution. In essence, DBR is not a specific research method but rather a methodological framework (Bakker, 2018). Although DBR research is becoming increasingly diverse, there are some common characteristics:

- DBR is based on the identification of discrepancies in educational practice, whether they are problems or goals that differ from current practice (McKenney and Reves 2012; Reinmann, 2022).
- DBR employs interventions to resolve this discrepancy. There are various types of interventions, such as educational programs, curricula, teaching methods, and technological tools (Bakker, 2018; Reinmann, 2022).
- DBR follows an iterative cycle that involves repeated design, development, testing, analysis, and evaluation (McKenney & Reeves, 2018; Bakker 2018; Reinmann, 2022).
- The development of interventions is closely collaborative between educational research and the respective educational practice (McKenney & Reeves, 2018).
- DBR is theoretically based, supporting design decisions, inspiring interventions, and providing a basis for evaluations and further development (van den Akker, 1999).
- DBR can include empirical research at all stages, including initial situation analysis and formative evaluations during design (McKenney & Reeves, 2012; Bakker, 2018).
- DBR results in both practical improvements in educational practices and theoretical insights (Anderson & Shattuck, 2012).
- To summarize: DBR is about education as it could be (Bakker, 2018).

3 Pilot-Projects: Vocational Education and Training for Sustainable Developement

Germany's pilot projects in VET were a significant forerunner and context for similar research approaches that evolved into Design-Based Research (DBR). Initially designed to promote innovation in education practice, pilot projects have been a part of German education policy since the 1970s (Holz, 1991; Euler and Sloane 1998. In pilot projects, medium-range innovations are tested in specific areas of VET and, upon success, are transferred to other areas (Euler and Sloane 1998). Although pilot projects and DBR emerged historically as distinct entities, their overlap and mutual influence in vocational education research have risen since the 2000s. Researchers such as Sloane, Euler, and Kremer have incorporated elements of the DBR discourse into their work (Sloane 2014; Euler 2014; Kremer et al. 2023). There are a wide range of methodological works and concrete studies that combine traditional design-oriented research in pilot projects with. This understanding is also reflected in various pilot projects on VETSD that have been funded by BIBB using resources from the Federal Ministry of Education and Research (BMBF) since 2010. In our next section, we will differentiate these pilot projects into three phases: (1) Initiation, (2) Consolidation, (3) Diffusion.

The first series of pilot projects specifically aimed at VETSD was conducted as part of the UN Decade (2004–2015). The first phase of the program can be referred to as the "initial phase." Through these pilot projects, solution concepts for designing and developing VETSD were developed in a range of industrial fields (metals/electrical, construction, housing, chemicals, and nutrition) with a focus on renewable energy. The goal was to raise awareness of sustainability among trainees and VET-educators. As a consolidation phase, the following period covers the pilot projects from 2015 to 2021. The focus shifted during this period to the development of concrete industry- and occupation-specific concepts and teaching/learning offers aimed at specifically improving trainees' sustainability-related occupational competence and the design of sustainable learning environments (Melzig et al. 2021). This was also accompanied by a significant expansion of the program structure, both in terms of content and organization. At the program level, cross-project scientific support was integrated to generate insights that will support the structural implementation of VETSD in Germany. In addition, an advisory board of academics, practitioners, and politicians was formed to disseminate the results of the VETSD pilot projects and initiate structural changes. In addition, an advisory board of academics, practitioners, and politicians was formed to disseminate the results of the VETSD pilot projects and initiate structural changes. Across three program lines, 45 university and research partners collaborated closely with over 280 practice and strategy partners. Program line 1 (Development of Sustainability Competencies in Commercial Occupations) concentrates on the promotion of sustainable competences in VET, particularly in wholesale, foreign trade, and logistics. VETSD was implemented in six pilot projects in companies where education settings are designed. As part of the cross project scientific support, a competence model and a basic didactic model for VETSD were developed (Casper et al. 2021). Program line 2 (Design of Sustainable Learning venues) aimed to develop and design sustainable learning environments in the workplace, focusing on the comprehensive transformation of educational settings. In six pilot projects, various approaches were pursued to implement VETSD within organizational development, learning and working environments, and teaching–learning situations. The scientific support was crucial for systematizing these approaches and developing indicators for sustainable learning environments (Hantsch and Weber 2023). Program

Line III (Developing Sustainability Competencies in the Food Sector) aims to develop sustainability-oriented concepts for competency development in the food industry. It supports the development and dissemination of didactic teaching and learning arrangements in specific occupational fields. Its purpose is to enable in-company trainers to develop VETSD learning arrangements. As a result, the competency framework and didactical recommendations from Program Line I can be compared and further developed. This enables the models to be tested and improved across sectors (Strotmann et al., 2023). The third period can be described as a "diffusion phase." During the "consolidation phase," efforts were made to facilitate the transfer of pilot project results, but achieving complete structural integration as sought by the UNESCO World Action Program "Education for Sustainable Development" proved challenging. It became evident that a timeframe and differentiated approach were necessary. It was challenging to apply insights gained from the pilot projects to a wide range of occupational fields and policy and structure aspects of VET. While some transfer activities succeeded, others faced barriers or lacked suitable starting points. In order to address this challenge, the program "VETSD Transfer" (BBNE Transfer) was designed to disseminate and sustain positive outcomes from pilot projects. In contrast to its predecessor, the program focused on practical implementation. However, universities and research institutions still played a vital role with their theoretical and design expertise. In this diffusion phase, the federal program "Sustainability in the Occupation" plays a significant role. In addition to drawing on insights gained from previous pilot projects, it no longer calls itself a pilot project. It is intended to integrate the new standard vocational training positions "Environmental Protection and Sustainability" into operational, inter-company, and extra-company training practices. As with the VETSD Transfer program, the goal here is to enhance the sustainability-related competencies of VET-educators through the development, implementation, and dissemination of VETSD qualifications and VETSD networks.

The following chapter introduces two DBR-based projects from the program, "Sustainability in Insurance and Consulting Professions" („NaVeBb: Nachhaltigkeit in Versicherungs- und Beratungsberufen") and "Sustainability in the Metal and Electrical Industry" (NiME: Nachhaltigkeit in der Metall- und Elektroindustrie).

4 DBR-Projects in the Federal Program "Sustainability in the Occupations"

The new standard vocational positions promote independent, reflective, and socially responsible individuals. The responsible use of digital media, data protection, and sustainable value chains are now integral parts of VET. This modernization is also based on the various pilot projects conducted by BIBB on behalf of the BMBF. In the federal program "Sustainable in the Occupation—Future-Oriented Training (NIB)," the standard vocational training position "Environmental Protection and Sustainability" is explicitly addressed. Through the program, sustainability will be promoted as an integral part of all occupational actions. It contributes to the future-oriented design of VET.

The two DBR-based projects "NaVeBb" and "NiME" are part of the program and will be discussed below. Special emphasis is placed on the interaction between science and practice as a necessary condition for designing innovation.

4.1 Sustainability in Insurance and Consulting Professions (NaVeBb)

There are two main focal areas of NaVeBb, distinguished by the occupations "insurance and financial clerks (KFV)" and "tax clerks (SFA)". The project pursues the following goals:

- Development of occupational-specific didactic materials for in-company training.
- Development, testing, and nationwide implementation of digital qualification modules for trainers and educators.
- Development of occupational-specific recommendations for designing exam tasks that integrate sustainability.
- Design, testing and nationwide implementation of digital qualification modules for examiners to design sustainability-related exam tasks.

Each focus area is managed by a tandem of a practice partner and a scientific partner (see Fig. 1). In the focus area "insurance and financial clerks", the German Association for Vocational Training in Insurance (BWV e.V.) acts as the practice partner and the University of Applied Labour Studies (HdBA) as the scientific partner. The focus area "tax clerks" is managed by the ECOVIS Academy as the industry partner, and the Institute for Business and Economic Education at the University of Rostock as the scientific partner. The project is also supported by a strategic and transfer advisory board. The strategic board consists of consortium members and industry experts who provide guidance, promote nationwide implementation, and ensure continuous evaluation. The transfer advisory board, which is composed of members of the strategic board and experts from other industries and political multipliers, serves as a platform for critical discussion and reflection. In addition, it provides guidance on how to adapt modified products and training programs to other occupations such as bank clerks.

In NaVeBb, science-practice collaboration is conceived as an "innovation partnership." Practice does not merely serve as the subject of research, but also as an equal partner in design and research. It emphasizes the importance of closely working with stakeholders rather than simply designing and developing for them. Despite the differences in their goals and focuses, science and practice benefit from one another's contributions. Using the DBR as a methodological framework improves both theoretical science and problem-solving skills.

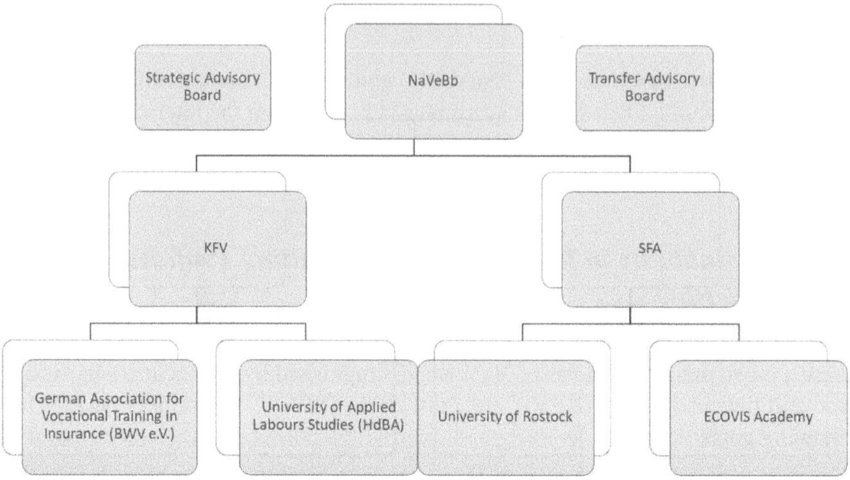

Fig. 1 Cooperative structure of NAaVeBb

Designers must investigate what they design and design what they investigate. The complexity of these systems cannot be separated in practice. There are several DBR models that help structuring, documenting, and communicating DBR projects. NaVeBb uses McKenney and Reeves' (2018) three-phase iterative process model, which consists of the following phases:

- Analysis and Exploration
- Design and Construction
- Evaluation and Reflection
- Furthermore, continuous implementation and dissemination are parallel dimensions.

In March 2024, the analysis and exploration phase of NaVeBb began. Analysis consists of systematically analyzing the issue, its causes, potential solutions, and conditions using an empirical, descriptive, and explanatory approach. Exploration involves learning from others, identifying best practices, and identifying effective solutions. The project employed various methods, including qualitative interviews, focus groups, and quantitative surveys with industry experts, trainers, and managers to gain a comprehensive understanding of local challenges and conditions. The strategy advisory board discussed findings, formulated strategic recommendations, and identified transfer potentials. During the exploration phase, the materials developed during the pilot projects were analyzed extensively for their adaptability to the two occupations. Additionally, sector-specific networks, such as those of sustainable tax consulting firms, were tapped for their extensive sustainability knowledge.

Design and construction involve determining the local requirements for a specific design and implementing it through a step-by-step revision process. Several general design principles are applied, such as various frameworks, tools, and patterns. Due

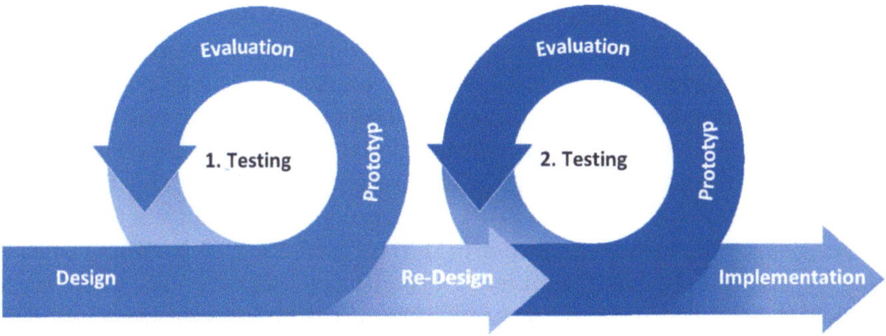

Fig. 2 Iterative design in NaVeBb

to the limited project duration, the design and construction phase runs parallel to the analysis and exploration phases. As a result of interviews and exchanges with practice partners, specific local requirements will be identified, leading to the creation of "personas" to help us better understand the needs, behaviors, and goals of the target group. This understanding will be incorporated into the prototype. Using Design-Based Research (DBR), two iterative loops are planned in total, as shown in Fig. 2. In the first testing session, scientific and practice partners will collaborate closely in tandem to test the prototype with VET trainers.

Each testing phase is followed by an evaluation phase to gain insight into the effectiveness of the approach. Methods such as feedback, meeting notes, and interviews are used to systematically capture participants' experiences and feedback. By analyzing this data, we can evaluate how well the prototype meets the defined objectives. In parallel, we are preparing for the national transfer following the lead of our practice partner. An effective public relations program is implemented, events are organized, and interactions with key stakeholders (e.g., associations and chambers) are promoted.

4.2 Sustainability in the Metal and Electrical Industry (NiME)

NiME addresses occupations in the metal and electrical industries, specifically focusing on training for "Industrial Clerk," "Industrial Mechanic," and "Electronics Technician for Automation Technology." In essence, NiME combines the learning task concept from the pilot project NaReLe with focusing on the occupation of "Food Technician," which also deals with sustainability in an industrial context. This concept should be applied to the metal and electrical industries, as these sectors play an important role in sustainable development because of their energy and emission intensity. NIME is based on a transformative view of VET. As such, VET is seen as a space for possibilities. It is associated with searches and negotiations for transitions from unsustainable routines to sustainable ones. The juxtaposition of these two routines creates an environment in which "transformation" occurs. In this space, two fundamental premises regarding occupational action are contrasted: economic and household-oriented occupational actions. Household-oriented actions serve as the target of sustainable occupational practice. Household is defined here as "preserving the ecological, economic, and social substance that society wants to utilize sustainably" (Müller-Christ, 2010, 179). By contrast, economic action pursues "acquisition for the sake of acquisition" (Müller-Christ 2001, 317) without considering its sustainability. We are currently experiencing contradictions, tensions, or conflicts of goals during the early phases of the sustainability transformation. This can catalyze the transformation process (WBGU, 2011). This requires, however, that the actors engaged in occupational practices have the ability to manage contradictions, tensions, and conflicts. It is in this sense that VET plays an important role, since it is directly related to the world of work, a central area of sustainability action (UNESCO, 2015).

NiME is based on an understanding of political education for sustainable development (VETSD). This approach is characterized by conflicting goals and contradictions in the workplace, such as the tension between economic efficiency and sustainability (Fischer et al. 2021; Singer-Brodowski and Holst 2022, 109). As Fischer et al. (2021, 93) describe it, managing contradictions is the "DNA" of VETSD. As a result, the political dimension is assigned a central role in VETSD. Building on this transformative, politically educative understanding of VETSD, the didactic concept of resonance-oriented learning tasks from the pilot project NaReLe is being adapted to NiME. This concept enables both VET-educators and trainees to experience resonance in relation to sustainable occupational actions. Resonance experiences, as defined by Rosa (2016) in his resonance theory, refer to interactions between subjects and the world that involve mutual responsiveness. Resonant subjects are not only influenced by the world; they actively participate in it and can thus potentially transform it or be transformed by it. In the context of (vocational) education, it is about creating educational opportunities for learners to engage reflectively and creatively with the question of how they locate themselves between the previously described household-oriented and economic occupational actions in their everyday work (Hantke, 2018).

Specifically, the NaReLe learning task concept to be transferred to occupations in the metal and electrical industry is as follows (Fig. 3).

NiME aims to provide education material-supported, inclusive training opportunities for VET-educators in the metal and electrical industries so that they can identify (organizational) action areas and learning situations that support sustainable transformation in the industry. The training opportunities will facilitate the creation of classroom and learning environments that foster participatory learning along with real-world, sustainable design options. It will be possible for vocational educators to complete the training courses both online and in person. IG Metall union representatives will also be trained to act as multipliers, following the train-the-trainer approach, to ensure long-term implementation of the in-person formats. In summary, NiME

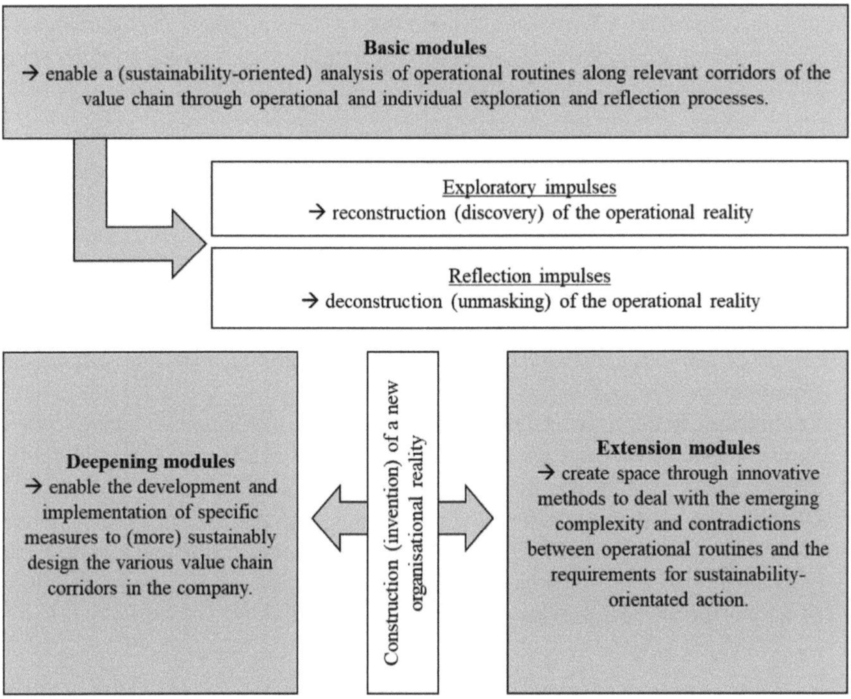

Fig. 3 Modular structure of the NaReLe learning task concept. *Source* Pranger and Hantke (2020, 88)

focuses on the following objectives. Promoting attitudes of VET-educators towards sustainability as an integral part of training.

- Supporting VET-educators in recognizing sustainability-related fields of action and their political significance both within and outside the company.
- Developing the didactic and methodological competencies of VET-educators to prepare sustainability-related learning situations.
- Assisting VET-educators in developing a sustainability-oriented training culture within their organizations.
- Qualiifying trainers (Train-the-Trainer) to further educate VET-educators in line with the above objectives.

Methodically, NiME operates as follows: A two-fold status quo analysis is conducted by NiME to assess the conditions, opportunities, and limitations of VETSD in metal and electrical companies. The needs and expectations, perspectives, and practical issues of stakeholders involved in training were first analyzed. Secondly, it examined the curricular and professional requirements of the occupations. By implementing the concept at both levels, it can be sustained in practice from an early stage. In collaboration with VET-educators in the metal and electrical industries, the didactic concept of resonance-oriented learning tasks will be adapted.

Through this approach, VET-educators are trained to be multipliers of sustainability-related occupational competencies, so that they can train others in these competencies as well. The project can be divided into the following steps (McKenney and Reeves 2018):

1. **Analysis and Exploration:** Specification of practical needs, including the analysis of potentially supportive and obstructive framework conditions from the perspectives of various groups involved in training, as well as the examination of curricular requirements.
2. **Adaptation and Implementation:** Customization of a material-supported, inclusively designed training program for and with gatekeepers, considering the results of the context analysis to promote political education for sustainable development in the metal and electrical industries.
3. **Evaluation, Reflection, and Subsequent Expansion:** Evaluation and reflection on the workshop concept and material components, with special consideration of participants' practical experiences.
4. **Distribution and Sustainability:** Distribution and institutionalization of the expanded workshop concept through Train-the-Trainer training for multipliers and educational representatives of the IG Metall union, as well as offering it as an online educational program.

5 Conclusion

The NaVeBb and NiME projects illustrate the substantial benefits of fostering collaboration between science and practice in vocational education and training (VET), especially in addressing sustainability challenges. These projects demonstrate that it is crucial to include practice as an equal partner in research in order to drive meaningful innovation in the field. Employing DBR as a methodological framework, both projects facilitate this collaboration by ensuring that theoretical models are continuously aligned with practical needs. This is done through iterative processes of design, implementation, and evaluation. DBR provides a robust structure for the ongoing enhancement of sustainability-related competencies, effectively leveraging scientific inquiry and vocational practice. This collaborative approach not only increases the practical relevance of research findings but also ensures their adaptability and transferability across diverse sectors and contexts.

A critical aspect of these DBR projects is participants' commitment. Higher engagement levels correlate with more relevant and practical outcomes. "Real" design, however, often deviates from the conventional demands and constraints of traditional scientific systems. The production of scientific knowledge and the development of practice-oriented materials must be balanced in order to generate real value for all stakeholders. The potential obstacles to such collaboration should not be underestimated. Bringing together scientists, practitioners, and policymakers requires ongoing dialogue and mutual understanding. This heterogeneity can be both a challenge and an opportunity for science-practice cooperation. While it fosters innovative perspectives, it also necessitates clear communicative structures to prevent misunderstandings and ensure alignment of both scientific and practical perspectives.

In projects like NaVeBb and NiME, researchers must reassess their roles and embrace active participation. In order to make this shift, we must develop a new self-concept that transcends traditional "ivory tower" thinking and fosters a close connection between theory and practice. The challenge lies in finding an equilibrium between theoretical distance and practical proximity in order to effectively navigate between theoretical constructs and their applications.

In summary, the partnership between science and practice within DBR projects such as NiME and NaVeBb not only contributes to the development of sustainable educational concepts but also enhances educational actors' capacity to proactively tackle the challenges posed by an evolving educational landscape. This form of collaboration, encompassing both challenges and opportunities, is crucial for embedding sustainable development in VET and could serve as a guiding principle for future initiatives.

References

Anderson T, Shattuck J (2012) Design-based research: a decade of progress in education research? Educ Res 41(1):16–25. https://doi.org/10.3102/0013189X11428813

Bakker A (2018) Design research in education: a practical guide for early career researchers (1st ed.). Routledge. https://doi.org/10.4324/9780203701010

Bereiter C, Scardamalia M (2003) Learning to work creatively with knowledge. In: Corte ED, Verschaffel L, Entwistle N, Merriënboer JV (eds) Powerful learning environments: unravelling basic components and dimensions. Elsevier Science, pp 73–78

Bogner A (2021) Die Epistemisierung des Politischen: Wie die Macht des Wissens die Demokratie gefährdet. Reclam

Casper M, Schütt-Sayed S, Vollmer T (2021) Nachhaltigkeitsbezogene Gestaltungskompetenz in kaufmännischen Berufen des Handels. In: Melzig C, Kuhlmeier W, Kretschmer S (eds) Berufsbildung für nachhaltige Entwicklung: Die Modellversuche 2015–2019 auf dem Weg vom Projekt zur Struktur. Budrich, pp 179–199. https://www.bibb.de/dienst/veroeffentlichungen/de/publication/show/16974

Diettrich A, Hahne K, Winzier D (2007) Berufsbildung für eine nachhaltige Entwicklung. Hintergründe, Aktivitäten, erste Ergebnisse. Berufsbildung in Wissenschaft und Praxis 36(5):7–12

Dilger B, Euler D (2018) Wissenschaft und Praxis in der gestaltungsorientierten Forschung – ziemlich beste Freunde. bwp@ Berufs- und Wirtschaftspädagogik – online, Ausgabe 33:1–18. http://www.bwpat.de/ausgabe33/dilger_euler_bwpat33.pdf

Euler D (2014) Design-Research–a paradigm under development. In: Euler D, Sloane PFE (Hrsg.). Design-Based Research (15–44). Stuttgart: Franz Steiner Verlag

Euler D, Sloane PFE (1998) Implementation als Problem der Modellversuchsforschung. Unterrichtswissenschaft (26), Artikel 4

Fischer A, Hantke H, Roth J-J (2021) Innovatives Lernen zwischen betrieblichen Anforderungen und nachhaltigen Herausforderungen. In Melzig C, Kuhlmeier W, Kretschmer S (eds) Berufsbildung für nachhaltige Entwicklung. Die Modellversuche 2015–2019 auf dem Weg vom Projekt zur Struktur. Budrich, pp 85–107

Gibbons M, Limoges C, Nowotny H, Schwartzman S, Scott P, Trow M (1994) The new production of knowledge: the dynamics of science and research in contemporary societies. SAGE

Hantke H (2018) „Resonanzräume des Subpolitischen" als wirtschaftsdidaktische Antwort auf ökonomisierte (wirtschafts-)betriebliche Lebenssituationen – eine Forschungsheuristik vor dem Hintergrund der Nachhaltigkeitsidee. *bwp@*, Ausgabe 35. http://www.bwpat.de/ausgabe35/hantke_bwpat35.pdf
Hantsch R, Weber H (2023) Gestaltungsansätze zur Entwicklung nachhaltiger Lernorte in betrieblichen Kontexten – Ansätze und Erfahrungen aus BMBF/BIBB-Modellversuchen. In: Pfeiffer I, Weber H (eds) Zum Konzept der Nachhaltigkeit in Arbeit, Beruf und Bildung – Stand in Forschung und Praxis. Budrich, pp 91–109
Hauff V (1987) Unsere gemeinsame Zukunft. Der Brundtland-Bericht der Weltkommission für Umwelt und Entwicklung. Eggenkamp
Hemkes, B, Srbeny C, Vogel C, Zaviska C (2017) Zum Selbstverständnis gestaltungsorientierter Forschung in der Berufsbildung – Eine methodologische und methodische Reflexion. bwp@, Ausgabe 33. http://www.bwpat.de/ausgabe/33/hemkes-etal
Holz H (1991) Modellversuche als angewandte Berufsbildungsforschung. In: Meyer-Dohm P, Schneider P (eds) Berufliche Bildung im lernenden Unternehmen. Klett-Verl. für Wissen und Bildung, pp 77–82
Kahan DM, Jenkins-Smith H, Braman D (2010) Cultural cognition of scientific consensus. J Risk Res 14(2):147–174. https://doi.org/10.1080/13669877.2010.511246
Kremer H, Ertl H, Sloane P (Hrsg.) (2023) Wissenschaft trifft Praxis—Designbasierte Forschung in der beruflichen Bildung. Bonn: Verlag Barbara Budrich
McKenney, S., & Reeves, T. (2018). Conducting Educational Design Research (2nd ed.). Routledge. https://doi.org/10.4324/9781315105642
McKenney S, Reeves T (2012) Conducting educational design research. Routledge
Mede NG, Schäfer MS, Füchslin T (2020) Wissenschaftsskepsis: Ursachen und Folgen. Publizistik
Melzig C, Kuhlmeier W, Kretschmer S (eds) (2021) Berufsbildung für nachhaltige Entwicklung. Die Modellversuche 2015–2019 auf dem Weg vom Projekt zur Struktur. Budrich. https://www.bibb.de/dienst/veroeffentlichungen/de/publication/show/16974
Müller C, Pranger J, Reißland J (eds) (2023) Nachhaltigkeitsorientierte Weiterbildungsdidaktik: Die doppelte Multiplikatorenqualifizierung. wbv Publikation. https://doi.org/10.3278/9783763974382
Müller-Christ G (2001) Nachhaltiges Ressourcenmanagement: Eine wirtschaftsökologische Fundierung. Metropolis
Müller-Christ G (2010) Nachhaltiges Management. Einführung in Ressourcenorientierung und widersprüchliche Managementrationalitäten. Nomos
Nowotny H, Scott P, Gibbons M (2001) Re-thinking science: knowledge and the public in an age of uncertainty. Polity Press
Parthey-Müller J, Eberl J-M, Kritzinger S, Bogner A (2023) Wissenschaftsskepsis. In Bogner A (ed) Nach Corona: Reflexionen für zukünftige Krisen – Ergebnisse aus dem Corona-Aufarbeitungsprozess, pp 88–116. https://doi.org/10.1553/978OEAW95696
Pranger J, Hantke H (2020) Die Wertschöpfungskette der Lebensmittelindustrie als Resonanzraum – Ein offenes Lernaufgabenkonzept im betrieblichen Einsatz. Haushalt in Bildung & Forschung (HiBiFo) 3:81–98
Reinmann G (2005) Innovation ohne Forschung? Ein Plädoyer für den Design-Based Research-Ansatz in der Lehr-Lernforschung. Unterrichtswissenschaft 1:52–69
Reinmann G (2022) Was macht Design-Based Research zu Forschung? Die Debatte um Standards und die vernachlässigte Rolle des Designs. EDeR 6(2). https://doi.org/10.15460/eder.6.2.1909
Rosa H (2016) Resonanz: Eine Soziologie der Weltbeziehung. Suhrkamp
Singer-Brodowski M, Holst J (2022) Nachhaltigkeit: Zielkonflikte und Widersprüche – (k)ein Thema für die politische BBNE? In: Hemkes B, Rudolf K, Zurstrassen B (eds) Nachhaltigkeit in der Berufsbildung. Politische Bildung als Gestaltungsaufgabe. Wochenschau, pp 109–117
Sloane PF (2014) Wissensgenese in design-based-research projekten. In: Euler D, Sloane PF (Hrsg.), Design-Based Research (1. Aufl., Beiheft 27, S. 113–139). Franz Steiner Verlag

Strotmann C, Kastrup J, Casper M, Kuhlmeier W, Nölle-Krug M, Kähler A-F (2023) Ein Modell zur Strukturierung und Beschreibung nachhaltigkeitsbezogener Kompetenzen für Auszubildende in Lebensmittelhandwerk und -industrie. In: Ansmann M, Kastrup J, Kuhlmeier W (eds) Berufliche Handlungskompetenz für nachhaltige Entwicklung: Die Modellversuche in Lebensmittelhandwerk und -industrie. Budrich, pp 159–179

UNESCO (2015) Unleashing the potential: transforming technical and vocational education and training. UNESCO

van den Akker J (1999) Principles and methods of development research. In: van den Akker J, Branch R, Gustafson K, Nieveen N, Plomp T (eds) Design approaches and tools in education and training. Kluwer Academic Publishers, pp 1–14

Wissenschaftlicher Beirat der Bundesregierung Globale Umweltveränderungen (WBGU) (2011) Welt im Wandel: Gesellschaftsvertrag für eine Große Transformation. WBGU

Integration of Sustainability Competencies in Creative Disciplines Training. Colombian Case

Lina María Suárez-Vásquez, Ana Elena Builes-Vélez, Juliana Restrepo Jaramillo, and Natalia Pérez-Orrego

Abstract As part of the global policies around education for sustainable development, we could indicate how UNESCO proposed to incorporate principles, values, and practices of Sustainable Development (SD) into education; among the multiple statements, the Nagoya Declaration (2014) confirms the responsibility of HEIs in SD education, the UN asks to connect sustainability with curricula and specifically for its 2030 Agenda, it is requested to prioritize education as a national investment, developing Education for Sustainable Development Policies. With this scenario, the research project "Education for sustainability in the creative disciplines," also called "Study for competencies for sustainable development in higher education specifically in art, architecture, and design," is proposed. This research carried out in the city of Medellín between 2021 and 2023, is an interinstitutional project among three Colombian universities that aims to recognize how we can contribute to resolve global crises from creative disciplines education and identify teaching strategies for the formation of competencies in sustainability in creative programs in the city of Medellin. In design and architecture education, sustainability, sustainable development, and climate change have been a challenge for decades. In this article, in addition to presenting the research's SustainUs toolkit results, we aim to recognize conditions of favorability or possibility and also limit conditions that have emerged in the research, institutions, disciplines, and curricula, which influence how easy it can be to integrate these competencies.

L. M. Suárez-Vásquez (✉) · N. Pérez-Orrego
Production and Design Faculty, Institución Universitaria Pascual Bravo, Medellín, Colombia
e-mail: l.suarezva@pascualbravo.edu.co

N. Pérez-Orrego
e-mail: Natalia.perezorrego@upb.edu.co

A. E. Builes-Vélez · J. R. Jaramillo · N. Pérez-Orrego
School of Architecture and Design, Universidad Pontificia Bolviariana, Medellín, Colombia
e-mail: ana.builes@upb.edu.co

J. R. Jaramillo
e-mail: juliana.restrepo@upb.edu.co

W. Leal Filho et al. (eds.), *Education for Sustainable Development: The Contribution of Universities*, World Sustainability Series,
https://doi.org/10.1007/978-3-031-86985-3_12

Keywords Education for Sustainability · Pedagogical Strategies · Creative Discipline · Sustainability Competences

1 Introduction

Education in sustainability (EFS), Education for Sustainable Development (ESD), and Sustainable Learning Education (SLE) have been essential issues for UNESCO (2005). All these educational approaches have been supported in multiple conferences and global agreements, such as the de Nagoya Declaration (2014), where governments and non-profit organizations recognize the responsibility of higher education Institutions (HEIs) to promote and achieve a sustainable future. HEIs are essential in encouraging sustainability and climate change education to future professionals to become agents of change (Sengupta, E. et al., 2020; Svanstrom, M. et al., 2008; Moore, 2005).

Education for sustainable development (ESD) grows with student-centered learning methods, such as project-based learning (PBL), problem-based learning (PBL), concept-based curriculum (CBL) or Universal Design for Learning (UDL); these methods are constantly used in creative disciplines and they promote critical thinking, the construction of new knowledge and learning new skills while working on wicked problems (Kolko, 2012), skills that have been proven to be necessary on solving sustainability and climate change issues. In architecture and design schools and faculties, these methods are used mainly in project and research courses because they allow students to comprehend and produce innovative solutions by considering various stakeholders' ideas and reflections on the problem they are trying to solve. In design and architecture education, sustainability, sustainable development, and climate change have been a challenge for decades. In order to address these challenges, the radical transformation of the world's development model is urgent. In the late 90 s, Manzini and Vezzoli (1998) argued the importance of changing our development model and moving towards a society where well-being is measured in terms of living better with less.

ESD should be considered a priority in architecture and design schools worldwide. With this framework, our project emerged as an answer to the constant questions inside the research team. From here, the requirement of a pedagogical strategy that focused on the integration and implementation of topics related to sustainability in professional training programs in creative disciplines emerged, and that is how SustainUs came about.

Through the methodological development and our project analysis, we can then account for these conditions of favorability or possibility and also limiting conditions that have emerged in the institutions, the discipline, and the curricula, which influence how easy it can be to integrate these competencies. The toolkit simultaneously was another resource or input for comprehending these conditions.

Table 1 Programs and Institutions where the research project was conducted

Institution	Programs					
Universidad Pontificia Bolivariana	Clothing Design	Graphic Design	Industrial Design	Architecture		
Universidad de Medellín	Space Design and Management	Fashion and Textile Design and Management	Product Design and Management			
Institución Universitaria Pascual Bravo	Clothing Design	Fashion Design	Graphic Design (Technology)	Graphic Design (Professional)	Digital Animation	Management Design

2 Methodological Approach

2.1 *Cases Study*

The research was conducted in three higher education institutions in Medellín, Colombia: two private and one public. It was centered on the faculties of architecture and design of all institutions. The study considers the content of teaching delivered to students and the pedagogical strategies used in class by teachers (Table 1).

2.2 *Methods*

The baseline study addressed the topics of curricula and SDGs in Colombia, especially in Medellín. The research team based in three design faculties reviewed programs such as graphic, industrial, clothing design, and architecture (12 programs revolving around creative disciplines). From a qualitative interpretative approach, the research team applied tools such as bibliographic analysis, semi-structured interviews (15 teachers were consulted), participatory observation, and an online questionnaire sent to approximately 700 creative discipline students, suited the goals of this study. Answers provided a "snapshot" of current teaching and Education for Sustainability practices across creative disciplines in the three institutions. From here, emerged the requirement of a pedagogical strategy that focuses on the integration and implementation of topics related to sustainability in professional training programs in creative disciplines. The challenge was how to include the sustainability criteria competencies in curricula.

After a documentary tracking of toolkits (global scale) used in creative disciplines and/or around teaching about sustainability, the research team developed an analysis grid to register the relevant elements found in the readings allowing their subsequent location through the following categories: complete bibliography (APA), brief description of the text (a paragraph or short ideas), textual citation or reference to the place in the text (example: pages 16 to 18), page, relevance to the project or the article.

Following the analysis of each toolkit, one of our conclusions was that according to the design process of creative disciplines, we used the techniques of these toolkits, linking them to the stages of the project; this vision allows us to begin to link the techniques in different combinations according to the multiple sustainability competences that students must acquire and the different purposes that the courses or projects require. To identify the techniques that would contribute most to this exercise, we designed another grid that allowed us to identify the stages of the process in which they performed most effectively (Preparation, Construction of meaning and Ideation, Sustainable Interaction Design, and Validation), their name and the toolkit of origin, the relevant actors or agents in the application of the technique and finally the beneficiaries or on which area we would see the impact of the result to the solution to the problem or the opportunity for transformation, among them: communities (c), Industrial, spaces, product (i), clothing, apparel, fashion (v), architecture, planning, landscape (a), graphic, communication, advertising (G).

2.3 Data Analysis

- Insight 1. Interviews with teachers

The teachers interviewed suggest that they participate in a "hidden curriculum" through which they influence their students beyond the classroom or subject. For example, they choose specific texts or projects that allow them to analyze the context and provide elements to discuss sustainability, climate change, and ecological justice, among other topics.

In general, the documents the universities propose as a guide to develop their courses do not explicitly state the need to develop competencies for sustainability.

- Insight 2. Review of curricula

The review of undergraduate and graduate curricula reveals a diversity of conceptions of sustainability and how to face the challenges it implies. Few programs approach the problem in a systemic and systematic way, that is, through a structured theoretical corpus that gives unity and coherence to the approach, and this doesn't occur in a sustained manner throughout the training process.

- Insight 3. Surveys with students

The surveys revealed aspects related to sustainability, for which more than half of the students indicated that they did not have enough information: bibliography, sustainable materials or processes, regional laws or policies related to sustainability, the local or regional environmental situation, best business or social practices for SD, and traditional knowledge for the care of nature.

After a documentary review, nine (9) toolkits were identified as containing the most relevant tools for design projects. Those selected were Frog Design - Collective Action Toolkit, The Sustainable Toolkit, DDC - Design your Circular Transition, SDG Fortune Teller, Nike: Circularity, VentureWell: Inventing Green Toolkits, Sustainable Communities Toolkit, Wonderland Studio Sustainable Digital Design, and The Circular Design Guide. Among these, 53 techniques were chosen and distributed among the four (4) defined stages and twelve (12) substages that could contain the vast journey of design projects.

3 Theoretical Background

3.1 Education for Sustainable Development (ESD) and Sustainable Learning Education (SLE)

Agenda 21 in Chapter 36 highlighted the importance of connecting sustainability and curricula (UNESCO, 2010). After this declaration, UNESCO launched the UN Decade of Education for Sustainable Development program, aiming to incorporate "principles, values, and practices of sustainable development in education" (UNESCO, 2005). Later, the UN created the 2030 Agenda for Sustainable Development, which includes the SDG, 17 goals, and 169 targets intended to promote actions by international governments to achieve sustainable development by balancing social, environmental, and economic dimensions. The Agenda explicitly mentions Education for Sustainability in SDG 4, which seeks to ensure equitable quality education around the globe and promotes sustainable learning.

Education for Sustainable Development (ESD) aims at enabling people to "not only acquire and generate knowledge but also to reflect on further effects and the complexity of behavior and decisions in a future-oriented and global perspective of responsibility" (Rieckmann, 2012, p. 128). These skills and attitudes may derive from individual competencies for sustainable development. Understanding competencies as enabling successful task performance and problem-solving concerning real-world problems, challenges, and/or opportunities (Barth, C. et al., 2007).

Sustainable Development Goals (SDGs) replaced the Millennium Development Goals in 2015. This objective is one of the critical aspects of education. Goal 4 of the SDGs calls for the world to "ensure inclusive and equitable quality education and promote lifelong learning opportunities for all" (United Nations, 2015). In addition, Target 4.7 demands the development of skills and knowledge "through education for sustainable development and sustainable lifestyles" (United Nations, 2015).

Education for Sustainable Development (ESD) requires different approaches and methods because of the complexity of the concept of sustainability; the understanding and definition of sustainability requires context and subjectivity, which makes traditional ways of teaching inadequate for exploring sustainability and sustainable development complexity (Builes-Vélez et al., 2023). When teaching ESD, a complete perspective of sustainable development that integrates ecological, economic, and social aspects is necessary. Borg, C. et al. (2012) recognize that the literature on ESD and Environmental Education (EE) emphasizes the importance of an interdisciplinary and holistic understanding of sustainable development.

SD and ESD raise controversial issues regarding which today's society is divided. SD is controversial because it has multiple perceptions and definitions within the environmental, social, and economic debate worldwide, as well as conflicts of interest. After all, social groups and governments tend to perceive and consider environmental issues differently. Regarding ESD, teachers may feel unprepared to teach sustainable and climate crisis issues due to a lack of knowledge, institutional support, and the absence of pedagogical strategies to reach sustainable development and climate change issues (Borg, C. et al., 2012).

Teachers must comprehend sustainability and sustainable development and must have a receptive and adaptative attitude towards sustainable development and climate change to connect students with the subject. Studies show that teachers' commitment to teaching ESD is relevant for overcoming barriers in the implementation of sustainable development (Timm, J. M., & Barth, M., 2021; Nousheen, A. et al., 2020; Hernández-Díaz, P. M. et al., 2021). Teachers from different educational backgrounds interpret sustainable development and climate change differently, which becomes a barrier when implementing ESD and SLE strategies. For instance, in some disciplines, sustainable development is seen as marginal, but it becomes a priority in others.

In their book *Design for Sustainability*, Cheschin and Gaziulusoy reviewed the different responses to sustainability and climate change challenges. In Multi-level Framework from Products to Socio-technical Systems (2020), regarding education, authors tracked the link between design and ecology (sustainability) back to the 1960s. Victor Papanek´s work *Design for the Real World: Human Ecology and Social Change* (1971) is considered the first to introduce an environmental and ethical reflection around the design profession; many others continued Papanek´s line of work, like Ezio Manzini and Carlo Vezzoli (1998). Some terms of the 1970s, such as 'design for need' or 'alternative design,' are precursors of the green design or ecodesign of the 1980s. The early 1990s focused on a more systematic response from design regarding sustainability; based on a redesign approach, the efforts examined how to reduce environmental impacts and increase efficiency. In the new millennium, ecodesign arrives with an expanded look at the complete life cycle of products. In the decade's second half, the answer takes another perspective with emotionally durable design.

New approaches to sustainable product design are aligned with the Sustainable Development Goals (2005) as they use a triple-line approach to understanding the 3Ps: planet, profit, and people to try to balance social responsibility, economic profit, and environmental impacts (Watkins, M. et al., 2021). Regarding creative disciplines, designers' and architects' decisions are responsible for the impact of the products, goods, buildings, and services we consume; that is why creative professionals involved in design and architecture play an important role in helping reduce the negative social, environmental, and economic impacts.

Education for Sustainable Development in design and architecture is fundamental in all stages of education; future creative professionals need to deal with wicked problems, crises, and transition issues during their training to become designers and architects with environmental and social consciousness so that when they start their carriers they design, products or spaces, thinking in environmental issues, economic profit and social responsibility. Persov, E. et al. (2017) argues that.

Educating designers for social and environmental sustainability within the disciplinary structure has proved to be a challenging task. In order to encourage creation of feeling, values and skills for sustainable design, a certain level of relevant sustainable literacy must be achieved (p. S1225).

As a challenge, ESD must be considered a priority in architecture and design schools worldwide. Designers and architects have proven they can work on transitioning from social, environmental, or economic crises.

Thus, we have seen that the perspective on sustainability approaches from the design and creative disciplines has been broad and diverse. Therefore, it was essential to determine the key variables that would be evaluated in the toolkits for this research. First, we adhere to the approach of UNESCO, which defines sustainability as "[…] development that satisfies the needs of the present generations without compromising the capacity of the future generations of satisfying their own needs" (1987). Within this frame, directly addressing this subject is difficult because of its broadness. Sustainability is an emergent state of equilibrium that contextual conditions can favor, some of which were identified by UNESCO (2015). Using this proposal as a basis, we interpreted these contextual elements as variables that could be tracked in the higher education context, as listed:

Curricular and/or technological resources used to teach sustainability.

- Information and/or training on local, national, and international laws, agreements, and policies to reduce the environmental impact of products and processes.
- Information about the local or regional situation linked to consumption and/or environmental protection.
- Knowledge of the best business or social practices for sustainable development. Social challenges linked to sustainable development.
- Culture of sustainability: ancestral, indigenous, or native knowledge related to the diversity, care, protection, and opportunities of natural resources. Frequent prejudices and misunderstandings about sustainable development.
- Integration of policies for ESD at the institutional level. As mentioned in this document, some instruments already exist to diagnose the quality of training processes, coverage, infrastructure, etc.

It is important to clarify that these variables could be tracked at different levels of development in higher education. In the Colombian context, competencies are assumed to have different levels of development. To recognize the strategies and pedagogical tools used for teaching sustainability in the creative disciplines' context, the planning of all courses in 12 programs was reviewed, searching for keywords linked to the variables referred to earlier in this text.

3.2 Pedagogical and Didactic Strategies

Pedagogical strategies are a series of procedures carried out by the teacher with the concept to be taught. On the other hand, didactic strategies are tools from which the teaching–learning process will be carried out from the teacher to the student. They make it easier for the teacher to transmit his knowledge and experiences. In addition, they will help to build new knowledge through the exchange of information between both parties and the knowledge base of the learners.

Pedagogical Strategy can be seen as having three different levels:

- A holistic approach where the principles of teaching and learning and theoretical concepts inform pedagogical strategies.
- At the programmatic or curriculum level, a general instructional strategy covers the various aspects of the sequencing and organizing of content, specifying learning activities, and deciding how to deliver this course content and associated activities.
- Pedagogical Strategy can be examined through the lens of how instructional design impacts the delivery of an individual lesson or module. This focuses on which specific instructional methods are best suited for the course content, modality, and learner.

According to Lozano (2010), a pedagogical strategy presents adverse elements (elements to be transformed). Its foundations are philosophical, pedagogical, sociological, and psychological objectives, stages (diagnosis, design, execution, and evaluation), and ways of improving teachers' professional pedagogical performance.

To gain a more refined understanding of the practical and theoretical aspects of teaching and learning sustainability and sustainable development, researchers did an extensive review of different pedagogical tools and strategies used in project courses in design and architecture programs to recognize the different ways in which tools are used to enhance creativity, develop new ideas, and implement actions for sustainability. The most relevant ones included in our proposal are described below.

The Collective Action Toolkit (CAT) proposed by Frog Design allows the team (designers or stakeholders) to be taken to a space of exploration to make things happen. The toolkit has five areas, each one with several group activities, that converge in a constant adjustment to team goals called "clarify". The moments are The Converge stage, the Build Your Group stage, the Seek New Understanding stage, the Imagine More Ideas stage, the Make Something Real stage, and lastly, the Plan for Action stage.

Nike Circularity initiative, arranged on a guide and its workbook, is structured based on principles that support a universal call to action for the industry regarding sustainability. Each of these principles has an introduction, some questions to ask during the process, and resources such as books, videos, documents, and examples of good practices. The principles are material choices, cyclability, waste avoidance, disassembly, green chemistry, refurbishment, versatility, durability, circular packaging, and new models.

VentureWell's Inventing Green: A toolkit for entrepreneurs is a set of resources to help inventors develop products that will improve people's lives and generate profits. The Lemelson Foundation designed it, and VentureWell partnered with NESsT and Presidio Graduate School. *VentureWell's Inventing Green: A Toolkit for Sustainable Design* helps early-stage inventors understand how the lifecycle of their products will affect the environment. The toolkit includes a video series and several resources that can be used within short workshops, multi-day accelerators, or as part of a university-level engineering or design course. It is divided into three areas that cover the general stages of product development: Supply Chain, Product Use, and End of Life.

Wonderland. Wonderland Studio proposed the Sustainable Digital Design initiative; the guide looks for a warning to creative communities in terms of digital production and consumption impacts. The initiative compiles a series of posts for designers, developers, and creators and helps them to reduce their work's carbon footprint. The guide includes minor changes such as resizing images, to full workflow reworks for static site generation. It is divided into three parts: size matters, tracking tech, and energy efficiency and consumption.

Living Better Using Less, a Sustainable Communities toolkit for community and voluntary organizations is an initiative from The Wheel (Ireland's national association of community and voluntary organizations, charities, and social enterprises) on behalf of the Environmental Protection Agency (EPA). This toolkit attempts to support local communities to stimulate consciousness and participation according to the sustainable premise of "living better by using less". The toolkit is divided into five themes: questions, case studies, and links to different resources. The topics are Ecological Integrity, Health and well-being, Participation and engagement, Culture and heritage, and Economic Resilience. They emphasize that Communities will look different and have different priorities depending on the community.

Designing Your Circular Transition Toolkit, created by The Danish Design Center, contains eight tools that guide you through a circular innovation process and provide you with the necessary tools. It is created as a resource for companies and organizations who want to work more creatively and collaboratively in their team or organization towards circularity. The methods are divided into three categories: Exploring the problem space, (Co)-creating new circular solutions, and Making the future concrete.

The analysis of these toolkits and strategies revealed the importance of seeking multiple perspectives on how to approach design or creative projects, the epistemological vision of complexity and the loop path, the opportunity that each stage of the project represents to relate with sustainability, and most importantly, that sustainability is in the hands of multiple agents, not only the users, not only the communities, not only the designers or the academy, everyone. From these findings, along with the justification and objective of each strategy, we developed the proposal of our meta-toolkit, a pedagogical proposal called SustainUs.

4 Pedagogical Strategy - SustainUs Toolkit

We will focus on the toolkit that, as a didactic strategy, provides the basis for constructing a pedagogical strategy that allows teachers and institutions to approach the ways and means that others have instituted to integrate the values and skills needed to teach sustainability into their exercises.

Subsequently, according to the information provided by the tracking and with the framework of action provided by the concept of didactic strategy (tools, knowledge, explanation, exchange of information, objectives, and stages), considering the possible relations with multiple typologies of a didactic strategy and the plurality of elements that allow its composition, the following methodological structure was considered for its development, which would take the form of a Toolkit. In view that one of the significant requirements detected is the lack of knowledge of the available resources for the design or creative exercise with sustainability items, the proposed strategy focuses on the recognition of the multiple existing tools, with emphasis on the actors (principal stakeholders, design/technical team and community) and possible beneficiaries in the application of such instruments, methods or methodologies (communities (c), Industrial, spaces, product (i), clothing, apparel, fashion (v), architecture, planning, landscape (a), graphic, communication, advertising (G)).

Taking into consideration that the development of the projective or creative exercise is the transversal element to the professional training in the analyzed disciplines, the following categories were considered in order to maintain the relevant elements of design and the creative in the construction of the strategy:

- Stage of the process (moment of the exercise where the tool, method, or methodology could be used).
- Name of the technique
- Function
- Toolkit of origin
- Beneficiaries (C) (I) (V) (G) (A)

After the analysis of each toolkit reviewed, including other worldwide experiences in the integration of sustainability into the creative exercise in the training of professionals, the research team integrated this information by developing a deck of 50 cards and a manual/handbook that will provide the necessary formats and indications for the development of the methodologies. The cards are classified into four categories according to each stage of the design process, in a general way: Preparation, Construction of meaning and Ideation, Sustainable Interaction Design, and Validation. Each card contains a tool, method, or methodology likely to be used in a sustainability-related project; it will indicate its name and a brief description of its application, making also reference to the specific toolkit from which it comes for further information on the user (as seen on Fig. 1).

The cards have several icons providing information to the user concerning the phase of the design process, actors that can or should participate, the role of the methodology, etc., allowing an effortless and agile approach from the graphic

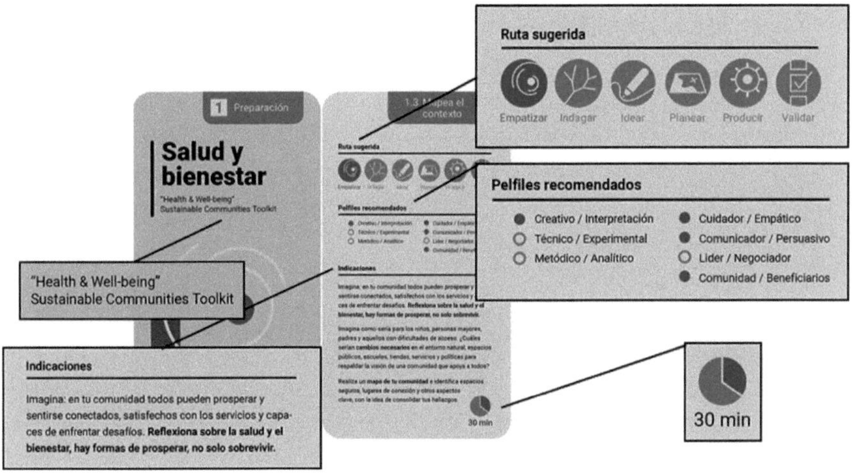

Fig. 1 Elements and symbols are included in a SustainUs card

construction of the material (as seen in Fig. 2). Thinking of an instrument like this, which may be useful for the creative disciplines also requires design considerations for its formatting. From the background training of the researchers, several key elements have been provided that will contribute to a better reading and use of the deck of cards.

Fig. 2 The final layout of SustainUs toolkit cards

4.1 Possible Routes for SustainUs

Paths to use the Toolkit depend on the type of course in which it would be used and the problem, need, or issue it addresses. So, the toolkit has multiple opportunities to be used. For example, a possible route for a research course can be:

a) Preparation – Phase 1
b) Construction of Meaning– Phase 2
c) Validation – Phase 4

A hypothetical scenario of using the deck of cards would initially seek to understand the moment in which the cards are being used (a strategy of dissemination of these tools or the application within a project with a situation that seeks to be transformed); whoever guides the activity could look for a diagnostic card (Preparation Stage) that allows visualizing the state of the situation (e.g., Review of Objectives). During the use of this card and the application of the activity proposed there, the conventions used in the deck could be explained along with the four possible phases of a creative project where the instrument can be used: Preparation stage (for a diagnosis stage), Construction of Meaning and Ideation, Sustainable Interaction Design (the execution moment), Validation (the evaluation phase). In a possible application with teachers or instructors, situations that could occur with students during the development of a project in the design laboratory (as a pedagogical figure within the design formation) could be considered.

The cards could be distributed, and each group could identify whether or not the cards they had could be used according to the situation presented. Each group could share what they have and why they think the strategy presented by their selected card works.

Finally, professions in areas other than the creative, such as education, medicine, engineering, etc., can use the cards to integrate sustainability into their teaching or workflow.

4.2 Testing of SustainUs Toolkit

Following the analysis of each toolkit, one of our conclusions was that, according to the design process of creative disciplines, we could use each of the techniques of these toolkits in relation to the stages of the project; this vision allows us to begin linking the techniques in different combinations according to the multiple sustainability competences that students must acquire and the different purposes that the courses or projects require.

The main emphasis of the SustainUs is to show the students how to include sustainable criteria within the projects, even in a pedagogical exercise. For the application of the strategy, in terms of previous knowledge, whether people are in the creative area is not required. Subsequently, this will become a condition of possibility because it does not establish biases or limitations for the people who can access this information. The typology of projects where the toolkit can be applied can be either of material or intangible characteristics; the type of artifact does not constitute an important degree of affectation for SustainUs.

When writing this article, the toolkit was tested with teachers of different programs. However, researchers have designed a workshop to do at the LATI Learning, Technology, and Innovation Laboratory, a space for multidisciplinary knowledge (UPB). They will be tested with teachers and administrative staff of the Universidad Pontificia Bolivariana. Also, researchers will test the deck, SustainUs, in the first semester of 2024 in two courses: a research course on Waterfront Landscape and Climate Change and a course on Creative Writing.

Furthermore, researchers will conduct a workshop with universities linked to the LENS network (Learning Network on Sustainability, node Colombia) to test the toolkit in a projective and inter-institutional exercise. Medellín, Bogotá, and Bucaramanga, as well as students from the multiple universities that are part of the network, will participate.

Within the exercises and internal applications, we conclude that the strategies can appear iterated. It does not matter the complexity of the project, the number of problems, transformation opportunities, or dimensions.

5 Conclusions – About Conditions of Possibility and/or Limitations

When we consider conditions of possibility, we refer to those structural elements or the nature of the context, which we find allow the integration of these competencies into the curricula. Among the conditions of possibility, according to our findings, there are undoubtedly the visions and political lines of the professors, the questions and issues that they have explored in their lectures, the pedagogical practices that are established in the academic modules, even when there is no explicit content in the descriptive letters or micro-curricula of the courses.

On the other hand, both the Clothing Design program of the Universidad Pontificia Bolivariana and the programs of the Department of Design of the Institución Universitaria Pascual Bravo have developed a curricular transformation proposal that has gathered concepts and performance criteria regarding sustainability competencies in their professional design project; in the future, this will allow, that naturally, teachers can include in their pedagogical discourse notions and scenarios of learning environments linked to Sustainability. In this sense, we are in congruence with the vision of the design theorist Tony Fry (2018), who talks about the need to restructure design education as a real education, leading to the creation of a new community of design educators:

(…) Recasting design education as an actual education: An absolute priority and in the recognition that de-signer needs to be educated in a worldly sense, and that designer who don't think critically are dangerous. By implication this means the creation of a new community of design educators (of which there are scatterings) and new communities of desig(ing) (Fry, 2018, p.175).

From another perspective, we can include the structural circumstances of the institution, initially in terms of the discourse to which each university is inscribed, for example in the Pascual Bravo, there is an expression of "green and intelligent campus," and this has even allowed me to be here because there are processes that unquestionably have the institutional support; in the same way, there is installed as well in terms of educational means and infrastructure, for example libraries, access to multiple databases, devices in the classroom to connect to the internet (allowing mirror classes for example).

In terms of infrastructure, there are also technical laboratories such as clothes-making machine rooms where students can make prototypes without any limitations, proving, testing, materializing by their own hands, and controlling conditions of production that allow them to understand in a more optimal way the impacts, the possibilities, the possible advantages of the processes in their different instances and materials (in the same sense would be understood the jewelry workshops, wood workshop, graphic techniques workshop, textile intervention, or even foundry laboratories). In this regard, the students achieve to be able to decide on their design projects in depth and with greater certainty about sustainability.

Relatedly, when we talk about limiting conditions, we can mention a structure that falls short in providing possibilities to broaden and deepen inquiries and explorations. This includes issues such as time requirements, whereby everything must be completed in less than four months.

On the other hand, the presence of characteristics associated with a traditional practice of design as a discipline or traditional ideas regarding materialization implies disadvantages for the integration of competencies. The programs circulate on a double thread; undoubtedly, within the dynamics of the academic modules themselves, there are a series of harmful issues, such as the imperious introduction of the concept of p (thought based on marketing or engineering, without questioning the modes, the use of resources, etc.), the theoretical affinity with questionable productive models (considered as success cases), it is still believed that for students to "get a job," classroom practices must be adjusted to the way industry operates, etc.

Also, there are issues of an economic nature regarding the students themselves, who generally have to pay for everything related to their projects, as well as those people or organizations involved in them. There are processes, materials, and resources that today, although they have a lower impact, are not economically accessible to students, are not scalable, or are not even available in the Colombian market.

At a more macro level, we can include what we will call institutional pressures concerning the relations that the faculties and programs must establish with the university itself or other institutions and governmental entities of a local and national nature, promoting specific actions based on this relationship.

It should not be forgotten that, due to historical trajectory, design programs in Colombia began to be created in response to an industrial need to differentiate themselves from imported products; therefore, it is reasonable to include chairs around productive practices of the industry, both local and global. The connection between the institutions, the local industry, and the market is evident through the presence and support of different interrelation instances. Therefore, with these institutional pressures, it is very important to have directives supporting critical designers' learning spaces.

Time is another limitation since the short four-month duration of an academic period makes it impossible to systematize, record, and archive the educational and pedagogical experiences that take place, and even more, the continuation of research and exploratory exercises (this is why the toolkit is so pertinent in the possibility of using different strategies, with different application times).

Nevertheless, all these circumstances respond to limitations, but we deem them opportunities to generate more processes and projects built from a critical discourse with the industry itself, with the way of thinking about artifacts, design projects, or any other creative discipline.

References

Barth M, Godemann J, Rieckmann M, Stoltenberg U (2007) Developing key competencies for sustainable development in higher education. Int J Sustain High Educ 8(4):416–430. https://doi.org/10.1108/14676370710823582

Borg C, Gericke N, Höglund HO, Bergman E (2012) The barriers encountered by teachers implementing education for sustainable development: Discipline-bound differences and teaching traditions. Res Sci Technol Educ 30(2):185–207. https://doi.org/10.1080/02635143.2012.699891

Builes-Vélez, A. E., Jaramillo, J. R., Pérez-Orrego, N., & Suárez-Vásquez, L. M. (2023). Teaching strategies for sustainability in creative disciplines. In *Sustainability in practice: Addressing challenges and creating opportunities in Latin America* (pp. 75–91). Cham: Springer Nature Switzerland.

Fry T (2018) Design, a philosophy of liberation and ten considerations. Strategic Design Research Journal 11(2):174–176. https://doi.org/10.4013/sdrj.2018.112.16

Hernández-Díaz PM, Polanco JA, Escobar-Sierra M, Leal Filho W (2021) Holistic integration of sustainability at universities: Evidences from Colombia. J Clean Prod 305:127145. https://doi.org/10.1016/j.jclepro.2021.127145

Kolko, J. (2012). *Wicked problems: Problems worth solving: A handbook & a call to action.* AC4D.

Lozano R (2010) Diffusion of sustainable development in universities' curricula: An empirical example from Cardiff University. J Clean Prod 18(7):637–644. https://doi.org/10.1016/j.jclepro.2009.07.005

Manzini, E., & Vezzoli, C. (1998). *Lo sviluppo di prodotti sostenibili: I requisiti ambientali dei prodotti industriali.* Rimini: Maggioli Editore.

Moore J (2005) Seven recommendations for creating sustainability education at the university level: A guide for change agents. Int J Sustain High Educ 6(4):326–339. https://doi.org/10.1108/14676370510623829

Nousheen A, Zai SAY, Waseem M, Khan SA (2020) Education for sustainable development (ESD): Effects of sustainability education on pre-service teachers' attitude towards sustainable development (SD). J Clean Prod 250:119537. https://doi.org/10.1016/j.jclepro.2019.119537

Papanek, V. (1971). *Design for the real world: Human ecology and social change.* Bantam Books.

Persov, E., Udyavar Yehuda, R., Kantor, R., & Pelman, B. (2017). PBL 2.0. Interdisciplinary, multicultural transformative sustainable education for design students. *The Design Journal, 20*(1). https://doi.org/10.1080/14606925.2017.1352652

Rieckmann M (2012) Future-oriented higher education: Which key competencies should be fostered through university teaching and learning? Futures 44(2):127–135. https://doi.org/10.1016/j.futures.2011.09.005

Sengupta, E., Blessinger, P., & Yamin, T. S. (2020). Introduction to integrating sustainability into the curriculum. In *Integrating sustainable development into the curriculum* (pp. 3–14). Emerald Publishing Limited. https://doi.org/10.1108/978-1-83909-437-720201002

Svanstrom M, Lozano-Garcia FJ, Rowe D (2008) Learning outcomes for sustainable development in higher education. Int J Sustain High Educ 9(3):339–351. https://doi.org/10.1108/14676370810885925

Timm JM, Barth M (2021) Making education for sustainable development happen in elementary schools: The role of teachers. Environ Educ Res 27(1):50–66. https://doi.org/10.1080/13504622.2020.1825637

UNESCO. (2005). *UN decade of education for sustainable development: The DESD at a glance.* http://unesdoc.unesco.org/images/0014/001416/141629e.pdf

UNESCO. (2010). *Education for sustainable development in action: Learning & training tools no. 2.* https://unesdoc.unesco.org/ark:/48223/pf0000190898

UNESCO. (2015). *Replantear la educación: ¿Hacia un bien común mundial?* http://unesdoc.unesco.org/images/0023/002326/232697s.pdf

United Nations. (2014). *Nagoya Declaration on Higher Education for Sustainable Development.* https://sdgs.un.org/documents/aichi-nagoya-declaration-education-sustaina-20607

United Nations. (2015). *The Sustainable Development Goals.* https://sdgs.un.org/es/goals

Watkins M, Casamayor JL, Ramiréz M, Moreno M, Faludi J, Pigosso DC (2021) Sustainable product design education: Current practice. She Ji: the Journal of Design, Economics, and Innovation 7(4):611–637. https://doi.org/10.1016/j.sheji.2021.11.003

Nourishing Change Agents by Undergraduate Program for Sustainable Futures at a Japanese University

Hideki Maruyama

Abstract This chapter examines the role of Japanese universities in promoting Education for Sustainable Development (ESD) through a case study of Sophia University's "Sustainable Futures" program for undergraduate students. Positioned within Japan's evolving socio-educational context, the study explores how the program integrates interdisciplinary learning, global–local connectivity, and sustainability competencies. It highlights innovative pedagogical approaches, including transformative and systems thinking, while addressing challenges posed by demographic shifts and the COVID-19 pandemic. The research underscores the significance of the "4Cs" framework—curriculum, campus, community, and culture—in creating a sustainable university environment. By analyzing Sophia University's efforts to align education, research, and societal contributions with sustainable development goals, this paper offers insights into how higher education institutions can foster socially conscious graduates and adapt to global challenges. It emphasizes the need for holistic approaches to education that address local and global sustainability issues in interconnected and transformative ways.

Keywords Education for sustainable development (ESD) · Interdisciplinary education · Transformative learning · Sustainable universities · Systems thinking

1 Introduction

Among the diverse functions of universities, some are local in nature, while others are universal, transcending national boundaries. Although universities have sometimes been derided as "ivory towers," since the onset of prominent globalization, there has been an increasing demand for universities to contribute to society. This shift has led to a reevaluation of the university's role in addressing broader societal challenges and engaging with communities beyond academia. Brubacher (1982)

H. Maruyama (✉)
Sophia University, Tokyo, Japan
e-mail: hideki@sophia.ac.jp

W. Leal Filho et al. (eds.), *Education for Sustainable Development: The Contribution of Universities*, World Sustainability Series,
https://doi.org/10.1007/978-3-031-86985-3_13

suggests two philosophies underlying the functions of the modern university: One is epistemological in nature and states that university's purpose is to answer the great questions of human existence, meaning that universities seek only knowledge and truth. The other is political philosophy of education states that universities not only seek knowledge but also apply knowledge to solve the complex problems of society. Furthermore, universities may need to engage in more self-criticism regarding their educational outcomes. This is because there are even assertions that those who have been highly educated by modern educational institutions are the ones making the Earth unsustainable. For example, Orr (2004) concludes that the sustainability crisis is driven by graduates educated in modernist notions of progress, economic growth, and the urge to dominate nature (p. 201).

More specifically, while global standards were prioritized as the "correct" approach throughout the twentieth century, with universities focusing on producing graduates equipped to excel in the global economy and competition, recent perspectives have challenged this narrow view. However, scholars such as Escobar (2018) emphasize that true diversity encompasses marginalized voices that often go unheard. For a sustainable future, it is imperative to foster an inclusive concept of "we" that extends beyond the realm of higher education. This necessitates a profound acknowledgment that the collective "we" integral to addressing global challenges is not limited to those with university degrees. This sentiment is echoed in UNESCO (2021) publications, underscoring the importance of diverse perspectives and inclusive participation in shaping a sustainable future.

Considering these global developments, what kind of social expectations are Japanese universities confronted with? How are they positioned to respond to these evolving demands within their unique cultural and educational context? Basic Act on Education (Act No. 120 of December 22, 2006) says, "Universities, as the core of scholarly activities, are to contribute to the development of society by cultivating advanced knowledge and specialized skills, inquiring deeply into the truth to create new knowledge, and broadly offering the fruits of these endeavors to society (Chapter 2, Article 7) (MEXT 2006)". In 2022, the Japan Business Federation (*Keidanren*) predicted the advent of a human-centered society that balances economic development with the resolution of social issues, premised on a future where digitalization and reality merge. In preparation for such a society, Keidanren proposed that universities should play crucial roles from three perspectives: education, research, and social contribution. Let us briefly summarize these three aspects as follows:

- [*Education*] While academic aspects are still included, the focus is increasingly shifting toward skill development that is considered "more practical" or "useful to society." There is a growing demand "to nurture and produce human resources who will lead a new era and play active roles at the core of society by cultivating the ability to explore and solve social issues through the acquisition of a wide range of knowledge, skills, and expertise." This is because it is believed that fostering and producing individuals who can drive innovation in future society will contribute to strengthening the competitiveness of both the nation and Japanese companies.

As the number of university applicants continues to decline due to the decreasing birthrate, some private university administrators are positioning their institutions as "education-focused" and using this as a means to attract prospective students.[1]

- [*Research*] It is frequently reported that Japan's research capabilities have been declining, as seen in the decrease in the number of highly cited papers.[2] Therefore, there are growing expectations for universities "to play an active role in creating innovations that contribute to solving social issues." However, in the face of challenges in university management due to declining student enrollment—especially for private universities located outside major metropolitan areas—there is an increasing emphasis on education over research.
- [*Social Contribution*] Through excellence in education and research, universities can contribute to society. In addition, it has been suggested that universities actively engage in interaction, dialogue, and collaboration with various stakeholders, such as businesses and communities, to consolidate and share knowledge. This is particularly expected in non-urban areas, where specific contributions are increasingly demanded from public universities, which are established and funded by local governments, rather than national universities.

Japan has 813 universities, of which about 77% are private universities (MEXT, 2024, August 28). The numbers of universities, excluding junior colleges, technical colleges, and vocational schools, include 86 national universities, which receive relatively large subsidies from the government; 103 public universities, which are established by the administrative units of 47 prefectures; and 624 private universities established by various corporations. In a situation where students continue to decline, MEXT is urging each university to leverage its strengths to secure student entry. Thus, universities are free to allocate their internal resources in whatever balance they see fit across the three aforementioned areas. However, for many private universities, which do not have expensive science faculties such as medical schools, it is difficult to focus on research only. As a result, they tend to place emphasis on either education or social contribution. Moreover, while universities in larger urban areas still have some flexibility in management, those in other areas face challenges due to smaller young populations. The situation can be illustrated as shown in Fig. 1, and the universities in Japan can hardly adopt the same approach, as they face quite different underlying challenges.

When focusing on the university's contribution to sustainability, all three aspects are important and interconnected because universal and global issues, such as sustainable development, are directly linked to local challenges. In other words,

[1] In 2024, the number of private universities with enrolled students fewer than their capacity reached a record high of 354 (up from 320 the previous year) out of 624 institutions, accounting for 59.2% of the total (up from 53.3%). Against a total capacity of 503,874 students, the number of enrolled students was 494,730. The population of 18-year-olds has decreased by approximately 34,000 from the previous year. https://www.asahi.com/articles/ASS9F1VKDS9FUTIL01ZM.html Asahi NP on 2024/9/13.

[2] From 1997 to 1999, Japan ranked second in the number of academic papers, following the United States. However, from 2017 to 2019, it fell to fourth place, behind China and Germany. The world rankings of Japanese universities are also on a declining trend.

	Urban Area		Other Areas	
	National & Public	Private	National & Public	Private
Research	***	**	***	*
Education	**	***	**	***
Social contribution	*	*	***	***

Fig. 1 Emphasis on three areas by university background

sustainability-related research conducted by universities is synergistically linked to education and addresses both global and local issues. Concentrating a university's resources on only one of the three aspects cannot be considered a holistic approach.

This paper will describe how a Japanese university is implementing ESD and contributing to the construction of a sustainable society, with a particular focus on education, through a case study. The featured private university is located in central Tokyo and upholds the philosophy of "For Others, With Others" for its education. It is also the first university in Japan to establish an undergraduate-level sustainability education program. Through examining the background, outcomes, and challenges of this case, the paper discusses the role of university education in a Japanese society facing difficulties in sustainability due to population decline. The university is ranked among the top in Japan and has a reputation for having outstanding students. However, even such a prestigious university faces challenges in fully addressing all three aspects.

2 Learning in University and Sustainability

2.1 Social and Transformative Learning as Higher Education Approach

In ESD, it is not expected to rely solely on textbooks that follow predetermined educational goals or standardized content, resulting in the same outcomes every year. Instead, it encourages taking topics from the immediate living environment and linking local issues with global challenges. Given the unpredictability of the future, the framework proposed by Scott and Gough (2003) is easy to understand. The first approach views environmental issues as challenges that can be resolved through appropriate science and technology. The second focuses on the social and political aspects of the challenges, using accumulated knowledge and wisdom to address symptoms rather than the root causes. The third point is that our knowledge and tools for the future are inherently insufficient, and thus any possibility should be seen as an opportunity, requiring lifelong learning. This third approach necessitates social learning and collaborative learning based on the assumption of uncertainty and complexity observed in the real world.

less transmission and delivery	more transformation and critical inquiry
less teaching organization	more learning organization
less discipline based	more inter- and transdisciplinary
less managerial	more participative

Fig. 2 Higher Educational Institutes need to become… *Note* Sterling (2013, p. 27)

Since the term ESD may implicitly suggest education as a tool for achieving sustainable development, it is necessary for higher education institutions to fundamentally transform education itself through ESD. The term "sustainable development" always carries the implication that economic growth is both possible and desirable, which inherently involves the contradiction of promoting development through economic growth and wealth redistribution while maintaining ecosystem health (Jickling 1992; Banerjee 2003; Kopnina 2012). It is important to avoid reducing education to a mere tool for this purpose. Wals (2011) has cautioned that ESD runs the risk of overemphasizing "correct" and "incorrect" behaviors, which could result in big-brother sustainability or eco-totalitarianism, instead of focusing on ways of thinking and problem solving. Sterling (2007) shows three learning levels: basic learning as doing things better, meta-learning as doing better things, and epistemic learning as seeing things differently. He also shows that inter- and trans-disciplinarity are common, there is an emphasis on real-life issues, and the boundaries between institution and community are fluid. Sterling (2013) illustrates the mainstream of higher education may largely be educating for a future that "no longer exists" and educational institutions need to be transformed (Fig. 2).

For universities to provide future-oriented education, both students and faculty will need to experience transformative learning. The proponents of transformative learning suggest frameworks such as Freire (1996), who highlighted the psychological aspects of transformative learning through the process of conscientization, and Mezirow (2000), who emphasized the perspective transformation by critical reflection to encourage individual self-actualization.

Additionally, when considering university facilities and environment, it can be organized into the following "four Cs" based on Selby's (2009) argument (Sterling 2013): A holistic "4C" approach to sustainable universities integrates Curriculum, Campus, Community, and Culture. This model sees these four elements as interconnected spheres that enhance each other and contribute to the learning experience of students and faculties. Curriculum covers learning, teaching, and related research; Campus involves physical changes; Community refers to engagement beyond the university; and Culture encompasses institutional values, policies, and practices.

2.2 Pedagogical Value of the Social Ecology Model in a Society with Population Decline

Generally, societies with a large working-age population experience what is known as a "demographic bonus" period, leading to significant production and consumption activities, often resulting in relatively high economic growth. However, as societies mature and the population balance stabilizes, consumption activities tend to become more modest. In an aging society, not only do consumption activities shrink, but production activities also decrease. Japan has developed various social infrastructure and systems during its period of high economic growth and is still trying to maintain many of them today. In elections, which signify political participation, policies for the elderly are prioritized. However, the world-leading pace of aging is changing the social context, rapidly marginalizing young people into a social minority, making it difficult for them to assert their presence even if they wish to participate in society. Nonetheless, while the elderly do not prioritize envisioning the future, awareness of environmental issues is rising among young people, who are directly connected to their future. The elderly, as the majority, often cannot escape from their past successful experiences or find it hard to accept changes in rules, whereas young people are more interested in sustainability. The process of moving from the periphery of society to the center is referred to as "legitimate peripheral participation (Lave and Wenger 1991)." They emphasize how learners become part of a community through increasing involvement in its sociocultural practices. Wenger further developed this theory, exploring concepts of meaning, community, and identity in relation to learning (Wenger 1998). It can be said that university education is important for having programs that encourage students' social participation.

In discussions of sustainable development, while the preservation of the natural environment, economic opportunities, and social inclusivity are often taken for granted, there exists a Social Ecology model that incorporates humans instead of economy. Social Ecology is a discipline that comprehensively considers environmental issues and social challenges. Bookchin (2022) captures the fundamental interconnectedness between nature and human society, arguing that everything is interdependent. Additionally, it has been noted that traditional education tends to be strongly anthropocentric, making it necessary to aim for experiences that foster reverence for nature. Hill (2020) presents an approach that addresses environmental issues and social challenges, focusing on the transformations of individuals, society and culture, and the environment and spirit. All these perspectives are different from claims that seek to make economic growth sustainable; they do not advocate for human development alone but emphasize interdependent relationships that include nature and demonstrate the potential of education. The Social Ecology model helps to bring the person into view, which has strong pedagogical merit (Mulligan 2017). The model highlights personal perspectives, countering the notion that sustainability is solely for experts. It helps alleviate despair over global challenges by focusing on individual action, as personal efforts can lead to broader social engagement. However, individual action is just a starting point for addressing issues that span from local to global scales. When related to universities, sustainability that focuses on the well-being of students and staff is sought (Sayce et al. 2013).

3 A Case: Sophia University

3.1 General Backgrounds

Sophia University has long been recognized as a global university in Japan. In 1549, Jesuit missionary Francis Xavier, who represented Western civilization, traveled all the way from Europe to Japan, marking the beginning of the introduction of Christianity. In a letter to a member of the Jesuits in Goa, he wrote, "The people of this country … are the greatest citizens; you will not find a superior people to the Japanese." During his mission, he traveled to the capital *Miyako* [present Tokyo] and planned to establish a university there, similar to the one in Paris where he had studied alongside Ignatius of Loyola. His vision was for it to be a great hub for exchanging both European and Japanese ideas and cultures (Sophia University 2023). Since its establishment as a modern university, Sophia University has a history of 111 years in education and has maintained connections with foreign partners, benefiting from opportunities for international exchange in education and research, particularly through the Jesuit network. In recent years, it has developed numerous collaborations with international organizations, including the United Nations programs.

For many years, Japanese universities provided education primarily in Japanese, but in response to international competition, there has been a growing emphasis on education conducted mainly in English (Rose and McKinley 2018). MEXT started emphasizing the internationalization of universities through the Top Global University Project (2014–2023). While various projects had existed prior to this, the past decade has seen a significant impact from this initiative. In order to compete for students amid a declining population, universities have actively utilized MEXT subsidies. However, these subsidies are competitive funds, and whether a university is selected or not has implications for its survival.

The Top Global University Project provided two types of subsidies to the universities. Type A offers support of up to 105 million yen (approximately 0.66 million euros) in ten years to universities aiming for the top 100 in global rankings, with 13 universities selected. On the other hand, Type B supports universities that drive the globalization of society, with a smaller amount of 47 million yen (approximately 0.3 million euros), and 24 universities were chosen for this category (Table 1).

Sophia University is a private university that applied for Type B and had its educational program proposal approved. Since it was competitive funding, a promising and appealing educational program that could convince the evaluators was required. Sophia University had a strong advantage in international education compared to other universities, as it had been conducting education in English. However, it was obvious that merely using English as the medium of instruction would not suffice

Table 1 The number of selected top global universities

	National and public	Private
Type A	11	2
Type B	12	12*

Note * The present case belongs here

to compete with foreign universities. Nevertheless, within Japan, the university was able to gain significant information and cooperation from international organizations. Against this backdrop, Sophia University connected its educational initiatives to the UN's SDGs and particularly emphasized proactive efforts towards sustainable development. This led to the creation of the "Sophia Program for Sustainable Futures: SPSF" to be explained in the next section.

3.2 Program Design

The SPSF is a four-year bachelor's program that started in September 2020. The medium of instruction is entirely in English, and one-third of the students in the program are international students who have enrolled from a wide range of regions, including the Americas, Australia, Europe, Southeast Asia, and the Asia–Pacific, in addition to East Asian students often seen in many universities in Japan. The next one-third consists of Japanese nationals who have graduated from high schools other than Japanese schools, while the final one-third is made up of students who have graduated from Japanese high schools and excel in English. Incoming students belong to one of six departments (Economics, Education, Global Studies, Journalism, Management, and Sociology) and receive one of seven degrees (Bachelor in Area Studies, Economics, Education, International Relations, Journalism, Management, or Sociology) upon graduation. This program is Japan's first undergraduate course that directly addresses sustainability.[3]

In the SPSF, in addition to generic skill courses such as academic writing and presentation, there are unique classes. All students take the same subjects in their first semester and in the first semester of their third year, reflecting on their experiences both on and off campus during their third year (Fig. 3). In the first semester of the first year, all students are required to take "*Lectures in/about Sustainable Futures*." In this mandatory course, the coordinator uses the Social Ecology model to teach sustainability and the latest global and local trends, while faculty from the six departments teach about sustainable futures based on their respective disciplines. As first-year students, they focus on acquiring foundational knowledge about sustainability, creating an opportunity for them to design the content they will learn throughout their four years at university. From the latter half of the first year, students

[3] Although there were seminars and classes focusing on sustainability at other universities before this, it marks the first establishment of a structured Bachelor course in this area at Sophia University.

	Year 1		Year 2		Year 3		Year 4	
	Autumn	Spring	Autumn	Spring	Autumn	Spring	Autumn	Spring
SPSF Core	*Lectures in/about Sustainable Futures*				*Seminar for Sustainable Futures*			
Dept. Ed	Economics						Bachelor's thesis	
	Education						Bachelor's thesis	
	Global Studies		Interdisciplinary learning				Bachelor's thesis	
	Journalism						Bachelor's thesis	
	Management						Bachelor's thesis	
	Sociology						Bachelor's thesis	
General Ed	Compulsory & elective courses							
Others	Study abroad, internships, social engagement programmes, extracellular programmes							

Fig. 3 Four-year design in SPSF. *Note* Modified from Sophia University (n.d.)

are free to choose their classes, although they are encouraged to intentionally take courses from other departments. Many students begin studying abroad,[4] participating in extracurricular activities, and internships at this stage. As a result, first and second-year students also attend classes outside their own department, broadening their perspectives as they engage in cross- and multi-disciplinary learning.

In the first semester of the third year, "*Seminar for Sustainable Futures*" is a mandatory course, and all students gather again. In this seminar, students work in groups to define sustainability based on previous research, discuss challenges, create specific projects, and implement these projects through fieldwork and investigations. To capture specific issues from multiple perspectives, it becomes essential to design projects in collaboration with students from other departments, making an interdisciplinary approach important. The most distinctive aspect of the third-year seminar in the SPSF program is that students first grasp the situation regarding sustainability through Systems Thinking. They recognize that individual issues cannot be solved by one solution, that solutions may create another problem, and that seemingly unrelated factors can influence one another (Stroh 2015; Meadows 2008). After identifying leverage points as a group, students incorporate the concept-based methods of Erickson et al. (2017) while following the lesson flow proposed by Spencer and Juliani (2017) to initiate and implement projects using Design Thinking procedures, ultimately presenting their results. Examples of past projects include initiatives aimed at social impact, such as reusing food to reduce food waste (*e.g.* selling fruit jams) and establishing a system for donating unwanted items to organizations supporting low-income families. Through this project, students realize that they possess the power to change society.

In the fourth year, the students engage in graduation research for their respective departments. The graduation research and thesis writing process is like that of the regular Japanese track; however, SPSF students, building on their experiences from the projects, become aware of the importance of not only their research subjects but also how their research can contribute to society. Here, there is an awareness of transforming traditional academic systems to include a transdisciplinary approach,

[4] Sophia University has exchange agreements with 800 universities worldwide.

but the faculty understands that this cannot be fully achieved within just four years of undergraduate education.

3.3 *Turning COVID-19 into Learning Opportunities*

The launch of the SPSF was anything but smooth sailing. This was because, around the time the first student cohort[5] enrolled in September 2020, the world was during the COVID-19 pandemic. Faced with unknown challenges and uncertain outcomes, the Japanese government began refusing entry to overseas travelers at the end of January 2020, and as more countries closed their borders, international mobility of international students became halted. The strictest response came with the complete suspension of entry for all foreigners starting January 13, 2021. By the summer of that year, vaccination became common in Europe and international travel was starting to resume; however, entry into Japan was postponed until November 8, when the number of infections decreased. Nevertheless, with the emergence of the Omicron variant, from November 30, 2021, to March 1, 2022, all new entries for foreigners were effectively suspended once again.

Schools across Japan began closing and transitioning to online classes in April 2020, starting time of the academic year in Japan. Sophia University also shifted entirely to online education at that time. The faculty members struggled to adapt to this unfamiliar mode of teaching, gathering in classrooms while wearing masks and participating in training sessions to conduct classes. The University has approximately 12,000 students and has accepted about 2000 international students from over 90 countries each year.[6] Due to the impact of the pandemic, international students who were unable to come to Japan attended online classes. However, for those in North America, the time difference meant that classes were held from 2 a.m. to 5 a.m. Japan time, causing significant mental distress for the students.[7] Additionally, the suspension of the university's study abroad programs led some students to choose to take a leave of absence or withdraw from their studies.

The faculty of the SPSF program also faced challenges, but fortunately, the preparation and experience gained from the Japanese track from April to August 2020 were applied to the English courses that started in September. As a result, the use of technology, including ICT by faculty members with about six months of experience, was relatively smooth, and the new students adapted quickly to online tools. In 2020, in addition to classes, regular online gatherings were held where students and faculty could come to chat, with multiple faculty members supporting students' mental

[5] Only four disciplines—Economics, Education, Global Studies, and Sociology—were involved initially. In 2021, the Management department began admitting students, and in 2022, the Journalism department started accepting students as well, completing the interdisciplinary course with six disciplines.

[6] https://www.sophia.ac.jp/jpn/aboutsophia/sophia_disclosure/info/sophia_disclosure_5/

[7] https://www3.nhk.or.jp/news/special/coronavirus/restrictions/detail/detail_22.html.

health. Joint classes and projects with universities from other countries, which would have been impossible before, were conducted online. For example, students were able to interview university management and discuss ways to achieve a sustainable campus (one of 4Cs). While mainstream media reported on the emptiness of once-crowded streets, creating a sense of urgency, this situation instead provided a setting for framing learning challenges. Students deepened their discussions on sustainability, realizing that addressing common human challenges was directly linked to sustainable development, that they were both witnessing and creating history, and that they could approach these issues as a system. Through confronting shared global and universal difficulties, students and faculty developed a sense of unity. As of April 2023, Japanese universities started to accept international students in a manner similar to before the pandemic.

4 Contributing to Sustainable Development

4.1 Competencies Set as Goals and Actual Potential

Through the SPSF course, students have acquired sustainability core competencies as Leicht et al. (2018) demonstrate (Maruyama 2025). This includes competencies tailored to specific situations and contexts, encompassing not only the designed learning outcomes but also the aspects of social and emotional learning that have gained attention recently, as well as the outcomes of non-formal education that include incidental learning (Maruyama 2019). Since sustainability often tends to fall into formal egalitarianism, it is necessary to discuss it within specific and individualized frameworks (Robertson 2021) and to emphasize informal learning, social practice, cultural contexts, and everyday experiences (Blewitt 2006). Research encourages co-creation of knowledge, requiring multiple perspectives and critical dialogue related to relevant concepts, along with pluralistic and liberating forms of education (Gough and Scott 2008). It is essential to always keep individual contexts and validity in focus.

While some faculty members pointed out a lack of knowledge in specific disciplines, the academic advisors who guided the graduation research in the six departments respected the research themes set by the students. Although true transdisciplinary research was not achieved, in June 2024, ten students conducted poster presentations of their graduation research publicly on campus, and in September, the first cohort of 18 SPSF students graduated. At this stage, we can only assess the learning outcomes of those who have completed the course, but interactions between SPSF student cohorts and students from other English courses have increased, and more diverse learning outcomes are expected in the future.

Additionally, it should be noted that students engaged in highly autonomous activities in extracurricular contexts. For example, they participated in circle activities that contribute to local environmental conservation and sustainable campus initiatives, as well as workshops offered by Stanford University in the United States of American, demonstrating the university's function as a living lab (Leal et al. 2020; Short et al. 2023). Furthermore, there was a strong emphasis on reclaiming and creating sustainability from everyday life (Washington 2015).

4.2 Faculty Development and Impacts on Campus

The professional development of faculty and staff was also an achievement. Typically, educational and administrative systems are established by department, making it extremely rare for curriculum arrangements across departments for SPSF. As a result, curriculum coordination was made across disciplines with sustainable futures as the core theme, often requiring responses that deviated from usual practices during the preparation. Additionally, some departments expressed resistance to the adjustments being made. However, the SPSF Office, established as a hub within the university, served as a foundation for dedicated coordination, enabling the program to move forward. Although the field of sustainability is often ambiguous and not prioritized as an academic discipline, it intersects with various fields of study. As Ratner (2004) suggests, this led to dialogue among faculty members.

Moreover, faculty members have begun to prioritize their own well-being. In Japanese universities, it is common for faculty to have significant responsibilities beyond the several 14-week teaching per semester, particularly the entrance exams that occur approximately many times a year. The entrance exams typically include a traditional written exam held once a year, for which faculty are responsible for creating the questions and grading the answers submitted by applicants. Additionally, there are recommendation exams, which include designated slots for high schools chosen by the university, as well as recommendations from schools affiliated with religious or educational organizations and general applications or self-recommendations. Faculty must review application materials, grade essays, and conduct interviews with applicants. Furthermore, faculty are expected to conduct trial lessons at high schools once or twice a year. Since all faculty are involved in these activities, they take time away from the three expected functions of education, research, and social contribution. Under such circumstances, the establishment of new educational courses has felt burdensome for faculty. Particularly during the pandemic, many unexpected tasks emerged, contributing to this burden. These experiences have unexpectedly fostered a "culture" of sustainability among faculty as one of 4Cs.

As it was the first attempt of its kind in Japan, SPSF was also an experimental initiative for the University. The president and vice president, who took the initiative during the preparation phase, stated that the experiences gained from SPSF, including its failures, would be beneficial for the university. Later, the vice president established a center that introduced a required course with an inclusive approach like SPSF in the Japanese track, using the operational insights from the SPSF as a reference. Additionally, the president emphasized the importance of SPSF both within and outside the university, establishing a Sustainability Promotion Headquarters to advance sustainability efforts campus wide as one of 4Cs.

However, regarding the community aspect as one of the 4Cs, significant progress has not been observed due to the reduced physical interactions caused by the pandemic. In the initial design of SPSF before the pandemic, students were expected to gain internship experiences with NGOs, international organizations, or private companies both in Japan and abroad, which would further their pursuit of sustainability. Although the severe restrictions on movement due to the pandemic have hindered this plan from proceeding as initially intended, there are expectations for future developments.

5 Conclusion: Holistic View on Education, Research and Social Contribution

The case study of Sophia University's "Sustainable Futures" program highlights the critical role that universities play in advancing ESD at higher education in Japan. By integrating interdisciplinary learning, systems thinking, and a focus on both local and global sustainability challenges, the program exemplifies how higher education can contribute to societal and learner's transformation. Despite facing significant challenges, particularly those arising from demographic decline and the COVID-19 pandemic, the program has successfully created a platform for students to engage in transformative learning experiences that extend beyond traditional academic boundaries. The pandemic, which occurred at an accidental timing, was not a favorable event for humanity, resulting in the loss of many lives. The impact of the pandemic on our daily lives may have been a negative factor for the campus life of undergraduate students, who have only four years. However, experiencing a common challenge in human history represents sustainable development also provided an opportunity to reflect on the role of research in society. This period fundamentally questioned the practice of Sophia University's motto, "For Others, With Others," prompting researchers to consider how their work connects with education and contributes to society. In this sense, many research efforts have reached a transdisciplinary moment. While this paper primarily focuses on education, the three expected roles of the university are interconnected. The emphasis on the "4Cs"—curriculum, campus, community, and culture—has proven essential in cultivating a holistic approach to sustainability within the university environment. As universities continue to navigate the complexities of sustainability, the University's efforts provide valuable insights into how institutions can evolve to foster socially responsible graduates as change agents for sustainable futures. Moving forward, it is essential to further strengthen the interconnectedness between education, research, and community engagement to create a sustainable future for all.

References

Banerjee SB (2003) Who sustains whose development? Sustainable development and the reinvention of nature. Organ Stud 24(1):143–180. https://doi.org/10.1177/0170840603024001341

Blewitt J (2006) The ecology of learning: sustainability, lifelong learning, and everyday life. Routledge

Bookchin M (2022) The philosophy of social ecology: essays on dialectical naturalism. AK Press

Brubacher JS (1982) On the philosophy of higher education. Jossey-Bass

Erickson HL, Lanning LA, French R (2017) Concept-based curriculum and instruction for the thinking classroom. Corwin

Escobar A (2018) Designs for the pluriverse: radical interdependence, autonomy, and the making of worlds. Duke University Press

Freire P (1996) Pedagogy of the oppressed. Penguin

Gough SR, Scott WAH (2008) Higher education and sustainable development: paradox and possibility. Routledge

Hill SB (2020) Transformative learning priorities. In: Wright D, Hill SB (eds) Social ecology and education: transforming worldviews and practices. Routledge, pp 28–40

Keidanren - Japan Business Federation (2022) Teigen: atarashii jidai ni taioushita daigaku kyouiku kaikaku no suishin—shutaiteki na gakushuu wo tsuujita tayouna jinzai no ikusei ni mukete [Suggestion: Promoting university education reform for the new age—Towards diverse human resource by active learning]. Keidanren

Jickling B (1992) Viewpoint: why I don't want my children to be educated for sustainable development. J Environ Educ 23(5):5–8. https://doi.org/10.1080/00958964.1992.9942801

Kopnina H (2012) Education for sustainable development (ESD): the turn away from "environment" in environmental education? Environ Educ Res 18(5):699–717. https://doi.org/10.1080/13504622.2012.658028

Lave J, Wenger E (1991) Situated learning: legitimate peripheral participation. Cambridge University Press

Leal WLF, Salvia AL, Pretorius RW, Brandli LL, Manolas E, Alves F, Azeiteiro U, Rogers J, Shiel C, Paco AD (eds) (2020) Universities as living labs for sustainable development. Springer. https://doi.org/10.1007/978-3-030-15604-6

Leicht A, Heiss J, Byun WJ (eds) (2018) Issues and trends in education for sustainable development. UNESCO

Maruyama H (ed) (2019) Cross-bordering dynamics in education and lifelong learning: a perspective from non-formal education. Routledge

Maruyama H (2025) SPSF 5 nenkan no keiken: Shunin kara miru seika to kadai [The five years of SPSF: Achievements and challenges from the chief coordinator]. Sophia University

Meadows DH (2008) Thinking in systems: a primer. Chelsea Green Publishing

Ministry of Education, Culture, Sports, Science, and Technology (MEXT) (2024) *Reiwa 6 nendo gakkou kihon chousa (sokuhouchi) ni tsuite kouhyou shimasu* [Announcing preliminary basic school survey 2024]. Retrieved 30 Sept 2024, from https://www.mext.go.jp/content/20240821-mxt_chousa01-000037551_001.pdf

MEXT (2006) Basic act on education (Act No. 120 of December 22, 2006). Retrieved from https://www.mext.go.jp

Mezirow J (2000) Learning as transformation: critical perspectives on a theory in progress. Jossey-Bass

Mulligan M (2017) An introduction to sustainability: environmental, social, and personal perspectives. Routledge

Orr D (2004) Earth in mind. Island Press

Ratner BD (2004) Sustainability as a dialogue of values: challenges to the sociology of development. Sociol Inq 74(1):50–69. https://doi.org/10.1111/j.1475-682X.2004.00079.x

Robertson M (2021) Sustainability principles and practice, 3rd edn. Routledge

Rose H, McKinley J (2018) Japan's English-medium instruction initiatives and the globalization of higher education. High Educ 75(1):111–129. https://doi.org/10.1007/s10734-017-0125-1
Sayce S, Bradley JF, Ritson J, Quinn F (2013) Well-being: what does it mean for the sustainable university? In: Sterling S, Maxey L, Luna H (eds) The sustainable university: progress and prospects. Routledge, pp 211–233
Scott WAH, Gough SR (2003) Sustainable development and learning: framing the issues. Routledge
Selby D (2009) Towards a sustainability university. J Educ Sustain Dev 3(1):103–106. https://doi.org/10.1177/097340820900300119
Short M, Nagai G, Maruyama H (2023) Engaging students in co-creating a sustainable campus as a living lab. Global Concern 5:158–175
Sophia University (2023). In Sophia University's roots—together with Francis Xavier's dreams and aspirations. Retrieved 30 Sept 2024, from https://www.sophia.ac.jp
Sophia University. (n.d.) Sophia program for sustainable futures (SPSF). Retrieved 30 Sept 2024, from https://spsf.sophia.ac.jp
Spencer J, Juliani AJ (2017) Empower: what happens when students own their learning. IMPress, LP
Sterling S (2007) Higher education, sustainability, and the role of systemic learning. In: Corcoran PB, Wals AJ (eds) Higher education and the challenge of sustainability. Springer, pp 49–70
Sterling S, Maxy L, Luna H (eds) (2013. The sustainable university: progress and prospects, Routledge.
Stroh DP (2015) Systems thinking for social change: a practical guide to solving complex problems, avoiding unintended consequences, and achieving lasting results. Chelsea Green Publishing
UNESCO (2021) Reimagining our futures together: a new social contract for education. UNESCO
Washington H (2015) Demystifying sustainability: towards real solutions. Routledge
Wals AEJ (2011) Learning our way to sustainability. J Educ Sustain Dev 5(2):177–186. https://doi.org/10.1177/097340821100500208
Wenger E (1998) Communities of practice: learning, meaning, and identity. Cambridge University Press

Preferences and Effectiveness of Sustainability Communication: A Study Amongst Maltese Generation X and Y

Clinton Cassar

Abstract Sustainability communication plays a crucial role in our society, influencing a wide range of domains and actors. Central to its role, is the dissemination of accurate information about pertinent global issues whilst serving as a driver to motivate individuals towards responsible behaviour. The main objective of this study is to investigate the preferences and effectiveness of communication methods among eighteen adult students, belonging to Generation X and Y, undertaking a study unit about sustainability at the University of Malta. Upon completion of the study unit, through purposive sampling, a questionnaire was administered to obtain participants' opinions about sustainability communication methods, with a particular focus on those utilised during the study unit, which includes, amongst others, comics, songs, and storytelling. Findings reveal that participants consider storytelling to be the most effective method which impacts significantly the participants' knowledge, attitudes and lifestyle. Communication methods related to the arts and literature such as theatre, poetry and puppetry are not preferred. While findings strongly reveal that the direct effectiveness of the sustainability methods as a learning tool is crucial during their educational experience, their applicability to personal choices and lifestyle is less evident. The Mann–Whitney U Test also revealed no differences between the two generations regarding: the level of understanding of sustainability communication; the influence of communication methods on knowledge, attitudes and lifestyle; the impact of the communication methods used during the course; and the characteristics of effective communication methods. The major contribution of this research is 'The Tripartite Model of Communication Effectiveness', which serves as a mapping tool to visually contextualize and plot the effectiveness of communication methods vis-à-vis knowledge, attitudes and lifestyle. This study, being part of a wider PhD study, warrants further investigation given its small sample size. However, it provides new insights into sustainability communication, indicating methods preferred among Generation X and Y participants, which may be relevant for stakeholders such as

C. Cassar (✉)
University of Malta, Msida, Malta
e-mail: clinton.cassar@um.edu.mt

W. Leal Filho et al. (eds.), *Education for Sustainable Development: The Contribution of Universities*, World Sustainability Series,
https://doi.org/10.1007/978-3-031-86985-3_14

policymakers, educators, and communication practitioners in designing effective strategies to foster behavioural change.

Keywords Sustainability · Sustainability communication · Education for sustainable development · Generation · Preferences

1 Introduction

Education for Sustainable Development (ESD) caters for different categories or groups of individuals, driving the need for sustainability communication across generations. Efforts are tailor-made to the needs and level of engagement of the targeted audience (Lamm et al. 2019; Warner et al. 2017), which in this study specifically includes the Maltese adult population, belonging to Generation X and Y. Generations X and Y possess distinct characteristics in a variety of domains. Generation X is a generation perceived as a transitional one or in the shadow of the previous generation (Baby Boomers) and therefore not considered as a driver for social change. The breakdown of traditional familial structures, rising divorce rates, and increased female workforce formed part of societal changes in this generation (Roberts and Manolis 2000). Generation Y individuals were born to the Baby Boom Generation. These individuals are self-confident, individualistic, consumeristic and demonstrate entrepreneurship. Technology plays an important part in the daily lives of this generation.

Even though such diverse factors have shaped these generations, a common essential function for each generation is communication, which serves a dual role: that of providing information and also optimizing how people think and perceive particular issues (Pelletier and Sharp 2008). Sustainability communication refers to the use of communicative processes and resources to promote effective policies, public participation, and the launching of projects oriented towards environmental sustainability (OECD 1999). Being a rather emerging field which has gained momentum over the past few years (Adomßent and Godemann 2011; Arief et al. 2022), this study provides a useful contribution to this field from a generational perspective.

Following an introduction of the research area, the next section elaborates on the study's objectives and research question.

2 Research Scope

The increasing prevalence of climate change, habitat degradation, economic recession and poverty have become a critical societal concern. Motivated by the need to tackle such issues, the demand for behaviour change through ESD strategies, including effective communication, has become crucial. Research has shown that communication can be challenging since it requires different strategies, channels, engagement methods and individual or group preferences (Fuchs 2017; Lamm et al. 2019). As the former postulates, there is a gap in the literature when addressing audience segmentation and communication channel considerations. Comprehensive studies focusing on the influence of sustainability communication on Generations X and Y and their response is still lacking. This research aims to address this gap and examines the extent to which sustainability communication impacts one's knowledge, attitudes and lifestyles. Given the different characteristics of each generation, understanding their preferences is essential when formulating effective communication strategies.

The research context takes place in the Maltese islands, the smallest EU member state in the middle of the Mediterranean Sea, with over 500,000 inhabitants in an area of 316 km^2. Being a microstate makes it more vulnerable to a variety of sustainability issues which need to be considered and communicated effectively, making this study pertinent in exposing the importance of sustainability communication. The main objective of this study is to identify the preferences of Maltese adult students, segmented into Generation X and Y, vis-à-vis various sustainability communication methods.

This study is exploratory in nature since it forms part of preliminary research embedded within a PhD study, aimed at gathering initial insights and gauging the appropriateness of the methods used. The crux of this phase is to ensure methodological refinement and shape a robust research design, yielding the likelihood of relevant findings within the entire PhD study. Fuelled by these objectives, this study seeks to answer the following core research question:

Which sustainability communication methods mostly impact Generation X and Y's knowledge, attitudes and behaviour?

To address this research gap and answer the aforementioned research question, a quantitative approach will be adopted via a questionnaire to accentuate the perspectives of the target population.

This research chapter is divided into a number of sections. The next section delineates the theoretical underpinnings followed by the methodological framework adopted. Following this, the successive section will roll out the findings to pave the way for the last section which will include the concluding remarks and recommendations.

3 Literature Review

3.1 Effectiveness of Sustainability Communication

Moser (2006:3) defines effective communication as "any form of public engagement that actually facilitates an intended behavioural, organizational, political or other social change consistent with identified mitigation and adaptation goals". This resonates with the objectives of Education for Sustainable Development (ESD), where behavioural change might be initiated through sustainability communication.

The effectiveness of communication as a driver for behavioural change has long been argued, since having a high level of education does not necessarily lead to adopting sustainable practices due to inconveniences and other psychological barriers (de Vries et al. 2019). This line of thought aligns with the linear rational choice models such as the Shannon–Weaver model of communication (Shannon and Weaver 1949), which postulates that knowledge directly translates into behavioural change, neglecting any underlying reasons which hinder this change such as cognitive biases, including denial, defensive avoidance, and emotions (fear and hope). Expectations about communication methods and the influence of other individual's behaviours should equally be acknowledged (de Vries 2019).

According to Moser (2016), the effectiveness of sustainability communication does not rest exclusively on awareness-raising or fearmongering but more on active engagement. Weder et al. (2021) consider the urgency to adopt a change in approach from communication about sustainability to communication for sustainability. This transition requires the inclusion of three different facets of sustainability communication: (i) communication about sustainability aims at providing information, interpretations, and opinions about sustainability; (ii) communication of sustainability, usually includes experts or practitioners who provide information to educate the audience to take action regarding pertinent issues; and, (iii) communication for sustainability emphasises on the normative influence of sustainable development, focusing more on its transformative power determined through measurable action.

McQuail (2010) asserts that the effectiveness of sustainability communication depends on the deliverance of the right message, to the right audience, through the right channel, at the right time. As postulated by the theory of motivated reasoning (Kunda 1990), individual predispositions such as attitudes, beliefs, defensive motivations and accuracy motivations should be considered (Chaiken et al. 1989). Rather than adopting a one-size-fits-all approach, messages should be tailored to the specific characteristics of each age group, increasing communication effectiveness (Corner and Randall 2011). This can be achieved through segmentation analysis, which involves the divisions of individuals into subgroups, usually involving factors such as geographic (region, population density, climate), socio-demographic (age, gender, class), psychological (values, attitudes) and behavioural patterns.

Various studies have implemented audience segmentation, the most renowned is that of Maibach et al. (2011) who identified the 'Six Americas' concerning climate change. This study segmented the wide perceptions of Americans about climate change into six categories: alarmed, concerned, cautious, disengaged, doubtful, and dismissive. These categories were predictors for nine policies about greenhouse gas reduction. Such a concept was also diffused in Australia, where Hine et al. (2013) proposed five other categories: alarmed, concerned, uncertain, doubtful, and dismissive. Despite these different classifications, the studies indicate similar attitudes among Americans and Australians.

Furthermore, a synthesis of the literature determines the following factors (categorised into two strands) as crucial for the effectiveness of sustainability communication (Corner et al. 2018; Kayaalp and Toprak 2017; Muradova et al. 2020; Primožič and Kutnar 2022):

The first strand focuses on the design of communication channels which are direct and easy for the audience to understand, based on: (i) transparency, where interested individuals need to be aware of the processes, measures, methods, and sources used during the communication process; (ii) appropriateness focuses on the relevance of information that meets the needs of the audience. Complex issues need to be simplified through action-oriented processes; (iii) credibility, emphasises that the message needs to be conveyed honestly and fairly; (iv) coherence stresses that values and actions should be aligned to expected outcomes.

The second strand encourages that communication methods initiate relatedness with the audience in a variety of ways, namely: (i) responsiveness, where the communication process addresses the needs of the concerned individuals and responds to them accordingly; (ii) connection focuses on the relevance of the message. According to Nisbet (2009), carefully selected visual or verbal metaphors or real-world analogies can help the audience to relate, especially when explaining distant or complex issues such as climate change; (iii) personalisation, focuses on messages that are tailor-made to the target audience; (iv) narration emphasizes on the presentation of meaningful experiences rather than statistical data; (v) concreteness stresses on real and tangible issues rather than abstract ones. Real people and their stories rather than staged photos are more effective.

These two strands fuse clarity with relatedness since the need to narrate, relate and personalize can make the message clearer and simpler to any audience. According to Cox (2013), the selection of the communication method can positively alter individual attitudes and behaviours, if it resonates with the target audience.

3.2 Generational Preferences in Communication Styles

An important element in sustainability communication is the channel and source through which the information is disseminated. The former refers to the method through which information is communicated whereas the latter refers to the individual or organization behind the message. Even though preferences are individualised, these might change as new means of communication are developed or individuals are exposed to more accessible methods.

Effective sustainability communication rests on personalisation, which is determined by demographics, location and the accessibility of the selected method (Wilkins et al. 2018). Differences in communication preferences depend on demographics—visual and printed communication is popular with adults while the younger generation prefers online communication. Also, personal experiences are a top communication method, which needs to be interactive, hands-on and conversational. Online means of communication seem prevalent in many studies such as Takahashi and Tandoc (2016) and Lamm et al. (2019). Ejaz et al. (2022) assert that the most important channel is television, closely followed by online sources such as websites and social media.

Previous studies about generational preferences for sustainability communication channels are limited, except for a Generational Sustainability Survey carried out in 2023. It reveals that at least 30% of every generation ranked social media as one of their top three sources, despite being their third or fourth most trusted source of information (EY and JA worldwide 2023). The use of technology is prevalent with Generation Y, who enjoy face-to-face communication but to a lesser degree than Generation X, since they are digital natives (Venter 2017), with the media consuming approximately 33–38 h of their time per week (Bennett et al. 2006). Despite being technological experts, the members of this generation struggle with oral, written and interpersonal skills (Hartman and McCambridge 2011). A study by Schroth (2019), confirms that, together with Generation Z, 74% of respondents from this generation communicate digitally while 26% prefer in person. They enjoy constant connections with family and friends via social networks (Black 2010), creating different types of communication styles. Liu et al. (2012) describe these styles as: instant, playful, collaborative, expressive, responsive and flexible. Such characteristics may imply openness to collaboration with other generations, with online means of communication being the preferred mode.

Generation X is considered a digital immigrant since they were born before the digital age and had to familiarise themselves with technological advancements. Fox (2014) describes this generation as stuck between the analogue and the digital generation. Since Generation X is not accustomed to technological exposure, they perceive traditional communication—face-to-face, letter and phone communication—as threatened (Haserot 2009). However, Martin (2011) warns that despite having more variety in communication the older generation still reverts to familiar forms.

In conclusion, the right combination of interactive strategies, channels and media formats as put forward by Lamm et al. (2019) can contribute to effective sustainability communication to specific audiences, including Generation X and Y, who like any other members in the community have their preferences. This shall be explored in more detail in the successive section, where the methodology is rolled out.

4 Methodology

As briefly outlined in the research scope, this exploratory study was conducted, as asserted by Aithal and Aithal (2023), to identify, understand and interpret a problem by providing possible solutions, which in this case focuses on the generational preferences of sustainability communication methods. Through the findings, as suggested by Robson (2004), new initial insights and ideas are generated to leverage methodological rigour rather than garner generalizable findings.

Table 1 Segmentation of the population sample by generation

	Male	Female	Total
Generation X	6	5	11
Generation Y	4	3	7
Total	10	8	18

Since this exploratory research was carried out in the early stages of the PhD, it was crucial to: (i) identify the main factors that require further investigation; (ii) test the feasibility of the data collection process and sustainability communication methods utilised; (iii) evaluate the logical considerations impacting the feasibility of such methods; and, (iv) finding adequate ways how to approach members from Generation X and Y.

Based on such justifications, this first stage of the preliminary study involves the acquisition of quantitative data based on Generation X and Y's preferences for communication methods through the use of a questionnaire.

4.1 Participants and Design

The population sample, selected through purposive sampling, consisted of 18 adult students who attended a module about sustainability delivered at the University of Malta from February to May 2024. As seen in Table 1, a total of 10 males and 8 females participated in the study, with 11 participants belonging to Generation X and 7 participants forming part of Generation Y.

4.2 Materials

The main tool utilised was a questionnaire which included 10 questions, divided into three sections, as follows:

i. Section A: Demographic Characteristics—contained three questions which focused on the age, gender, and locality of respondents;
ii. Section B: Knowledge about Sustainability Communication—The main question asked about the participants' level of understanding of sustainability communication.
iii. Section C: The Effectiveness of Sustainability Communication—Participants were asked to rank the impact of various methods on their knowledge, attitudes, and lifestyle, with one question focusing on methods used during the course whereas another question displayed methods not used during the study unit. Other questions: presented a series of statements about the relevance of the communication methods used during the course, to be rated on a 5-point

Likert scale (Highly agree, Agree, Neither agree nor disagree, Disagree, Highly disagree); addressed the challenges that hinder the effectiveness of communication methods; and, displayed a set of statements about what makes sustainability communication methods effective.

4.3 Procedure

The following sustainability communication methods were utilised during the course:

i. Storytelling was utilized to narrate the life of factory workers in Bangladesh and the lack of social justice and quality of life amongst people.
ii. Newspaper articles were utilized to create discussion about environmental justice based on climate change and pollution amongst different social groups.
iii. Songs such as 'The Earth Song' by Michael Jackson in 1995 and 'Where is the Love?' by Black Eyed Peas in 2003 were utilized to highlight different sustainability challenges and the trajectory of sustainable development throughout the years.
iv. Comics were utilized to create a discussion in highlighting different sustainability challenges such as climate change, acid rain, deforestation, poverty and development.
v. Photographs were utilized depicting different forms of development, eliciting characteristics belonging to ecocentric and anthropocentric approaches.

Upon completion of the module, students were asked to complete an online self-administered and anonymous questionnaire. The questionnaire was specifically designed to assess the impact of the aforementioned methods and other possible ones in impacting the respondents' knowledge, attitudes and behaviour.

4.4 Limitations and Challenges

A major challenge of this study was accessing the population, by specifically delivering an educational experience involving Generation X and Y. Being the sole researcher of the study, with limited time and resources, lecturing a module at the University of Malta facilitated this challenge. Ethical considerations were respected given the researcher-educator role and participants' anonymity and confidentiality were assured. As asserted by Daniel (2012:238), in an exploratory study "the researcher is not attempting to make conclusive analyses, and a small sample size may suffice". Therefore, the main focus of this study involves the exploration of potential methodological avenues rather than the acquisition of generalizable findings that guide the wider subsequent PhD data collection process.

In the light of this, larger data sets including quantitative data from other generations (Baby Boomers, Generation Z and Alpha) and qualitative data from focus groups will be included in the subsequent phases of this preliminary study. While it is understood that the results may lack replicability and that relationships between variables are limited, findings will be interpreted with caution. The appropriate statistical test, the Mann U Whitney Test, was carefully selected, to cater for relationships created within this small sampling size.

5 Findings and Discussion

This section presents the findings obtained through the questionnaire both in a descriptive and inferential manner, followed by a discussion which leads to answering the research question.

5.1 Knowledge About Sustainability Communication

Participants were asked to rate their level of understanding of sustainability communication, through a rating scale ranging from no understanding (1) to extensive understanding of sustainability communication (5). Figure 1 demonstrates that there are some differences in the frequency of ratings. The lowest ranking (1) was only scored by Generation Y whereas the highest ranking (5) was generated only by Generation X. The highest frequency for both generations is 3, which is a moderate understanding of sustainability communication (Table 2).

A Mann–Whitney U test was run to investigate if there are any differences between generations. H1 hypothesised a relationship between 'Generation (Gen X vs. Gen Y)' and 'Knowledge of communication for sustainability' [KnowComm]. The distribution scores of KnowComm for Gen X and Gen Y were similar as assessed by visual inspection. KnowComm scores for Gen X (Mean = 3.73) and Gen Y (Mean = 2.86) were not statistically different, $U = 22.5$, $z = -1.514$, $p = 0.15$. Therefore, no support was provided for H1 and the null hypothesis of no relationship was retained.

5.2 Preferences and Effectiveness of Sustainability Communication Methods

Participants were asked to rank on a scale from 1 to 5 (1 being the least ranking while 5 being the highest ranking) the effectiveness of methods used during the course on their knowledge, attitudes and lifestyle. The methods included storytelling, newspaper articles, songs, comics and photographs. Another question then asked participants to rate other methods not used during the course on their knowledge,

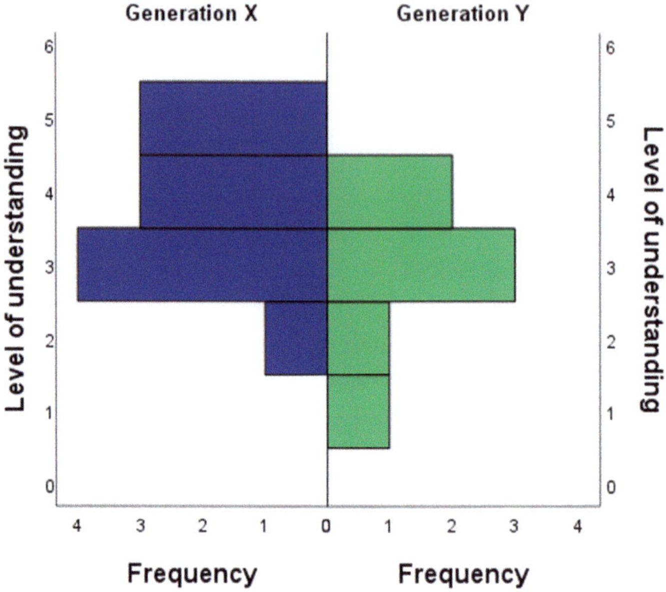

Fig. 1 The level of understanding of sustainability communication by generation

Table 2 The level of understanding across generation X and Y

Hypothesis	*p*-value	Mean	
		Gen X	Gen Y
The level of understanding is the same across Generation X and Y	0.151	3.73	2.86

attitudes and lifestyle, which included billboards, online material, brochures, theatre, cartoons, movies, puppetry and poems.

The mean rating score for the methods used in the course is found in Table 3, with the highest scores indicated in bold whereas the lowest scores are underlined. Storytelling emerged as the most effective method for influencing knowledge, attitudes, and lifestyle, achieving the highest average rating in all three domains. It has the greatest impact on attitudes, with a mean score of 4.67, the highest recorded in Table 3. In contrast, newspaper articles received the lowest mean score for knowledge, attitudes, and lifestyle. Notably, this method, along with songs and comics, shared the lowest mean score of 3.67 for their impact on lifestyle.

Table 4 represents the median score for the impact of communication methods used during the course, which is above 3, for all three aspects. There is a strong consensus amongst participants that storytelling, songs, and photographs are impactful on knowledge and attitudes, all garnering a median score of 5. This means that more than half of the respondents rated these methods as effective. On the other hand, the impact on lifestyle was not as strong with all methods obtaining a median score of 4. The lowest median score of 3.5, is the impact of newspaper articles on

Table 3 Mean rating score for the impact of communication methods used during the course

	Storytelling	Newspaper article	Songs	Comics	Photographs	Total mean rating score
Impact on knowledge	4.56	<u>3.89</u>	3.94	4.33	4.44	**21.16**
Impact on attitudes	**4.67**	<u>3.78</u>	4.06	4.11	4.39	**21.01**
Impact on lifestyle	<u>4.06</u>	<u>3.67</u>	<u>3.67</u>	<u>3.67</u>	3.83	**18.9**

lifestyle. This indicates that some respondents may have rated highly while others do not, suggesting a weaker impact on lifestyle.

Table 5 presents the mode for all communication methods presented during the course, representing the most frequently occurring rank in the dataset. Except for the impact of newspaper articles on lifestyle, which had a mode of 3, all other methods received high scores. Storytelling and songs achieved a mode of 5 across all three domains: knowledge, attitudes, and lifestyle. Notably, as seen in the previous Table 4, the impact on lifestyle shows the lowest mode, with a score of 3 appearing in newspaper articles, comics, and photographs.

The mean rating scores for the methods not used in the course are shown in Table 6. The highest scores are highlighted in bold, while the lowest are underlined. Online material achieved the top ratings for its impact on knowledge, attitudes, and lifestyle. Together with movies, it also obtained the highest score for influencing knowledge (4.17). In contrast, puppetry garnered the lowest mean scores across all three domains. Interestingly, puppetry, along with theatre and poetry, all received a mean score of 2 in each domain.

When compared to Table 4, Table 7 shows lower median scores ranging from 1.5 up to 4. No median score of 5 is displayed. While online materials are consistently effective in influencing all three domains, achieving a high media score of 4, puppetry

Table 4 The median score for the impact of communication methods used during the course

	Storytelling	Newspaper article	Songs	Comics	Photographs
Impact on knowledge	5	4	5	5	5
Impact on attitudes	5	4	5	4	5
Impact on lifestyle	4	3.5	4	4	4

Table 5 The mode for the impact of communication methods used during the course

	Storytelling	Newspaper article	Songs	Comics	Photographs
Impact on knowledge	5	4, 5	5	5	5
Impact on attitudes	5	4	5	4	5
Impact on lifestyle	5	3	5	3, 4	3, 4

Table 6 Mean rating score for the impact of communication methods not used during the course

	Billboards	Online material	Brochures	Theatre	Cartoons	Movies	Puppetry	Poetry	Total mean rating score
Impact on knowledge	3.28	4.17	3.11	2.67	3.39	4.17	2	2.2	24.99
Impact on attitudes	3.17	4	3.17	2.78	3.44	3.83	2.11	2.44	24.94
Impact on lifestyle	3.28	4.06	3.278	2.61	3.17	3.72	2.11	2.67	24.898

has the lowest score across all three categories, implying it is the least effective method according to participants. On the other hand, movies and cartoons are also effective. Poetry also obtained a low median score but fared slightly better than puppetry.

Table 8 presents the mode for all communication methods not included in the course. Online material shows the highest mode across all three domains, alongside movies, which achieved a mode of 5 for its impact on knowledge. In contrast, puppetry obtained a mode of 1 in all three domains, with poetry following closely, reflecting its low impact on knowledge and attitudes.

In Table 9, H1 hypothesised a relationship between 'Generation (Gen X vs. Gen Y)' and 'Influence of communication for sustainability' [InfluenceComm]. A Mann–Whitney U test was run to investigate if there are any differences between generations. The distribution scores of InfluenceComm for Gen X and Gen Y were similar as assessed by visual inspection. InfluenceComm scores for Gen X and Gen Y were not statistically different. Therefore, no support was provided for H1 and the null hypothesis of no relationship was retained.

Interestingly, the impact of songs and comics on knowledge garnered a p-value of 1.00. This is the highest value in Table 9, suggesting the biggest discrepancy between the aforementioned generations. This is followed by the impact of newspaper articles on attitudes and lifestyle, obtaining *p*-values of 0.930 and 0.860 respectively.

Participants were asked to rank various statements about the communication methods used during the course as seen in Table 10. H1 hypothesised a relationship between 'Generation (Gen X vs. Gen Y)' and 'Impact of communication for

Table 7 Median score for the impact of communication methods not used during the course

	Billboards	Online material	Brochures	Theatre	Cartoons	Movies	Puppetry	Poetry
Impact on knowledge	3	4	3	3	4	4	1.5	2
Impact on attitudes	3	4	3	3	3.5	3.5	1.5	2.5
Impact on lifestyle	3	4	3	3	3	4	2	3

Table 8 The mode for the impact of communication methods not used during the course

	Billboards	Online material	Brochures	Theatre	Cartoons	Movies	Puppetry	Poetry
Impact on knowledge	3	5	3	3, 1	4	5	1	1
Impact on attitudes	3	5	3	1	3	3	1	1
Impact on lifestyle	3	4	4, 3	3	3	5, 3	1	3

Table 9 The influence of various communication methods used during the course on Generation X and Y

	Knowledge			Attitudes			Lifestyle		
	p-value	u	Mean	*p*-value	u	Mean	*p*-value	u	Mean
Storytelling	0.375	48.5		0.536	43.5		0.425	47.5	
Gen X			4.45			3.73			3.73
Gen Y			4.71			4.14			4.29
Newspaper article	0.479	46.5		0.930	37		0.860	36.5	
Gen X			4.27			4.45			4.72
Gen Y			4.43			4.43			4.57
Songs	1.00	38		0.479	46.5		0.659	43.5	
Gen X			3.73			3.91			4.09
Gen Y			3.86			4.29			4.14
Comics	1.00	38.5		0.425	47.5		0.285	50.5	
Gen X			4.36			3.91			3.45
Gen Y			4.43			4.29			4.00
Photographs	0.479	47		0.479	46.5		0.596	44.5	
Gen X			3.45			3.55			3.73
Gen Y			4			3.86			4.00

sustainability' [ImpactComm]. A Mann–Whitney U test was run to investigate if there are any differences between generations. The distribution scores of ImpactComm for Gen X and Gen Y were similar as assessed by visual inspection. ImpactComm scores for Gen X and Gen Y were not statistically different, $U = 28$, $z = -1.214$. Therefore, no support was provided for H1 and the null hypothesis of no relationship was retained .

Participants were required to select three challenges from a list of options displayed in Fig. 2, focusing on the implementation of sustainability communication. The top three challenges chosen by respondents were: Lack of interest by audience (11), Lack of funds (10) and Lack of resources (materials, items, devices) (8). Two respondents chose 'other' by adding their responses. These included: the lack of knowledge conveyed to the public and the lack of seriousness in implementing sustainability.

Participants also had to rate the following statements about what makes a sustainability communication method effective. The Mann–Whitney U Test was used to statistically determine whether the following statements about communication methods varies across Generation X and Y. H1 hypothesised a relationship between 'Generation (Gen X vs. Gen Y)' and 'Effectiveness of communication for sustainability' [EffectivenessComm]. A Mann–Whitney U test was run to investigate if there are any differences between generations. The distribution scores of EffectivenessComm for Gen X and Gen Y were similar as assessed by visual inspection. EffectivenessComm scores for Gen X and Gen Y were not statistically different.

Table 10 The impact of communication methods used during the course

Statement	*p*-value	U	Mean	
			Gen X	Gen Y
(a) The various communication methods were enjoyable	0.375	28	4.82	4.14
(b) The communication methods were useful to learn about sustainability	0.596	45	4.64	4.43
(c) The communication methods helped me to understand difficult concepts	0.069	18	4.45	3.57
(d) The communication methods help me to express my opinion	0.085	19	4.91	3.86
(e) The communication methods encouraged us to interact and participate together	0.375	28	4.82	4.14
(f) The communication methods helped me think about the impact of my choices on sustainability issues	0.179	23	4.45	3.71
g) The communication methods helped me make good choices in everyday life	0.375	28	4.45	3.86

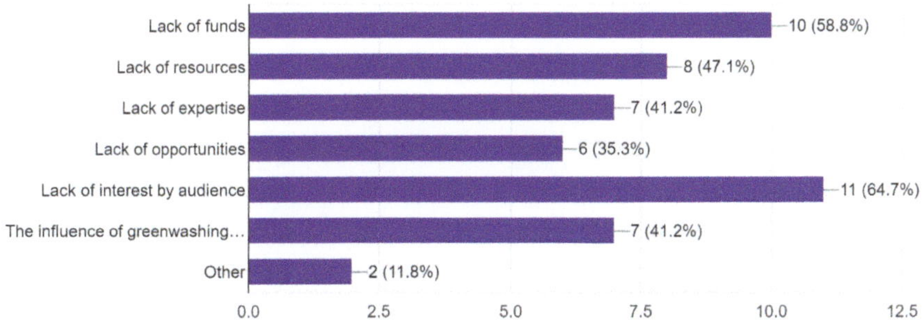

Fig. 2 The challenges in implementing sustainability communication

The level of significance for each statement exceeds 0.05. Therefore, no support was provided for H1 and the null hypothesis of no relationship was retained.

It appears that both Generation X and Y participants agree on Statement D since it scored 1.00, whereas Statement H slightly exceeded the level of significance, achieving a value of 0.056. This indicates that the similarity of beliefs in this regard is extremely weak (Table 11).

6 Discussion

The current study produced findings which need to be viewed within its parameters, primarily identifying patterns and evaluating the suitability of communication methods. It also demonstrates no significant differences between the communication

Table 11 Effective characteristics of communication methods

Statement	*p*-value	U	Mean	
			Gen X	Gen Y
i. The communication method presents the message clearly	0.860	36.5	4.73	4.29
ii. The communication method presents the message in simple language.	0.724	34	4.64	4.14
iii. The communication method presents messages with updated information.	0.724	43	4.18	4.14
iv. The communication method presents messages through effective visuals.	1.000	39	4.82	4.43
v. The communication method presents messages relevant to the world's needs	0.126	21	4.55	3.57
vi. The communication method presents messages relevant to the people's needs	0.659	33.5	4.09	3.71
vii. The communication method presents messages which are easy to remember	0.151	22	4.64	3.86
viii. The communication method is effective as the speaker is effective.	0.056	17.5	4.27	3.29
ix. The communication method is effective as the speaker and audience participate effectively.	0.596	32	4.64	4.14

preferences of Generation X and Y. This is confirmed by The Mann–Whitney U Test. In all cases when this was utilised, the null hypothesis was always accepted, which states that there are no differences between the two generations in: the level of understanding of sustainability communication; the influence of communication methods on knowledge, attitudes and lifestyle; the impact of the communication methods used during the course; and the characteristics of effective communication methods.

This shows that the communication methods used had the same degree of effectiveness for both generations uniformly and impacted them similarly. Overall, the sampling population of this study consider storytelling as the most effective method, similarly impacting knowledge, attitudes, and lifestyle. This conforms to the strand of relatedness, as discussed in the literature review, implying that participants can connect with this communication method.

Communication methods related to the arts and literature (theatre, poetry, and puppetry) are not preferred channels as much as online means. It can be assumed that online methods can be easily accessed and are considered more appealing. This conforms to studies mentioned in the literature review, such as Takahashi and Tandoc (2016) and Lamm et al. (2019).

A discrepancy is visible in the applicability of communication methods between knowledge and lifestyle. This might indicate a knowledge-behaviour gap which can be addressed through appropriate educational opportunities, considering Weder et al. (2021) shift towards communication for sustainability to initiate behavioural change.

These preliminary results have been framed within the scope of this study, however such insights warrant further investigation, which will inform the next stage of the data collection: (i) the preferences amongst Generation X and Y participants, will be utilised to perform future comparative quantitative and qualitative analysis with other generations, namely: Baby Boomers, Generation Z and Alpha and, (ii) the familiarisation with a range of communication techniques, helped the researcher understand which approaches are most effective with these two generations, which will also be applied to the other generations as well. This will ensure that suitable communication methods will be used for the actual data collection process.

A major contribution of this study includes an original model entitled 'The Tripartite Model of Communication Effectiveness'. This tool is derived from the median rating scores garnered in Tables 4 and 7. Inspired by these findings, a mapping tool was devised to visually represent the interconnected impact of a communication method on knowledge, attitudes and lifestyle. This is a mapping model which consists of three triangles each representing knowledge, attitudes, and lifestyle. Each triangle displays the impact of a communication method, ranging from low impact (1) to high impact (5). It is an important model to: visually plot the effectiveness of a communication method; examine the strengths and weaknesses of each method; and categorise methods which are most effective and least effective for each population group. Based on the parameters of this study, this tool sheds light on visually mapping the communication preferences of different generations. It will assist in choosing methods which are well-suited to the different generations, within the wider scope of the PhD study (Fig. 3).

7 Conclusion

This study has set out to explore the preferences and effectiveness of communication methods amongst Maltese adult learners at the University of Malta. Given that no generalizability is possible through such a limited sample, this study still provides an understanding of how Generation X and Y respond to particular issues and which methods might be used for an intended audience. Rather than garnering definitive conclusions, this study is constructive by informing the next stages of the PhD study, ensuring that they are carefully planned and methodologically sound.

Ultimately, the study has shed light on the need for reorienting educational opportunities by utilizing adequate communication approaches to meet the needs of the target audience which might encourage behavioural change. Other researchers and policymakers might use this case study to further investigate how communication

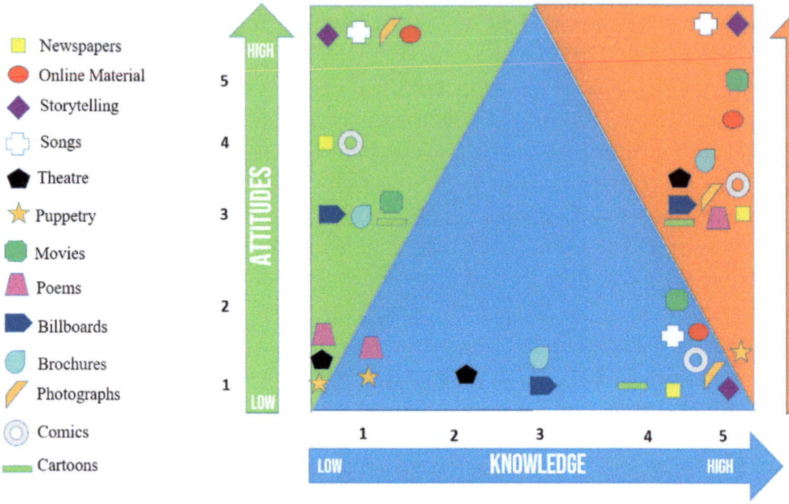

Fig. 3 The Tripartite model of communication effectiveness

methods can be utilized within formal, informal and non-formal education contexts. Further research is required to analyse other audience segments as regards sustainability communication to ensure that such educational experiences are of benefit to students and the wider community.

References

Adomßent M and Godemann J (2011) Sustainability communication: an integrative approach. In: Godemann J, Michelsen G (eds) Sustainability communication: interdisciplinary perspectives and theoretical foundation. Springer, Dordrecht, The Netherlands, New York, NY, USA, pp 27–37

Aithal PS, Aithal S (2023) New research models under exploratory research method. In: Paul PK et al (ed) Emergence and research in interdisciplinary management and information technology. New Delhi Publishers, New Delhi, India

Arief NN, Famiola M, Pratama AP, Anggahegari P, Putri ANA (2022) Sustainability communication through bio-based experiential learning. Sustainability 14:5204. https://doi.org/10.3390/su14095204

Bennett G, Sagas M, Dees W (2006) Media preferences of action sports consumers: differences between Generation X and Y. Sport Mark Q 15(1):40–49

Black A (2010) Gen Y: Who they are and how they learn. Educ Horizons 88(2):92–101

Chaiken S, Liberman A, Eagly AH (1989) Heuristic and systematic information processing within and beyond the persuasion context. In: Uleman JS, Bargh JA (eds) Unintended thought. Guilford Press, New York, pp 212–252

Corner A, Randall A (2011) Selling climate change? The limitations of social marketing as a strategy for climate change public engagement. Glob Environ Chang 21:1005–1014

Corner A, Shaw C, Clarke J, Wang S (2018) Communicating environmental and sustainability science: challenges, opportunities, and the changing political context. Climate Outreach

Cox R (2013) Environmental communication and the public sphere. Sage Publications, Thousand Oaks, CA
Daniel J (2012) Choosing the size of the sample. In Daniel JN (ed) Sampling essentials: practical guidelines for making sampling choices. SAGE Publications, Inc., pp 236–253. https://doi.org/10.4135/9781452272047.n
de Vries G (2019) Public communication as a tool to implement environmental policies. Soc Iss Policy Rev 14(1):244–272. https://doi.org/10.1111/sipr.12061
de Vries G, Rietkerk M, Kooger R (2019) The hassle factor as a psychological barrier to a green home. J Consum Policy 43:345–352. https://doi.org/10.1007/s10603-019-09410-7
Ejaz W, Mukherjee M, Fletcher R, Nielsen RK (2022) How we follow climate change: climate news use and attitudes in eight countries. Reuters Institute for the Study of Journalism, Oxford
EY and JA worldwide (2023) How can we empower the next generations to build a more sustainable future? Generational Sustainability Survey. https://assets.ey.com/content/dam/ey-sites/ey-com/en_gl/topics/corporate-responsibility/ey-ja-2023-sustainability-report-27-july-2023.pdf
Fox P (2014) GenX and social media: stuck in the middle. Business2Community. https://www.business2community.com/social-media-articles/gen-x-social-media-stuck-middle0939476#cxho5TVy2EhIqX%20GZ.97hotelmarketing.com/index.php/content/article/online_hotel_%20booking_market_estimated_to_exceed_100_billion_in
Fuchs C (2017) Social media: a critical introduction, 2nd edn. Sage Publications Ltd., Los Angeles, CA
Hartman JL, McCambridge J (2011) Optimizing millennials' communication styles. Bus Commun Q 74(1):22–44
Haserot PW (2009) The keys to maintaining professionalism & harmony across the generations at work. Prof Lawyer 19(3):10425675
Hine DW, Reser JP, Phillips WJ, Cooksey R, Marks ADG, Nunn P, Watt SE, Bradley GL, Glendon AI (2013) Identifying climate change interpretive communities in a large Australian sample. J Environ Psychol 36:229–239
Kayaalp T, Toprak D (2017) Communication strategies for sustainability (TC Ministry of Urbanization and Environment, Communication for Sustainability). Sustainable production and consumption publications—VII (ET 1.10.2021). https://rec.org.tr/wp-content/uploads/2017/02/
Kunda Z (1990) The case for motivated reasoning. Psychol Bull 108(3):480–498
Lamm KW, Borron A., Holt J, Lamm AJ (2019) Communication channel preferences: a descriptive audience segmentation evaluation. J Appl Commun 100(3). https://doi.org/10.4148/1051-0834.2238
Liu W, Pasman G, Stappers PJ, Taal-Fokker J (2012) Making the office catch up: Comparing Generation Y interactions at home and at work. Newcastle, DIS 2012, pp 349–352
Martin EMP (2011) *D*igital natives and digital immigrants: teaching with technology. Unpublished Doctoral dissertation, Northeastern University
Maibach EW, Leiserowitz A, Roser-Renouf C, Mertz CK (2011) Identifying like-minded audiences for global warming public engagement campaigns: an audience segmentation analysis and tool development. PLoS ONE 6(3):e17571. https://doi.org/10.1371/journal.pone.0017571
McQuail D (2010) McQuail's mass communication theory, 6th edn. SAGE, London
Moser SC (2006) Talk of the city: engaging urbanites on climate change. Environ Res Lett 1(1):014006
Moser SC (2016) Reflections on climate change communication research and practice in the second decade of the 21st century: what more is there to say? WIREs—Clim Change 7(3):345–369. https://doi.org/10.1002/wcc.403
Muradova L, Walker H, Colli F (2020) Climate change communication and public engagement in interpersonal deliberative settings: evidence from the Irish citizens' assembly. Clim Policy 20(10):1322–1335. https://doi.org/10.1080/14693062.2020.1777928
Nisbet MC (2009) Communicating climate change: why frames matter for public engagement. Environ: Sci Policy Sustain Dev 51(2):12–23

OECD (1999) Environmental communication. In: Applying communication tools towards towards sustainable development, Paris. OCDE

Pelletier LG, Sharp E (2008) Persuasive communication and pro-environmental behaviours: How message tailoring and message framing can improve the integration of behaviours through self-determined motivation. Can Psychol/psychologie Canadienne 49(3):210–217. https://doi.org/10.1037/a0012755

Primožič L, Kutnar A (2022) Sustainability communication in global consumer brands. Sustainability 14:13586. https://doi.org/10.3390/su142013586

Roberts JA, Manolis C (2000) Baby boomers and busters: an exploratory investigation of attitudes toward marketing, advertising and consumerism. J Consum Mark 17(6):481–497. https://doi.org/10.1108/07363760010349911

Robson C (2004) Real world research, 2nd edn. Blackwell Publishing, London

Schroth H (2019) Are you ready for Gen Z in the workplace? Calif Manage Rev 61(3):5–18. https://doi.org/10.1177/0008125619841006

Shannon CE, Weaver W (1949) The mathematical theory of communication. The University of Illinois Press, Urbana, IL, pp 1–117

Takahashi B, Tandoc EC (2016) Media sources, credibility, and perceptions of science: learning about how people learn about science. Public Underst Sci 25(6):674–690. https://doi.org/10.1177/0963662515574986

Venter E (2017) Bridging the communication gap between Generation Y and the Baby Boomer generation. Int J Adolesc Youth 22(4):497–507. https://doi.org/10.1080/02673843.2016.1267022

Warner LA, Chaudhary AK, Rumble JN, Lamm AJ, Momol E (2017) Using audience segmentation to tailor residential irrigation water conservation programs. J Agric Educ 58(1):313–333. https://doi.org/10.5032/jae.2017.01313

Weder F, Karmasin M, Krainer L, Voci D (2021) Sustainability communication as critical perspective in media and communication studies—an introduction. In: Weder F, Krainer L, Karmasin M (eds) The sustainability communication reader. Springer VS, Wiesbaden. https://doi.org/10.1007/978-3-658-31883-3_1

Wilkins EJ, Miller HM, Tilak E, Schuster RM (2018) Communicating information on nature related topics: preferred information channels and trust in sources. PLoS ONE 13(12):e0209013. https://doi.org/10.1371/journal.pone.0209013

Shaping University Teaching Programs on Sustainability: Solidifying the Fluid Concept of Sustainability

Martijn Pieter van der Steen, Kristina Linke, Paula Dirks, and Nazim Hussain

Abstract This chapter examines the development and implementation of a sustainability focused university program. Given the complexity and fluidity of sustainability as a teaching subject, the creation of such a program is fraught with challenges. Based on a retrospective qualitative analysis of the implementation of such a program at the University of Groningen (UG), the chapter documents that the design choices for such a program are particularly difficult to make. These choices entail fitting the complex and fluid subject-matter of sustainability into a mostly stable university infrastructure. The analysis reveals that, in doing so, universities may help constrain and solidify an otherwise multi-layered and fluid concept. As such, universities can have an important role in making sustainability actionable for current and future generations of students. This is, the chapter argues, the educational challenge for the years to come: embedding the concept of sustainability in university teaching practices in such a way that it enables graduates to address the moving target that sustainability currently is.

1 Introduction

Over time, sustainability has unfolded into a concept with a multiplicity of meanings and implications. As early as 1987, the Brundtland report, published by the UN World Commission on Environment and Development, highlighted the difficulties of providing a single definition of what sustainability constitutes, and noted that sustainable development "meets the needs of the present without compromising the ability of the future generations to meet their own needs" (p. 16). More recently, alternative definitions have been coined, including the following: "sustainability is the ability to achieve sustained economic prosperity over time, while protecting the natural systems of the planet and providing a high quality of life for people" (as cited by Piacentini and Della Ceca 2017). This definition aims to capture the breath of the

M. P. van der Steen (✉) · K. Linke · P. Dirks · N. Hussain
Faculty of Economics and Business, University of Groningen, Groningen, Netherlands
e-mail: m.p.van.der.steen@rug.nl

W. Leal Filho et al. (eds.), *Education for Sustainable Development: The Contribution of Universities*, World Sustainability Series,
https://doi.org/10.1007/978-3-031-86985-3_15

concept of sustainability, which arguably comes at the expense of representing its essence and its implications at more fundamental levels (Schulz et al. 2018; Sherren 2005).

These definitional difficulties notwithstanding, the perceived need for sustainability has already greatly impacted economic and social actors, such as governments, companies, consumers and citizens. The sense of urgency has preceded societies' abilities to delineate, conceptualize and ground their understanding of sustainability, and to render it actionable for future generations. It is against this background that institutions of higher education in general, and universities in particular, play an important role. Part of universities' social remit is based on their ability to provide higher-order teaching to generations that are expected to encounter problems that are unique to their generation—one of which is related to sustainability (Leal et al. 2019). Sometimes put in shorthand as a "wicked problem", sustainability constitutes a wide variety of unique problems at all levels of society and for all actors inhabiting this society (Hahn et al. 2014; Toman 2006). However, as the fluid and multi-dimensional concept of sustainability has yet to stabilize into more established permutations, it may be timely and relevant to explore if and how universities can develop their teaching on sustainability and how they can integrate this into their established curricula.

This is the objective of this chapter. Drawing on a case study at the University of Groningen (UG), this chapter examines the development and implementation of a new master's program on sustainability management and documents some of the associated concerns. Apart from the provisional "lessons learnt", the chapter also considers how the incorporation of these programs at university levels contributes to making sustainability "actionable". By conceptualizing, defining, classifying and embedding sustainability as a subject into the functional domains taught, we argue, university programs on sustainability are key to stabilizing sustainability in ways that render it actionable.

To this end, the next section will briefly introduce the literature on the fluidity of the concept of sustainability and the potential problems that this may incite. Then, the UG case is discussed. The first subsection introduces the UG as the site of our case. The second subsection examines the process of developing a teaching program on sustainability as a subject-in-flux. The third subsection argues that, in the process of affixing such a program to the university infrastructure, many choices need to be made that delineate the nature of sustainability and classify it into categories that ultimately render sustainability more manageable. Consequently, we argue that the mere practice of program development is impactful for the ways sustainability is experienced by students, the actionability of the knowledge about sustainability the students obtain, and the future forms of this knowledge that graduates may put into practice. Finally, the chapter ends in a conclusion and recaps and discusses the arguments brought forward.

2 The Fluidity of Sustainability and the Consequences for University Teaching

It is not self-evident how sustainability needs to be taught because the mere concept of sustainability is fluid, ambiguous and volatile (Hahn et al. 2014; Toman 2006). This is the case for several reasons. First, the generation of knowledge in sustainability spans a variety of disciplines, including environmental sciences, economics, and social and policy studies (Mayer et al. 2004). By their nature, sustainability issues surpass these domains, and this interdisciplinarity requires a conceptualization of sustainability that accommodates different meanings from these different domains (Edwards 2019). These difficulties are visible in attempts to carve out different dimensions of sustainability, such as environmental, social, and economic sustainability (Carew and Mitchell 2008; Santoyo-Castelazo and Azapagic 2014). Although these attempts have mobilized communities around the different dimensions, they may be problematic when they meet, for instance by privileging some aspects over others (see e.g. Halme et al. 2004). To illustrate, it is considered notoriously difficult to understand, let alone quantify, economic sustainability in the face of environmental sustainability (Norton and Toman 1997). Second, sustainability, in most of its potential meanings, is an inherently long-term phenomenon (Kates and Parris 2003). This makes the concept unpredictable and, consequently, difficult to specify. For instance, it concerns a phenomenon that needs to be considered in the perspective of a future that is largely unknown.

Third, what is considered *sustainable* varies between cultures and contexts (Horlings 2015). Practices that are considered sustainable in specific parts of the world are not in others, depending on climate, natural resources, population, historical development and ecological state (Giddings et al. 2002; Kronenberg and Andersson 2019). For example, many Europeans are concerned about climate change, and by extension about CO_2 emissions, as they are increasingly confronted with floodings and other weather-induced anomalies. By contrast, large-scale deforestation in South America is at the root of sustainability concerns there, because of the rapid destruction of the traditional rain-forest ecology. Fourth, sustainability is a relatively young domain in its own right. Although there is interest dating back to the 1950's, larger interest in the topic has only evolved over the last three decades. Although many welcome this upscaling of sustainability as a domain of interest, it has consequences for the relevance of the concept, as it needs to accommodate an increasing number of problems, issues and viewpoints (Giddings et al. 2002). Consequently, the concept of sustainability is expanded to accommodate these new problem areas (for example, see Ertz and Leblanc-Proulx (2018) for a review of the expanding literature about the collaborative economy—a subcategory of the sustainability literature). Finally, sustainability does not only encompass factual states. It includes judgments about what is desirable for the future of ecologies, populations and individuals (Lélé and Norgaard 1996). These judgments are not fixed, but, rather, moving panels, which respond to wider political, economic demographic and social changes. Highlighting the essence of the above arguments, Giddings et al. (2002, p. 188) note:

> *"It is clear from all the debates about sustainable development that there is no common philosophy. There are so many interpretations of sustainable development that it is safe to say that there is no such thing as sustainable development-ism, in contrast to the schools of neo-liberalism, feminism, deep ecology or socialism. Rather, the existing worldviews of people and organizations flow into their conception of sustainable development"*

Consequently, definitions of sustainability often emphasize flexibility to allow for adaptation and reinterpretation according to specific contexts and evolving knowledge. This flexibility and fluidity can be both a strength, allowing for customized discourses, targeted at specific cross-sections of stakeholders, as well as a challenge, making it harder to establish a clear scope of sustainability, especially when it transcends disciplinary boundaries.

This flexibility makes teaching sustainability in higher education particularly challenging (Schulz et al. 2018; Sherren 2005). Given its emerging nature, there is only a limited tradition in teaching sustainability at institutions of higher education. The relatively blank slate in teaching, combined with the observed fluidity and multi-dimensionality of sustainability, yields the important question of whose needs are to be served by effective teaching about sustainability. This question is not only relevant for *teaching* sustainability, but also for *creating teaching programs on sustainability*. Another challenge related to creating teaching programs is that, although universities are often involved in the creation of innovative products and processes, they tend to be quite inert when it comes to their own organizational learning (White and Weathersby 2005). The creation of new teaching programs on emerging subjects, i.e. subjects that are still in flux, may therefore be problematic, because this requires the mobilization of new teaching practices and organizational structures and resources, all of which may be fraught with risk, at least from the perspective of the institution. In the case of teaching a full program on the subject of sustainability, it is quite unclear who will be the prime societal benefactors of these programs. Related questions are how to best teach such a topic that is still in flux, or where in the university hierarchy to embed a subject, which, by its very nature, is multi-disciplinary. In a university as a political stakeholder organization (Bleiklie and Kogan 2007), the embedding of such a new initiative is a delicate undertaking. Therefore, the remainder of this chapter will discuss the creation of a program in sustainability, the associated pitfalls and lessons learned, and suggestions for other institutions who wish to follow suit. To this end, the next subsection examines the introduction of a master's program on sustainability: Msc. in Accountancy and Controlling—Sustainability Management and Controlling.

3 MSc. in Accountancy and Controlling—Sustainability Management and Controlling

To examine how UG's implementation of Sustainability Management and Controlling (SMC) brought about a more solidified concept of sustainability, we employed the analytical model depicted in Fig. 1. The top half depicts the creative construction of a curriculum on sustainability. Subsequently, this curriculum was fitted into the university infrastructure as depicted in the lower half of the figure. We argue that, at this stage, the possibilities *and* impossibilities afforded by the university infrastructure condition the concept of sustainability and are restrictive to the attributes that can be taught to students. We conclude by claiming that, over time, this more restricted concept of sustainability becomes solidified for new generations of students, and consequently, more actionable. The subsections below first introduce the UG and its new program on sustainability entitled SMC. Then, the subsequent subsections explain the process of creative construction that was at the basis of SMC, followed by a discussion of the university infrastructure that made the program possible, but, simultaneously, constrained what it could deliver.

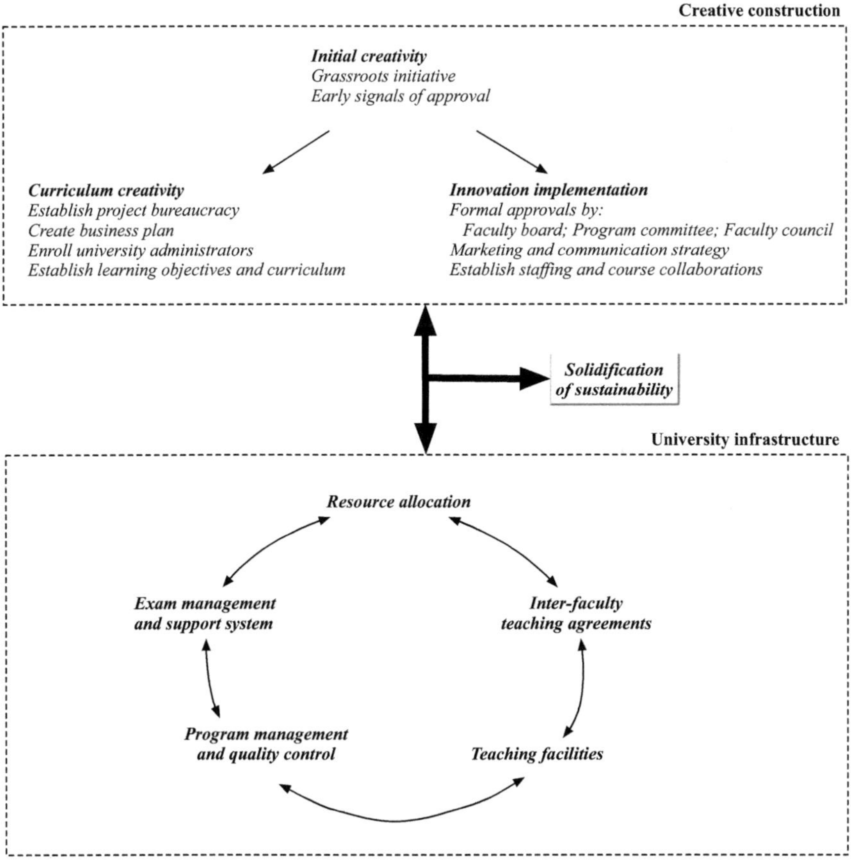

Fig. 1 Analytical framework

3.1 Sustainability and UG

UG, located in the city of Groningen in the Netherlands, is a multidisciplinary university with eleven faculties, including the Faculty of Economics and Business, which is responsible for the new SMC program. The 60 EC, one-year program is situated in the Msc. Accountancy and Controlling and consists of eight courses and a graduate thesis in sustainability management, accounting and control.

In the current strategic plan of UG, entitled 'making connections', societal outreach is an important theme, considered relevant for both teaching and research. According to this strategy, UG needs to better articulate its responsibility towards society—a society that is facing challenges in combating climate change and in development towards a more sustainable way of working and living. UG also has the ambition to encourage interdisciplinary teaching by bringing together various disciplines in several so-called schools. One of these schools is focused on sustainable governance, politics, and a sustainable society, more generally.

Sustainability is also a recurring strategic theme at the Faculty of Economics and Business. Its importance is underlined by providing a so-called *focus area of sustainability* that students can participate in by following courses that include specific sustainability topics. Moreover, several programs at the faculty have integrated sustainability (and energy)-related themes in the courses they offer. While much progress has been made, the faculty also acknowledges that still much more work needs to be done. A survey[1] in 2022 indicated that students were insufficiently aware about how sustainability was included in each of their courses. In general, they indicated that they would prefer receiving more sustainability teaching in their curriculum, either in a degree program or through thematic minors. Against this background, the development of the new formal master program focusing on sustainability was initiated to cater to these needs.

3.2 Teaching Sustainability

Although it was clear that a program on sustainability would cater to students' needs, there was no formal assignment or implementation process of the SMC program. Instead, the fluid and flexible character of sustainability provided the soil for grassroot initiatives in curriculum development. In the past two decades, grassroot initiatives played a central role in the development toward sustainable production and consumption solutions (Hossain 2020). At the Faculty of Economics and Business, the project team applied a grassroot innovation approach to create the new sustainability curriculum. Amabile and Pratt (2016) divide grassroot initiatives into two

[1] *Source* Annual sustainability report (2022). https://www.rug.nl/about-ug/profile/facts-and-figures/duurzaamheid/documenten/annual-report-sustainability-2022.pdf. Date accessed/13/09/2024.

stages. First, creativity is needed to produce novel and useful ideas by either individuals or small groups. Second, innovation can also be defined as successful implementation of creative ideas in organizations. The stages of creativity and innovation implementation serve as structure to describe the case, situated in the department of Accounting and Auditing (A&A).

3.2.1 Initial Creativity

Currently, the field of accounting is marked by the introduction of the EU Directive Corporate Sustainability Reporting (2022) that requires about 49,000 firms in Europe to report extensively about sustainability related matters (KPMG 2021). This creates a high demand for accounting and auditing professionals with specific knowledge about sustainability, and, at the same time, the traditional activities of management accountants and auditors seem to become less attractive to students (Burke and Polimeni 2023). In the A&A department, the project team noticed this decline of students in the more traditionally oriented programs. Horne et al. (2024) found that individuals are more interested in becoming accountants when sustainability assurance positions are advertised, appealing to a more prosocial value orientation. In addition, Dirks et al. (2024) argue that accounting and finance professionals are potentially positioned well to participate in the organizational path towards sustainability, which may make a sustainability-oriented accounting program especially attractive.

Considering this, a member of the department floated the idea of a new master program that takes both legal requirements of corporations and changing student values into account. A diverse group of department members consisting of different research fields, hierarchical functions, genders and nationalities was contacted by email. An unstructured string of emails followed, concluding that a new master program would be desirable. Department management attempted to receive formal faculty board approval to proceed but learned that there is only one moment each year, including a strict application deadline of November 1st, where such requests can be incorporated in the examination rules and regulations. In turn, this incorporation was tied to resources allocated, and therefore, there was considerable time pressure to prepare such a formal application. However, the faculty board did signal a benevolent attitude toward such request but made sure to avoid making any commitments. Taking the risk of rejection into account, a project team was assembled, consisting of sustainability experts and program management.

3.2.2 Curriculum Creativity

When the project team started, two months' time was left to meet the application deadline. Although the deadline seemed unrealistic, a project delay of one year was not desirable either, especially since other institutions of higher education were also looking into developing a curriculum on sustainability, and a possible first-mover advantage would be lost. Weekly meetings were scheduled to discuss the curriculum, make decisions and distribute tasks to be completed by the next week. First, the scope needed clarification, and a decision was made to target management accountants since this profession is not as regulated as auditing professionals, and therefore, this group would be more amenable to adopting a broadly defined and fluid concept of sustainability.

In the absence of a blueprint for curriculum development, two key program features were defined: traditional management accounting knowledge and skills, and sustainability knowledge covering the pillars *environment*, *social*, *governance*, as well as interdisciplinary skills. Starting from these features, a search for suitable, already existing courses was conducted. The UG strategic plan appeared to have been fruitfully implemented by several faculties, so that almost the entire program could be constructed using existing courses. Such "curriculum bricolage" was possible because the new program could contain many different dimensions of sustainability. Courses taught by staff of diverse backgrounds, all of whom considered sustainability from their own points of view and expertise, were considered and adopted. The resulting multi-disciplinarity was both required by the current state of the art of sustainability and made possible by this approach.

Once the curriculum structure became clear, the team wrote a business plan comprising eight sections (curriculum description, curriculum embedding, graduate description, learning goals, market research, target student group, curriculum courses (mandatory/electives), development of new flagship course, and challenges). Importantly, in doing so, the project team restricted sustainability especially to a more restricted form: that of a market opportunity. For instance, in the documents, Fig. 2 related sustainability to numbers of vacancies in other domains, to provide a sense of the opportunities that were present, which made sustainability especially more relatable to the university administrators that needed to be convinced.

In the final phase of writing the proposal, it was important to take note of all formal faculty application requirements and to elicit regular feedback from a broader group of department members. In all, the application deadline could be met, because the fluidity and flexibility of the concept of sustainability allowed the project team to adopt many existing UG offerings as useful bases for curriculum development.

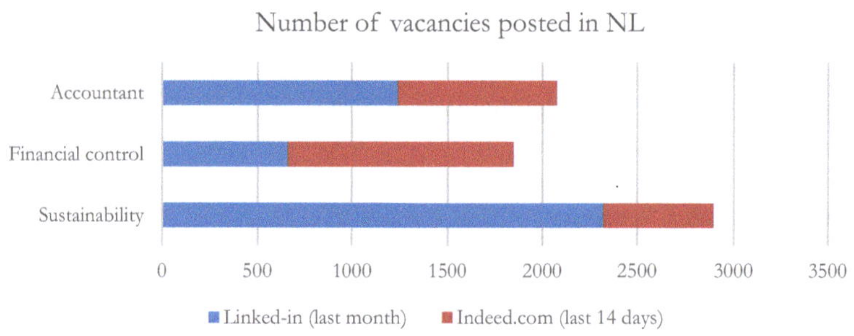

Fig. 2 Vacancies in sustainability and related domains (per 21-10-2023), provided to administrators

3.2.3 Innovation Implementation

The Faculty Board approved the curriculum application six weeks after submission. In that timeframe, service-level agreements with other faculties were negotiated (see the next section). One example is the course "Interdisciplinary Perspectives on Sustainability" in which staff of the faculties of Behavioral Sciences, Spatial Sciences and Science and Engineering cooperated and who were joined by staff of the faculty of Economics and Business. After the Faculty Board's approval, two additional approvals were required. The Program Committee approved the program as it was expected that it would improve graduates' labor market perspectives, and the Faculty Council provided its approval as its legal embedding in the examination regulations was considered satisfactory.

Subsequently, a marketing strategy with the communications department was developed, the entire program website was redesigned, and promotional materials were developed. After the *Master's Day*, the faculty's main promotional event, six weeks were left for further promotional activities until the Dutch central registration system closed for the subsequent academic year. In the meantime, faculty support departments made sure the new curriculum was ready for teaching (see *solidification* in the next section). Finally, staffing for new courses needed to be decided upon and course materials needed to be developed. Additionally, modifications of existing courses were also required to avoid overlap and to extend course content to include sustainability subjects of sufficient level and depth.

3.3 Slow Solidification of Sustainability at UG

In the subsequent phase, the UG has successfully developed its new program on sustainability, placed it under the umbrella of its MSc Accountancy and Controlling curriculum, and integrated it into the existing fabric of university teaching programs and infrastructure. This integration encompassed the solidification of the concept of sustainability into university resources and infrastructure. In this context, "solidification of sustainability" refers to the process of making the fluid concept of sustainability concrete and actionable within and outside the university's teaching frameworks. Below, several university processes are discussed which were very helpful in developing a program on sustainability, but, simultaneously, constrained the variations of sustainability that the university could bring into wider circulation. Through resource allocations to the program, inter-faculty teaching agreements, assignment of teaching facilities, program management and quality control, and exam management and support systems, the complexities of sustainability were channeled *and* reduced. This section therefore reveals how the use of a well-established infrastructure to incorporate sustainability in a teaching program has allowed the project team to quickly make available this highly fluid concept for teaching. Inevitably, however, this had to come at the price of reducing its multi-dimensionality and flexibility to manageable proportions. However, this also brought additional benefits, including the increased actionability of the knowledge students obtained in the program.

3.3.1 Resource Allocation

Extant educational innovation literature emphasizes the importance of resource allocation in curriculum development. For instance, Sahlberg (2006) highlights that successful educational reforms often require substantial investments in developing new teaching materials and training faculty members. By investing in the development of sustainability courses, UG aligns with best practices in educational innovation, but in doing so, it locks down which aspects of sustainability take center stage in the new program (see also, Gusc and van Veen-Dirks 2017). Every higher education institution faces scarcity of resources. Despite all constraints, UG's commitment to sustainability education is evident through its investment of resources in developing new courses and content. The allocation of dedicated resources has enabled the creation of comprehensive courses and modules addressing various dimensions of sustainability, including economic, environmental, social, and governance aspects. The alignment of these contents with the United Nations Sustainable Development Goals was important to university administrators, as it signaled the credibility of the program. However, fulfilling these requirements unavoidably restricted the variations of sustainability that could be addressed.

3.3.2 Inter-Faculty Teaching Agreements

The interdisciplinary approach in sustainability education is important, as highlighted by Annan-Diab and Molinari (2017). In line with prior sections, they argue that sustainability issues often span multiple disciplines, necessitating an educational approach that integrates knowledge from various fields. As a mature institution, UG has administrative frameworks in place which organize inter-faculty teaching collaborations. Such collaborations facilitate interdisciplinary teaching, which is especially important for understanding of the multidisciplinary nature of sustainability. For example, UG's institution-wide scheduling system facilitates this integration, as it enables the integration of sustainability courses across different programs. This approach allows students from various disciplines to engage with sustainability topics in ways that do justice to their fluidity. It is important to note, however, that although some of this infrastructure fostered sustainability as a multi-disciplinary subject, there were strict limits that were imposed. For example, students could attend sustainability courses in other faculties, but the integrated scheduling systems could not accommodate overlapping courses or student travel time. Consequently, students remain exposed to only a limited number of perspectives, outside their own faculty, as others are simply not practically accessible to them. Hence, the scheduling and course information systems of UG made it possible to offer an interdisciplinary program at the MSc level, but they equally restricted the number of variations of sustainability that students could be exposed to. We thus see in the case at UG that the way in which these multiple disciplines are mobilized depends on the offerings available in other faculties, their interest in sustainability, and the administrative frameworks in place to facilitate such collaboration. Consequently, the nature of inter-disciplinary course offerings on sustainability is largely dependent on the maturity of the institution and the variety of course offerings already present. This means that a program on sustainability at UG may be different from other institutions, at least, for as long as the fluidity and multi-disciplinarity of the topic of sustainability facilitates its teaching in many different ways.

3.3.3 State-of-the-Art Teaching Facilities

The use of advanced teaching facilities also improves students' learning experiences. According to Waheed et al. (2016), the effective use of technology in education can significantly enhance student engagement and learning outcomes. Therefore, by providing modern teaching facilities, the UG seeks to ensure that its students enhance their understanding of complex concepts such as sustainability. Like in many established institutions, UG's modern classrooms are equipped with advanced technological tools. These tools enable the use of interactive and engaging teaching methods, such as case studies, simulations, and group projects, which help teaching complex sustainability concepts (Brundiers et al. 2010). However, these facilities also constrain the dimensions of sustainability that can be taught. For instance, the plight of farmers facing increased regulatory pressures or changes in climate cannot be fully appreciated from the classroom alone. Hence, even state of the art teaching facilities present in institutions of higher education may limit the variations of sustainability that can be taught to the (im-)possibilities that these facilities offer.

3.3.4 Program Management and Quality Control

Program management's involvement in course (re-)development, continuous feedback and iterative course enhancement enable educators to make adjustments to curricula, thereby ensuring alignment with student needs and broader stakeholder demands. UG has a comprehensive quality control system in place designed to facilitate the continuous improvement of both its established programs and new offerings. This system gathers feedback from both students and faculty, which is then used to implement regular updates in the courses. Such a feedback mechanism is important for identifying areas in need of improvement, ensuring that the curriculum remains pertinent and effective (Bovill et al. 2011) and can, as such, also be helpful in addressing the evolving challenges of sustainability. The iterative process of feedback and curriculum refinement contributes to the gradual solidification of the otherwise fluid concept of sustainability within the university's academic framework, because it operationalizes only a handful of quality metrics. Among others, it measures lecturers' perceived proficiency in explaining sustainability, and students' impressions of the topical relevance. As lecturers are aware of these evaluations, they impact the ways that they approach these topics in the classroom. Hence, such systems may affect the teaching dynamics of sustainability in rather unpredictable ways.

3.3.5 Exam Management and Support System

According to Nicol and Macfarlane-Dick (2006), using diverse assessment methods provides a more comprehensive evaluation of student learning, catering to different learning styles and abilities. UG's exam management and support system embodies these principles, ensuring that students are assessed comprehensively. Although it facilitates various types of assessments, including written and digital exams, project reports, and presentations, which are crucial for a thorough understanding of complex issues (Banta and Palomba 2014), it puts further restrictions on the ways that sustainability is experienced and understood by students. As sustainability as a concept is routed through the university exam management system, it limits the nature and number of dimensions that can be evaluated. In addition, there is a risk that the stringent use of such a system may reduce sustainability to a singular set of checkboxes that students will be preoccupied to fill.

3.4 Relevance and Impact

The integration of sustainability into the MSc Accountancy and Controlling program at UG has a significant impact on both students and its wider academic community. By embedding sustainability into the core curriculum, the university impacts the relevance and applicability of its programs, by infusing the education of future graduates with specific understandings of sustainability. The experiences at UG illustrate that the design choices and the subsequent embedding of teaching programs in the university's educational structures and practices are both empowering as well as restrictive to what sustainability is and can become. These choices influence not only the quality of education but also the broader societal understanding of sustainability that graduates will mobilize in their future careers.

4 Discussion and Conclusion

This chapter has problematized the development of teaching programs in sustainability. Its obvious societal importance makes it important to deliver sustainability teaching in higher education, but in doing so, the organization of higher education also restricts the number of sustainability dimensions that can be taught. Furthermore, we argued that sustainability is highly fluid, but by exposing it to the context of a university, it gets increasingly constrained to fit the university infrastructure and, subsequently, it solidifies into a less malleable form. In other words, teaching sustainability unavoidably leads to limitations to what sustainability can be. In UG, these limitations included (1) the unavoidable privileging of sustainability in economic and business terms, as it is taught in a faculty and department with expertise in these areas; (2) teaching sustainability as an essentially administrative and managerial problem, not only due to the orientation of the program, but also due to the absence of laboratories that would enable more practical teaching methods; and (3) a structure of functionally separate courses, which constrains the emphasis on the interdisciplinary nature of sustainability.

The question is how much of a problem this may be. The initial fluidity of sustainability was particularly helpful to the development of the program. Due to its multidimensionality, the project team was able to quickly develop the program as it could seek out and integrate many of the dimensions of sustainability that were already in circulation throughout the organization. Additionally, the project team was able to switch between the various meanings of sustainability, depending on the audiences and interests it wanted to serve. However, as the program became integrated in the university structure, so did the concept of sustainability that the program teaches. We view this as beneficial for progress in the ultimate objective of combating climate change. Solidifying sustainability into more stable forms is required to have more uniform vocabularies that stakeholders can use to negotiate potential solutions. Such stability may also render these solutions more actionable, because then, the essence of sustainability, on which these solutions are based, is less likely to change halfway through their implementations. A constrained but well-established concept of sustainability is also beneficial for the universities involved. The monopolization of specific instances of sustainability provides an unbreakable association of the institution with the concept and can provide many reputational benefits. In other words, for new, fluid and flexible concepts, the first mover advantage counts.

There are, however, disadvantages associated with these outcomes of the introduction of university teaching on sustainability. If, indeed, the concept of sustainability becomes more constrained, there is an increasing risk that new developments, which may for very good reasons be associated with sustainability, be ignored. Such myopia may challenge graduates' ability to recognize and act upon new threats to sustainability. It is therefore important that institutions of higher education seek ways to retain some of the fluidity of the concept through their offering of higher-order teaching, which will teach future graduates to address sustainability in whatever form it will present itself to these generations.

References

Amabile TM, Pratt MG (2016) The dynamic componential model of creativity and innovation in organizations: making progress, making meaning. Res Organ Behav 36:157–183

Annan-Diab F, Molinari C (2017) Interdisciplinarity: practical approach to advancing education for sustainability and for the sustainable development goals. Int J Manage Educ 15(2):73–83
Banta TW, Palomba CA (2014) Assessment essentials: planning, implementing, and improving assessment in higher education. John Wiley & Sons, New York City, United States
Bleiklie I, Kogan M (2007) Organization and governance of universities. High Educ Pol 20(4):477–493
Bovill C, Cook-Sather A, Felten P (2011) Students as co-creators of teaching approaches, course design, and curricula: implications for academic developers. Int J Acad Dev 16(2):133–145
Brundiers K, Wiek A, Redman CL (2010) Real-world learning opportunities in sustainability: from classroom into the real world. Int J Sustain High Educ 11(4):308–324
Burke JA, Polimeni RS (2023) The accounting profession is in crisis: a partial solution to the shortage of accountants. CPA J 93(9/10):6–8
Carew AL, Mitchell CA (2008) Teaching sustainability as a contested concept: capitalizing on variation in engineering educators' conceptions of environmental, social and economic sustainability. J Clean Prod 16(1):105–115
Directive (EU) 2022/2464 of the European Parliament and of the Council of 14 December 2022 amending Regulation (EU) No 537/2014, Directive 2004/109/EC, Directive 2006/43/EC and Directive 2013/34/EU, as regards corporate sustainability reporting
Dirks P, Hussain N, Van der Steen MP (2024) The roles of the financial professional in combating climate change through business circularity: implications for education and research. In: University initiatives on climate change education and research, pp 1–22
Edwards GI (2019) Multidisciplinary approach to environmental problems and sustainability. In: Leal Filho W (ed) Encyclopedia of sustainability in higher education. Springer International Publishing, Cham, pp 1184–1189
Ertz M, Leblanc-Proulx S (2018) Sustainability in the collaborative economy: a bibliometric analysis reveals emerging interest. J Clean Prod 196:1073–1085
Giddings B, Hopwood B, O'Brien G (2002) Environment, economy and society: fitting them together into sustainable development. Sustain Dev 10(4):187–196
Gusc J, van Veen-Dirks P (2017) Accounting for sustainability: an active learning assignment. Int J Sustain High Educ 18(3):329–340
Hahn T, Preuss L, Pinkse J, Figge F (2014) Cognitive frames in corporate sustainability: managerial sensemaking with paradoxical and business case frames. Acad Manage Rev 39(4):463–487
Halme M, Jasch C, Scharp M (2004) Sustainable homeservices? Toward household services that enhance ecological, social and economic sustainability. Ecol Econ 51:125–138
Hossain M (2020) Frugal innovation: conception, development, diffusion, and outcome. J Clean Prod 262:121456
Horlings L (2015) The inner dimension of sustainability: personal and cultural values. Curr Opin Environ Sustain 14:163–169
Horne E, Loftus S, McCoy SS, Winn AM (2024) Attracting the next generation of accountants: the joint impact of sustainability emphasis and social value orientation on accounting career perceptions. Audit: J Pract Theory 43(1):191–209
Kates RW, Parris TM (2003) Long-term trends and a sustainability transition. Proc Natl Acad Sci 100(14):8062–8067
KPMG (2021). https://kpmg.com/nl/en/home/topics/environmental-social-governance/corporate-sustainability-reporting-directive.html. Visited on 22 Jul 2024
Kronenberg J, Andersson E (2019) Integrating social values with other value dimensions: parallel use vs. combination vs. full integration. Sustain Sci 14(5):1283–1295
Leal F, Walter S, Chris P, Arminda M, Mark Á, Veiga L, Brandli LL, Molthan-Hill P, Pace P, Azeiteiro UM, Vargas VR, Caeiro S (2019) Sustainable development goals and sustainability teaching at universities: falling behind or getting ahead of the pack? J Clean Prod 232:285–294
Lélé S, Norgaard RB (1996) Sustainability and the scientist's burden. Conserv Biol 10(2):354–365
Mayer AL, Thurston HW, Pawlowski CW (2004) The multidisciplinary influence of common sustainability indices. Front Ecol 2:419–426

Nicol DJ, Macfarlane-Dick D (2006) Formative assessment and self-regulated learning: a model and seven principles of good feedback practice. Stud High Educ 31(2):199–218
Norton BG, Toman MA (1997) Sustainability: ecological and economic perspectives. Land Econ 73(4):553–568
Piacentini RD, Della Ceca LS (2017) The use of environmental sustainability criteria in industrial processes. Drying Technology 35(1):1–3
Sahlberg P (2006) Education reform for raising economic competitiveness. J Educ Change 7:259–287
Santoyo-Castelazo E, Azapagic A (2014) Sustainability assessment of energy systems: integrating environmental, economic and social aspects. J Clean Prod 80:119–138
Schulz K-P, Finstad-Milion K, Janczak S (2018) Educating corporate sustainability—a multidisciplinary and practice-based approach to facilitate students' learning. J Clean Prod 198:996–1006
Sherren K (2005) Balancing the disciplines: a multidisciplinary perspective on sustainability curriculum content. Aust J Environ Educ 21:97–106
Toman MA (2006) The difficulty in defining sustainability. In: The RFF reader in environmental and resource policy, 2nd edn. Routledge
Waheed M, Kaur K, Ain N, Hussain N (2016) Perceived learning outcomes from Moodle: an empirical study of intrinsic and extrinsic motivating factors. Inf Dev 32(4):1001–1013
White J, Weathersby R (2005) Can universities become true learning organizations? Learn Organ 12(3):292–298

Sowing Seeds of Change: Cultivating Sustainability Literacy in TU Dublin's Language Studies Curriculum

Odette Gabaudan, Pilar Molina Taracena, and Alexandra Severino

Abstract Embedding sustainability literacy into the Higher Education curricula is imperative particularly considering the sobering findings of the 2023 SDG Progress Report (https://unstats.un.org/sdgs/report/2023/) which indicate that only 12% of the Sustainable Development Goal (SDG) targets are on track. Resulting from a collaboration between the Language Studies department at TU Dublin and UNITAR (United Nations Institute of Training and Research), this paper presents the fruit of a professional development (PD) initiative aimed at equipping Language Studies colleagues with practical strategies to integrate Education for Sustainable Development into their pedagogical practices. Recognizing the pivotal role of educators, as articulated by (Tribello et al. in Ir J Acad Pract 11(2), Article 6, 2024), the chapter details the stages of a PD process undertaken to empower and motivate lecturers to incorporate SDGs into their modules. The PD process started with an online roundtable

Disclaimer

The views and opinions expressed in this content are solely those of the authors and do not reflect the official policies, positions, or endorsements of the United Nations or its affiliates.

Additional Information Note

This paper was developed as part of the collaborative partnership between UNITAR and TU Dublin, emphasizing the critical role of research and collaboration in advancing the 2030 Agenda for Sustainable Development, particularly Sustainable Development Goal (SDG) 17: Partnerships for the Goals.

O. Gabaudan · P. M. Taracena (✉)
Language Studies, TU Dublin, Dublin, Ireland
e-mail: pilar.molina@tudublin.ie

O. Gabaudan
e-mail: odette.gabaudan@tudublin.ie

A. Severino
Online Learning and Education Unit, Division for Peace, UNITAR, Geneva, Switzerland
e-mail: alexandra.severino@unitar.org

W. Leal Filho et al. (eds.), *Education for Sustainable Development: The Contribution of Universities*, World Sustainability Series,
https://doi.org/10.1007/978-3-031-86985-3_16

discussion focusing on enhancing language lecturers' sustainability literacy specifically in the context of language learning and teaching. The roundtable discussion was followed by a webinar that delved into TU Dublin's Sustainability Education Framework. Finally, a day-long workshop facilitated by UNITAR on the topic of ESD for Language Studies further emphasised key cross-cutting areas of knowledge and competencies essential for fostering meaningful learning around UN's Agenda 2030 and the SDGs. The workshop, supported by self and group-led processes yielded valuable insights, equipping Language Studies lecturers with the knowledge and skills to integrate SDGs at programmatic and modular levels, aligning with TU Dublin's Strategic Intent 2030. This chapter contributes to providing educators and managers with strategies for embedding sustainability principles within the curriculum of Higher Education institutions.

Keywords Education for sustainable development · Sustainability literacy · Language studies · Languages

1 Introduction

Education for Sustainable Development (ESD) has gained increasing prominence in Higher Education both in Europe and beyond. TU Dublin in Ireland serves as a notable example. This paper outlines the initiatives undertaken to promote professional development in ESD among faculty in the Language Studies Department at TU Dublin. Offering lecturers the chance to engage with the United Nations' 17 Sustainable Development Goals in language teaching has proven to provide an effective approach to integrating sustainability issues into course content thereby educating students on these topics. A key innovation of this initiative is the consolidation of a multi-faceted collaboration with the United Nations Institute of Training and Research (UNITAR). This work is presented as a case study of a series of activities that culminated in a workshop by UNITAR for colleagues in Language Studies in TU Dublin. The professional development framework developed by the National Forum for the Enhancement of Teaching and Learning for Higher Education (Ireland) underpins and frames all the activities undertaken as part of this case study.

2 Context

The Education for sustainable development (ESD) initiative that is presented in this paper is grounded in Ireland's first technological university, namely TU Dublin. This university, the largest higher education institution in Ireland, offers courses in varying fields, from science and health to business and law, tourism, hospitality, and social sciences. ESD has become a fundamental tenet of TU Dublin's University Education

Model and it is also an essential pilar in UNITAR's vision. The United Nations Institute for Training and Research (UNITAR) is a United Nations agency focused on building global capacity for sustainable development, peace, and prosperity. Through innovative training and knowledge-sharing, UNITAR supports individuals, organizations, and governments in addressing global challenges. Its emphasis on ESD aligns with TU Dublin's efforts, reflecting a shared commitment to advancing sustainability through transformative education.

In March 2022 TU Dublin published a sustainability statement stating that 'through its three pillars of People, Planet, and Partnerships, sustainability drives our vision, mission, and values (…). TU Dublin is committed to embedding sustainability into all current University plans, strategies, policies, and everyday operations.' (TU Dublin statement 2022) As part of the TU Dublin strategic plan 2024–2028, sustainability education has two interlinked and complementary elements. Firstly, 'it involves engaging academic and professional staff through training, professional development, and education to build capacity, skills, and knowledge (…). Secondly, education involves infusing sustainability through TU Dublin's curricula to enable and empower students to progressively develop their sustainability knowledge and skills in a manner, which is clearly linked and applied to their programme.' (https://www.tudublin.ie/explore/about-the-university/sustainability/sustainability-education/#) It is within this broad context that our Language Studies faculty has embarked on professional development for ESD and that TU Dublin has established a collaboration with UNITAR.

Complementing TU Dublin's approach, UNITAR envisions a world 'in which individuals, institutions and organizations are equipped with the knowledge, skills and other capacities to overcome global challenges.' (https://unitar.org/about/unitar/mission-vision-and-core-functions) UNITAR supports this vision by developing capacities at all levels through high-quality learning solutions and knowledge products, enhancing decision-making and supporting country-level actions (https://unitar.org/about/unitar/mission-vision-and-core-functions). UNITAR's expertise in granting tailored educational resources and technology-based knowledge services is in strong alignment with TU Dublin's ESD goals (https://unitar.org/sustainable-development-goals). By facilitating innovative strategies, leveraging from its multi and transdisciplinary approaches to ESD, and promoting knowledge sharing, UNITAR fosters a collaborative environment for sustainability-focused curricula in general, and in the context of this initiative, for Language Studies curricula.

By being involved in this project through the participation of one of the authors of this chapter, the Online Learning and Education Unit (OLEU), part of UNITAR's Division for Peace, was able to build from its expertise and experience both in online learning and blended approaches, to support sustainability education in Language Studies (TU Dublin) by experimenting with methodologies and approaches from non-formal and informal education, as well as with transdisciplinary perspectives on curricula design and integration. One of the key objectives of UNITAR's involvement in its collaboration with TU Dublin was to offer a hands-on approach to include the SDGs in Language Studies curricula with special emphasis on global citizenship and sustainability awareness, and always within the framework of a partnership (in alignment with SDG 17).

3 Literature Review: ESD for Language Teaching and Learning

The integration of Education for Sustainable Development (ESD) into language education has gained momentum in recent years, highlighting an important intersection that might not always come across as evident, i.e. the nexus between sustainable literacy and language learning. de la Fuente (2022, 5) asserts that the ability to 'understand and tolerate culturally different points of view' and 'develop sensitivity to local cultures' are vital components of sustainability efforts and, in that measure, sustainable literacy in language education involves as one of its core elements the cultivation of critical awareness of sustainability issues, while promoting multilingualism and multiculturalism.

Sustainability literacy in language education is relatively new, with most research emerging since 2010. As noted by Molina Taracena (2022), the development of sustainability literacy is shaped by fields such as ecolinguistics, the ecology movement, and environmental humanities. Early research often focused on practice-based approaches, where educators sought to provide practical examples of integrating sustainability into language modules (Molina Taracena 2022, 14). These approaches aimed to inspire other colleagues by demonstrating concrete ways to inspire sustainability into language teaching approaches and practices.

Laurie et al. (2016) emphasise that ESD pedagogies have progressed beyond traditional instruction methods, focusing instead on learner-centered, participation-infused approaches. In that sense, ESD fosters active learning through problem-solving, critical thinking, and collaboration (Laurie et al. 2016, 231). According to de la Fuente (2022, 5), these principles are markedly embedded in problem-based and community-based learning environments in modern language teaching methods. Accordingly, the concerted effort between language education and ESD lies remarkably in the emphasis of real-world applicability and intercultural competence development.

Pedagogical strategies in ESD often involve issue-based and place-based learning approaches, which support students to engage with sustainability issues in their local context critically, highlight Laurie et al. (2016, 230–231) in their analysis of *The Education for Sustainable Development Sourcebook* by UNESCO (UNESCO 2012, 15). By creating the link between global sustainability matters and their microenvironment, i.e. a closer-to-home feeling, there is potential for bridging and creating space for a more encompassing and systemic understanding of sustainability, which is undeniably crucial (UNESCO 2012; UNESCO 2019). Tilbury (2011, 29, cited in

Laurie et al. 2016) identifies essential ESD learning processes that include encouraging learners to ask reflective questions, clarify values, and envision more sustainable futures. This student-centered approach mirrors the methodologies used in language education, where students and learners are often connected in project-based learning or task-based instruction that emulates real-life communication scenarios (Laurie et al. 2016, 228).

The multidisciplinary nature of ESD requires educators to facilitate connections between environmental, social, and economic aspects of sustainability, further highlights Barth (2015, 134). One way to facilitate these connections in language studies is by embedding sustainability topics into reading, writing, and speaking exercises, and by integrating them into other aspects of language study, e.g. cultural components of language learning. By incorporating these perspectives and approaches, there is an opportunity to enhance linguistic skills while also encouraging students to develop a deeper understanding of sustainability issues and their relevance to students' lives, as highlighted by Laurie et al. (2016).

Language education is uniquely positioned to foster global citizenship by teaching students how to sail across diverse cultural and linguistic landscapes and contexts, according to O'Boyle (2024, 14). Therefore, the addition of sustainability themes in language learning can develop and enhance students' critical thinking skills as well as their ability to reflect on global and local sustainability issues (Laurie et al. 2016, 230–231; UNESCO 2012), hopefully making them: i. more engaged in their communities; and ii. committed global citizens.

However, the successful integration of ESD into language studies requires significant professional development for educators. Also stressed by Laurie et al. (2016), there is a noticeable need for ongoing professional development to ensure that teachers (and instructors) are equipped to handle the interdisciplinary nature of ESD (2016, 237). Interdisciplinary teaching materials are also important to support educators who might feel overwhelmed by the multidisciplinary nature of ESD and who are unfamiliar, or less acquainted, with environmental and sustainability issues (Laurie et al. 2016, 240). Barth (2015) shares this view and emphasises that teachers need to develop new competencies to include and embed sustainability into their teaching practices (Barth 2015, 145). For language educators, specifically, this often means building knowledge and skills in areas outside their traditional expertise, namely environmental science or sustainable economics.

O'Boyle (2024) further adds that teacher capacity is critical when embedding sustainability topics into language education, as alongside the needs highlighted above, many educators may lack confidence in covering and addressing sustainability issues (20) through their discipline's curricula (O'Boyle 2024, 20). Professional development workshops, therefore, must focus on both enhancing language instructors' understanding of sustainability and equipping them with practical tools to introduce these topics into their teaching.

The needs identified above, are also addressed at policy level. The UNESCO Framework for Education for Sustainable Development Beyond 2019 (2019) stresses that 'educators must be supported through comprehensive training programs that foster their capacity to teach sustainability across all subjects.' (2019, 15) This shows that professional development initiatives for language teachers and educators require an ongoing systemic, robust, and interdisciplinary (or even transdisciplinary) approach to ensure long-term success and positive outcomes. In line with this analysis, the rationale and purpose of our workshop, along with the methodologies employed, were specifically designed to contribute to the field of ESD within language studies, addressing some of these needs and fostering a deeper integration of sustainability into language education in TU Dublin.

As it can be inferred by the analysis to this point, the integration of ESD into language education has encountered different challenges, primarily due to its multidisciplinary nature and characteristics. As noted by Molina Taracena (2022), many educators find it difficult to implement sustainability topics in beginner-level courses (CEFR A1-A2), as these themes are often perceived to be too complex for beginner-level learners (Molina Taracena 2022, 17). However, contrasting perspectives have been offered by scholars like Méndez Seijas and Parra (2021), who argue that even beginner courses should move beyond the simplistic topics often found in traditional language curricula. By advocating for a content-based instruction (CBI) model that directly addresses sustainability and social justice issues, Méndez Seijas and Parra (2021), propose a more immediate immersion into complex, meaningful content, which they consider essential for fostering critical thinking and global citizenship. These contrasting views demonstrate that there is no one-size-fits-all approach to integrating ESD into language studies. Instead, this process benefits from active learning strategies such as problem-solving, critical thinking, and collaboration among language educators and educational institutions themselves.

In a nutshell, this requires a flexible curriculum design that allows for the inclusion of sustainability topics in a meaningful way, which might imply policy and decision-making change and openness towards a lower comfort zone, namely at the higher education level. In other words, and regardless of the ways in which ESD is integrated in language studies curricula, what seems indisputable is that sustainability topics should not be canned as isolated themes but should be fully integrated into the language curriculum (Barth 2015, 132) through a systems thinking perspective (Laurie et al. 2016, 240), intrinsic to ESD.

While challenges as those analysed above can be identified, the integration of ESD into language studies also offers many opportunities for pedagogical innovation. As already mentioned, ESD promotes innovative teaching methodologies such as project-based learning and experiential education (Laurie et al. 2016, 238). When applied to languages' teaching, these methods encourage students to work collaboratively both on real-world problems—and matters that are relevant to them—such as environmental issues or social justice topics—while practicing their language skills. UNESCO (2019) also underlines the importance of interdisciplinary approaches in ESD, suggesting that subjects like language studies can serve as a platform for exploring wider sustainability issues, consequently enriching both the language curriculum and students' awareness of an increasing global world (UNESCO 2019, 12). The integration of sustainability topics into language learning offers students the opportunity to engage with resources that deal with global issues, thereby fostering critical global citizenship. As de la Fuente (2022) also asserts, the link between sustainability and language learning is a natural fit, advancing critical thinking, cultural sensitivity, alongside the aforementioned global citizenship dimensions, offering an excellent opportunity to boost the development and implementation of ESD via language studies.

While integrating these two fields requires substantial professional development for educators, involving both time and resource investment, this challenge can also be viewed as an opportunity. As language teachers engage in the professional development necessary to align with international policy frameworks, such as the United Nations 2030 Agenda and the Sustainable Development Goals (SDGs) and incorporate ESD principles into language pedagogy (UNESCO 2019), they can become true agents of change. In doing so, they play a crucial role in advancing sustainable development goals while simultaneously enhancing their students' language proficiency.

Mambu's (2023) case study is a very good example of how among the different challenges analysed above, ESD, and the integration of the SDGs specifically in language studies curricula, can be a promising opportunity to advance critical thinking, cultural awareness, and global citizenship while advancing linguistic education (Mambu 2023, 279). The case study explores the integration of SDGs into English language teacher education, focusing on how critical pedagogy can enhance students' critical thinking and social awareness. The study follows Pamela, a student in Mambu's Critical Pedagogy and Literacy (CP and L) course, as she engages with SDGs such as reducing inequality (SDG 10) and promoting quality education (SDG 4). Through a service-learning approaches, Pamela developed and applied lesson plans that addressed real-world challenges (inequality as addressed by SDG 10: Reduce Inequality Within and Among Countries) with a strong connection to their local context, i.e., socio-economic disparities in her Papuan community. This approach is good evidence of the bridging analysed above between global sustainability challenges and issues and the local context as an important element of the dialogue between language studies and SDG curricular integration (Laurie et al. 2016, 230–231; UNESCO 2012). Mambu's approach highlights participatory and experiential learning, encouraging students to critically analyse societal issues and apply these insights to language teaching. This approach and practice demonstrate how SDGs can serve as a framework for interdisciplinary education, enhancing language curricula while advancing global citizenship and critical analysis and insight among students (de la Fuente 2022). Mambu's findings highlight the transformative potential of connecting language education to sustainability goals and benchmarks, showcasing how such integration helps prepare students to address demanding global challenges while developing and enhancing their pedagogical and linguistic knowledge, skills and competencies.

Another interesting case study that embeds a good alignment with the rationale and approach to our Workshop, is Méndez Seijas and Parra's (2021) case study with a beginner Spanish curriculum. Their study, similarly, to Mambu's (2023) showcases how encompassing sustainability topics into language education can foster linguistic development and critical thinking skills with global citizenship as a sort of umbrella. Méndez Seijas and Parra's (2021) scope of study is within beginner

Spanish curriculum and the focus was on embedding themes of social, cultural, and environmental sustainability, with special emphasis on SDGs such as Quality Education (SDG 4) and Reducing Inequalities (SDG 10).

The curriculum redesign presented in the case study incorporates four thematic units that explore sustainability challenges in Spanish-speaking communities. These units embed and explore authentic content, such as multimodal and multimedia texts and project-based tasks (Méndez Seijas and Parra 2021, 96), to establish meaningful connections between language learning and real-world issues (Méndez Seijas and Parra 2021, 98). This enshrines the localisation element that was also so important in Mambu's study, where Pamela bridged global issues like inequality with her Papuan community. In line with this, in Méndez Seijas and Parra's (2021), students in this Spanish curriculum engaged with diverse sustainability topics embedded in the cultural contexts of the Spanish-speaking world, encouraging them to critically examine global challenges while advancing their language skills (Méndez Seijas and Parra 2021, 99–100). Students were actively immersed in discussions, collaborative group work, and multimedia-based tasks focused on sustainability themes, contributing to more familiarity and a deeper sense of engagement with societal and community challenges. In a sense, Méndez Seijas and Parra offered case studies as a key methodological choice within their own study, by focusing on the analysis of Spanish-speaking countries relevant issues, allowing students to connect language education with concrete, globally relevant issues. This content-driven approach is reflective of the principles seen in content-based instruction or CBI (Bula 2014). Interestingly the participants of our workshop, especially in the Re-drafting Modules Activity (with the proposal of embracing Backwards Design as the approach and methodology) were organically driven towards discussions around CBI as a crucial element when integrating the SDGs into their curricula (2014, 74).

The methodology adopted by Méndez Seijas and Parra (2021) aligns closely with Mambu's focus on participatory and experiential learning. This approach highlights the significance of integrating ESD, and the SDGs specifically, into language studies curricula, as demonstrated through the analysis of this section. Furthermore, it reflects key elements of non-formal education, which have also shaped the design and delivery of our workshop. Echoing the outcomes of Mambu's work, Méndez Seijas and Parra's work (2021) highlights the transformative potential of connecting language education to the SDGs, equipping students with critical awareness and cultural competence to address global challenges.

Our workshop served as a dynamic and informal platform that showcased many of the ideas and conclusions explored in these case studies. Our participants experimented with similar methodologies to the ones used, engaging in collaborative activities or reflecting on how to replicate these with their students. Both from the case studies and our workshop one main idea seems to emerge and occupy a central stage: the importance of sustainability literacy for both educators and students and the role it has in language education as an ally of sustainable development.

In this regard, this literature review has shown the potential of integrating ESD into language education but also the challenges associated with the interdisciplinary nature of ESD and therefore the professional development required to support educators in incorporating ESD in their teaching practices, but also opportunities, as showcased in the case studies discussed. The following section outlines the professional development framework used to underpin the set of activities aimed at promoting ESD in Language Studies.

4 Methodological Considerations

4.1 Professional Development Framework

Cultivating sustainability in the TU Dublin Language Studies curriculum is presented in this paper as a descriptive case study of a professional development (PD) activity implemented to support language colleagues at TU Dublin in incorporating sustainability into their curriculum. A descriptive case study explores and describes a case's unique characteristics, processes, and contextual details. Therefore, the focus of this chapter is on providing a detailed description of the activities conducted to help the reader understand the "what" and "how" aspects of the initiative, more specifically its structural design, its process, the participants' experiences and the facilitators' interventions. Data underpinning the description includes facilitators' observations, surveys about participants' knowledge and experiences, and preparatory documentation, which is patent in the different sections of this chapter Yin (2018).

The case study is based on a professional development initiative. Desimone (2009) defines professional development as the 'teachers' knowledge and skills that relate to changes in instructional practice, which increase students' learning and achievement'. This definition, unlike many other PD definitions (Sancar et al. 2021) is learner-centered and focuses on the student. Therefore, it aligns well with our PD objective for language lecturers.

Indeed, the ultimate goal of our series of PD activities is to equip our language students with ESD competencies. Our PD is based on Desimone's definition above and on the National Professional Development Framework (NPDF) articulated by the National Forum for the Enhancement of Teaching and Learning in Ireland (Infographic Draft 3 teachingandlearning.ie). The framework recognises that there are many forms of professional development (PD) and proposes a typology of (PD) activities that distinguishes between informal collaborative, unstructured or structured non-accredited activities and formal accredited activities.

The PD activities for ESD in Language Studies can be characterised as structured and non-accredited. The framework also identifies four different types of learning including new learning, consolidating learning, mentoring and leading. The first two types of learning were relevant to the PD activity for ESD, even if for some faculty members an element of mentoring and leading also came into play. The framework is composed of four domains, three of which were addressed in our PD activity, namely Professional Knowledge and Skills; Professional Communication and Dialogue; Professional Identity, Values and Development. The values underpinning the PD framework also underpin our PD initiative and include collaboration, authenticity, inclusivity, learner centeredness and scholarship.

The next section outlines in chronological order the stages of the PD initiative.

4.2 An Outline of the Activities of the Professional Development Initiative in Language Studies

Before narrating the pathway that unfolded in the professional development initiative undertaken by Language Studies in TU Dublin, the timeline in Fig. 1 provides a visual illustration.

The engagement of two of the Language Studies co-authors (Gabaudan and Molina Taracena) in ESD commenced in earnest in 2020 when they joined the newly formed TU Dublin SDG Literacy Community of Practice. As Wenger-Trayner puts it, a community of practice 'has now become the foundation of a perspective on knowing and learning that informs efforts to create learning systems' (2015). By bringing academics together, a community of practice can be a very effective means of supporting collaborative non-accredited learning, both in a structured and unstructured format as its activities take multiple formats. The SDG Literacy Community of Practice (CoP) aligns well to the principles of the NPDF as evidenced by its multiple activities, events and outputs. For example, resulting from their involvement with the CoP, the two Language Studies authors of this paper were invited in 2022 to edit a special section on sustainable development literacy in Language Learning and Teaching in the journal *Language Learning in Higher Education.* This process enabled them to connect with a network of Language Studies academics interested in Education for Sustainable Development. A new area of expertise within Language Studies was emerging and the TU Dublin authors of this paper were encouraged to build on this initial PD and engagement with ESD.

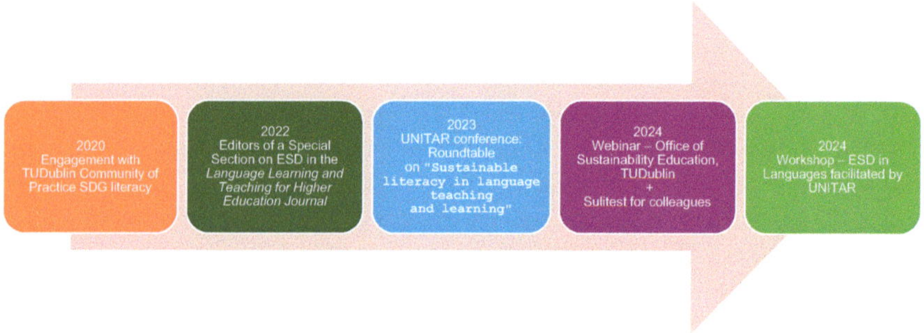

Fig. 1 Summary of PD activities for ESD in language studies

With TU Dublin aiming to embed the Sustainable Development Goals (SDGs) in the first year of all programs by September 2024, it was becoming increasingly urgent to support colleagues in reviewing their module learning outcomes, content, and assessments while ensuring that topics related to the UN's 2030 Agenda would be meaningfully and coherently integrated into their teaching. In October 2023, the TU Dublin authors secured a small amount of funding aimed at driving teaching and learning innovations and enhancements in the Higher Education sector in Ireland. This grant, known as SATLE (Strategic Alignment of Teaching and Learning Enhancement) Funding, included three strands, one of which focused on Education for Sustainable Development. The funding allowed the TU Dublin oauthors to organise two sets of events aimed at promoting and disseminating Sustainability literacy in Language teaching and learning.

The first event was a hybrid roundtable discussion on sustainable literacy in language teaching and learning held as part of an international research conference on Climate Change at the University of Bonn. Organised by TU Dublin in partnership with UNITAR, this structured event featured four experts in sustainability literacy in the context of language teaching and learning -María José de la Fuente, Kiley Kost, Owain Llewellyn and Seth Peabody-. Online participation was open, free of charge, to all language colleagues and it was attended by 35 participants. The roundtable was a follow-up to the afore mentioned publication of the Special Section on 'Sustainable development literacy in Language Learning and Teaching', in the Journal of *Language Learning in Higher Education* (2022). The discussion once again underscored the lack of research in this area and emphasized the need to raise awareness about the connections between language learning and sustainability literacy. Hosting the roundtable under the UNITAR umbrella, within the theme of Pedagogical Approaches for Climate Change Education, encouraged participants to reflect on how to incorporate sustainability topics into foreign language curricula.

The second activity was divided into two parts. The first part was a webinar for TU Dublin Language Studies staff, delivered in January 2024 by the Head of Sustainability Education in TU Dublin. Its purpose was to introduce TU Dublin's Sustainability Education Framework, the UN's Agenda 2030 with its SDGs as well as the cross-cutting competencies needed to achieve the SDGs (as outlined by UNESCO Education for Sustainable Development). This webinar was attended by 17 TU Dublin language lecturers representing Chinese, English, French, German, Italian, Irish and Spanish language teaching. The second part was a one-day workshop titled *Language Education for Sustainable Development: Integrating SDGs into Language Studies Curricula.* The workshop aimed to provide university language teachers with practical strategies for incorporating the SDGs into their curricula using non-formal and informal educational approaches. It was facilitated by UNITAR programme coordinator Alexandra Severino (one of the authors of this paper) who also designed the activities for the day as illustrated in Fig. 2. The workshop was promoted among the Third Level Institutions based in Dublin, however only Language Lecturers working in TU Dublin attended. 15 Language Lecturers teaching Chinese, English, French, German, Italian, and Spanish participated. All of them attended the previous

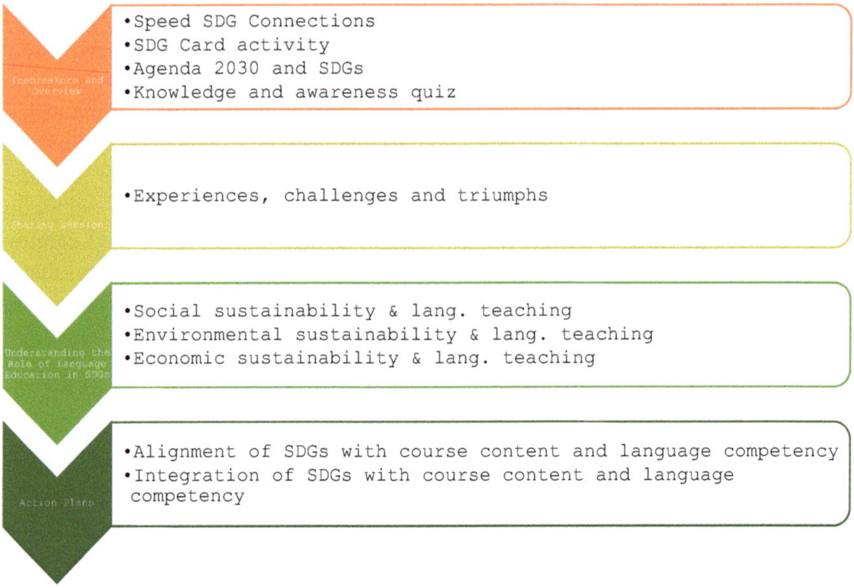

Fig. 2 Stages of the workshop

webinar on Education for Sustainable Development. The next section will examine the workshop in more detail.

5 Workshop on Language Education for Sustainable Development

The central piece to the series of professional development activities for colleagues in Language Studies was the workshop on language education for sustainable development. The learning methodologies used for the workshop will be considered, some detail on the sequencing of the workshop and its outcomes will be provided, all within the context of relevant literature and the National Forum PD Framework.

5.1 Learning Approaches to the Workshop

The approach and methodology used in the workshop was anchored in UNESCO's Education for Sustainable Development guidelines, which advocates integrating sustainability into all educational aspects. UNESCO guidelines promote critical thinking, problem-solving, and participatory teaching methods, aligning perfectly with the workshop's focus on the integration of SDGs into the Language Studies curricula. By incorporating ESD principles, the workshop aimed to foster awareness and action towards sustainability, enhancing global awareness among the participants (UNESCO 2020), leveraging from their valuable knowledge, skill set and experience to inspire action in terms of their curricula shift towards intentionally embedding the SDGs.

The workshop methodology employed active learning, and constructivist approaches to engage the TU Dublin's language teachers in meaningful and reflective discussions about the space that the SDGs occupy already in their work or the potential to encapsulate these further in the curriculum of their discipline—including at the learning objectives level. Active learning, as noted by Bonwell and Eison (1991), emphasises student engagement through activities that promote analysis, synthesis, and evaluation of class and course content. In the "Speed SDG Connections" exercise, those who participated engaged in dynamic and interactive activities that required them to move, communicate, and reflect on their experiences and connections to the SDGs. This also allowed the teachers to reflect and test potential ways in which, departing from the student perspective level, they can replicate the methods used in the workshop with their students for a joint and collaborative work on the integration of the SDGs in the curriculum.

The overall approach to the workshop is also in line with constructivist theory, which posits that learners construct knowledge through social interactions and experiences (Woolfolk 2019) in full harmony to the approach of OLEU to higher education from the delivery and recipient perspectives. Additionally, by facilitating peer-to-peer learning and discussion, the workshop allowed the teachers to collaboratively construct their understanding of the role of the SDGs in their curricula (Dicker et al. 1994, 40–45), aligned with the goals enshrined in the TU Dublin's Sustainability Statement. This approach seems to be particularly helpful in language education (Dicker et al. 1994, 40–45), and especially useful in making sure that the Leave no One Behind (LNOB) promise of the United Nations 2030 Agenda, in which the SDGs occupy a central stage and role, is embraced at the language studies level.

The methodology adopted in this workshop also drew on aspects of social learning theory, which underscores the importance of observational learning, imitation, and modelling. By encouraging participants to share their experiences and perspectives, the workshop fostered a collaborative learning environment, where teachers could benefit from observing and interacting with their peers. This was in line with the previous approaches adopted by TU Dublin, including the Conference session, and is especially beneficial in this area because it enhances communicative competence and cultural understanding, as different languages and different levels of studies might require different approaches, through peer interaction (Schunk 2012, 242–245).

The methodological goals of OLEU (UNITAR) in adopting these different approaches were to ensure that the activities were pedagogically suited to the Language Studies lecturers participating in the workshop but also aligned with broader educational goals of fostering sustainable development, which are part of the mandates of the two institutions—TU Dublin and UNITAR.

The authors are aware that there are some limitations to consider for the chosen methodological approaches. While the descriptive case study methodology offers in-depth insights into specific instances, its findings are often context-specific and may not be easily generalizable to other settings (Yin 2018). Additionally, as the approach relies on the (co)authors interpretation of data, there is an element of potential bias that is acknowledged (Merriam 2009). In an effort to address these limitations, there was an intentional focus on active participation, collaboration, and the exchange of diverse perspectives, which were reflected in this paper. This was also one of the reasons why utilising real-time engagement tools like Mentimeter and participatory activities, such as action plan development, were part of the methodological choices. While the specific outcomes are unique to each context, the adaptable nature of these workshops suggests that similar initiatives could inspire meaningful progress in diverse educational settings, even within the constraints of a single case study.

5.2 Execution and Analysis of the Workshop

The workshop began with "Speed SDG Connections," an engaging icebreaker designed to initiate discussions on how participants' courses could align with the Sustainable Development Goals (SDGs). This activity, where participants selected SDG cards that resonated with their teaching contexts, successfully fostered conversations on the relevance and integration of SDGs into language curricula. Through this exercise, teachers identified connections between their courses and specific SDGs, highlighting the interdisciplinary nature of both sustainability and language education (Tilbury 2011; Laurie et al. 2016; de la Fuente 2022; UNESCO 2012, 2019, 2020, Gabaudan 2022). This activity was developed and implemented with the active learning approaches promoted in Education for Sustainable Development (ESD) in mind. These approaches emphasise experiential and real-world engagement (UNESCO 2012; 2019a, 8) also stresses that learning environments should encourage participation, critical thinking, and collaboration, all of which were methods used to facilitate the activity but also outcomes of the activity itself (Laurie et al. 2016; UNESCO 2020).

By moving beyond traditional methods of instruction and fostering peer-to-peer dialogue, participants engaged in the kind of participatory learning central to effective ESD implementation (Laurie et al. 2016, 230). This choice and approach are in line with Laurie et al.'s (2016) findings that ESD pedagogies should emphasise and focus on learner involvement in meaningful contexts, helping participants make connections between sustainability and their academic practices (Fig. 3).

Following the SDG card activity, participants reflected on the patterns emerging from their choices, including common priorities such as environmental sustainability, social justice, and economic empowerment. This step aligns with the reflective processes highlighted by Tilbury (2011, 29) as critical in ESD, where learners must not only engage with sustainability concepts but also reflect on their values and the broader implications of their learning. These reflections had an important role in

Fig. 3 SDG card activity for speed connections

the overall workshop dynamic and helped participants recognise that sustainability encompasses more than just environmental concerns. Indeed, the UNESCO (2012, 10) highlights the holistic nature of sustainability, which includes economic and social dimensions (Acosta Castellanos and Queiruga-Dios 2022; Boar et al. 2020) and not only ecological and climate considerations.

To deepen their understanding of the broader framework, participants were introduced to Agenda 2030 and the SDGs through a Mentimeter quiz. This quiz revealed that while many participants were familiar with environmental aspects of sustainability, there was less awareness about the broader social and economic dimensions. This finding is consistent with research in ESD, which shows that while environmental sustainability is often prioritized, other dimensions of sustainability require further emphasis (Acosta Castellanos and Queiruga-Dios 2022; Boar et al. 2020). UNESCO (2012, 22) encourages educational settings to explore all aspects of sustainability, including human rights, gender equality, and poverty reduction (Fig. 4).

Subsequently, participants explored the role of language education in contributing to SDGs. The discussion was organised according to the thematic areas previously addressed, i.e., participants were divided into groups based on some of the different dimensions of sustainability: environmental, economic, and social (Acosta Castellanos and Queiruga-Dios 2022; Boar et al. 2020). These group discussions also helped to illustrate the interdisciplinary potential and components of language education to address various SDGs. For example, Group A focused on how eco-friendly vocabulary and environmental concepts can be integrated into language lessons, echoing the ecolinguistics movement, which has gained prominence in ESD discourse (Molina Taracena 2022). Group B explored economic sustainability by discussing ethical business practices and entrepreneurship in language studies—topics supported by UNESCO (2019, 14), which encourages the inclusion of economic sustainability within educational programs. Group C emphasized social sustainability, addressing how language education can promote inclusive and resilient communities, which is

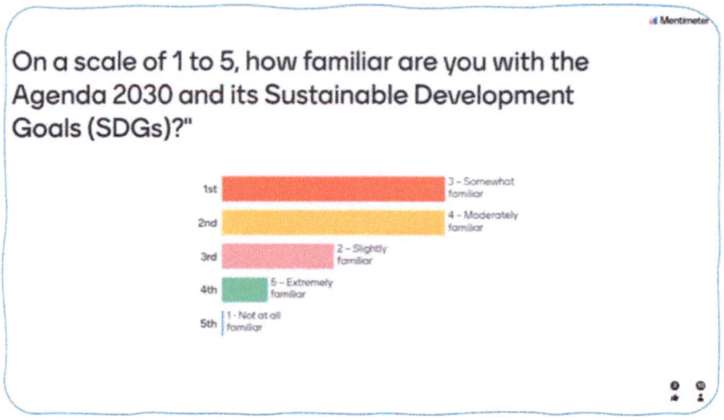

Fig. 4 Results of quiz on participants familiarity with Agenda 2030 and SDGs

a key pillar of the SDGs as also outlined by international policy frameworks relevant to ESD (UNESCO 2015, 10).

The next phase of the workshop involved developing action plans for integrating SDGs into language curricula. Teachers reviewed their existing course content through the lens of the Common European Framework of Reference for Languages (CEFR) and identified specific SDGs that aligned with their teaching goals. This process also mirrors the approach recommended by the UNESCO ESD Roadmap (2020), which encourages educators to incorporate sustainability into existing curricula rather than treating it as an add-on. Participants were also guided through the Backward Design framework (Drake and Reid 2018, 35–36), which supports the development of clear learning outcomes and assessments aligned with sustainability objectives and as a potential tool for curricula integration of the ESD framework and were invited to experiment and test with it. The backward design methodology also facilitated the incorporation of simplified sustainability themes into lower proficiency levels. Indeed, carefully designed and scaffolded activities can be an effective way to begin embedding ESD into entry-level language courses. One example provided was the topic of transportation, commonly included in many entry-level language courses. This topic can serve as a platform to introduce discussions on sustainable transportation practices, among other related sustainability topics.

In the concluding sessions, lecturers present reflected on their progress and discussed the practical implementation of their action plans. This step also contributed to reinforce the importance of continuous professional development, and adaptability, as highlighted in the UNESCO Education for Sustainable Development Roadmap (2020). This roadmap stresses the need for educators to adapt and refine their approaches to meet the evolving challenges that relate to sustainable development. Lecturers were also encouraged to practice these exercises, use the backward design methodology, and consider curricula adaptation as a co-creation process to be further developed with their student (Laurie et al. 2016).

Finally, the discussions also focused significantly on the collaborative aspect of ESD, with participants recognising the value of engaging with stakeholders, including students and local communities (UNESCO 2012, 15), to ensure the successful integration of SDGs into their teaching. The stakeholders identified by the teachers included TU Dublin services and departments, senior management, and other institutions and organizations. Participants in the workshop recognised that students must be involved in this process and participate actively in the curricula adaptation process to fully encompass the SGDs and ESD. Ideally some of the activities related to ESD in language learning should be done with the participation of local community members although this can be challenging to implement in an English-speaking environment.

Evaluation was conducted informally and in real time using tools like Mentimeter, as per the figures shared in this section, focusing primarily on participants' perceptions rather than employing a formal survey. However, we plan to conduct a follow-up survey in the future to assess progress and gather deeper insights into the long-term impact of the workshop. This follow-up will help us understand how participants have applied the concepts discussed, identify areas where further support might be needed, and evaluate how effectively the workshop has contributed to integrating sustainability into language studies curricula.

6 Conclusion

As we reflect on the framework that initially guided the PD initiative for Language Studies, the National Forum PD Framework proved to be a relevant and useful compass to guide our decisions along the PD journey for ESD. The types of PD activities were all collaborative, structured and non-accredited activities. They were facilitated by a range of supporting networks, both internally and externally, and by faculty members, champions of ESD within the discipline of Language Studies. As highlighted through the learning approaches in the workshop section, reflective peer discussion through active learning are most successful when aiming to understand and integrate into the curriculum the complex issues raised by ESD.

For all colleagues, the learning was at times new learning or a consolidation of existing learning. Through peer-to-peer discussions, an element of mentoring and leading occurred, according to participants' own areas of strength. A range of PD domains were addressed through the various activities and particularly the workshop. Through their engagement in the PD for language ESD, participants expanded their knowledge around the UN Agenda 2030 and were able to explicitly articulate the skills required to incorporate SDGs into the curriculum. Communication and dialogue were a welcome feature of the workshop as participants were given a space to collaborate and question their values and approaches to their language teaching practices (Fig. 4). Given the nature of ESD, all activities were underpinned by the value of authenticity. The focus was always on the development of the student's language competency and sustainability literacy. The approach was inclusive to facilitate engagement from participants with varying backgrounds e.g. cultural studies, linguistics, applied language teaching, literary studies.

Upon conclusion of the workshop, the facilitator (also co-author) noted that the participants demonstrated significant engagement and enthusiasm throughout the workshop activities. Moreover, the diverse perspectives and methodologies adopted by participants highlighted the flexibility and adaptability required in integrating SDGs into language curricula.

The facilitator's role in guiding and supporting self and group-led processes proved effective, contributing to positive feedback from participants. The quiz activities provided valuable insights into participants' familiarity, perceptions, and priorities related to Agenda 2030 and the SDGs. As mentioned by participants, the workshop served as a platform for meaningful exchange, fostering collaboration and sharing of best practices among language educators who otherwise seldom get the time and opportunity to delve into language education matters for an entire day. The closing session emphasised the importance of continued dialogue and networking for sustained progress in integrating sustainability into language studies education.

Through these activities seeds of change have been sown by cultivating Sustainability Literacy in TU Dublin's Language Studies Curriculum. The next step is to continue implementing the integration of SDGs into Language modules both within TU Dublin but also beyond by facilitating other language colleagues' journey. It is important to adopt a flexible methodological approach that can be adapted to different teaching contexts and respond to specific institutional priorities. We believe this study has the potential to be adapted in universities, institutes, and language centres that aim to incorporate ESD in their curricula and equip language educators with the much-needed ESD competencies.

References

Acosta Castellanos P, Queiruga-Dios A (2022) From environmental education to education for sustainable development in higher education: a systematic review. Int J Sustain High Educ 23(3):622–644. https://doi.org/10.1108/IJSHE-04-2021-0167

Barth M (2015) Implementing sustainability in higher education: learning in an age of transformation. Routledge Studies in Sustainable Development

Boar A, Bastida R, Marimon F (2020) A systematic literature review. Relationships between the sharing economy, sustainability and sustainable development goals. Sustainability 12(17):6744. https://doi.org/10.3390/su12176744

Bonwell CC, Eison JA (1991) Active learning: creating excitement in the classroom. ASHE-ERIC Higher Education Report No. 1. The George Washington University, School of Education and Human Development

Bula O (2014) Content-based instruction: a relevant approach of language teaching. Innovaciones Educativas 15:71. https://doi.org/10.22458/ie.v15i20.515

De la Fuente, MJ (ed) (2022) Education for sustainable development in foreign language learning: content-based instruction in college-level curricula. Routledge

Desimone LM (2009) Improving impact studies of teachers' professional development: toward better conceptualizations and measures. Educ Res 38(3):181–199. https://doi.org/10.3102/0013189X08331140

Dicker C, Chamot A, O'Malley J (1994) The Calla Handbook: implementing the cognitive academic language learning approach. Addison-Wesley Publishing Company

Drake SM, Reid JL (2018) Integrated curriculum as an effective way to teach 21st-century capabilities. Asia Pac J Educ Res 1(1):31–50. https://doi.org/10.30777/APJER.2018.1.1.03

Gabaudan O (2022) On a journey towards Education for Sustainable Development in the foreign language curriculum. Lang Learn High Educ 12(2):525–546. https://doi.org/10.1515/cercles-2022-2059

Laurie R, Nonoyama-Tarumi Y, Mckeown R, Hopkins C (2016) Contributions of education for sustainable development (ESD) to quality education: a synthesis of research. J Educ Sustain Dev 10(2):226–242. https://doi.org/10.1177/0973408216661442

Mambu JE (2023) Embracing sustainable development goals critically to explore life purposes in english language teacher education. TEFLIN J 34(2):264–282. https://doi.org/10.15639/teflinjournal.v34i2/264-282

Méndez Seijas J, Parra ML (2021) Engaging students with social, cultural, and environmental sustainability topics in the Spanish-speaking world: a reimagined beginner Spanish curriculum. In: De la Fuente MJ (ed) Education for sustainable development in foreign language learning: content-based instruction in college-level curricula. Routledge, pp 87–104

Merriam SB (2009) Qualitative research: a guide to design and implementation. Jossey-Bass, San Francisco, CA

Molina Taracena P (2022) Bibliographical review on sustainable literacy in language learning and teaching. Lang Learn High Educ 12(2):513–524. https://doi.org/10.1515/cercles-2022-2059

O'Boyle A (2024) Educating for sustainability in language education. In: PowerPoint presentation, closing plenary, equals annual international conference. Lisbon

Sancar R, Atal D, Deryakulu D (2021) A new framework for teachers' professional development. Teach Teach Educ 101. https://doi.org/10.1016/j.tate.2021.103305

Schunk DH (2012) Learning theories: an educational perspective. 6th ed. Pearson

Tilbury D (2011) Education for sustainable development: an expert review of processes and learning. UNESCO. ED-2010/WS/46

United Nations Educational, Scientific and Cultural Organization (UNESCO) (2020) Education for sustainable development: a roadmap. United Nations Educational, Scientific and Cultural Organisation

United Nations Educational, Scientific and Cultural Organization (UNESCO) (2019) Framework for the implementation of education for sustainable development (ESD) beyond 2019. UNESCO

United Nations Educational, Scientific and Cultural Organization (UNESCO) (2015) Global action programme on education for sustainable development. UNESCO

United Nations Educational, Scientific and Cultural Organization (UNESCO) (2012) The education for sustainable development sourcebook. Education for sustainable development in action, learning and training tools No. 4. UNESCO

United Nations Institute for Training and Research (UNITAR) (n.d.) Mission, vision and core functions. Available at: https://unitar.org/about/unitar/mission-vision-and-core-functions

Wenger-Trayner E, Wenger-Trayner B (2015) An introduction to communities of practice: a brief overview of the concept and its uses. Available from authors at https://www.wenger-trayner.com/introduction-to-communities-of-practice

Woolfolk A (2019) Educational psychology. 14th ed. Pearson

Yin, R. K. (2018) Case study research and applications: design and methods 6th edn. Thousand Oaks, CA. Sage

Sustainable Competencies of the Soon-to-be Leaders and Experts: Evidences from Higher Education System

Lazar Pavić, Mateja Čuček, and Matevž Obrecht

Abstract Although sustainable development is gaining increasing global attention, the actual level of knowledge about this topic among current experts and leaders remains limited. This issue is likely to persist among future leaders and experts unless higher education study programs incorporate sustainable content more effectively and accurately in the near future. The main objective of this paper is to examine the inclusion of sustainable content in accredited higher education study programs in Slovenia with the measure of the Sustainability Assessment Score (SAS) of individual study programs and to identify statistically significant differences in SAS between different groups of study programs based on factors such as the type of institution offering the program, the KLASIS-P16 classification, the study cycle of the program, and the presence of sustainability-related keywords in at least one course title within the program. Based on an analysis of all 956 Slovenian higher education study programs, the SAS was calculated for each program by searching for the occurrence of sustainability-related keywords in Slovenian, as identified by a focus group of renowned experts in sustainable development in Slovenia. These keywords included: sustainable development, green/environmental, environmental protection, circular economy, social responsibility, and ecology. Parametric tests (T-test and ANOVA) were applied to determine statistically significant differences in SAS across different groups of study programs. The results revealed that higher education study programs offered by private institutions and programs containing courses with sustainability-related keywords in their titles had higher SAS. Additionally, statistically significant differences in SAS were observed between study programs across different fields.

L. Pavić (✉) · M. Čuček · M. Obrecht
Faculty of Logistics, University of Maribor, Celje, Slovenia
e-mail: lazar.pavic@um.si

M. Čuček
e-mail: mateja.cucek1@um.si

M. Obrecht
e-mail: matevz.obrecht@um.si

W. Leal Filho et al. (eds.), *Education for Sustainable Development: The Contribution of Universities*, World Sustainability Series,
https://doi.org/10.1007/978-3-031-86985-3_17

Keywords Sustainable competence · Sustainable knowledge · Study programs · Higher education system

1 Introduction

Sustainability is no longer just a scientific concept but has become a central theme in daily news, politics, and business transformation. As its significance continues to grow, there is an urgent need to systematically explore and enhance education in the field of sustainability, ensuring it becomes a core subject at universities. Recent studies highlight a global shortfall in environmental and sustainability education, alongside a pressing need to develop innovative frameworks to address these gaps (Brudermann et al. 2019; Daub et al. 2020; Do 2020; Eppinga et al. 2020). This challenge is particularly pronounced in the Global South (Ulmer and Wydra 2020).

The climate crisis has become one of humanity's most pressing concerns; therefore, addressing sustainability-related challenges through appropriate knowledge and education will play a critical role. To advance sustainability, it is essential to educate future leaders and experts in sustainable development and environmental sustainability. This requires the study of various environmental protection concepts, particularly at the higher education level, as they are inherently complex and interdisciplinary (Daneshjoo et al. 2019). Sustainability science is widely recognized as a guiding framework for the sustainability paradigm and forms the core philosophy of national and international development agendas, such as the Paris Agreement. It is also embedded in the United Nations Sustainable Development Goals (SDGs), specifically within SDG 4 (Quality Education), which emphasizes the importance of education for sustainable development (SD). Target 4.7 of SDG 4 states that "all learners acquire the knowledge and skills needed to promote sustainable development, including, among others, through education for sustainable development and sustainable lifestyles, human rights, gender equality, promotion of a culture of peace and non-violence, global citizenship, and appreciation of cultural diversity and culture's contribution to sustainable development" (United Nations 2015).

Furthermore, sustainability education is a core competency for managing sustainability and Environmental, Social, and Corporate Governance (ESG) initiatives (Piza et al. 2018).

1.1 Theoretical Backgrounds

Higher education for sustainable development addresses highly complex real-world challenges, including pollution, climate change, resource scarcity, overproduction, biodiversity loss, inequitable wealth distribution, and emerging concepts like degrowth. These topics are critical for transforming both society and higher education institutions (HEIs) toward a more sustainable future (Priyadarshini and Abhilash 2020; Barth and Michelsen 2013; Benton-Short and Merrigan 2016).

While some challenges of economic growth are presumed solvable through decoupling growth from resource consumption, the prevailing "sustainability and growth" paradigm has often exacerbated inequalities, increased pressure on certain resources (e.g., for e-mobility), harmed biodiversity, and created additional social tensions. Consequently, the paradoxes of sustainable development must be integrated into education, particularly for future leaders. This includes adopting alternative pedagogical methods such as indigenous learning, eco-pedagogy, eco-centric education for a steady-state and circular economy, and approaches emphasizing planetary ethics, empowerment, and liberation (Viegas et al. 2016; Wals 2014). However, achieving these goals requires a systematic analysis of higher education to design effective transformation strategies.

Pioneering studies in sustainability education have tracked its evolution, from a focus on end-of-pipe pollution control to courses emphasizing sustainability (Obrecht et al. 2022). More recent studies reveal that integrating environmental education at the university level fosters environmentally conscious leadership, as most current and future leaders, engineers, and educators graduate from the university system (Glavič 2006). As sustainability increasingly becomes a top concern for future managers (Ayalon and Avnimelech 2007), higher education must adopt transformative approaches that challenge worldviews and societal values (Zorio-Grima 2020).

Significant changes are needed in the number of sustainability-related programs and courses, as well as in how these topics are taught, to provide more effective education for future managers, experts, and other societal stakeholders while also raising environmental awareness (Howlett et al. 2016; Kopnina 2020). The growing importance of sustainability education is reflected in changing university curricula and visions, with many institutions prioritizing sustainability not only for students but also for academics (Wamsler 2020).

For example, Brudermann et al. (2019) explored the theoretical foundations of education for sustainable development in Austria, where the first programs in this field were established two decades ago. Environmental management and sustainability-related courses are increasingly popular among students, signaling a need for educational programs to adapt to this demand (Wals 2014). Studies show that students across diverse fields consistently view sustainability as a critical component of their higher education (Dong et al. 2023).

However, research by Paço and Lavrador (Boarin et al. 2020) reveals no clear correlation between environmental knowledge, attitudes, and sustainable practices at universities or on their campuses (Paço and Lavrador 2017; Ozdemir et al. 2020). It is suggested that sustainability topics be better integrated into economics courses, such as marketing, which often promote greenwashing, overconsumption, and material accumulation as markers of success—despite evidence showing that these are not indicators of personal happiness or societal well-being (Roos and Guenther 2020).

Sustainability science as a distinct field has struggled to gain traction in educational settings. Skills such as foresight, advanced problem-solving, and interdisciplinarity remain underutilized (Kemper et al. 2020). This gap is especially critical given that small and medium-sized enterprises (SMEs)—which contribute over 60% of all environmental impacts—often lack the skills and empowerment necessary for sustainable management (Hermann and Bossle 2020). Lifelong learning (LLL) programs addressing sustainability challenges are urgently needed to support active professionals and the organizations they serve (Denac et al. 2018).

As Mulder emphasizes, sustainability issues should guide educational priorities, particularly for engineers (Hay and Eagle 2020). Achieving the transformation to an ecologically sound society and providing adequate educational programs requires a careful analysis of context, the role of education, existing competency gaps, and sustainability performance (Mulder 2017; Orlović-Lovren 2011; Wang et al. 2013).

This study focuses on the integration of sustainability in study programs within the Slovenian higher education system. Slovenia, located in Central Europe and a member of the European Union (EU), has been actively aligning its educational strategies with EU-wide policies such as the European Green Deal, the United Nations' Sustainable Development Goals (SDGs), and the Education for Sustainable Development (ESD) framework promoted by UNESCO. Higher education in Slovenia is structured around a dual system of public and private institutions, offering programs at professional, university, master's, and doctoral levels. Additionally, lifelong learning (LLL) seminars form a vital component of Slovenia's education ecosystem, aiming to provide opportunities for continuous education beyond formal degree programs.

Slovenian higher education institutions are categorized into public universities, private universities, and standalone higher education institutions. Public institutions dominate the landscape and are primarily funded by the government, while private institutions often rely on tuition fees and industry partnerships. These structural differences influence how curricula are designed, the degree of innovation in program offerings, and the responsiveness to emerging trends such as sustainability.

Study programs in Slovenia are classified using the KLASIUS-P16 framework, which is a standardized system for categorizing education programs based on fields of study and educational levels. This classification enables consistent evaluation and comparison of programs across disciplines, including arts and humanities, natural sciences, engineering, and health sciences, among others. Such a framework is not unique to Slovenia; similar systems are used internationally (e.g., UNESCO's ISCED classification), making the results of this study comparable on a global scale.

Lifelong learning is a cornerstone of Slovenian education policy, designed to support individuals in adapting to evolving societal and economic demands. LLL programs are typically shorter, non-degree programs targeting professionals and adults seeking to update their knowledge and skills. These programs play a critical role in advancing sustainability, as they offer tailored courses addressing specific challenges, such as renewable energy technologies, sustainable business practices, and environmental conservation.

Globally, LLL programs are recognized as essential tools for achieving SDG 4 ("Ensure inclusive and equitable quality education and promote lifelong learning opportunities for all"). In Slovenia, the integration of sustainability into LLL programs reflects broader international efforts to bridge knowledge gaps in adult education, particularly in the context of transitioning to greener economies.

2 Materials and Methods

Reviewed studies primarily focus on the perceptions of students and staff regarding sustainability education (Kopnina 2020; Emblen-Perry 2022; Bell et al. 2020; Sandri 2020), often overlooking the specific study programs and the content being taught. Some studies provide only a cursory review of study programs, based on their titles, educational outcomes (e.g., academic profile or degree title), or by examining a limited subset of programs (Brudermann et al. 2019; Daub et al. 2020; Obrecht et al. 2022; Kopnina 2020; Dong et al. 2023; Valderrama-Hernández et al. 2019). The findings suggest that study programs are sometimes evaluated based on expert opinions rather than through statistical analysis of curricula, such as in-depth factor analysis and content evaluation (Ulmer and Wydra 2020). Additionally, some research favors public institutions, failing to highlight differences between public and private institutions (Eppinga et al. 2020; Ulmer and Wydra 2020). Others rely heavily on university rankings as a basis for their conclusions (Obrecht et al. 2022; Fokdal et al. 2020).

The research focuses on a comprehensive overview of all 956 Slovenian higher education study programs that were available in the academic year 2021/2022. It measures the sustainability assessment score for all study programs by identifying occurrences of key terms, which were previously determined based on focus group results from renowned experts in the field of sustainable development in Slovenia. The identified keywords are: **sustainable development, green/environmental, environmental protection, circular economy, social responsibility,** and **ecology**.

Our primary research question was how well sustainability is integrated into these 956 study programs and what kind of data should be used for in-depth statistical analysis, forming the foundation for data-driven decision-making in the development of new programs, curricula, and content modifications. The goal is to enhance education for future environmental experts, managers, and leaders. Therefore, our objective was to statistically analyze a set of study programs and address statistically significant differences using various statistical tests and approaches. This research continues previous studies focused on course content and the interconnectivity of sustainability topics (Boarin et al. 2020).

Gaining in-depth insight is viewed as the first step toward creating guidelines and strategies for more intensive integration of sustainability into higher education. This includes designing and implementing lifelong learning (LLL) seminars for experts who have already completed their formal education. The research team is also motivated to develop an LLL education platform to bridge the gap between sustainability knowledge and the requirements set by EU legislation (such as the

EU taxonomy and sustainability reporting directive), as well as the climate crisis, which is driving us toward ecological overshoot. Data-driven assessment can help improve the quality and sustainability of education systems, making it crucial to examine and assess the national higher education system in terms of its integration of environmental sustainability (Kovačič Lukman et al. 2021; Alvarez-Lopez and Matarranz 2019; Vagnoni and Cavicchi 2015). This study gives the broad directions to the process of designing sustainable education curricula specifically tailored to address the needs of students, lifelong learners and alumni in fostering environmental and social responsibility.

The course content of 956 Slovenian higher education study programs was examined. In the next phase, the situation in the domain of higher education environmental education was thoroughly analyzed through comparisons, examining course topics and content using specific keywords. A panel of sustainable development experts determined the sustainability assessment score for each program. After conducting two iterative cycles of the Delphi method, involving a panel of experts in sustainability and education, six keywords were identified and selected as representative of the core essence of sustainability within study programs. These keywords were chosen based on a consensus among the experts, ensuring they accurately encapsulate the multifaceted concept of sustainability and its relevance to educational curricula. The keywords are: (1) "green/environmental"; (2) "environmental protection"; (3) "sustainable development/sustainability"; (4) "circular economy"; (5) "ecology/ecological"; and (6) "social responsibility," along with related acronyms and synonyms.

The comparisons were made based on study area, institution type, study level, and subject name inclusion. Parametric tests were performed to establish statistical significance. The three study levels—bachelor's, master's, and doctoral degrees—were examined, and the following KLASIUS-P16, which is a compulsory national standard to be used in the recording, collection, processing, analysis, transmission and presentation of statistical and analytical data relevant for monitoring socio-economic and demographic conditions and trends in the Republic of Slovenia, study fields were considered: arts and humanities; social, business, management, and law studies; computer engineering, mathematics, and natural sciences; mechanics, production technologies, and construction; health and social care; agriculture, forestry, fisheries, and veterinary studies; and educational sciences and teacher training. Data were collected from public universities, institutions of applied sciences, private colleges, and secondary schools.

Courses within the study programs were further analyzed to identify the three most frequent and three least frequent themes related to sustainability. This helped determine which sustainability issues are currently considered a priority in sustainability-related subjects. Additionally, cross-sectional analyses were conducted to identify significant links between topics and those that are not yet connected. The six keywords defined by the expert panel were chosen because they represent the essence of sustainability topics and directly relate to learning outcomes in sustainable development.

Through detailed analysis of course content, it was determined that some programs also included courses without the identified keywords, which still reflected indirect relationships with learning outcomes related to sustainable development. However, due to the indirect nature of these relationships and the fact that these courses accounted for less than 3% of the study programs, they were excluded from further analysis.

3 Results

In the first phase, the sustainability assessment score (SAS) for all 956 programs was calculated by summing the occurrences of the six previously mentioned keywords. The mean SAS value for all 956 study programs is 2.65 (Table 1), indicating that the inclusion of sustainable content in higher education study programs in Slovenia is relatively low on average. In the next research phase, parametric tests were used to determine statistically significant differences between different groups of study programs, based on factors such as the type of institution where the program is taught, the KLASIUS-P16 category of the program, the study cycle of the individual program, and the inclusion of sustainability keywords in the name of at least one subject within the program. Although the Kolmogorov–Smirnov test results indicated that the data were not normally distributed ($p<0.05$), parametric tests were deemed appropriate for this analysis, as they are more robust and powerful for detecting differences when the assumption of normality is met.

The ANOVA test was used to investigate statistically significant differences between study programs belonging to different KLASIUS-P16 fields. The results of the ANOVA test, presented in Table 2, indicate significant differences in sustainability scores among various academic fields at the 0.01 level of significance. The field of Arts and Humanities achieved the highest mean sustainability score (4.1667), suggesting a strong emphasis on or integration of sustainability within these programs. Conversely, the field of Health and Social Work displayed the lowest mean score (1.5263), indicating a comparatively limited focus on sustainability in these curricula. Fields such as Services and Environmental Safety (3.2439) and Engineering, Manufacturing Technology, and Construction (2.8517) also scored relatively higher, reflecting their potential alignment with sustainability themes in practice-oriented or sector-specific contexts. In contrast, fields like Agriculture, Forestry, Veterinary, and Fishing (2.3306) and Natural Sciences, Mathematics, and Computer Science (2.4766) scored lower, which may be surprising given their natural connection to environmental topics, pointing to potential gaps in curriculum design. These

Table 1 Descriptive statistics test results for variable SAS score

Variable	Valid N	Min	Max	Mean
Sustainability assessment score	956	0	6	2.6569

results highlight the variability in the prioritization of sustainability education across disciplines and suggest a need for targeted curricular enhancements to integrate sustainability more evenly across all academic fields.

The ANOVA test was used to investigate statistically significant differences between different study programs at different study levels (professional, university, master and doctoral). The ANOVA test results in Table 3 reveal statistically significant differences in sustainability scores across different study levels ($p < 0.05$). Doctoral programs exhibited the highest mean sustainability score (2.7862), suggesting a stronger emphasis on sustainability at the advanced research and academic level. Professional programs also demonstrated a relatively high mean score (2.7510), potentially reflecting their applied nature and alignment with industry needs, where sustainability may be increasingly emphasized. Similarly, Master's programs scored moderately high (2.7055), indicating an integration of sustainability themes at the graduate level, likely as part of specialized or advanced study areas. In contrast, University programs at the undergraduate level had the lowest mean score (2.4257), pointing to a potential gap in incorporating sustainability topics within foundational academic curricula. These results highlight a trend where sustainability appears to gain greater focus at higher levels of study, emphasizing the need to strengthen sustainability integration at the undergraduate level to build foundational knowledge and awareness early in students' educational journeys.

The T-test was used to investigate statistically significant differences between study programs taught in private and public institutions. Statistically significant differences were found ($p < 0.05$). Based on the mean values for the different groups of study programs, we can conclude that study programs taught in private institutions include significantly more sustainable content than those taught in public institutions (Table 4). This discrepancy may reflect differing institutional priorities, with private institutions potentially responding more dynamically to industry demands

Table 2 ANOVA test results: field of study

ANOVA				
Sustainability score				
	N	Mean	F	Sig.
Education	22	2.4545	8.298	0.000**
Arts and humanities	12	4.1667		
Social business law and administrative sciences	149	2.7919		
Natural sciences, mathematics, computer science	342	2.4766		
Engineering, manufacturing tech. and construction	209	2.8517		
Agriculture, forestry, veterinary, fishing	121	2.3306		
Health and social work	19	1.5263		
Services, environmental safety	82	3.2439		
Total	956	2.6569		

** Statistical significance at 0.01 level

Table 3 ANOVA test results: level of study

ANOVA				
Sustainability score				
	N	Mean	F	Sig.
Professional	253	2.7510	2.996	0.030*
University	249	2.4257		
Master	309	2.7055		
Ph.D.	145	2.7862		
Total	956	2.6569		

* Statistical significance at 0.05 level

and global sustainability trends due to their closer ties to stakeholders or market pressures. Public institutions, on the other hand, may face challenges such as rigid administrative structures or a slower pace of curricular reform, which could limit the integration of sustainability-related content.

This finding underscores the importance of addressing these gaps by encouraging public institutions to adopt best practices from private institutions, such as engaging with external partners and aligning curricula with emerging sustainability standards. Additionally, further research could explore whether the observed differences are influenced by factors like funding, institutional autonomy, or faculty expertise in sustainability topics. These insights can help foster a more balanced approach to sustainability education across both types of institutions.

The T-test was used to investigate statistically significant differences between study programs that included at least one subject with at least one keyword in the subject's name. Statistically significant differences were found ($p \leq 0.01$). Based on the mean values for the different groups of study programs, we can conclude that study programs with at least one subject containing a previously defined sustainability keyword include significantly more sustainable content than those without such subjects (Table 5). This finding underscores the critical role that explicit labeling and naming of sustainability-related subjects play in embedding sustainable content within educational curricula.

The results suggest that including sustainability-related keywords in subject names likely signals a deliberate effort to integrate sustainability concepts into the curriculum, either as a primary focus or a key theme. Conversely, the lower scores

Table 4 T-test results: type of institution

T-test				
	Type of institution	N	Mean	Sig.
Sustainability score	Private	185	2.8757	0.022*
	Public	771	2.6044	

* Statistical significance at 0.05 level

Table 5 T-test results: inclusion in the name of at least one subject

T-test				
	Inclusion in the name of the subject	N	Mean	Sig.
Sustainability score	NO	415	2.1373	0.001**
	YES	541	3.0555	

** Statistical significance at 0.01 level

for programs without such subjects indicate a potential lack of structured emphasis on sustainability or an absence of explicit integration into the curriculum.

This observation highlights the importance of intentional curriculum design in promoting sustainability education. Institutions could enhance their sustainability profiles by reviewing and updating subject offerings to explicitly reflect sustainability themes in course titles and content. Further research might investigate the extent to which this naming practice influences students' awareness and engagement with sustainability concepts, as well as the quality and depth of sustainability content delivered in these programs.

4 Discussion

The previously presented results show that study programs with a higher SAS score tend to be those taught in private institutions and those with subjects that include sustainable development keywords in the name of at least one course. The results also indicate statistically significant differences between fields of study according to KLASIUS-P16, as well as between study programs (professional, university, master's, and doctoral). These results appear logical based on the content analysis, which will be presented below.

Firstly, statistically significant differences were found when comparing by KLASIUS-P16 fields. This is likely because the keywords used to calculate the SAS score in our study were primarily focused on environmental and basic components of sustainability (e.g., green/environmental, environmental protection, sustainable development/sustainability, circular economy, ecology, and social responsibility). Specific aspects of sustainability related to particular SDGs were not included. Therefore, it can be assumed that study programs in fields like health and social work included keywords related to specific SDGs, such as health, well-being, and healthcare, along with other keywords more focused on the social aspects of sustainability.

Secondly, statistically significant differences were found between study programs. This may be due to the fact that some study programs are more focused on academic and management skills, which tend to address less specific topics. This could explain why these programs show less integration of sustainability compared to others. It can also be assumed that higher SAS scores are found in upper-level programs (master's and doctoral) because these programs are shorter and can adapt more quickly to change.

Thirdly, study programs taught in private institutions have significantly higher SAS scores compared to those in public institutions. This can be explained by the fact that many private institutions focus primarily on economics, business, and management programs, which generally do not require specialized laboratories and may not have the same traditional structure as other programs. As a result, these programs are more flexible in adapting to market demands and can easily incorporate sustainable business practices as a primary field of focus and competitive advantage.

Finally, study programs that include at least one subject with one of the defined keywords often use this as a promotional tool, as it allows them to easily demonstrate their sustainability orientation.

Education for sustainable development is a strategic focus of EU education. Sustainability will continue to be developed as a strategic part of study curricula, but first, researchers need to identify and develop the skills required for future sustainability experts, who will be capable of distinguishing greenwashing from genuine green prosperity. One of the primary goals of the EU, in its path to meeting climate targets by 2030 in a fair, cost-effective, and competitive manner, is to make Europe the first climate-neutral continent. This goal involves preparing all sectors of the EU economy to meet this challenge. As a result, greater efforts will be made towards sustainable investments and jobs related to emission reduction, responsible growth, addressing energy poverty, reducing external energy dependency, and improving health and well-being.

In this context, the European Union Recovery and Resilience Facility for Education, for the period 2023–2025, aims to achieve curricular renewal and changes for a green, resilient, sustainable, and digitally connected educational system. Various methods and forms of education for sustainable development will be introduced and tested through pilot projects. The Facility will simultaneously contribute to achieving two national strategic goals of sustainable education: empowering youth and adults to work and live in a sustainable, environmentally responsible society, and transitioning to a low-carbon circular economy, as well as improving environmental literacy as a key component of functional literacy.

The planned content should be linked to the United Nations Sustainable Development Goals, which call on all learners to acquire the knowledge and skills necessary to promote sustainable development by 2030 and beyond. Integrating environmental and climate issues into future leaders' education, alongside the use of digital technologies as powerful enablers of the green transition, will contribute to sustainable behavior and the development and use of digital solutions. This aligns with the EU industrial strategy. Recognizing that universities are key stakeholders in addressing major social challenges, the proposed transition must align with the objectives of the Guidelines for the Renewal of Higher Education Professional Education. This includes a plan for implementation, specifically focused on the green transition, the digital transition, and the area of lifelong learning (LLL) and micro-credentials.

LLL and micro-credentials are of special importance because current students will be significantly better educated in sustainability than their predecessors, who hold top managerial positions in companies that need to start sustainability reporting (e.g., large companies by 2024, medium-sized companies by 2026) but lack such knowledge and education. LLL is thus essential to reducing the mismatch between graduates' knowledge and labor market needs, as it addresses the harmonization of expectations regarding the knowledge required by the labor market and future professions, on one hand, and the competencies that graduates should acquire through higher education, on the other. However, these competencies may still be lacking in the generation of older or more experienced professionals. Therefore, LLL programs will allow them to acquire valuable skills.

Thinking strategically, on a national level, it is necessary to establish and manage sustainable supply chains with a particular focus on knowledge management (creation, storage, transfer, and utilization) and its impact on supply network performance. These supply chains should include experts with varying priorities (social, technical, natural) focused on sustainable development. Such managed supply chains should facilitate the flow of materials, information, financial resources, and more, between different actors so that knowledge sharing can significantly improve the level of skills and expertise across the entire supply chain. This is especially critical if the EU, as a whole, wants to become more resilient and competitive. Future experts will be required to solve environmental challenges as one of their daily priorities. In the near future, a visible shift from classical economic theory toward greater inclusion of environmental and social performance will also be promoted through the assessment of the sustainability of investments based on the EU taxonomy.

The findings of this study have broader implications for global higher education systems. Slovenia's integration of sustainability serves as a microcosm for understanding how institutional and program-level factors shape the adoption of sustainability principles. By situating these results within the context of EU initiatives and global frameworks, this study contributes to international discussions on best practices for embedding sustainability in education.

The emphasis on lifelong learning aligns with UNESCO's advocacy for continuous education as a driver for sustainable development. Similarly, the insights into institutional differences and curriculum design are relevant for countries seeking to balance public and private education while addressing the challenges of sustainability integration. By examining Slovenia as a case study, this research offers valuable lessons for policymakers, educators, and institutions worldwide.

4.1 Limitations and Study Implications

The study design has several limitations that should be considered when interpreting the results. One potential limitation is keyword selection bias. The study relies on a predefined set of sustainability-related keywords, selected through expert consensus. While this process was systematic, it may have excluded other relevant terms that could impact the accuracy of the Sustainability Assessment Score (SAS). As a result, the study's focus on these specific keywords might not fully capture the diversity of sustainability topics integrated into the curriculum.

Another limitation is the study's geographic focus on Slovenia. While this provides valuable insights into higher education in Slovenia, the findings may not be directly applicable to other countries with different educational systems, sustainability priorities, and curricula. Educational frameworks and sustainability integration practices can vary significantly across regions, which may limit the generalizability of the results.

The data collection method also presents limitations. The study relies on course titles and descriptions to assess the sustainability content of programs. However, this method may not fully capture the depth of sustainability integration, as some courses may incorporate significant sustainability content without explicitly using the predefined keywords. Therefore, the true extent of sustainability within the curriculum might be underestimated.

Additionally, the study utilizes a cross-sectional design, offering a snapshot of sustainability integration at a single point in time. While this provides useful insights, it does not allow for the assessment of trends or changes over time. Longitudinal studies would be necessary to explore how sustainability education evolves and how curricula adapt to emerging challenges and opportunities.

When considering the limitations of the study's results, it's important to note the issue of generalizability. While the findings offer valuable insights into sustainability in Slovenian higher education, they may not directly apply to other countries or educational systems without considering local policies, cultural contexts, and institutional structures. Further studies in different regions or countries would be required to verify the broader applicability of the conclusions.

The depth of sustainability integration is another limitation. The SAS provides a quantitative measure of sustainability content based on keyword frequency, but it does not assess the quality or depth of sustainability coverage in individual courses. It is possible that some courses with high keyword frequency may still vary in how comprehensively they address sustainability issues. A more qualitative assessment would be needed to fully understand how well sustainability is integrated into curricula.

The study also does not address the impact of sustainability education on students. While it provides an assessment of the presence of sustainability content, it does not measure its effectiveness in changing students' knowledge, attitudes, or behaviors. Future research could explore how sustainability education influences student outcomes and contributes to the development of sustainability competencies.

While the study highlights differences between public and private institutions in terms of sustainability integration, it does not delve deeply into the factors driving these differences. Institutional characteristics such as funding models, faculty expertise, and institutional priorities could play a significant role in shaping sustainability curricula. Further investigation into these factors would provide a more nuanced understanding of why these differences exist.

To improve the specificity of the study's aims, several revisions could be made. First, the scope of the study should be clearly stated, emphasizing that the aim is to evaluate the presence and extent of sustainability content within Slovenian higher education programs, with a focus on identifying gaps and opportunities for improvement. Key terms such as "sustainability," "sustainability assessment score," and "sustainability-related keywords" should be defined clearly to ensure consistency and clarity throughout the study. The methodology used to select keywords, collect data, and analyze results should be described in more detail, including the rationale behind the choice of statistical tests and how they address the research questions. Finally, the practical implications of the findings should be highlighted, discussing how the results can inform policy decisions, curriculum development, and strategies at both institutional and national levels to enhance the integration of sustainability in higher education.

This study emphasizes the need for more comprehensive integration of sustainability content across various fields of study. Programs with higher Sustainability Assessment Scores (SAS) often include sustainability-related keywords in course titles, suggesting that incorporating these keywords could signal the importance of sustainability and ensure it becomes a core component of education. Fields such as Health and Social Work, which scored lower, could benefit from targeted curricular enhancements, such as revising existing courses or introducing new ones that focus on sustainability's environmental and social dimensions.

Significant differences between public and private institutions indicate that private institutions may be more responsive to sustainability trends. Public institutions could learn from private ones by forming partnerships with external organizations and aligning curricula with global sustainability standards. Policymakers should also consider providing additional support to public institutions, such as funding for curriculum redesign and faculty development, to help them better integrate sustainability into their programs.

The study highlights the importance of preparing students to address sustainability challenges. Programs with higher SAS scores likely produce graduates who are more knowledgeable in sustainability practices, meeting the growing demand for professionals in this field. Lifelong learning (LLL) programs also play a key role in providing current professionals with the skills they need to address sustainability challenges, particularly for those who did not receive formal sustainability education.

While this study provides a snapshot of the current state of sustainability education, future research should track trends over time through longitudinal studies. Assessing the effectiveness of sustainability education in shaping students' knowledge and behaviors is also crucial, as it would provide insights into how teaching methods and curricula can be improved. Comparative studies across countries could identify best practices for integrating sustainability, while international collaboration would allow institutions worldwide to share resources and strategies, further enhancing sustainability education.

In conclusion, addressing these issues will help higher education institutions advance sustainability and prepare students to effectively tackle environmental and social challenges. By refining curriculum development, strengthening institutional strategies, and fostering global collaboration, higher education can play a critical role in building a sustainable future.

Acknowledgements The research was funded by the European Union-Next Generation EU & The Ministry of Higher Education, Science and Innovation. Research was carried out within the project entitled "Establishing an environment for green and digital logistics and supply chain education".

Funding Please add: "This research was funded by the European Union-Next Generation EU & The Ministry of Higher Education, Science and Innovation.

Conflicts of Interest The authors declare no conflict of interest.

References

Álvarez-López G, Matarranz M (2019) Quality and evaluation as global educational political trends: comparative study of national evaluation agencies in compulsory education in Europe. Revista Complutense De Educación 31(1):85–95. https://doi.org/10.5209/rced.61865

Ayalon O, Avnimelech Y (2007) The role of an academic institute in setting national environmental policy: the case of Israel. Energy 32(6):927–934. https://doi.org/10.1016/j.energy.2006.12.004

Barth M, Michelsen G (2013) Learning for change: an educational contribution to sustainability science. Sustain Sci 8(1):103–119. https://doi.org/10.1007/s11625-012-0181-5

Bell K, Cash B, Boetto H, Thampi K (2020) International study abroad programmes: exploring global south student perspectives, reciprocity and sustainability. Soc Work Educ 1–13. https://doi.org/10.1080/02615479.2020.1711881

Benton-Short L, Merrigan KA (2016) Beyond interdisciplinary: how sustainability creates opportunities for pan-university efforts. J Environ Stud Sci 6(2):387–398. https://doi.org/10.1007/s13412-015-0341-x

Boarin P, Martinez-Molina A, Juan-Ferruses I (2020) Understanding students' perception of sustainability in architecture education: a comparison among universities in three different continents. J Clean Prod 248:119237. https://doi.org/10.1016/j.jclepro.2019.119237

Brudermann T, Aschemann R, Füllsack M, Posch A (2019) Education for sustainable development 4.0: lessons learned from the University of Graz, Austria. Sustainability 11(8):2347. https://doi.org/10.3390/su11082347

Daneshjoo K, Motevalli Haghighi H, Talaei M (2019) Community space and the school ground A 3D book for the concepts of sustainable development. MANZAR, Sci J Landsc 11(49):38–47

Daub CH, Hasler M, Verkuil AH, Milow U (2020) Universities talk, students walk: promoting innovative sustainability projects. Int J Sustain High Educ 21(1):97–111. https://doi.org/10.1108/IJSHE-04-2019-0149

Denac M, Obrecht M, Radonjič G (2018) Current and potential eco-design integration in small and medium enterprises: construction and related industries. Bus Strateg Environ. https://doi.org/10.1002/bse.2034

Do NM (2020) An overview of strategic responses of Vietnamese higher education institutions. Int J Educ Manag 34(1):35–44. https://doi.org/10.1108/IJEM-02-2019-0074

Dong X, Zhang X, Zhang C, Bi C (2023) Building sustainability education for green recovery in the energy resource sector: a cross country analysis. Resour Policy 81:1–8. https://doi.org/10.1016/j.resourpol.2023.103385

Emblen-Perry K (2022) Auditing a case study: enhancing case-based learning in education for sustainability. J Clean Prod 381(1):1–12. https://doi.org/10.1016/j.jclepro.2022.134944

Eppinga MB, Lozano-Cosme J, de Scisciolo T, Arens P, Santos MJ, Mijts EN (2020) Putting sustainability research into practice on the university campus. Int J Sustain High Educ 21(1):54–75. https://doi.org/10.1108/IJSHE-03-2019-0131

Fokdal J, Čolić R, Milovanović Rodić D (2020) Integrating sustainability in higher planning education through international cooperation. Int J Sustain High Educ 21(1):1–17. https://doi.org/10.1108/IJSHE-01-2019-0045

Glavič P (2006) Sustainability engineering education. Clean Technol Environ Policy 8(1):24–30. https://doi.org/10.1007/s10098-005-0025-4

Hay R, Eagle L (2020) Impact of integrated sustainability content into undergraduate business education. Int J Sustain High Educ 21(1):131–143. https://doi.org/10.1108/IJSHE-05-2019-0174

Hermann RR, Bossle MB (2020) Bringing an entrepreneurial focus to sustainability education: a teaching framework based on content analysis. J Clean Prod 246. https://doi.org/10.1016/j.jclepro.2019.119038

Howlett C, Ferreira J-A, Blomfield J (2016) Teaching sustainable development in higher education. Int J Sustain High Educ 17(3):305–321. https://doi.org/10.1108/IJSHE-07-2014-0102

Kemper JA, Ballantine PW, Hall CM (2020) The role that marketing academics play in advancing sustainability education and research. J Clean Prod 248:119229. https://doi.org/10.1016/j.jclepro.2019.119229

Kopnina H (2020) Education for the future? Critical evaluation of education for sustainable development goals. J Environ Educ 1–12. https://doi.org/10.1080/00958964.2019.1710444

Lukman RK, Omahne V, Sheikh LTe, Glavič P (2021) Integrating sustainability into logistics oriented education in Europe. Sustainability 13(4):1667. https://doi.org/10.3390/su13041667

Mulder KF (2017) Strategic competences for concrete action towards sustainability: an oxymoron? Engineering education for a sustainable future. Renew Sustain Energy Rev 68:1106–1111. https://doi.org/10.1016/j.rser.2016.03.038

Obrecht M, Feodorova Z, Rosi M (2022) Assessment of environmental sustainability integration into higher education for future experts and leaders. J Environ Manag 316. https://doi.org/10.1016/j.jenvman.2022.115223

Orlović-Lovren O (2011) The role of education in protected area sustainable governance. Manag Environ Qual 22(1):48–58. https://doi.org/10.1108/14777831111098471

Ozdemir Y, Kaya SK, Turhan E (2020) A scale to measure sustainable campus services in higher education: "Sustainable Service Quality." J Clean Prod 245:118839. https://doi.org/10.1016/j.jclepro.2019.118839

Paço A, Lavrador T (2017) Environmental knowledge and attitudes and behaviours towards energy consumption. J Environ Manage 197:384–392. https://doi.org/10.1016/j.jenvman.2017.03.100

Piza V, Aparicio J, Rodríguez C, Marín R, Beltrán J, Bedolla R (2018) Sustainability in higher education: a didactic strategy for environmental mainstreaming. Sustainability 10(12):4556. https://doi.org/10.3390/su10124556

Priyadarshini P, Abhilash PC (2020) From piecemeal to holistic: introducing sustainability science in Indian Universities to attain UN-sustainable development goals. J Clean Prod 247:119133. https://doi.org/10.1016/j.jclepro.2019.119133

Roos N, Guenther E (2020) Sustainability management control systems in higher education institutions from measurement to management. Int J Sustain High Educ 21(1):144–160. https://doi.org/10.1108/IJSHE-01-2019-0030

Sandri O (2020) What do we mean by 'pedagogy' in sustainability education? Teach High Educ 1–16. https://doi.org/10.1080/13562517.2019.1699528

Ulmer N, Wydra K (2020) Sustainability in African higher education institutions (HEIs). Int J Sustain High Educ 21(1):18–33. https://doi.org/10.1108/IJSHE-03-2019-0106

United Nations (2015) Sustainable development goals. https://www.un.org/sustainabledevelopment/energy/. Accessed 10 Sept 2021

Vagnoni E, Cavicchi C (2015) An exploratory study of sustainable development at Italian universities. Int J Sustain High Educ 16(2):217–236. https://doi.org/10.1108/IJSHE-03-2013-0028

Valderrama-Hernández R, Alcántara Rubio L, Sánchez Carracedo F, Caballero Franco D, Serrate González S, Gil-Doménech D, Vidal-Raméntol S, Miñano R (2019) Does the Spanish university system teach sustainability? Perception of the students of four universities. Educ XX1 23(1). https://doi.org/10.5944/educxx1.23420

Viegas CV, Bond AJ, Vaz CR, Borchardt M, Pereira GM, Selig PM, Varvakis G (2016) Critical attributes of Sustainability in Higher Education: a categorisation from literature review. J Clean Prod 126:260–276. https://doi.org/10.1016/j.jclepro.2016.02.106

Wals AEJ (2014) Sustainability in higher education in the context of the UN DESD: a review of learning and institutionalization processes. J Clean Prod 62:8–15. https://doi.org/10.1016/j.jclepro.2013.06.007

Wamsler C (2020) Education for sustainability: fostering a more conscious society and transformation towards sustainability. Int J Sustain High Educ 21(1):112–130. https://doi.org/10.1108/IJSHE-04-2019-0152

Wang Y, Shi H, Sun M, Huisingh D, Hansson L, Wang R (2013) Moving towards an ecologically sound society? Starting from green universities and environmental higher education. J Clean Prod 61:1–5. https://doi.org/10.1016/j.jclepro.2013.09.038

Zorio-Grima A (2020) Driving factors for having visibility of sustainability contents in university degree titles. J Clean Prod 242:114746. https://doi.org/10.1016/j.jclepro.2018.10.344

The Social Capital of a Higher Education Institution and the 2030 Agenda

Gabriela Daiana Christ, Lair Barroso Arraes Rocha, Moacir Piffer, and Jandir Pauli

Abstract The 2030 Agenda for Sustainable Development, which encompasses the Sustainable Development Goals (SDGs), calls for collective action for a more sustainable world, engaging governments and civil society. This research focused on how the Western Paraná State University (Unioeste) is engaging with this agenda. Through qualitative research with 14 management leaders of Unioeste and analysis of institutional documents, this study examined how the institution mobilizes social capital to implement the SDGs. The conclusions suggest that Unioeste has social capital, but not enough to fully implement the SDGs, so there is room for improvement. Thus, placing social capital at the center of this effort is a possible approach for such implementation. This research provides valuable information on the complex process of integrating the SDGs into universities and contributes to the theoretical understanding of SDG implementation in academic settings. Furthermore, this study has practical implications for helping decision-makers to manage sustainability issues.

Keywords Sustainable development goals · Social capital · University · Brazil

G. D. Christ (✉) · M. Piffer
Universidade Estadual do Oeste do Paraná, Programa de Pós-Graduação em Desenvolvimento Regional e Agronegócio, Toledo, Paraná, Brazil
e-mail: gabriela.christ@unioeste.br

M. Piffer
e-mail: moacir.piffer@unioeste.br

L. B. A. Rocha
Universidade Estadual de Maringá, Programa de Pós-graduação em Administração, Maringá, Brazil
e-mail: lair_rocha@uel.br

J. Pauli
Atitus Educação, Programa de Pó-graduação em Administração, Passo Fundo, Rio Grande do Sul, Brazil
e-mail: jandir.pauli@atitus.edu.br

W. Leal Filho et al. (eds.), *Education for Sustainable Development: The Contribution of Universities*, World Sustainability Series,
https://doi.org/10.1007/978-3-031-86985-3_18

1 Introduction

Sustainability can be understood as the right of future generations to enjoy equal or better conditions based on present conditions (UN 1985). In September 2015, the UN approved the Sustainable Development Goals (SDGs) at its 70th General Assembly to promote sustainability and address global challenges (UN 2015). The 2030 Agenda includes 17 goals and 169 targets to be achieved by 2030. These goals aim to benefit people, the planet, and prosperity (Prieto-Jiménez et al. 2021).

This study shows that the SDGs implementation needs to consider the social capital of a region, particularly strategic actions for problem-solving, and institutions embedded in the geographical spaces of the regions (Putnam 2006; Sen 2000). The key players in development are representatives of civil society; local, regional, and federal governments; the private sector, and academia (Ferrera de Lima and Alves 2012).

In this context, Universities are considered strategic actors for implementing the SDGs (SDSN Australia/Pacific 2017). When investigating implementation strategies in Higher Education Institutions, Christ and Piffer (2022) identified a research gap regarding the study of social capital in universities to contribute to the implementation of the 2030 Agenda and regarding sustainability in such institutions. University outreach, in particular, has the role of bringing the knowledge developed in academic spaces to the community and providing assistance to the population, whether through culture, free care at a school clinic, or legal advice. These actions impact the economic and social environment of the places where they are inserted (Brown et al. 2006; Leal Filho et al. 2019).

The present study analyzes the mobilization of the social capital of a University regarding the implementation of the 2030 Agenda. It aims to understand the role of the University in mobilizing institutional, human, synergetic, and civic capital for the implementation of the SDGs. To this end, an empirical study was conducted through a case study, triangulating data from interviews and institutional documents.

2 Social Capital

According to Putnam, social capital involves the networks of social relationships, patterns of trust, and cooperation that exist within a community or society. These networks can help people achieve goals together, solve problems, and also improve how society functions in general.

Social scientists have offered a vast array of definitions for the concept of social capital (see Table 1), with the idea being widely used in regional development (Bourdieu 1980; Coleman 1988; Fukuyama 1996; Nahapiet; Ghoshal 1998; Putnam 2006).

Furthermore, social capital can be a tool for improving the living conditions of a community or region, being run by leaders (Melo et al. 2015; Piacenti 2016;

Table 1 Concepts of social capital

Source	Concept of social capital	Perspective
Bourdieu (1980, p. 2) tradução dos autores	"Le capital social est l'ensemble des ressources actuelles ou potentielles qui sont liées à la possession d'un réseau durable de relations plus ou moins intitutionnalisées d'interconnaissance et d'interreconnaissance; ou, en d'autres termes, à l'appartenance à un groupe, comme ensable d'agents qui ne sont pas seulement dotés de propriétés communes (susceptibles d'être perçues par l'observateur, par les autres ou par eux-mêmes) mais sont aussi unis par des liaisons permanents et utiles"	Structuralism
Coleman (1988, p. 98)	"Social capital is defined by its function. It is not a single entity but a variety of different entities, with two elements in common: they all consist of some aspect of social structures, and they facilitate certain actions of actors—whether persons or corporate actors—within the structure"	Utilitarianism
Putnam (2006, p. 177)	"Social capital here refers to features of social organization, such as trust, norms, and networks, that can improve the efficiency of society by facilitating coordinated actions"	Traditionalist
Fukuyama (1996, p. 41, 2000, p. 28)	"A capability that arises from the prevalence of trust in a society or in certain parts of it" "the existence of a certain set of informal values or norms shared among members of a group that permit cooperation among them"	Utilitarianism
Nahapiet and Ghoshal (1998, p. 243)	"The sum of the actual and potential resources embedded within, available through, and derived from the network of relationships possessed by an individual or social unit. Social capital thus comprises both the network and the assets that may be mobilized through that network" Dimensions of the concept: structural, cognitive, and relational	Functionalist

Source Developed by the authors based on (Boeira and Borba 2006; Melo et al. 2015)

Sen 2000). Thus, like institutional capital (public and private institutions or organizations), human capital (stock of knowledge and skills), synergetic capital (real or latent capacity for community coordination), and civic capital (democratic practices) (Boisier 2000), social capital is an intangible capital and is fundamental for regional development (Duarte and Schneider 2019; Piacenti 2016).

Regardless of the perspective or epistemological approach, three elements connect the several definitions of social capital (Bourdieu 1980; Coleman 1988; Fukuyama 1996; Nahapiet and Ghoshal 1998; Putnam 2006): (i) people (actors, agents, etc.); (ii) network (organizations, groups, social structures, etc.); and (iii) actions (trust, commitment, collaboration, shared culture, ties, cooperation, etc.). Putnam's understanding of social capital, developed in his extensive two-decade work (from 1970 to 1989), is a prominent theoretical basis in the present study due to its relationship with regional development.

In addition to the understanding of how certain social actors behave, social capital concerns "trust, norms, and networks, that can improve the efficiency of society by facilitating coordinated actions" (Putnam 2006, p. 167). Moreover, a strong and active civil society is essential for strengthening and consolidating democracy. The quality of public life and the performance of institutions are strongly influenced by norms and networks of civic engagement (Ostrom 1990; Putnam 1995, 2006; Sander and Putnam 2010; Wubah et al. 2021).

Putnam (1995) argues that civic engagement norms and networks improve government performance. Trust among members fosters development, social capital, and civic engagement (Pizzutilo and Venezia 2021). Identifying effective organizations and networks is essential for generating social capital and promoting regional development.

In this regard, an important step for regional analysis can be to creatively explore how public policies or even actions that shape the region's productive or political structure affect or can affect the development of social capital (Chibiaqui 2020; Corrêa 2021; Goebel and Bidarra 2019). Finally, trust is also fundamental to building social capital: "effective collaborative institutions require interpersonal skills and trust, but those skills and that trust are also inculcated and reinforced by organized collaboration" (Putnam 2006, p. 180).

3 Sustainable Development and the 2030 Agenda

Sustainability represents an alternative form of economic growth that does not harm environmental development, human rights, or future generations (UN 1985). This path includes the search for processes that are less environmentally degrading and more socially responsible. In organizational contexts, the concept of sustainability is essential due to the need to maintain business continuity (Brown et al. 2006). The economic, social, and environmental dimensions are represented by the acronym TBL (Triple Bottom Line), which refers to the primary definition of corporate sustainability (Dias et al. 2021). Issues related to the environment, which in the 1980s were limited to technical and scientific matters, have now permeated the political, economic, and social spheres (Barbieri 2011; Lago 2013).

"Sustainable development" gained relevance with the 1987 Brundtland Report by the UN. It became prominent in the twenty-first century due to ecological imbalance and social inequalities, highlighting the unsustainability of current production and consumption modes (Gallo and Setti 2014). Vries and Petersen (2009) define it as enhancing the quality of life, requiring a transdisciplinary approach.

In 2015, the UN proposed the 2030 Agenda during its 70th General Assembly, focusing on Sustainable Development. The Agenda includes 17 SDGs and 169 targets to guide global development efforts until 2030 (Managi et al. 2021). These goals legitimize regional and national efforts, empowering communities to take control of their futures.

Therefore, the 2030 Agenda can be considered an action plan for Sustainable Development. This proposal requires support in updating legal frameworks to promote favorable institutional conditions and financing conditions that can provide economic value to the proposed actions (Prieto-Jiménez et al. 2021).

4 Methodological Procedures

This study examines how a university uses social capital to implement the 2030 Agenda and the SDGs. It explores the role of institutional, human, synergetic, and civic capital, emphasizing the importance of context, norms, and processes. A qualitative approach was used to understand the events and social processes involved (Bardin 2011; Creswell 2007; Oliveira 2007; Van Maanen 1979).

Regarding the objectives, this research is exploratory since it seeks to identify the factors that determine or contribute to the occurrence of phenomena (Creswell 2007; Oliveira 2007; Rea and Parker 2000). In terms of the nature of the research, this study can be considered mostly descriptive, as it aims to describe in depth a certain reality, obtaining detailed information about the defined problem (Triviños 2009). A basic characteristic of this type of study is using standardized data collection techniques (Triviños 2009).

Regarding the environment, this is a field study, in which the selected university becomes a unit of context, with a cross-sectional time dimension, considering that the research was based on a specific period (Oliveira 2007; Rea and Parker 2000).

Fourteen (14) people who occupied management and leadership positions at the university were interviewed. This choice was due to these people holding decision-making power over the university's choices, given the institution's governance model (Fig. 1). The interviewed leaders comprised two groups: the rectory (8 interviewees) and the campus directory (6 interviewees) (Fig. 1). Data collection consisted of approximately 10 h of recording/interviewing and 165 transcribed pages. The interviews were carried out online via videoconference, using Microsoft Teams and Google Meet.

Table 2 summarizes the interviews of the research.

Figure 1 displays the research sample considering the organization chart of the institution adapted for this purpose.

Content analysis (Bardin 2011) was used to evaluate the data. The three stages were followed: pre-analysis, material exploration, as well as treatment and interpretation of the results. Field notes, authorized recordings of interviews, and materials mainly found on Unioeste's website (mainly the Institutional Development Plan—IDP) were also used for data analysis.

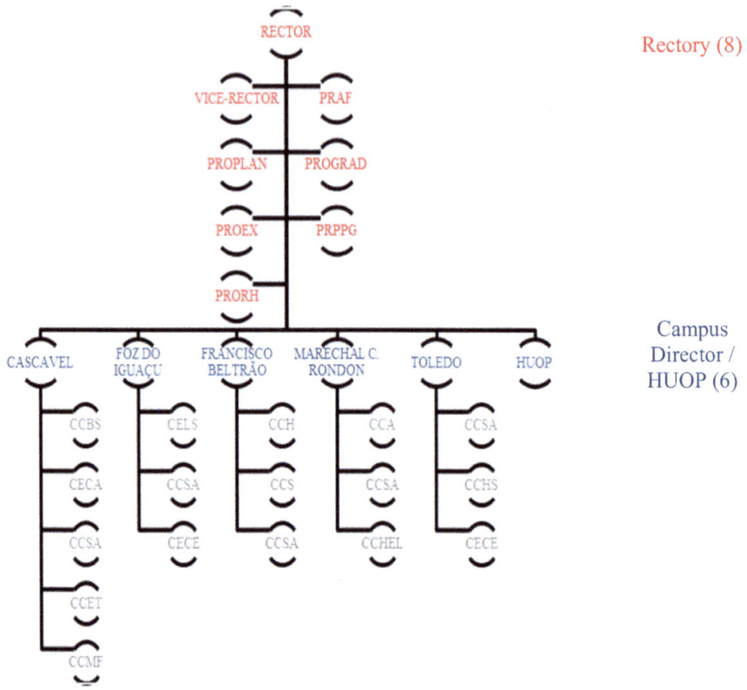

Fig. 1 Research sample: adapted organization chart. *Source* Developed based on (CHRIST2022; UNIOESTE 2021)

Table 2 Interviews conducted during the research

#	Group	Interviewee code	Date	Format	Interview length	No. of pages transcribed
1	Rectory	G_1	08/10/2021	Virtual	00:30:09	10
2	Rectory	G_2	11/10/2021	Virtual	00:51:30	16
3	Rectory	G_3	19/10/2021	Virtual	00:53:34	14
4	Rectory	G_4	22/10/2021	Virtual	00:28:36	10
5	Rectory	G_5	25/10/2021	Virtual	00:48:12	12
6	Rectory	G_6	25/10/2021	Virtual	00:39:23	14
7	Rectory	G_7	25/10/2021	Virtual	01:09:49	18
8	Directory	G_8	26/10/2021	Virtual	00:28:31	9
9	Rectory	G_9	27/10/2021	Virtual	00:26:05	7
10	Directory	G_10	27/10/2021	Virtual	01:13:14	14

(continued)

Table 2 (continued)

#	Group	Interviewee code	Date	Format	Interview length	No. of pages transcribed
11	Directory	G_11	05/11/2021	Virtual	01:16:56	17
12	Directory	G_12	09/11/2021	Virtual	00:36:55	12
13	Directory	G_13	09/11/2021	Virtual	00:18:01	4
14	Directory	G_14	12/11/2021	Virtual	00:26:57	8
Total					10:07:52	165

Source Adapted from (CHRIST 2022)

5 Discussions and Results

5.1 Western Paraná State University (Unioeste): Brief Notes

Unioeste is a state autonomous, non-profit university with a multi-campus structure, headquartered in Cascavel, Paraná, Brazil. It is linked to the General Superintendence of Science, Technology, and Higher Education (SETI) and authorized by State Law No. 8680 of 1987. Unioeste has campuses in Cascavel, Toledo, Marechal Cândido Rondon, Foz do Iguaçu, and Francisco Beltrão.

Unioeste's academic community comprises teachers, university staff, and students. Graph 1 indicates that over the last 17 years (2006–2023), while the number of students increased by 17.22%, the number of teachers also rose by 11.02%.

The total number of people directly connected to the institution was 16,279 (Graph 1), of which 12,900 were students (81% enrolled in undergraduate courses, 5% in specialization courses, and 14% in master's and doctoral courses) and 3379 civil servants (37.29% teaching staff and 62.71% university agents). The university hospital has 1415 staff (67% of the HEI's staff).

5.2 Unioeste's Social Capital and the Implementation of the SDGS

Regarding the "social capital" category, and based on the interviews, 23 codes were assigned, which were identified 402 times in the interviews. As for the "SDG" category, 15 codes were assigned, and they were mentioned 159 times during the interviews. Among the social capital codes most mentioned by managers are: cooperation (66 times), norms (66 times), trust (36 times), systems (28 times), and governance (27 times).

In addition to the code "Unioeste Forward Action", used to categorize the messages of future actions that managers mentioned regarding the implementation of the SDGs at Unioeste (a code that was used 43 times), the codes most frequently found

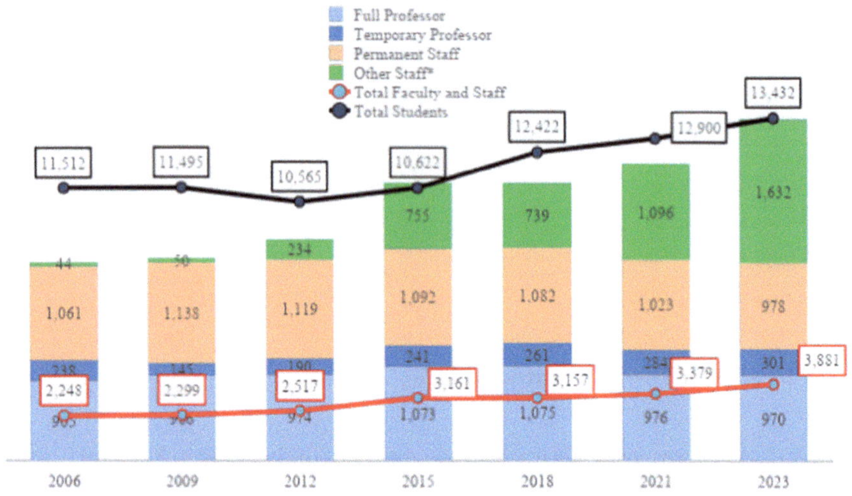

Graph 1 Academic community of Unioeste (2006/2023). *Source* Developed based on (CHRIST 2022; UNIOESTE 2023). Note *Other Staff include: CRES,[1] apprentice, trainee, and outsourced

were: Unioeste Action (42 times); International Office (10 times)—Unioeste's International Relations Office for the implementation of the SDGs, SDG 3 good health and well-being (10 times), SDG NI Not Identified (9 times) and SDG 2 zero hunger and sustainable agriculture (8 times).

Figure 2 evidences the relationship between the analyzed categories (social capital and SDGs) and was elaborated by connecting the interviewees' knowledge of the SDGs at Unioeste (in red) with the social capital codes (in blue).

To understand the connection between social capital and the SDGs, each topic is explored in detail. Unioeste's administration includes a University Council and holds elections for key positions, with participation from the entire academic community. The University's governance model was frequently mentioned and positively viewed by interviewees.

> [...] I look inward at how the university works, its decision-making model, its organization, everything, it is, in my view, it has a functional model, it has a model, from society's perspective, that is very representative (Manager 3).

> [...] and if we observe how the CEP [Teaching, Research and Extension Council], the University Council, are structured nowadays, we need only to look at the infrastructure. We have an auditorium. That in itself indicates that, look, not only the members representing the community will be here, this space, the councils, are open to the participation of the academic community and the external community (Manager 11).

---

[1] Fixed-term contracts occur as Special Regime Contracts (CRES), also known as PSS. These temporary contracts should only be used to fill vacancies as well as temporary and emergency absences at educational institutions.

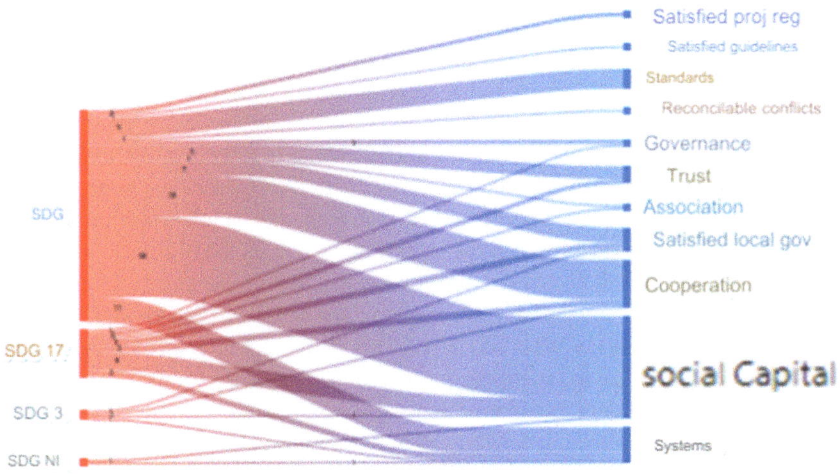

Fig. 2 Relationship between SDGs and Social Capital. *Source* adapted from (CHRIST 2022). *Note* *SDG NI (SDG Not Identified)

The responses highlight the importance of councils and individual empowerment within the university's civic context. All staff, students, and the external community are involved in debates, enhancing social capital. According to Putnam (2006), democratic environments foster citizenship and social capital, encouraging cooperation and participation. Governance is another fundamental aspect of any organization since it has a set of procedures that establishes the relationships among the institution's various stakeholders (Lugoboni and Marques 2021). All the interviewees were aware of the governance mechanisms, such as the Institutional Development Plan (IDP).

When questioned about social capital—which is the ability of actors to secure benefits through integration into social networks (Melo et al. 2015)—many interviewees mentioned outreach activities, research groups, actions by study centers, institutional practices, and other actions related to the course collegiates of both undergraduate and graduate courses (Unioeste 2021).

It was possible to identify that the set of characteristics: norms, systems, and trust are linked to social capital as perceived by the interviewees. In other words, shared norms (in which Unioeste's governance is the modus operandi) are associated with the system (that is part of the satisfaction with the local government and dialogue with the community where Unioeste is inserted), which in turn leads to relationships of trust that are part of cooperation and belonging. These relationships can be considered a virtuous cycle (Putnam 2006), meaning that the more they are used, the greater their value.

However, the opposite is also true. According to the reports of some managers, it was possible to verify what Putnam (2006) attributed to the lack of mutual and trusting commitment, in which all parties who could benefit from cooperation would rather behave opportunistically. As observed in the interview with a research subject:

"…the university is a very beautiful environment, it's a very beautiful ambiance, but it is also an environment with much dispute and conflict" (Manager 11, directory). This evidence can lead to distrust and rupture in the virtuous cycle of social capital, turning it into a vicious circle.

Nevertheless, it is possible to envision social capital as a form of dialogue, coexistence, trust, and decision-making through the institution's collegiate body. In other words, there is social capital, but it is not fully activated. The relationships found in the social capital category based on the interviews are depicted in Fig. 3.

Figure 3 shows a link between managers' responses and a sense of belonging, which is crucial for regional development and creating social capital (Putnam 2006). Unioeste, located in West and Southwest Paraná, fosters local development and empowerment. With international partnerships and borders with Paraguay and Argentina, it has a national and international impact.

From document analysis and manager interviews, we found that student organizations at Unioeste are not mentioned in the institution's IDP but are crucial for outreach. Unioeste has a social capital framework for regional development, including private foundations, clinics, a school pharmacy, centers, outreach activities, research projects, and groups. An example is the American Institute of Chemical Engineers (AIChE) at the Toledo campus, active since 2012. Encouraging and supporting new student organizations should be a priority for the university.

The majority of centers at Unioeste establish coordination with other actors in the region in their statutes, such as municipal offices, civil society, and other partner organizations. The relationship between Unioeste and the community as well as its surroundings was mentioned by some managers:

> […] We are part of an innovation ecosystem in which we realize our importance, and within innovation, we realize our importance, but within regional development, we also realize our importance and we cannot accomplish it on our own. (Manager 1, rectory).
>
> This year, the coordination with the city councilors, along with the mayor, with the deputy mayor, is such that they visit us constantly, they come to ask us for help, the city hall, they come to ask us for help constantly, and the opposite too, we often seek them for help too, with the municipal secretaries, with the mayor himself, so I have never seen such good rapport (Manager 12, directory).

Regarding the implementation of the 2030 Agenda at Unioeste, some managers demonstrated the importance of being aware of the SDGs as follows:

> […] He [the officer] had been emphasizing for a long time, and I think you were at ARI [International Relations Office], we need to establish the relationship between our projects and the SDGs, because in the forms, in the proposals that we have to formulate for the whole world, there must be a relationship, there must be indicators and we do not have them (Manager 1, rectory).
>
> […] The SDGs were discussed precisely because of the reports, those proofs, which are carried out by the international relations office, precisely because of the rankings and indicators, but our role is to implement the actions that we have been implementing and raise awareness among our colleagues, right?" (Manager 6, rectory).

In other words, the demand for the search and systematization of information was perceived to have emerged from the need to submit information to international

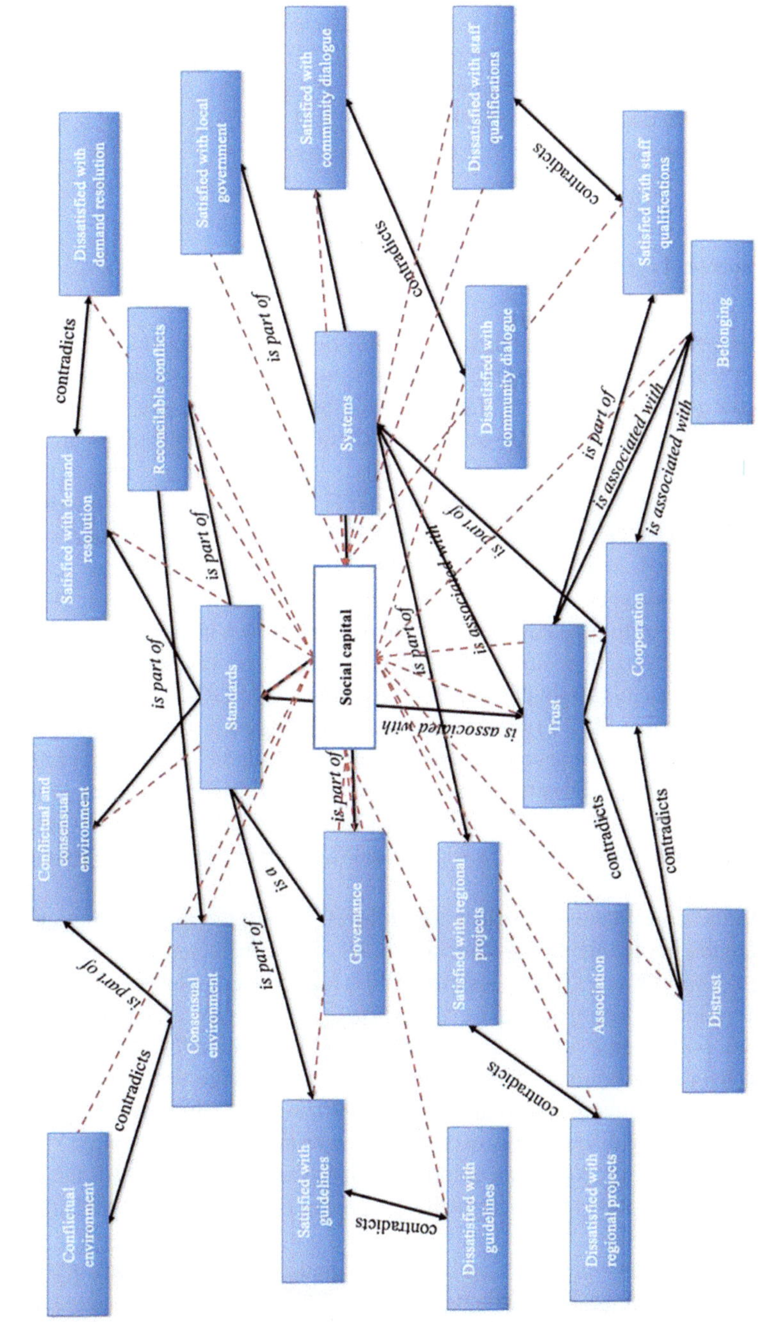

Fig. 3 The connection network of the social capital category. *Source* adapted from (CHRIST 2022)

rankings, i.e., it was exogenous. This exogenous demand observed in the submission to international rankings was also noticed in other HEIs (De La Poza et al. 2021; Torabian 2019). In this context, Unioeste is attentive to exogenous demands and is among the 1201 + best universities in the world according to Times Higher Education (THE). In 2021, it provided data to compare the performance of more than 1600 institutions from 99 countries (THE 2021) in the global ranking. However, it was in 2019 that THE launched its University Impact Rankings.

5.3 *Unioeste Actions Towards the SDGS*

The rankings are presented as evaluation tools following predefined criteria. The THE Impact Ranking shows how HEIs are working towards achieving the SDGs and allows for a series of analyses. Table 3 shows both: how Unioeste was ranked regarding the SDGs, as well as the leading universities in the evaluated years and their overall position.

According to Table 3, in 2019 (concerning 2017), Unioeste stood out in 9 of the 11 SDGs evaluated by THE, with a focus on SDGs 3, 4, and 16. In 2020 (concerning 2018) THE evaluated all 17 SDGs. At this time, Unioeste remained prominent in SDGs 3 and 16. In that year, SDG 6 was also highlighted. In 2021 (concerning 2019), Unioeste's best-performing SDGs were still 3 and 16. However, in this instance, SDG 8 also featured in the top 3 position in the ranking. In the last year evaluated (2022 concerning 2020), the highlights for the 2030 Agenda were SDGs 2, 6, and 8.

Among the 17 SDGs, SDG 3 (good health) and SDG 16 (peace, justice) stood out at Unioeste in the initial three years evaluated. SDG 3 was highlighted due to the services provided by the University Hospital and school clinics, while SDG 16 was notable for the work of the Legal Practice Centers and efforts to combat violence against women. These activities exemplify Unioeste's social capital practices extending beyond the University.

Based on the understanding of sustainable development, the implementation of the 2030 Agenda in organizations, especially in HEIs, is considered an institutional imperative, not simply a desirable possibility in the face of the challenges of contemporaneity. In addition to impacting all aspects of life in the city, on campuses and elsewhere, adopting public policies that foster the 2030 Agenda influences the entire external pattern of institutional reference through its partnerships and relationships.

Ultimately, both Putnam (2006) and North (1990) write about how the form of colonization influenced development in countries over time, notably the United States in comparison with Latin American countries. History cannot be changed, but it is certainly possible to plan with incremental changes. This is also reflected in the history of Unioeste.

The HEI has a webpage on its website dedicated to the SDGs (https://www.unioeste.br/portal/ods-unioeste), which is very important for communicating and publicizing its actions. Another action carried out by Unioeste is to indicate in its newsfeed the respective SDG pertaining to the content of the published article, using

Table 3 Unioeste in THE Impact Rankings (2019/2022)

SDG	2019	2020	2021	2022
SDG 1: No poverty	*Not assessed	301 +	401+	401–600
SDG 2: Zero hunger	*Not assessed	201 +	201–300	**101–200**
SDG 3: Good health and well-being	**201–300**	**401–600**	**401–600**	801–1000
SDG 4: Quality education	**201–300**	401–600	401–600	1001 +
SDG 5: Gender equality	201–300	401+	401–600	601–800
SDG 6: Clean water and sanitation	*Not assessed	**201–300**	201–300	**101–200**
SDG 7: Affordable and clean energy	*Not assessed	301+	301–400	301–400
SDG 8: Decent work and economic growth	201+	401+	**401–600**	**401–600**
SDG 9: Industry, innovation and infrastructure	Information not submitted*	401+	601 +	601–800
SDG 10: Reduce inequalities	201+	301–400	601 +	801 +
SDG 11: Sustainable cities and communities	201+	401+	401–600	601–800
SDG 12: Responsible consumption and production	Information not submitted*	301+	401+	601+
SDG 13: Climate action	101–200	301+	301–400	401–600
SDG 14: Life below water	*Not assessed	201+	101–200	101–200
SDG 15: Life on land	*Not assessed	201+	201–300	201–300
SDG 16: Peace, justice, and strong institutions	**101–200**	**201–300**	**301–400**	401–600
SDG 17: Partnerships for the goals	301+	601+	401–600	301–400

(continued)

Table 3 (continued)

SDG	2019	2020	2021	2022
Leading university in the overall ranking (score)	University of Auckland (97.2)	University of Auckland (98.5)	University of Manchester (98.8)	Western Sydney University (99.4)
Unioeste's position in the overall ranking (score)	301+ (23.8–53.6)	601+ (9.5–46.6)	601–800 (47.6–56.5)	601–800 (59.7–66.7)
Universities assessed (countries)	450 (76 countries)	768 (85 countries)	1.115 (94 countries)	1.591 (112 countries)
Brazilian universities	15	30	38	47
Universities in Paraná	4	6	6	6

Source Developed based on the THE University Impact Rankings (THE 2023)

a graphic representation (image) of the SDG below the news article. Such actions demonstrate the university's concern to publicize the classification of the action according to the 2030 Agenda.

Furthermore, recently (October 2023) Unioeste published its first Sustainability Report (UNIOESTE 2023) for 2022. This report reflects the importance of the issue for the institution and aims to diagnose actions carried out at the institutional level, based on social, environmental, and economic projects and programs, which are aligned with achieving the SDGs. Finally, not only providing information on the website is very important, but it is also essential to proceed with plans to communicate and disseminate this document among the organization's leadership, teachers, students, and technical staff, as well as stakeholders.

6 Conclusions

The 2030 Agenda for Sustainable Development draws attention to learning opportunities throughout its implementation process. This research aimed to analyze the implementation of the SDGs at Unioeste influenced by social capital. To meet this objective, we interviewed 14 managers in leadership positions at the University, as well as triangulated and analyzed secondary data such as THE Impact Rankings, the IDP, and the institutional website.

We found that Unioeste has been implementing the SDGs, although reactively, and has an idle social capital, diagnosed mainly by the heavy workload of the university's agents and professors. As new demands emerge, such as the 2030 Agenda, connecting actors, promoting discussion forums and networking must stem from Unioeste.

Regarding the implementation of the SDGs at Unioeste, we found that the university seeks to meet the demands of agents or factors exogenous to the institution. Outdated structures will not ensure the results needed for a sustainable world. Therefore, it is important to emphasize that it is possible to shift this scenario towards proactivity, as suggested by the present study since Unioeste is a significant institution for local and regional development.

It was possible to observe that submitting information to the Times Higher Education Impact Ranking is complex and requires strategic efforts. The fact that Unioeste has been ranked since the launch of the Ranking reflects the university's pioneering spirit, its concern for sustainability, and its accountability to the community. However, this ranking is underutilized because, among other factors, it allows benchmarking with other institutions, such as the State University of Londrina (UEL) and the State University of Maringá (UEM), which are better placed in the ranking.

As a recommendation for future research, we suggest further exploring the report, the opportunities identified in this paper, and the diagnosis found in the report (UNIOESTE, 2023). This can be accomplished through an effective action plan. However, the best practices adopted at other HEIs similar to UNIOESTE should be observed.

It is worth highlighting that agendas represent local, national, or international commitment instruments, as the 2030 Agenda, and are directed toward a common goal. Based on the theoretical framework and literature review presented in this study, social capital can be a mechanism for achieving the goals proposed by the agendas.

Finally, Unioeste can set an example in the implementation of the SDGs, not only as an isolated institution within its walls but also as a protagonist in the coordination with other bodies and in the ecosystem in which it is inserted. In other words, Unioeste has social capital, but it is not sufficient to implement the SDGs within the institution. Therefore, it is necessary to cooperate to achieve the Sustainable Development Goals!

Acknowledgements The authors would like to thank the Western Academic Writing, Translation and Revision Center (Centro de Escrita Acadêmica, Tradução e Revisão do Oeste, CETRO—https://www.unioeste.br/portal/centros-prppg/cetro) of the Western Paraná State University (UNIOESTE) for assistance with English language translation.

This study was financed in part by the Coordenação de Aperfeiçoamento de Pessoal de Nível Superior—Brasil (CAPES)—Finance Code 001.

References

Barbieri JC (2011) Gestão ambiental empresarial, 3rd edn. Saraiva, São Paulo

Bardin L (2011) Análise de conteúdo, 2nd edn. São Paulo: Edições 70

Barlete ARL (2019) A cultural political economy account of higher education in Mercosur (Doctoral thesis, Wolfson College, Faculty of Education). University of Cambridge, Cambridge

Boeira SL, Borba J (2006) Os fundamentos teóricos do capital social. Ambiente Sociedade 9(1):187–193

Boisier S (2000) El desarrollo territorial a partir de la construcción de capital sinergético. Revista Brasileira De Estudos Urbanos e Regionais 2:39
Bourdieu P (1980) Le capital social. Actes De La Recherche En Sciences Sociales 31:2–3
Brown D, Dillard J, Marshall RS (2006) Triple bottom line: a business metaphor for a social construct. Departament d'Economia de l'Empresa—Universitat Autònoma de Barcelona, 06/2(March) 1–39
Chibiaqui EDP (2020) O Programa Oeste em Desenvolvimento e seus impactos na governança da região Oeste do Paraná (Doctoral thesis, Programa de Pós-graduação em Desenvolvimento Regional e Agronegócio). Universidade Estadual do Oeste do Paraná, Toledo
Christ GD (2022) Capital social e desenvolvimento sustentável: uma análise da implementação da Agenda 2030 na Unioeste (Master's dissertation). Universidade Estadual do Oeste do Paraná, Toledo
Christ GD, Piffer M (2022) Rumo à sustentabilidade: uma análise da implementação dos Objetivos do Desenvolvimento Sustentável em Instituições de Ensino Superior. In XLVI Encontro da ANPAD - EnANPAD. Anais. Retrieved from http://anpad.com.br/uploads/articles/120/approved/7e6ff0205749bc6025b51155e26f6ced.pdf
Coleman JS (1988) Social capital in the creation of human capital. Am J Sociol 94:95–120
Corrêa CP (2021) O papel da Universidade Estadual do Oeste do Paraná no ecossistema de inovação da região Oeste do Paraná (Master's dissertation, Programa de Pós-graduação em Administração). Universidade Estadual do Oeste do Paraná, Cascavel
Creswell JW (2007) Projeto de pesquisa: método qualitativo, quantitativo e misto, 2nd edn. Artmed, Porto Alegre
De la Poza E et al (2021) Universities' reporting on SDGs: using the impact rankings to model and measure their contribution to sustainability. Sustainability (Switzerland) 13(4):1–30
Dias GP, Silva ME, Viana FLE (2021) Reviewing the role of social capital towards supply chain sustainability practices. In: XLV Encontro da ANPAD. Anais. Maringá: Associação Nacional de Pós-Graduação e Pesquisa em Administração
Duarte VN, Schneider MB (2019) Capitais intangíveis e desenvolvimento territorial: algumas considerações. In: XVIII ENANPUR. Anais. Natal: ENANPUR
Ferrera de Lima J, Alves LR (2012) Mecanismos e propostas para acelerar o desenvolvimento local e regional no Paraná. In: Raiher AP et al (eds) Economia paranaense: crescimento e desigualdades regionais. Editora UEPG, Ponta Grossa, pp 205–232
Fukuyama F (1996) Confiança: as virtudes sociais e a criação da prosperidade. Rocco, Rio de Janeiro
Fukuyama F (2000) A grande ruptura: a natureza humana e a reconstituição da ordem social. Rocco, Rio de Janeiro
Fundação Araucária (2021) Araucária promove quatro dias de debates com relação aos Objetivos de Desenvolvimento Sustentáveis (ODS). Retrieved from http://www.fappr.pr.gov.br/Noticia/Araucaria-promove-quatro-dias-de-debates-com-relacao-aos-Objetivos-de-Desenvolvimento
Gallo E, Setti AFF (2014) Território, intersetorialidade e escalas: Requisitos para a efetividade dos objetivos de desenvolvimento sustentável. Ciência e Saúde Coletiva 19(11):4383–4386
Goebel MA, Bidarra ZS (2019) Política pública e a tutela das relações comerciais brasileiras. In: XVI Seminário Internacional Demandas Sociais e Políticas Públicas na Sociedade Contemporânea
Lago AAC (2013) Conferências de Desenvolvimento Sustentável, vol 1. Brasília: Fundação Alexandre de Gusmão (FUNAG)
Leal Filho W et al (2019) Sustainable development goals and sustainability teaching at universities: falling behind or getting ahead of the pack? J Clean Prod 232:285–294
Lugoboni LF, Marques M (2021) Governança pública e sustentabilidade em instituições de ensino introdução. In: XXIV SEMEAD Seminários em Administração. Anais. PPGA FEA USP, São Paulo
Managi S, Lindner R, Stevens CC (2021) Technology policy for the sustainable development goals: from the global to the local level. Technol Forecast Soc Chang 162

Melo PTNB, Regis HP, van Bellen HM (2015) Princípios epistemológicos da teoria do capital social na área da administração. Cadernos EBAPE.BR 13(1):136–164
Nahapiet J, Ghoshal S (1998) Social capital, intellectual capital, and the organizational advantage. Acad Manag Rev 23(2):242–266
North D (1990) Institutions, institutional change and economic performance. Cambridge University Press, Cambridge
Oliveira MM (2007) Como fazer pesquisa qualitativa. Vozes, Petrópolis
ONU (n.d.) World Commission on Environment and Development (Brundtland Commission). In: The palgrave encyclopedia of global security studies. A/42/427: [s.n.], Nova Iorque
ONU (n.d.) Transforming our world: the 2030 Agenda for sustainable development. In: General assembly. A/RES/70/1: [s.n.], Nova Iorque
Ostrom E (1990) Governing the commons: the evolution of institutions for collective action. Cambridge University Press, Nova Iorque
Piacenti CA (2016) Indicadores do potencial de desenvolvimento endógeno dos municípios paranaenses no período 1999/2013, 1st edn. Editora Parque Itaipu, Foz do Iguaçu
Pizzutilo F, Venezia E (2021) On the maturity of social responsibility and sustainability integration in higher education institutions: descriptive criteria and conceptual framework. Int J Manag Educ 19(3):100515
Prieto-Jiménez E et al (2021) Sustainable development goals and education: a bibliometric mapping analysis. Sustainability (Switzerland) 13(4):1–20
Putnam RD (1995) Bowling alone: america's declining social capital. J Democr 6(1):65–78
Putnam RD (2006) Comunidade e democracia: a experiência da Itália moderna, 5th edn. Editora FGV, Rio de Janeiro
Rea LM, Parker RA (2000) Metodologia de pesquisa: do planejamento a execução, 1st edn. Pioneira Thomson Learning, São Paulo
Sander TH, Putnam RD (2010) Still bowling alone? The post-9/11 split. J Democr 21(1):9–16
SDSN Australia/Pacific (2017) Getting started with the SDGS in Universities: a guide to for Universities, Higher Education Institutions and the Academic Sector. Sustainable Development Solutions Network, Melbourne
Sen A (2000) Desenvolvimento como liberdade. Companhia das Letras, São Paulo
The Times Higher Education (2021) World university rankings. Retrieved from https://www.timeshighereducation.com/
The Times Higher Education (2023) Impact Rankings. Retrieved from https://www.timeshighereducation.com/impactrankings/
Torabian J (2019) Revisiting global university rankings and their indicators in the age of sustainable development. Sustainability (United States) 12(2):1–20
Triviños ANS (2009) Introdução à pesquisa em ciências sociais: a pesquisa qualitativa em educação. Atlas
UN (1985) World Commission on Environment and Development (Brundtland Commission) In: The palgrave encyclopedia of global security studies. A/42/427: [s.n.], New York
UN (2015) Transforming our world: the 2030 Agenda for sustainable development. In: General assembly. A/RES/70/1: [s.n.], New York
UNIOESTE (2021) Organograma. Disponível em: https://www.unioeste.br/portal/sobre-a-unioeste/organograma. Acesso em: 3 mar. 2021
UNIOESTE (2023) Estatística—Unioeste em números. Disponível em: https://www.unioeste.br/portal/acesso-aos-dados/estatistica. Acesso em: 31 out. 2023
UNIOESTE (2022) Unioeste promove Diálogo com Educadores sobre os Objetivos do Desenvolvimento Sustentável. Disponível em: https://www.unioeste.br/portal/central-de-noticias/55523-unioeste-promove-dialogo-com-educadores-sobre-os-objetivos-do-desenvolvimento-sustentavel. Acesso em: 5 mar. 2022
Van Maanen J (1979) Reclaiming qualitative methods for organizational research: a preface. Adm Sci Q 24:520–524

Vries BJM, Petersen AC (2009) Conceptualizing sustainable development: an assessment methodology connecting values, knowledge, worldviews and scenarios. Ecol Econ 68(4):1006–1019
Wubah D et al (2021) Funding community sustainable development using zero energy buildings. Int J Sustain High Educ 22(1):29–43

University Teachers' Competencies for Implementing ESD in Higher Education Curriculum: A Case Study of Malaysia

Sharifah Intan Sharina Syed-Abdullah, Edy Hafizan Mohd Shahali, and Najwa Huda Shaharil

Abstract Education for Sustainable Development (ESD) is a crucial approach to foster transformative change in education at higher education institutions (HEIs) . The implementation of ESD in Malaysia has not yet been mainstream. Prior to the extensive implementation of ESD at HEIs, it is imperative to guarantee that university teachers possess the requisite qualifications in competencies for effective instruction. This chapter presents a study that was conducted to assess the level of ESD-related competencies among university teachers. In particular, the competencies in this study were framed by the concept of key sustainability competencies (KSCs) and pedagogical content knowledge (PCK). The study additionally compares the competencies of university teachers according to their research fields, natural sciences or social sciences. The study employed a quantitative survey method, gathering data from 267 university teachers representing 20 Malaysian public universities. Descriptive statistics were used to summarise the overall competency levels, while inferential statistical tests, such as independent t-tests, were conducted to compare the competencies between teachers from different research fields. The findings revealed that Malaysian university teachers demonstrated moderate competency levels overall. A statistically significant difference was observed, with social science teachers exhibiting higher competency levels in both KSCs and PCK compared to those in natural sciences. These results underscore the importance of interdisciplinary approaches and highlight the evolving understanding of sustainable development as encompassing both natural and social science perspectives. The study concludes with recommendations

S. I. S. Syed-Abdullah (✉) · N. H. Shaharil
Faculty of Educational Studies, Universiti Putra Malaysia, Serdang, Malaysia
e-mail: sharifahintansharina@upm.edu.my

N. H. Shaharil
e-mail: najwa.shaharil93@gmail.com

E. H. Mohd Shahali
Faculty of Education, Universiti Malaya, Kuala Lumpur, Malaysia
e-mail: edyhafizan@um.edu.my

W. Leal Filho et al. (eds.), *Education for Sustainable Development: The Contribution of Universities*, World Sustainability Series,
https://doi.org/10.1007/978-3-031-86985-3_19

for targeted professional development programs, strategic institutional planning, and policy reforms to mainstream ESD effectively in Malaysian HEIs.

Keywords Education for sustainable development · University teachers · Key sustainability competencies · Pedagogical content knowledge · Research fields

1 Introduction

As climate change intensifies, global income inequality widens, social injustice persists, and migration movements continue, it is becoming increasingly important for higher education institutions (HEIs) to address these "Grand Challenges" by integrating sustainability into their curricula (Reid et al. 2010). The integration of Education for Sustainable Development (ESD) in higher education is significant because it aims to cultivate a new generation of professionals who engage in critical thinking, are committed to improving their local environments in their daily lives, and are equipped to understand the world in terms of interconnected relationships (Reid et al. 2010). It also instills the confidence that their actions can drive meaningful change in society (Christie et al. 2013).

HEIs play a pivotal role in advancing sustainability by integrating sustainable practices into their educational frameworks, research agendas, and community engagement efforts. Globally, HEIs are key drivers of the Sustainable Development Goals (SDGs), embedding them into curricula through interdisciplinary approaches and fostering collaborations with external stakeholders to enhance sustainability initiatives and policy formulation (Cembranel et al. 2024). By embedding ESD in HEIs' academic programs, it aims to go beyond traditional knowledge transmission to actively engage students with real-world situations as well as to promote reflection on values such as justice, solidarity, equity, ethics, and respect for diversity (Tilbury 2012). This approach not only equips graduates with the competencies needed to tackle complex sustainability challenges but also fosters a culture of social responsibility and ethical decision-making (Barth 2015).

In addition to research and education, HEIs implement campus sustainability initiatives that address environmental management alongside social and economic dimensions. Programs such as energy efficiency, waste reduction, and sustainable transportation exemplify efforts to create greener campuses while fostering sustainability consciousness among students and staff (Shetty 2024). However, challenges persist, particularly in achieving a more comprehensive integration of social and economic aspects of sustainability practices. This underscores the evolving role of HEIs in driving holistic sustainability efforts that balance environmental, social, and economic imperatives.

While many HEIs globally have made significant strides in incorporating sustainability practices into their campus operations and resource management (Dawodu et al. 2022), the progress in embedding sustainability into research and teaching has been comparatively slow (Tian et al. 2024; Tilbury 2012). In Malaysia, not all

HEIs have established specialised centers for sustainability. The majority of those established were founded post-2013. The establishment of sustainability centers is very new and can be regarded as tardy, especially in light of the enduring history of the sustainability concept. Sustainability has historical origins spanning several centuries, with its contemporary definition emerging in the mid-twentieth century, notably marked by the 1987 Brundtland Report and the establishment of the United Nations' Sustainable Development Goals (SDGs) in 2015. Given this historical context, the late establishment of sustainability-focused centers in Malaysian HEIs highlights a missed opportunity to proactively align with global sustainability efforts earlier. This underscores the need for accelerated and more strategic integration of sustainability into institutional structures to address contemporary environmental and developmental challenges.

Now, many universities have introduced sustainability courses, often adding them to existing, sometimes fragmented, degree programs. This approach represents a promising initial step toward integrating ESD into the curriculum (Barth 2015). However, relatively few universities have taken the further step of embedding ESD comprehensively across all their degree programs and fostering a campus-wide culture of sustainability (Singer-Brodowski 2016).

In Malaysia, the efforts of HEIs to promote and implement ESD were believed to be insufficient. The dedication of HEIs in Malaysia to the promotion and implementation of ESD is under-researched. Relevant available research focuses, for instance, on implementing ESD not at the university level in Malaysia but rather in schools (Hoque et al. 2022). While research conducted in the HEI context in Malaysia exists, it indirectly or inexplicitly links to ESD (Yassin et al. 2020; Chinedu et al. 2023), as it is more about education about sustainable development rather than education for sustainable development. Furthermore, the concept of sustainable development in these institutions appears to be narrowly focused on environmental issues. In addition, the concept of sustainable development appears to be restricted to the environmental sphere. Engineering programs, influenced by national engineering professional bodies, are among the very few programs that systematically align their courses with sustainability-related outcomes, leaving other disciplines with limited integration of sustainability into their curricula.

The implementation of Education for Sustainable Development (ESD) in higher education requires university teachers to possess specific competencies that align with UNESCO's framework for sustainable development. These competencies integrate knowledge, skills, values, and attitudes essential for fostering transformative learning experiences and addressing sustainability challenges. UNESCO identifies key sustainability competencies (KSCs) such as systems thinking, futures thinking, strategic action, and values-based thinking, which guide educators in equipping learners to engage with complex global issues. These competencies are further reinforced by frameworks like those proposed by Osman et al. (2017), which emphasise cognitive, affective, and psychomotor domains and their alignment with the Sustainable Development Goals (SDGs). Understanding and cultivating these competencies is vital for embedding ESD in higher education curricula, fostering resilient, responsible global citizens equipped to navigate and contribute to a sustainable future.

University teachers, who are the focus of this study, hold a pivotal role in influencing students' intellectual, emotional, moral development, and behaviours toward sustainability. They serve as champions of sustainable practices and advocates for principles that promote renewable and responsible living. As they guide students to become change agents, university teachers must inspire and nurture the unique passions of each individual, fostering critical thinking and compassionate actions to help students integrate sustainability into their personal and professional lives (Gopalakrishnan et al. 2019).

In the context of higher education, university teachers are often divided into specific disciplines, and it is essential that they tailor sustainability competencies based on their field while maintaining an interdisciplinary approach. By aligning their teaching methods with sustainability principles and collaborating across disciplines, university teachers could create a holistic learning environment that prepares students to tackle complex global challenges (Sterling 2010; Tilbury 2012). Additionally, university teachers act as role models by demonstrating ethical decision-making and sustainable behaviours, which reinforces the transformative role of higher education in promoting sustainability. Their influence extends beyond the classroom as they champion sustainability within the broader academic community and professional networks, ensuring that the principles of sustainable development are embedded both in curricula and in society at large (Brundtland Commission 1987).

Given the significant role of university teachers outlined above, they are anticipated to possess a certain set of competencies to guide their students in becoming capable contributors to sustainable development problem-solving. Therefore, a study was conducted to assess the level of ESD-related competencies among university teachers from their own perspectives. This study focuses on the university teachers' competencies, namely key sustainability competencies (KSCs) (Annelin and Boström 2023) and ESD Pedagogical Content Knowledge (ESD-PCK) (Lee 2016). Given that the concept of sustainability originates from environmental concerns, it raises the question of whether university teachers from various disciplines possess differing levels of skills. Consequently, the study also compares the university teachers' competencies based on their research fields. The study specifically analyses the competencies derived from the fields of natural science and social science education.

By examining these competencies across different research domains, such as natural science and social science education, the study was expected to reveal potential disparities in the ability of university teachers to integrate sustainability into their teaching practices. This knowledge is crucial for informing targeted professional development programs that address specific gaps in ESD competencies. In addition, the findings could contribute to the broader discussion on interdisciplinary approaches to sustainability education, highlighting the need for tailored strategies based on disciplinary backgrounds. This has implications for curriculum design, teacher training, and institutional policies aiming to enhance the role of higher education in advancing sustainable development goals.

2 Competencies for Implementing ESD

Based on an analysis of several policy documents, most HEIs around the world implement it across curricula, rather than through stand-alone courses. For those who take this approach, several scholars have provided guidance on its implementation. Osman et al. (2017) have developed a curricular framework for integrating ESD within the curriculum to aid curriculum developers in aligning with the Sustainable Development Goals (SDGs). Published by the Commonwealth Secretariat, the framework is said to help states produce resilient, effective, and responsible citizens who embrace the Commonwealth Charter's basic values and principles while pursuing sustainable development.

This framework serves two purposes. The curriculum framework was designed as a flexible, non-prescriptive instrument that adheres to a competency development paradigm integrating knowledge, skills, values, and attitudes. When developing the framework, they asserted that it should entail the identification and delineation of integrated competencies learners must attain, focusing on learning outcomes. Besides serving as a guide for ESD content or the learning outcomes that students must attain, it also addresses the competencies required of teachers.

The integrated competencies and learning outcomes are delineated into three categories: knowledge and understanding; skills and application, and values and attitudes (Osman et al. 2017). These three categories reflect the three learning domains, namely cognitive, affective and psychomotor. They also mirror the three areas of the division of Global Development Education/ESD competencies, which are recognising (Erkennen), assessing (Bewerten), and acting (Handeln) (Schreiber and Siege 2016).

The framework sought an integrated set of competencies derived from key sustainability competencies (KSCs) that learners must cultivate for active and responsible engagement in various life domains and for the implementation of the SDGs. The list of KSCs that Osman et al. (2017) list for each category of the ESD curriculum framework is shown in Table 1.

While the framework is rooted in the Commonwealth Charter, the values and strategies it outlines are adaptable to various contexts, making them applicable and beneficial to non-Commonwealth nations seeking similar outcomes. Resilient and responsible citizens are critical for any nation's progress, aligning with global citizenship education (GCED) principles endorsed by UNESCO, which advocates for shared responsibilities in a globalised world.

On the other hand, Qablan (2018) identified four KSCs. First, "learning to know" involves comprehending the challenges that both local and global communities face and understanding the roles educators and learners can play. Second, "learning to do" focuses on acquiring practical skills and competencies necessary for action related to ESD. Third, "learning to live together" emphasises fostering partnerships and valuing interdependence, diversity, mutual understanding, and peace. Finally, "learning to be" pertains to developing personal qualities and the capacity to act with greater autonomy, judgement, and responsibility regarding sustainable development.

Table 1 Description of integrated competencies and learning outcomes based on categories

Categories	Integrated competencies and learning outcomes
Knowledge and understanding	• Understanding of key social, environmental and economic challenges and complexities • Understanding of local, national and global challenges and complexities • Knowledge to respond to these challenges and complexities • Multiple literacies • Lifeskills • Understanding of key socio-political challenges and conflicts • Differential and multiscalar understanding • Systems theory and reflexivity
Values and attitude	• Sense of purpose and hope • Commitment to justice • Confidence, resilience and adaptability • Openness, respect for diversity • Communication, engagement and integration • Responsible, active, productive and engaged citizens • Duty bearers • Commitment to community engagement for constructive responses to societal issues • Self-esteem, self-understanding and clearer sense of identity
Skills and application	• Analysis • Fact finding • Listening, observing and problem solving • Independent learning and critical thinking • Goal-setting skills • Planning and decision-making • Ability to identify and solve problems, and to set goals • Effective communication and social integration skills • Capacity to think independently • Self-reflective and reflective skills

The four competencies align with the earlier categories of curriculum mentioned. "Learning to know" corresponds to knowledge and understanding—cognitive or recognising. "Learning to do" aligns with skills and application, psychomotor or acting, while "learning to live together" and "learning to be" reflect values and attitudes, affective or assessing.

The aspect of knowledge and understanding, apart from those integrated competencies listed by Osman et al. (2017) in Table 1, can be directly linked to the 17 SDGs. Each goal provides a specific context for learners to deepen their understanding of global challenges, such as poverty, inequality, and climate change. By grounding the competency of "learning to know" in the SDGs, university teachers can cultivate a more relevant and impactful educational experience that not only highlights the importance of these goals but also empowers learners to engage critically with the issues they address. This connection reinforces the significance of developing informed and responsible citizens who can navigate the complexities of sustainable development in a globalised world.

The aspect of skills in the framework aligns with varying sets of key sustainability competencies (KSCs) articulated by some scholars as necessary for ESD. These scholars emphasise the importance of developing practical skills that enable individuals to effectively address sustainability challenges. Wiek et al. (2015) proposed six core competencies, emphasising futures thinking, values-based thinking, and strategic action. Osman et al. (2017) highlight five core competencies: envisioning, critical thinking and reflection, systemic thinking, building partnerships, and participation in decision-making. Meanwhile, Rieckman et al. (2017) and Hoffman and Siege (2018) expand this to eight competencies, including systems thinking, anticipatory, normative, strategic, collaboration, self-awareness, and integrated problem-solving. More recently, Annelin and Boström (2023) have restructured these competencies into a more streamlined model of seven, focusing on systems thinking, strategic thinking, futures thinking, values thinking, interpersonal, intrapersonal, and implementational competencies (Table 2). This evolution of frameworks reflects a broader attempt to balance comprehensive skill sets with practical application, signalling a shift towards a more holistic and action-oriented understanding of sustainability education. The varied sets of KSCs reflect ongoing efforts to refine and prioritise these competencies. In addition, it underscores the challenge of distilling a definitive set of competencies, as different frameworks emphasise distinct aspects of sustainability, depending on their educational focus.

However, while the competencies they list appear to align primarily with "learning to do" or "learning to be," some authors have mapped these competencies to encompass all three principal categories: knowledge, values, and action. This broader mapping underscores that effective skill development is not isolated but interconnected with understanding the underlying knowledge and fostering the values necessary for responsible action. By recognising these interconnections, educators can create a more integrated approach that ensures learners are well-equipped to engage comprehensively with sustainability issues. For instance, KSCs such as systems thinking, anticipatory skills, and normative skills, all of which are crucial for engaging in sustainable practices. By integrating these competencies into the "learning to do" aspect of the framework, university teachers can better equip learners with the tools necessary for proactive participation in sustainability initiatives. This connection highlights the critical role of skill development in fostering responsible and capable individuals who can contribute to the achievement of the SDGs.

The progress of global universities in embedding sustainability competencies into curricula is marked by both advancements and challenges. While there is a growing recognition of the importance of sustainability education, many institutions still face significant barriers to its effective integration. Many universities are beginning to align their curricula with the Sustainable Development Goals (SDGs), with some institutions successfully integrating up to 14 out of the 17 SDGs into their programs (Cleveland et al. 2023). Interdisciplinary approaches are increasingly emphasised as essential for fostering sustainability education, as they encourage collaboration

Table 2 Description of key sustainability competencies

Competencies of sustainability	Descriptions
Futures thinking	The ability to grasp and evaluate many futures—possible, probable, and desirable—create one's own visions, apply the precautionary principle, analyse implications, and deal with risks and changes
Values thinking	The ability to comprehend and reflect on one's norms and values to negotiate sustainable values, principles, goals, and targets in the face of disputes, trade-offs, uncertainty, and inconsistencies
Intrapersonal thinking	The ability to interrogate norms, practices, and opinions, to contemplate one's own values, perceptions, and behaviours, and to adopt a stance in the sustainability dialogue
Systems thinking	The ability to understand relationships, analysing complex systems, thinking about how systems are integrated in different domains and scales, and dealing with uncertainty
Interpersonal thinking	The ability to acquire knowledge from others, comprehend and honour their needs, viewpoints, and behaviours, to connect with and exhibit sensitivity towards others, to manage group conflicts, and to promote collaborative and participatory problem-solving
Strategic thinking	The ability to collaborate on creative sustainability measures at the local and global level
Implementational thinking	The ability to employ various problem-solving frameworks for intricate sustainability issues, to formulate feasible, inclusive, and equitable solution alternatives that advance sustainable development, and to synthesise the aforementioned competencies

across various fields and disciplines (Abo-Khalil 2024). Despite these efforts, challenges persist. Resistance to change within institutions, limited resources, and a lack of awareness among students continue to hinder the effective integration of sustainability competencies (Park 2024). Furthermore, the absence of a unified policy framework for embedding sustainability into curricula remains a significant barrier, limiting the consistency and depth of sustainability education across institutions (Khare and MacNeil 2024).

In Malaysia, efforts to integrate sustainability competencies into higher education curricula are steadily gaining traction, reflecting both progress and persistent challenges that align with the global context. Malaysian universities have begun prioritising sustainability within their managerial agendas, demonstrating alignment with global trends in organisational adoption. This focus highlights the critical role of institutional leadership in driving effective implementation of sustainability initiatives (Syed-Abdullah et al. 2023). Additionally, several institutions have adopted holistic, pluralistic, and action-oriented methods in education for sustainable development (ESD), mirroring international best practices aimed at fostering students' sustainability consciousness (Saleem et al. 2022). However, despite these advancements, research specifically examining sustainability competencies in Malaysian higher education remains limited, leaving a critical gap in understanding the extent and effectiveness of such efforts.

While studies explicitly addressing sustainability competencies in Malaysian universities are sparse, a wealth of research exists that focuses on individual competencies or uses alternative educational frameworks. For example, numerous studies have explored teacher competencies in environmental education, climate literacy, and other sustainability-related areas. These often employ frameworks such as 21st-century skills (e.g. Sulaiman and Ismail 2020), pedagogical content knowledge (PCK) (e.g. Abdullah and Halim 2012; Shafie et al. 2019), and other pedagogical models. Such research provides valuable insights into the knowledge and skills educators need to effectively incorporate sustainability concepts into their teaching. Although these studies do not always adopt a holistic approach to sustainability competencies, they contribute significantly to understanding the foundational elements required for successful ESD integration. This interplay between broader institutional efforts and specific competency-focused research underscores the need for a more unified and comprehensive approach to sustainability education in Malaysia.

2.1 Pedagogical Content Knowledge

Pedagogical content knowledge (PCK), a framework proposed by Shulman (1987), elucidates the competencies required by educators including university teachers for teaching a specific subject. In the framework, he recommended two essential competencies required by educators: subject knowledge and pedagogical knowledge.

PCK represents a topic-specific knowledge that integrates content and pedagogy, in which instructors use both their subject matter knowledge and their knowledge of students to represent appropriate content to be learned. In other words, the PCK is an amalgam of subject-specific and pedagogical knowledge that includes an awareness of how students comprehend and process certain learning topics and difficulties (Singer-Brodowski 2016).

In addition, PCK highlights common challenges encountered during the learning process and builds upon them to suggest effective teaching practices that educators might implement (Shulman 1987). Shulman (1987) maintains that simply knowing material and general pedagogy is insufficient for effective instruction. Rather, effective teaching requires *"the blending of content and pedagogy into an understanding of how particular topics, problems, or issues are organised, represented, and adapted to the diverse interests and abilities of learners, and presented for instruction"* (Shulman 1987, p. 8).

Since its conception over three decades ago, the PCK construct has sparked a great deal of study interest as it strives to explain the complicated, nuanced, and sophisticated process of teaching. The PCK models utilised by researchers differ in their incorporation of PCK components. Chan and Hume's (2019) review identified the most prevalent PCK components used in the literature, finding that, while nomenclature varies across the PCK literature, the majority of PCK components are derived from three frameworks: Shulman (1987), Grossman (1990), and Magnusson et al. (1999).

Grossman (1990) extended Shulman's original concept of PCK by incorporating curricular knowledge, which is determined by the instructor's notions of purpose for teaching certain topics, along with knowledge about students' understanding and successful instructional practices. Grossman (1990) delineated the domains of instructor knowledge and identified four indispensable knowledge areas constituting professional knowledge for teaching: general pedagogical knowledge (GPK), subject matter knowledge (SMK), knowledge of context (KC), and pedagogical content knowledge (PCK). In Grossman's work (1990), GPK includes prescriptive knowledge, beliefs, and skills for effective instruction related to promoting learner's learning, classroom management, and curriculum design. SMK represents content knowledge of a subject area referring to the nature of the subject matter, including substantive and syntactic structures as a prerequisite for the development of PCK. KC includes knowledge of individual students and the demands of the community or districts in which instructors work. Using Shulman's and Grossman's work as an inspiration, Magnusson et al. (1999) then proposed two additional PCK components: instructors' orientations towards teaching and assessment knowledge. More recent conceptualisations draw on these previous models, consolidating contributions to reconceptualise models of PCK (e.g., Carlson et al. 2019; Gess-Newsome 2015).

In short, PCK can be defined as both a single domain of instructors' knowledge as well as an overarching domain that both integrates and transforms other knowledge domains. As such, researchers can approach understanding instructors' knowledge for particular content area teaching through the lens of PCK, which can be defined as an instructor's professional knowledge to transform what he or she knows into teaching strategies that make the knowledge accessible to learners.

The idea of PCK that has been developed further by various scientists (Barnett and Hodson 2001) provides an important foundation for quality teaching not only in schools but also in higher education (e.g. Trigwell and Shale 2004). PCK is underpinned by the substantive and syntactic content knowledge of subject area (Smith 1999), so it is necessary to elicit specific components of PCK in ESD. In the context of ESD, we revised Lee's (2016) PCK framework to propose competencies required by university teachers to implement ESD. Lee (2016) introduced ePCK as a framework for analysing the professional development of elementary teachers inside a graduate-level Interdisciplinary Environmental Education program. In the framework, there are seven subdomains of PCK were proposed, namely (1) knowledge of subject matter, (2) knowledge of curriculum, (3) knowledge of learner, (4) knowledge of instructional strategies, (5) knowledge of environmental context, (6) knowledge of educational goal, and (7) knowledge of evaluation.

Based on this framework and through a literature review and by conducting interviews, we elicit specific components of each knowledge of the PCK domain, which we refer to as a whole as ESD-PCK. Table 3 presents the modified definition of the components to suit the context and need for ESD.

Table 3 Description of ESD-PCK domains

Component	Definition
Knowledge of subject matter	Teacher's knowledge about sustainability subject content and the processes involved in teaching for sustainable development
Knowledge of curriculum	The knowledge required to understand the system of curricula related to sustainability education and to effectively implement and deliver sustainability-focused curricula
Knowledge of learner	Teacher's collective knowledge of learners who participate in ESD classes
Knowledge of instructional strategies	The teacher's collective knowledge about the appropriate teaching and learning methods suitable for the purposes of ESD
Knowledge of environmental context	The teacher's knowledge about the various social, cultural, and tangible resources that can be utilised for teaching and learning in ESD
Knowledge of educational goal	The teacher's knowledge required to establish the purpose of ESD and to set appropriate learning goals accordingly
Knowledge of evaluation	The teacher's knowledge required to apply assessment methods that are suitable for the purposes of ESD and to effectively utilise the results

3 Methodology

A quantitative survey was performed to assess the competencies of Malaysian university teachers in integrating ESD into the higher education curriculum. 267 respondents from a total population of 31,508 participated in the survey. From this number, 136 of them are experts in natural science fields, while the rest are experts in social sciences. The sample size was insufficient to meet the minimal requirement for the population size. This is attributable to the constraint that the data was gathered during the MCO, a period when the entire world grappled with adapting to life amid COVID-19. Nevertheless, this number was considered sufficient as many statisticians concur that a size of 100 is the minimum number of samples needed for meaningful results (Memon et al. 2020).

This study focused exclusively on university teachers at 20 public universities in Malaysia, which represent a critical component of the country's higher education landscape. Public universities in Malaysia are highly influential in shaping national educational strategies, particularly those aligned with the United Nations' SDGs. These institutions have a unique mandate to drive societal development by fostering sustainable practices and policies, given their close ties to government directives and their role in producing a large proportion of the nation's graduates. Additionally, Malaysian public universities operate under the unified governance of the Ministry of Higher Education (MOHE), which provides a standardised framework for policy implementation, including those related to ESD. This centralised structure facilitates a consistent approach to embedding sustainability into curricula across institutions, making public universities an ideal setting for studying ESD-related competencies. By focusing on public universities allows this study to conduct a more consistent analysis of ESD competencies under similar regulatory conditions.

While this strategy provides valuable insights into the state of ESD-related competencies in public HEIs, it also means that the findings may not fully represent the broader landscape of university teachers in private institutions or other types of educational organisations. Thus, the finding is limited to university teachers in public universities and cannot be generalised to all educators in higher institutions in the country. Therefore, caution is advised when generalising the results beyond public universities.

The competencies were evaluated according to Annelin and Boström's (2023) sustainability competencies and Lee's (2016) framework of pedagogical content knowledge (PCK). To recap, sustainability competencies encompass future thinking, values thinking, intrapersonal thinking, systems thinking, interpersonal thinking, strategic thinking, and implementational thinking. The questionnaire comprises 23 items designed to assess KSCs. Conversely, Lee's (2016) PCK framework comprises seven subconstructs: knowledge of subject matter, knowledge of curriculum, knowledge of learner, knowledge of instructional strategies, knowledge of environmental context, knowledge of educational goals, and knowledge of evaluation. There are 38 items to measure these competencies of university teachers. All construct measurements were modified from previous studies that have shown adequate psychometric consistency. All constructs were measured on a five-point Likert-type scale where 'strongly agree' was marked as 5 and 'strongly disagree' was marked as 1.

Data collection for this study took place during the global COVID-19 pandemic, which necessitated the adoption of an online method to gather survey responses. Due to restrictions on in-person interactions and the shift toward remote work, the survey was distributed electronically to research respondents via email and online platforms. This approach was thought would allow for broader geographic reach and ensure the safety of respondents while maintaining the integrity of the data collection process. The online method also should have enabled respondents to respond at their convenience, which may have contributed to a higher response rate despite the challenging circumstances of the pandemic. However, as previously elucidated, the turnover rate was rather low.

The data collected for this study was analysed using SPSS software. First, descriptive analysis was conducted to summarise the demographic characteristics of the respondents and to assess the general level of educators' sustainability competencies and ESD Pedagogical Content Knowledge (ESD-PCK). Descriptive statistics, such as mean and standard deviation, were used to provide an overview of the data. Following this, an independent t-test was performed to compare the sustainability competencies and ESD-PCK of the university teachers based on a specific variable, which is their academic discipline (natural sciences and social sciences). The t-test allowed for the identification of significant differences between the groups, providing insight into how these factors may influence the university teachers' competencies. A total of 267 Malaysian university instructors responded to the questionnaire. Based on instructors' teaching areas, 136 (50.9%) of the respondents are natural science instructors, while the remaining 131 (49.06%) of them are social science instructors.

4 Findings

Table 4 shows the means and standard deviations for each domain of sustainability competency. Overall, the level of sustainability competencies for Malaysian university teachers is at a moderate level (M = 3.27, S.D = 0.50). Findings from the results of the descriptive analysis show that only one domain of sustainability competency recorded an overall mean reading that gave a high level of interpretation, namely interpersonal competency (M = 3.42, S.D = 0.55), while for the other domains, they are at a moderate level (refer Table 4). Based on instructors' teaching area of expertise, only social science instructors showed a high level of competency for this domain of interpersonal competence (M = 3.52, S.D = 0.55) while for natural science instructors' are at a moderate level (M = 3.34, S.D = 0.55). The lowest mean score of sustainability competencies based on subdomain for both natural science (M = 3.11, S.D = 0.57) and social science (M = 3.23, S.D = 0.64) university teachers is for normative competency.

Table 5 presents the mean and standard deviation of Malaysian university teachers' ESD-PCK. The findings showed that overall ESD-PCK for Malaysian university teachers' is at a moderate level (M = 3.05, S.D = 0.61). Based on the ESD-PCK sub-domain, the analysis revealed that the highest mean is for subdomain knowledge of sustainable development context for both natural science (M = 3.19, S.D = 0.69) and social science (M = 3.33, S.D = 0.69) instructors. The lowest mean score of ESD-PCK based on subdomains is for subdomain knowledge of curriculum for both natural science (M = 2.76, S.D = 0,70) and social sciences (M = 3.05, S.D = 0.70) instructors.

Table 6 presents the results of the t-test analysis for the sustainability competency and ESD-PCK level of Malaysian university teachers, which are based on their academic area of expertise. The t-test analysis results show that there was a significant difference ($p < 0.05$) in the scores for social science (M = 3.341, S.D = 0.510) and natural science (M = 3.194, S.D = 0.487) sustainability competency; $t(265) = 2.40$, $p = 0.017$. The same findings were also observed for the university teachers' ESD-PCK where the findings revealed that, for ESD-PCK, t-test analysis results showed that there was a significant difference ($p < 0.05$) in the scores for social science (M = 3.341, S.D = 0.510) and natural science university teachers (M = 3.194, S.D = 0.487) for ESD PCK; $t(193) = 2.90$, $p = 0.004$.

Table 4 Means and standard deviations for the Malaysian university teachers sustainability competencies (N = 267)

Sustainability competencies	Overall			Natural sciences			Social sciences		
	Mean	SD	Interpretation	Mean	SD	Interpretation	Mean	SD	Interpretation
Futures thinking	3.19	0.54	Moderate	3.11	0.51	Moderate	3.27	0.56	Moderate
Values thinking	3.17	0.61	Moderate	3.11	0.57	Moderate	3.23	0.64	Moderate
Intrapersonal thinking	3.27	0.57	Moderate	3.20	0.55	Moderate	3.35	0.58	Moderate
Systems thinking	3.22	0.59	Moderate	3.19	0.57	Moderate	3.26	0.60	Moderate
Interpersonal thinking	3.42	0.55	High	3.34	0.55	Moderate	3.52	0.55	High
Strategic thinking	3.23	0.65	Moderate	3.13	0.63	Moderate	3.34	0.65	Moderate
Implementational thinking	3.25	0.61	Moderate	3.17	0.58	Moderate	3.33	0.63	Moderate
Overall	3.27	0.50	Moderate	3.19	0.49	Moderate	3.34	0.51	Moderate

Table 5 Means and standard deviations for the Malaysian university teachers ESD PCK (N = 267)

ESD PCK	Overall			Natural sciences			Social sciences		
	Mean	S.D	Interpretation	Mean	S.D	Interpretation	Mean	S.D	Interpretation
Knowledge of subject matter	2.97	0.68	Moderate	2.85	0.65	Moderate	3.09	0.69	Moderate
Knowledge of curriculum	2.90	0.70	Moderate	2.76	0.67	Moderate	3.05	0.70	Moderate
Knowledge of learner	3.01	0.67	Moderate	2.90	0.65	Moderate	3.13	0.68	Moderate
Knowledge of instructional strategies	3.07	0.76	Moderate	2.94	0.75	Moderate	3.20	0.75	Moderate
Knowledge of SD context	3.26	0.70	Moderate	3.19	0.69	Moderate	3.33	0.69	Moderate
Knowledge of educational goal	3.05	0.68	Moderate	2.91	0.67	Moderate	3.19	0.67	Moderate
Knowledge of assessment	2.96	0.70	Moderate	2.83	0.67	Moderate	3.11	0.71	Moderate
Overall	3.05	0.61	Moderate	2.93	0.59	Moderate	3.18	0.60	Moderate

Table 6 Malaysian university teachers' sustainability competency and ESD-PCK level based on area of academic expertise

	t	df	p
Sustainability competencies	2.40	265	0.017
ESD PCK	2.90	193	0.004

5 Discussion and Conclusion

The moderate level of sustainability competencies and ESD-PCK among university teachers presents a significant challenge for effectively implementing ESD at HEIs. These competencies are essential for preparing future generations to address complex sustainability challenges (Annelin and Boström 2023; Wiek et al. 2015). However, disparities between the competencies of university teachers in the social and scientific fields open opportunities to explore and address these differences in greater detail.

Historically, sustainable development has been associated predominantly with environmental issues, as the term was originally introduced to address concerns about ecological degradation. This environmental focus remains prominent in many scientific fields. However, the contemporary framing of the SDGs emphasises the interconnectedness of natural science and social science issues, underscoring the importance of addressing not only environmental concerns but also social and economic challenges. SDGs highlight the need to solve interrelated issues, such as poverty, inequality, climate change, and resource management, which requires an integration of both scientific and social knowledge.

The disparity in sustainability competencies between university teachers in the natural science and social science fields is noteworthy. The higher KSCs and ESD-PCK level observed among social science university teachers could be attributed to the human-centred focus inherent in social science disciplines. PCK in general involves not only mastery of subject matter but also the ability to teach it effectively by connecting content to students' lived experiences and societal issues, a skill often emphasised in social sciences. Given that social sciences engage directly with human behaviour, ethics, and societal structures, these educators are likely more familiar with interdisciplinary approaches and the incorporation of human and social dimensions of sustainability, such as equity, justice, and governance. Teaching skills, which are at the core of PCK, align closely with the human and relational skills more deeply embedded in social sciences, allowing these teachers to integrate sustainability in a way that resonates with both the subject matter and students' realities. This is supported by studies suggesting that educators in social science fields may have stronger foundations in addressing complex societal challenges, which are central to ESD (Rieckmann 2018).

Therefore, it is important to ensure university teachers in natural science fields are aware of the critical need to "put the social into sustainability science" as suggested by Kargan and Burton (2018). Sustainability science that neglects social perspectives is theoretically unfit and empirically unable to fully address the complex crises facing humanity. For instance, without considering social factors such as equity, cultural values, and governance, efforts to tackle environmental issues like climate change may fail to consider the societal drivers of these problems. A holistic approach, which blends natural and social sciences, is necessary to devise solutions that are not only technically sound but also socially acceptable and sustainable in practice.

Alternatively, Leal Filho (2022) proposes an interdisciplinary approach as a viable solution to this issue. He underscores the necessity of multidisciplinary activity to facilitate sustainable development that incorporates insights from various industries and disciplines. By integrating interdisciplinary approaches, university teachers can establish a comprehensive learning environment that equips students to address intricate global issues (Sterling 2010; Tilbury 2012). The findings of this study indicate that both sample groups exhibit a high degree of interpersonal thinking, suggesting the potential for collaboration between university teachers from the natural sciences and social sciences to mainstream ESD in Malaysian HEIs.

In conclusion, addressing the moderate level of sustainability competencies and PCK among university teachers requires a concerted effort to balance the focus on natural and social science perspectives. Sustainable development is not solely about environmental concerns; it encompasses social and economic dimensions that are equally important. Therefore, integrating the 'social' into sustainability science is critical for both theoretical coherence and practical efficacy in dealing with global crises. By enhancing interdisciplinary competencies in sustainability education, educators can better prepare students to engage with the complex, interrelated challenges of the SDGs, ultimately fostering a more sustainable future.

References

Abdullah SISS, Halim L (2012) Influence of teaching option and teaching experience on science teachers' pedagogical content knowledge of environmental education. J Korean Assoc Sci Educ 32(8):1378–1389

Abo-Khalil AG (2024) Integrating sustainability into higher education challenges and opportunities for universities worldwide. Heliyon 10:e29946

Annelin A, Boström G-O (2023) An assessment of key sustainability competencies: a review of scales and propositions for validation. Int J Sustain High Educ 24(9):53–69

Barnett J, Hodson D (2001) Pedagogical context knowledge: toward a fuller understanding of what good science teachers know. Sci Educ 85(4):426–453

Barth M (2015) Implementing sustainability in higher education: learning in an age of transformation. Routledge. https://doi.org/10.4324/9780203488355

Brundtland GH (1987) Our common future: report of the world commission on environment and development. UN-Document A/42/427. http://www.un-documents.net/ocf-ov.htm

Carlson J, Daehler KR, Alonzo AC, Barendsen E, Berry A, Borowski A, Wilson CD (2019) The refined consensus model of pedagogical content knowledge in science education. In: Repositioning pedagogical content knowledge in teachers' knowledge for teaching science, pp 77–94

Cembranel P, Gewehr L, Dal Moro L, Fuchs PG, Birch RS, Andrade Guerra JBSODA (2024) The pivotal role of higher education institutions in cultivating a sustainable development goals-centric culture. Int J Sustain High Educ 25(7):1385–1411

Chan KKH, Hume A (2019) Towards a consensus model: literature review of how science teachers' pedagogical content knowledge is investigated in empirical studies. In: Repositioning pedagogical content knowledge in teachers' knowledge for teaching science, pp 3–76

Chinedu CC, Saleem A, Wan Muda WHN (2023) Teaching and learning approaches: curriculum framework for sustainability literacy for technical and vocational teacher training programmes in Malaysia. Sustainability 15(3):2543

Christie B, Miller K, Cooke R, White J (2013) The sustainability education and dynamic teaching strategies. Int J Sustain High Educ 14(4):404–417

Cleveland D, Van Thang LN, Chi CNTM (2023) Integrating sustainability in higher education curricula: a transformative approach towards marketing education and brand social responsibility. ASCILITE Publications, pp 345–350

Dawodu A, Dai H, Zou T, Zhou H, Lian W, Oladejo J, Osebor F (2022) Campus sustainability research: Indicators and dimensions to consider for the design and assessment of a sustainable campus. Heliyon 8(12):e11864

Gess-Newsome J (2015) A model of teacher professional knowledge and skill including PCK: results of the thinking from the PCK summit. Routledge, pp 28–42

Gopalakrishnan AC, Bhaskaran S, Ambujakshi S (2019) Relationship between proactivity and personal pedagogical knowledge in ESD among pre-service teachers: sub-sampling on social media usage and subject discipline. J Teach Educ Sustain 21(2):73–90

Grossman P (1990) The making of a teacher. Teacher's College Press, New York

Hoffmann T, Siege H (2018) What is education for sustainable development (ESD). Hum Dev 1(8):1–6

Hoque F, Yasin RM, Sopian K (2022) Revisiting education for sustainable development: methods to inspire secondary school students toward renewable energy. Sustainability 14(14):8296
Kargan C, Burton MH (2018) Putting the 'social' into sustainability science. In: Handbook of sustainability science and research, pp 285–298
Khare A, MacNeil P (2024) Adopting sustainability competencies in management education-a scoping review of progress. https://doi.org/10.20944/preprints202410.0652.v1
Leal Filho W (2022) Pursuing sustainable development across disciplines. Handbook of best practices in sustainable development at university level. Springer International Publishing, Cham, pp 495–504
Lee YJ (2016) Using ePCK as a framework for examining elementary teachers' professionalization of teaching in a graduate-level interdisciplinary environmental education program (Doctoral dissertation, Seoul National University)
Magnusson S, Krajcik J, Borko H (1999) Nature, sources, and development of pedagogical content knowledge for science teaching. In: Gess-Newsome J, Lederman NG (eds) Examining pedagogical content knowledge: the construct and its implications for science education. Kluwer, Boston, pp 95–132
Memon MA, Ting H, Cheah JH, Thurasamy R, Chuah F, Cham TH (2020) Sample size for survey research: review and recommendations. J Appl Struct Equ Model 4(2):1–20
Osman A, Ladhani S, Findlater E, McKay V (2017) Curriculum framework for the sustainable development goals. Commonwealth Secretariat. Available online: https://www.thecommonwealth-educationhub.net/wp-content/uploads/2017/01/Curriculum_Framework_for_SDGs_July_2017.pdf
Park Y (2024) Sustainable education practices: voices from higher education institutions. PREPRINT (Version 1) available at Research Square. https://doi.org/10.21203/rs.3.rs-3923563/v1
Qablan A (2018) Building capacities of educators and trainers. In: Issues and trends in education for sustainable development, vol 133
Reid W, Chen D, Goldfarb L, Hackmann H, Lee Y, Mokhele K, Ostrom E, Raivio K, Rockström J, Schellnhuber H, Whyte A (2010) Earth system science for global sustainability: grand challenges. Science 330(6006):916–917
Rieckmann M, Mindt L, Gardiner S (2017) Education for sustainable development goals: learning objectives. UNESCO publishing
Rieckmann M (2018) Learning to transform the world: Key competencies in education for sustainable development. In: Leicht A, Heiss J, Byun WJ (eds) Issues and trends in education for sustainable development goals. UNESCO publishing, pp 39–59
Saleem A, Aslam S, Sang G, Dare PS, Zhang T (2022) Education for sustainable development and sustainability consciousness: evidence from Malaysian universities. Int J Sustain High Educ 24(1):193–211
Schreiber JR, Siege H (2016) Curriculum framework education for sustainable development. In: Standing conference of the ministers of education and cultural affairs; engagement global
Shafie H, Majid FA, Ismail IS (2019) Technological pedagogical content knowledge (TPACK) in teaching 21st century skills in the 21st century classroom. Asian J Univ Educ 15(3):24–33
Shetty S (2024) Higher education institutions as a catalyst for sustainability development. Goi EC (ed) Teaching and learning for a sustainable future: innovative strategies and best practices. IGI Global, pp 1–16. https://doi.org/10.4018/978-1-6684-9859-0.ch001
Shulman L (1987) Knowledge and teaching: foundations of the new reform. Harv Educ Rev 57(1):1–23
Singer-Brodowski M (2016) Students' competency development in the context of self-organised and project-oriented sustainability seminars: research at the interface between self-description and real-life action. In: Barth M, Michelsen G, Rieckmann M, Thomas I (eds) Routledge handbook of higher education for sustainable development. Routledge, pp 411–420

Smith DC (1999) Changing our teaching: the role of pedagogical content knowledge in elementary science. Examining pedagogical content knowledge: the construct and its implications for science education. Springer, Netherlands, Dordrecht, pp 163–197

Sterling S (2010) Transformative learning and sustainability: sketching the conceptual ground. Learn Teach High Educ 5:17–33

Sulaiman J, Ismail SN (2020) Teacher competence and 21st century skills in transformation schools 2025 (TS25). Univers J Educ Res 8(8):3536–3544

Syed-Abdullah SIS, Kushnir I, Abdrahim NA (2023) Narratives on education for sustainable development in Malaysian universities. Sustainability 15(17):13110

Tilbury, D. (2012). Assessing ESD experiences during the DESD: an expert review on processes and learning for ESD. UNESCO. http://unesdoc.unesco.org/images/0019/001914/191442e.pdf

Tian W, Ge J, Zheng X (2024) Understanding the landscape of education for sustainable development in China: a bibliometric review and trend analysis of multicluster topics (1998–2023). Hum Soc Sci Commun 11:1213. https://doi.org/10.1057/s41599-024-03713-y

Trigwell K, Shale S (2004) Student learning and the scholarship of university teaching. Stud High Educ 29(4):523–536

Wiek A, Bernstein MJ, Foley RW, Cohen M, Forrest N, Kuzdas C, Keeler LW (2015) Operationalising competencies in higher education for sustainable development. In: Routledge handbook of higher education for sustainable development. Routledge, pp 241–260

Yassin AA, Abdul Razak N, Qasem YAM, Saeed Mohammed MA (2020) Intercultural learning challenges affecting international students' sustainable learning in Malaysian higher education institutions. Sustainability 12(18):7490

A Digital Training Tool to Support the Industry 5.0 Transition—Case Description with Bachelor Students in a Brazilian University

João Victor Encide Salla, Ana Carolina Godoy Albino, Daiane Vitória da Silva, Antônio Carlos Farrapo Jr, Yovana María Barrera Saavedra, and Diogo Aparecido Lopes Silva

Abstract Industry 5.0 (I5.0) is a transformative shift in business models, emphasizing technology and customization within ecological boundaries. Despite its importance, there is a scarcity of research on evaluating digital tools that support training for the I5.0 transition. This study aims to fill this gap by assessing the quality and effectiveness of the I-GO Assistant tool from the perspective of bachelor students in Brazil. This self-assessment digital tool can support companies towards the I5.0 transition. To measure the student's expectations and perceptions of the use of the tool, the SERVQUAL model served as the instrument for comparing the I-GO Assistant application and quality gap analysis, involving the profiling of 148 participants. Our findings reveal that, while user expectations generally exceeded their perceptions across the five service quality dimensions, statistical analysis showed a high level of satisfaction, with responses predominantly falling within the 4–5 range on

J. V. E. Salla (✉) · A. C. G. Albino · D. V. da Silva · A. C. Farrapo Jr · D. A. L. Silva
Research Group on Sustainability Engineering (EngS Group), Department of Production Engineering, School of Management and Technology, Federal University of São Carlos, Sorocaba,, São Paulo, Brazil
e-mail: vesjoao@estudante.ufscar.br

A. C. G. Albino
e-mail: carolinagodoyalbino@gmail.com

D. V. da Silva
e-mail: daianevitoria@estudante.ufscar.br

A. C. Farrapo Jr
e-mail: acfarrapo@ufscar.br

D. A. L. Silva
e-mail: diogo.apls@ufscar.br

Y. M. B. Saavedra
Natural Sciences Center—CCN Ufscar-Lagoa do Sino, Federal University of São Carlos, UFSCar, Buri, São Paulo, Brazil
e-mail: yovana.saavedra@ufscar.br

W. Leal Filho et al. (eds.), *Education for Sustainable Development: The Contribution of Universities*, World Sustainability Series,
https://doi.org/10.1007/978-3-031-86985-3_20

the Likert scale. Recommendations for improvement included incorporating social aspects into the next version of the I-GO Assistant, one of the pillars of I5.0. This study contributes to the limited body of knowledge on the evaluation of digital tools for training in I5.0 and provides insights for improving such tools to better support organizations in the transition to I5.0.

Keywords SERVQUAL · Expectations · Perception · Sustainability · Industry 5.0

1 Introduction

The technological development fostered by the industrial revolution over the years has provided the optimized resources to support sustainable development. (David et al. 2022). For example, there is economic growth with job creation, innovation in operational processes, and societal lifestyle changes. The Fourth Industrial Revolution (IR4) was a revolutionary change of horizontal expansion in which information technology proliferated across industries (Lee et al. 2018). However, there are still concerns ahead of IR4 about the impacts of these technologies on water, energy, and food to the human economy (David et al. 2022). These three resources are global issues discussed in 2030 Agenda to achieve the Sustainable Development Goals (SDGs), such as sustainable consumption and production patterns (SDG 12).

This concern about management of vital resources is related to the transition from IR4.0 to the Fifth Industrial Revolution (IR5.0). Among the features of I5.0 are cooperation between humans and machines, improved quality of life at work which, and planetary boundaries which ensure the availability of natural resources over the years (European Commission 2021). Because of this, an approach is being developed aimed at creating factories of the future, called the human cyber-physical system, which fosters the involvement of humans with technological transformations while respecting the environment conservation (Saniuk et al. 2022).

A recent initiative by the I5.0 Community of Practices highlights this concern, aiming to unite stakeholders in implementing I5.0 principles and reshaping traditional business models to generate value (European Commission 2023). In this context, digital technologies—such as the Internet of Things, artificial intelligence, databases, and simulators—play a critical role. Among these, the I-GO Assistant tool (detailed in Sect. 2) facilitates work customization and training, aligning with the core pillars of I5.0 (The Green Growth Knowledge Partnership—GGKP 2022).

The 5 Gaps Model is a business analysis model that aims to identify and close gaps between customer expectations and company deliverables (Stylidis et al. 2019; Tumsekcali et al. 2021). Parasuraman et al. (1985) and Parasuraman et al. (1988) joined these attributes into five service quality dimensions: tangibility, responsiveness, reliability, assurance, and empathy.

Recently, Delgado et al. (2022) explained the difficulties on properly managing quality of digital services focused on sustainability issues based on the 05 Gaps Analysis Model application in a case study with an e-learning platform. Hussain et al. (2023) studied the perspectives of hospitality educators about the concerns relating to career sustainability. On the other hand, Nayeri et al. (2023) investigated the responsive supply chain problem based on the I5.0 dimensions.

However, in the context of I5.0 there is still a lack of publications in the topic of quality assessments of current digital tools. Hein-Pensel et al. (2023) highlights the existence of a gap in prescriptive maturity models that can assist organizations, especially for small and medium-sized enterprises (SMEs), which can be addressed by the I-GO Assistant tool.

Therefore, this study aims to fill this gap by assessing the quality and effectiveness of the I-GO Assistant from the perspective of potential users, specifically bachelor students in Brazil. The research question formulated is: "Does the I-GO Assistant tool have the quality and functionality to serve as an effective support within organizations aligned with the context of Industry 5.0?".

By evaluating the expectations and perceptions of these students using the SERVQUAL model, we seek to provide insights into the strengths and areas for improvement of the I-GO Assistant, contributing to its development and better alignment with the principles of I5.0. The results can inform both the tool developers and organizations seeking digital solutions to support their transition to I5.0.

The findings of this study are expected to offer practical recommendations for enhancing the I-GO Assistant tool, particularly in incorporating social aspects, which is one of the pillars of I5.0. Furthermore, this research contributes to the limited body of knowledge on evaluating digital tools in the context of I5.0, thereby filling a gap in the current literature.

2 Methodology

2.1 Object Under Investigation

The primary objective of this study is to evaluate the quality and effectiveness of the I-GO Assistant tool—a self-assessment platform developed by the Green Growth Knowledge Partnership (GGKP) to support the sustainable development of small and medium-sized enterprises (SMEs). The I-GO Assistant aims to enhance resource efficiency in SMEs by allowing them to self-assess their maturity levels in key categories such as resource efficiency strategy, energy management, water conservation, material efficiency, sustainable chemicals, and waste management. Its regional pilot application took place in three regions (Eastern Europe and Central Asia, Africa, and Latin America & the Caribbean during 2022) with the goal of improving resource efficiency in SMEs. To provide more technical and tailored support, the GGKP hosted regional workshops and meet-and-greet events (GGKP 2022). The I-GO Assistant will also introduce an 'Implementation Manager' feature to help SMEs develop and manage resource efficiency projects for lasting sustainability (GGKP 2022).

In this study, we administered the I-GO Assistant to a cohort of 148 bachelor students at the Federal University of São Carlos (UFSCar) in Brazil, representing diverse academic backgrounds. The students engaged with the platform between 2021 and 2022, involving 52 participating companies, 60% of which were SMEs. The participants were enrolled in undergraduate courses directly or indirectly related to sustainable resource efficiency.

By utilizing the SERVQUAL model to measure the expectations and perceptions of these students regarding the I-GO Assistant, our aim is to assess whether the tool possesses the quality and functionality to effectively support organizations, particularly SMEs, in their transition towards Industry 5.0 (I5.0). This evaluation seeks to identify the strengths and weaknesses of the I-GO Assistant, providing insights for its improvement and better alignment with the principles of I5.0. Ultimately, the study aims to contribute to the development of effective digital tools that facilitate the I5.0 transition, addressing a gap in current research and offering practical recommendations for enhancing such tools.

2.2 Research Instrument

For the analysis of the I-GO Assistant use, it developed an adaptation of the SERVQUAL instrument, a widely used method for evaluating service quality (Sumi and Kabir 2021). The study was conducted based on Fig. 1.

(1) **Questionnaire formulation**: Drawing from a comprehensive literature review, a set of questions was crafted to assess the gap between expectations and perceptions concerning the service delivery of the I-GO Assistant. Respondents provided their feedback using a Likert scale, wherein scores ranged from 1 (strongly disagree) to 5 (strongly agree). Additionally, qualitative and descriptive questions were included to explore participants' demographic characteristics.

(2) **Questionnaire application**: The survey was conducted with students enrolled in undergraduate courses such as 'Energy Resource Management', 'Environmental Management', 'Special Topics in Production Engineering', and 'Life Cycle Engineering and Management of Products'. Utilizing the Google Forms platform, the questionnaire was meticulously structured and subsequently made accessible for students to participate and provide their responses based on the I-GO use with SMEs.

(3) **Data analysis**:

   a. Data collection and tabulation: Following the distribution of the questionnaire, the responses were collected and tabulated in spreadsheets to facilitate calculations and the analysis in the subsequent stage.
   b. Statistical analysis: The sample underwent descriptive statistical tools and qualitative examination to identify patterns in relation to control variables.

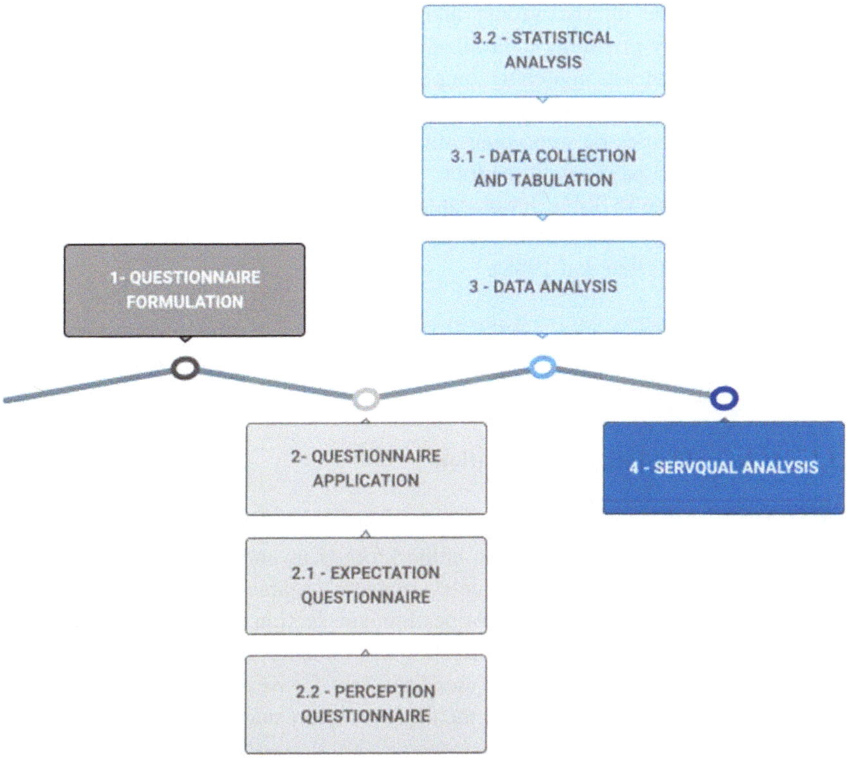

Fig. 1 Methodology overview. *Source* own authorship

The process was supported by statistical analyses using Cronbach's Alpha and Kurtosis. Further elaboration on these analyses can be found in Sect. 3.2.

(4) **SERVQUAL analysis**: Subsequent to this, the data underwent processing involving the evaluation of expectations, perception, satisfaction levels, and the comprehensive identification of gaps within the five dimensions of service quality provided by the I-GO Assistant tool. This instrument facilitates the measurement of quality through the lens of service gaps, quantifying the disparities between customer expectations of the service and their perceptions following its actual delivery. Consequently, the scale operates within two spheres: the expectation sphere, wherein attribute importance is evaluated, and the perception sphere, which allows users to appraise their real-time assessment of each attribute after engaging with it.

Throughout the undergraduate courses, the responsible professors worked with the students on the importance of the I-GO Assistant for application in SMEs in the context of I5.0. Additionally, they provided instructions on utilizing the SERVQUAL instrument to assess the quality of the digital tool. Students from all the courses were encouraged to create internal groups and select a SME of their choice to undertake a concise study involving the utilization of the I-GO Assistant. As part of the process, each student was invited to complete both an expectation questionnaire and a perception questionnaire regarding the platform application. The objective was to pinpoint any disparities between their initial expectations and the actual outcomes experienced after engaging with the tool.

The questionnaire in Table 1 was categorized into the five distinct dimensions: tangibility, reliability, responsiveness, assurance, and empathy, as established by Parasuraman et al. (1985 and 1988).

2.3 *Data Analysis*

2.3.1 Data Collection and Tabulation

The questionnaire encompassed inquiries designed to capture demographic and behavioral insights. In this way, the authors opted to craft questions that would provide insights into both the respondents' (students') characteristics and their chosen SMEs (in groups of 3–6 students). The questionnaire section dedicated to profiling the students included queries aimed at collecting respondents' email addresses, enrolled courses, current semester, and gender (see Sect. 3.1). Conversely, another set of questions was directed at comprehending the focal company under investigation by each group of students.

The university-level Research Ethics Committee granted approval to the survey, as mandated by Brazilian regulations (Certificate of Ethical Appreciation Presentation CAAE—63204122.0.0000.5504).

2.3.2 Statistical Analysis

Statistical analysis was conducted to better understand the questionnaire data. To achieve this, simpler analyses were performed to map the profile of the respondents (course and gender) and the studied companies (size and sector). Subsequently, analyses were conducted to examine the frequency of responses using Likert scale indicators (1–5) in the expectation and perception questionnaires. Finally, more robust statistical analyses were carried out, including dendrograms, Cronbach's alpha, and distribution curves, to identify potential trends in the data, search for outliers, and even identify similarities between questions (clustering).

Table 1 Research tool (adaptation of the SERVQUAL instrument)

Dimension	Cod	Expectation	Perception
Tangibility	T1	The I-GO Assistant tool must have a platform that is intuitive	The I-GO Assistant tool features an intuitive platform
	T2	The I-GO Assistant tool must have an interface that is modern and visually appealing to users	The I-GO Assistant tool boasts a modern interface that is visually appealing to users
	T3	The contact channels provided by the I-GO Assistant tool must be accessible and user-friendly	The contact channels provided by the I-GO Assistant tool are accessible and user-friendly
Reliability	C1	The I-GO Assistant tool must offer reliability and the ability to meet resource efficiency commitments	The I-GO Assistant tool offers reliability and the ability to meet resource efficiency commitments
	C2	The explanations provided for each question in the I-GO Assistant tool must be sufficiently explanatory to avoid interpretation errors	The explanations provided for each question in the I-GO Assistant tool are sufficiently explanatory to avoid interpretation errors
	C3	The support materials (weblinks, articles, tools) provided by I-GO Assistant must be relevant for a thorough understanding of resource efficiency	The support materials (weblinks, articles, tools) provided by I-GO Assistant are relevant to ensure a thorough understanding of resource efficiency
Responsiveness	P1	Access (login/registration) to the I-GO Assistant tool must be fast to appeal to users	Access (login/registration) to the I-GO Assistant tool is swift to capture users' favor
	P2	The I-GO Assistant tool must have an effective problem resolution system and internet-based response system	The I-GO Assistant tool includes an effective problem resolution system and internet based response mechanism
	P3	The I-GO Assistant tool must provide a quick results system for SMEs to understand their status in terms of efficient resource use	The I-GO Assistant tool features a rapid results system for SMEs to comprehend their position in terms of efficient resource use
Assurance	S1	Files/information provided by companies must be kept secure by the I-GO Assistant tool	Files/information provided by companies are securely maintained by the I-GO Assistant tool
	S2	The I-GO Assistant tool must have a section for privacy policies and data protection provided by users	The I-GO Assistant tool includes a section dedicated to privacy policies and protection of user-provided data
	S3	The I-GO Assistant tool must have interfaces in different languages to reach SMEs in other countries	The I-GO Assistant tool offers interfaces in various languages to reach SMEs in different countries

(continued)

Table 1 (continued)

Dimension	Cod	Expectation	Perception
Empathy	E1	The I-GO Assistant tool must offer individualized support based on each company's needs	The I-GO Assistant tool provides individualized support based on each company's needs
	E2	The I-GO Assistant tool must have channels that allow users to receive individualized electronic assistance	The I-GO Assistant tool features channels that allow users to receive personalized electronic assistance
	E3	The I-GO Assistant tool appears to care about and address the needs of SMEs	The I-GO Assistant tool is committed to addressing the needs of SMEs

Source own authorship, adaptation based on Farrapo Júnior (2019)

2.4 Statistical Analysis

When consumers make purchases of goods, they often rely on various tangible cues to assess quality, including attributes like style, durability, color, labels, and packaging. In contrast, when obtaining services, the indicators of quality tend to be less tangible. Typically, tangible evidence in service contexts is confined to physical facilities, equipment, and the personnel of the service provider (Parasuraman et al. 1985). This challenge extends to the realm of I5.0, where software and other digital technologies also offer services that often possess intangible characteristics from the user's perspective (Delgado et al. 2022). In this context, in the absence of tangible evidence to assess service quality, consumers and service providers are left to various aspects to measure the quality of received services (in the case of consumers) and provided services (in the case of providers), resulting in 5 typical gaps (Fig. 2).

Gap 1, delves into the disparities between the perceptions of managers and the expectations of customers. Gap 2, known as the design gap, emerges from the disconnect between customer expectations and the actual design of the service. Gap 3, also recognized as the delivery gap, assesses the distinction between the planned service design and its real-world execution for customers. Gap 4, referred to as the communication gap, gauges the contrast between what is conveyed to customers and what is effectively delivered. Lastly, Gap 5, referred to as the perception gap, permits an examination of the variance between customers' perceptions of the received service and their original expectations set before the service implementation. To tackle this challenge, the significance of the SERVQUAL scale introduced by Parasuraman et al. (1985) becomes evident, as it offers a framework for comprehending Gap 5. This instrument facilitates the measurement of quality through the lens of service gaps, quantifying the disparities between customer expectations of the service and their perceptions following its actual delivery. Consequently, the scale operates within two spheres: the expectation sphere, wherein attribute importance is evaluated, and the perception sphere, which allows users to appraise their real-time assessment of each attribute after engaging with it.

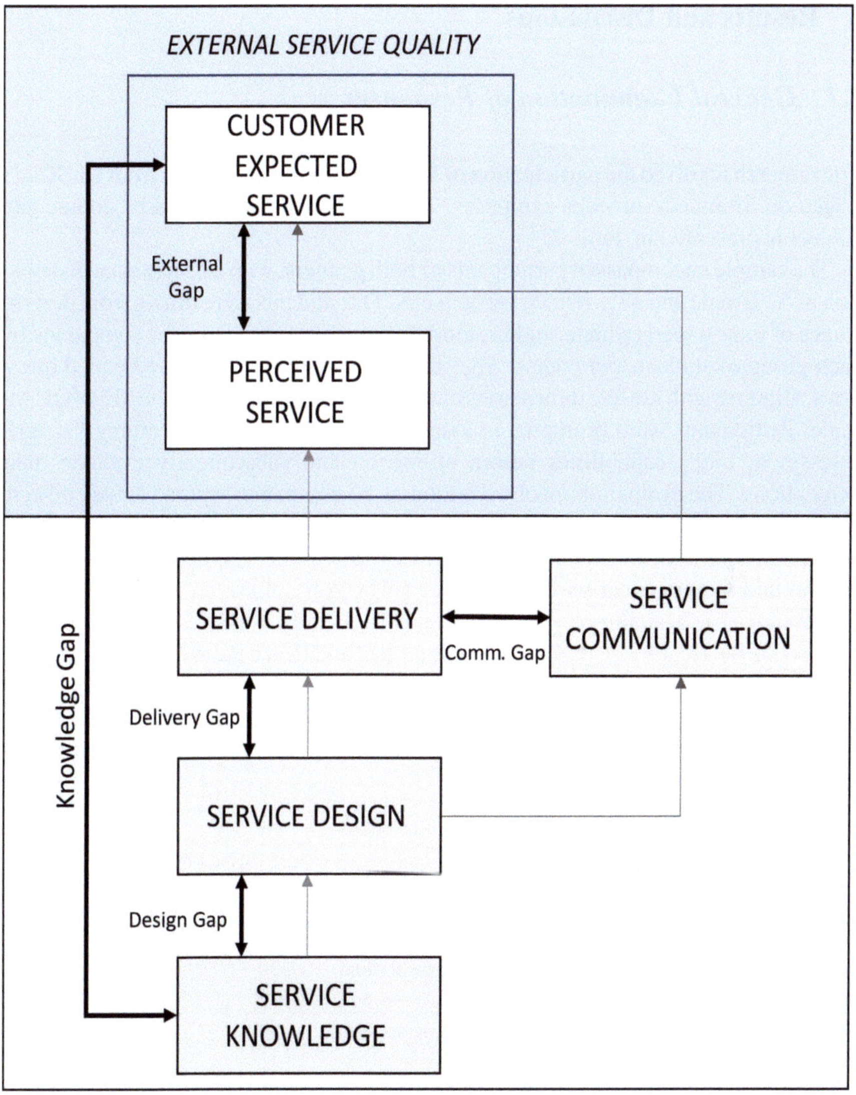

Fig. 2 The 5 gaps of service quality. *Adapted from* Gajanan (2016)

Both the Five Gaps model from Fig. 2 and the SERVQUAL scale served as the foundation for analyzing the quality of the I-GO Assistant tool.

3 Results and Discussions

3.1 General Examination of Responses

The research involved the participation of 148 undergraduate students from UFSCar's Lagoa do Sino and Sorocaba campuses. The number of participants by course and subject is presented in Table 2.

The sample encompassed participants of both genders, with a nearly even distribution of 52% male and 48% female participants. The students were drawn from diverse stages of their undergraduate studies, aiming to encompass a range of companies by each group of students per course. The survey encompassed 15 closed-ended questions, aligning with the five dimensions of service quality outlined in the SERVQUAL scale. Participants were prompted to assess their expectations concerning the self-assessment tool's capabilities before utilization and subsequently evaluate their perceptions. The evaluation involved a total of 52 companies, spread across diverse sectors such as Wholesale and Retail (AV), Food and Beverage Processing (PAB), Manufacturing (M), Restaurant (R), Civil Construction (CC), Logistics (L), Information and Communication Technologies (ITI), Agriculture (A), Forest Products Processing (F), Metallurgy (M), Finance (FI), Beauty Products Distribution for Salons (DPB), Healthcare (CS), Ornamental Plant and Peach Palm Production (PPO), Paper and Packaging (PE), Energy Generation and Supply Services (SGFE), Social Actions (AS), Automotive (AUTO), Candy Distribution (DD), and Waste Management and Recycling (GRR). The sector-wise distribution of the evaluated companies can be observed in Fig. 3.

Table 2 Descriptive statistics on respondents' profile

Undergraduate course	Undergraduate subject/subtopic	UFSCar campus	Number of students	%
Administration	Energy Resources Management	Lagoa do Sino	30	20
	Environmental Management	Lagoa do Sino	36	24
Production Engineering	Special Topics in Production Engineering	Sorocaba	38	26
Food Engineering	Food Engineering	Lagoa do Sino	25	17
Environmental Engineering	Product Life Cycle Management and Engineering	Lagoa do Sino	17	12
Agricultural Engineering		Lagoa do Sino	2	1
			148	100

Fig. 3 Number of evaluated companies per sector. *Source* own authorship

Out of these companies, 60% are classified as small-sized companies (<50 employees), 27% as medium-sized companies (<250 employees), and 13% as large-sized companies (>250 employees).

3.2 *Statistical Analysis of Responses*

For the statistical evaluation of the responses, we first choose to analyze the students' expectations and perceptions for each of the questions. To do so, two graphs were initially plotted based on the frequency of responses using the Likert scale (from '1–Strongly Disagree' to '5—Strongly Agree') in Fig. 4. The responses of the 148 people were filtered to ensure that only the responses of those who had answered both questionnaires (expectation and perception) were analyzed, resulting in 101 valid responses.

It is evident that the maximum score (5) for expectations was more frequently observed in comparison to perception responses. Additionally, there was a notable trend where respondents commonly assigned a score of 4 on the Likert scale when evaluating their perceptions. This disparity between expectations and actual user perceptions underscores the presence of a quality gap. Consequently, the perception scores were lower when contrasted with responses related to expectations on the I-GO Assistant use.

Dendrogram principles were employed to analyze the existence of similarities for each question and among the five dimensions of service quality. The dimension of quality that held the most significance among respondents was assurance (S) (Fig. 5). Issues about the 'quality of datasets recommended for the assessment and study of

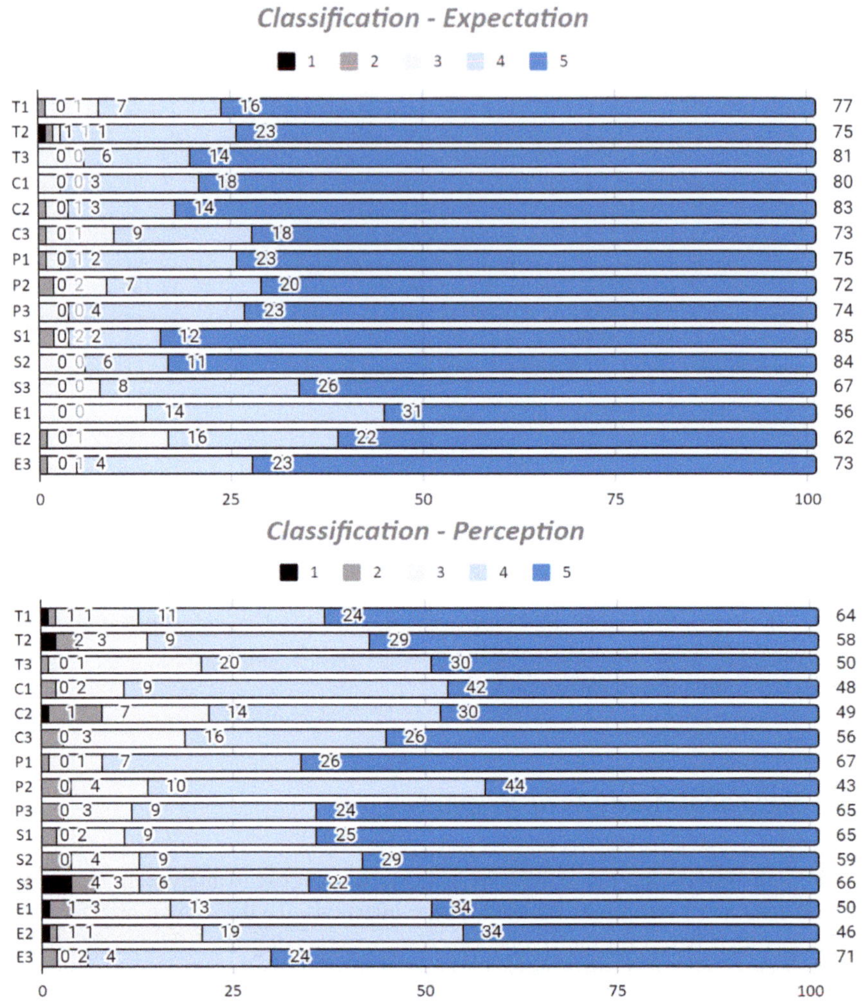

Fig. 4 Distribution of ratings: expectation and perception profiles. *Source* own authorship

SMEs provided by the I-GO Assistant platform'; 'the guarantee of personal data protection'; and the 'user interface of the tool being available in different languages' stood out. Questions related to reliability (C), i.e., 'how good the platform and its specific functions are', was the second-ranked dimension in Fig. 5. Finally, dimensions related to responsiveness (P) and tangibility (T) were equally relevant among respondents, with empathy (E) returning to a lesser extent.

A grouping of the perception questions outlined in Table 1 was performed. It can be observed some relevant clusters on the x-axis of Fig. 6, specifically at number 20:

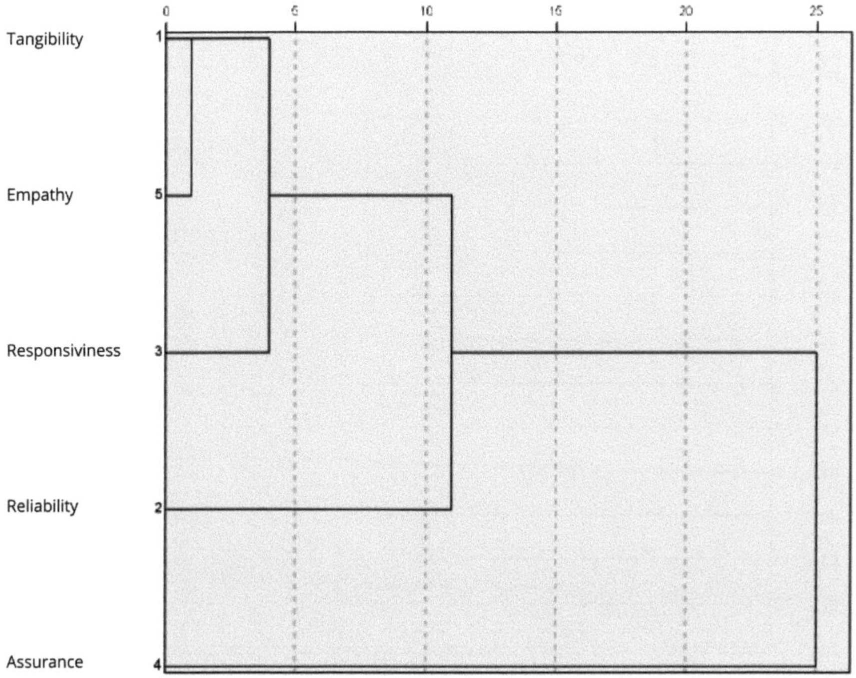

Fig. 5 Dendrogram of quality dimensions applied to the I-GO Assistant use. *Source* own authorship

- **Cluster 1**—User Relationship and Loyalty: The highlighted questions in this cluster pertain to three critical aspects: tangibility, empathy, and security (specifically, T3, E1, and S2). Among these, question T3, which addresses the accessibility and use of contact channels in the I-GO tool, emerged as the most significant. The other two questions, E1 and S2, corresponding to empathy and security, respectively, are associated with providing support to companies and ensuring the protection of personnel data and secure access to the platform. It's noteworthy that question E2, concerning channels that enable individualized electronic assistance, was a result of the integration of these two attributes (E1 and S2). This alignment is logical since both attributes are closely tied to supporting users, encompassing security and assistance aspects.
- **Cluster 2**—Platform Resources: the aspects of reliability and tangibility of the platform were most prominent (C1, C2, T2, and C3). All reliability-related questions are located within this cluster, indicating higher engagement among users. These questions include the ability to fulfill the tool's proposed tasks, the accuracy of questions to prevent misinterpretations, and the availability of supporting materials for users. Additionally, the question about the tangibility aspects referring to the I-GO Assistant's user interface (T2) was among the issues in this grouping. This indicates that the user experience with the platform was a relevant factor in the evaluation, increasing the likelihood of long-term usage.

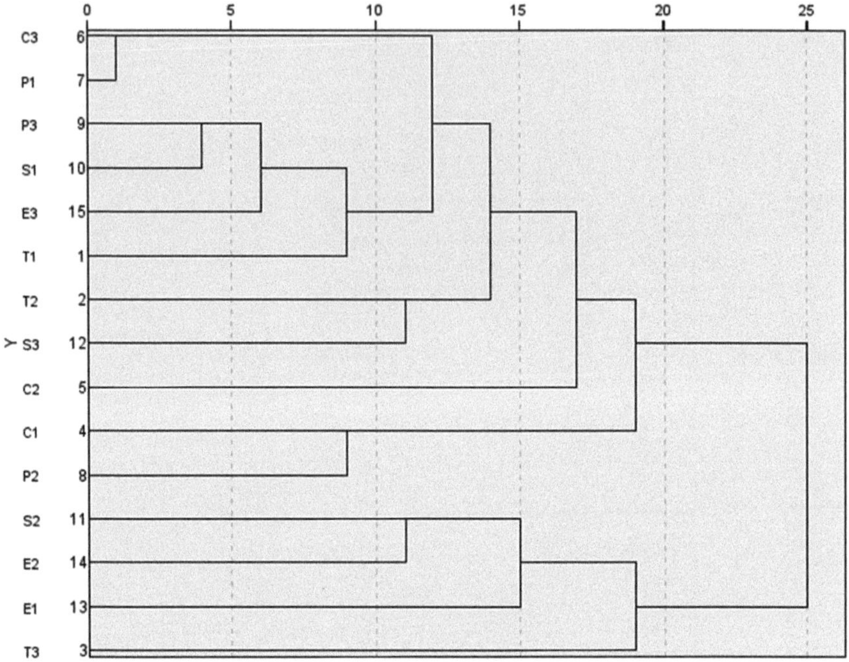

Fig. 6 Dendrogram of perceptions for each question. *Source* own authorship

In addition to the statistical assessment of individual responses, the data distribution of these responses was also explored. In this regard, statistical measures such as Cronbach's alpha, kurtosis, and skewness were used (Hair et al. 2009) and the results can be seen in Table 3.

From a statistical point of view, conventionally, values for Cronbach's Alpha above 0.8 are considered to indicate strong internal consistency (Hajjar et al. 2018; Taber 2018). Thus, the obtained result of 0.899 allows us to conclude that the questions have a high interrelation and that the research instrument used establishes a reliable measure of the underlying construct.

On the other hand, Fig. 7 shows the distribution of response frequencies regarding users' perception of using the I-GO Assistant tool for each of the 15 questions. It can be observed that all the graphs exhibit different behaviors from a normal distribution, meaning they tend to have negative skewness or left-skewed.

Table 3 Cronbach's alpha analysis

All combined results considering the total expectation and perception questions on the survey		
Cronbach's alpha	Cronbach's alpha based on standardized items	Number of items
0.899	0.907	30

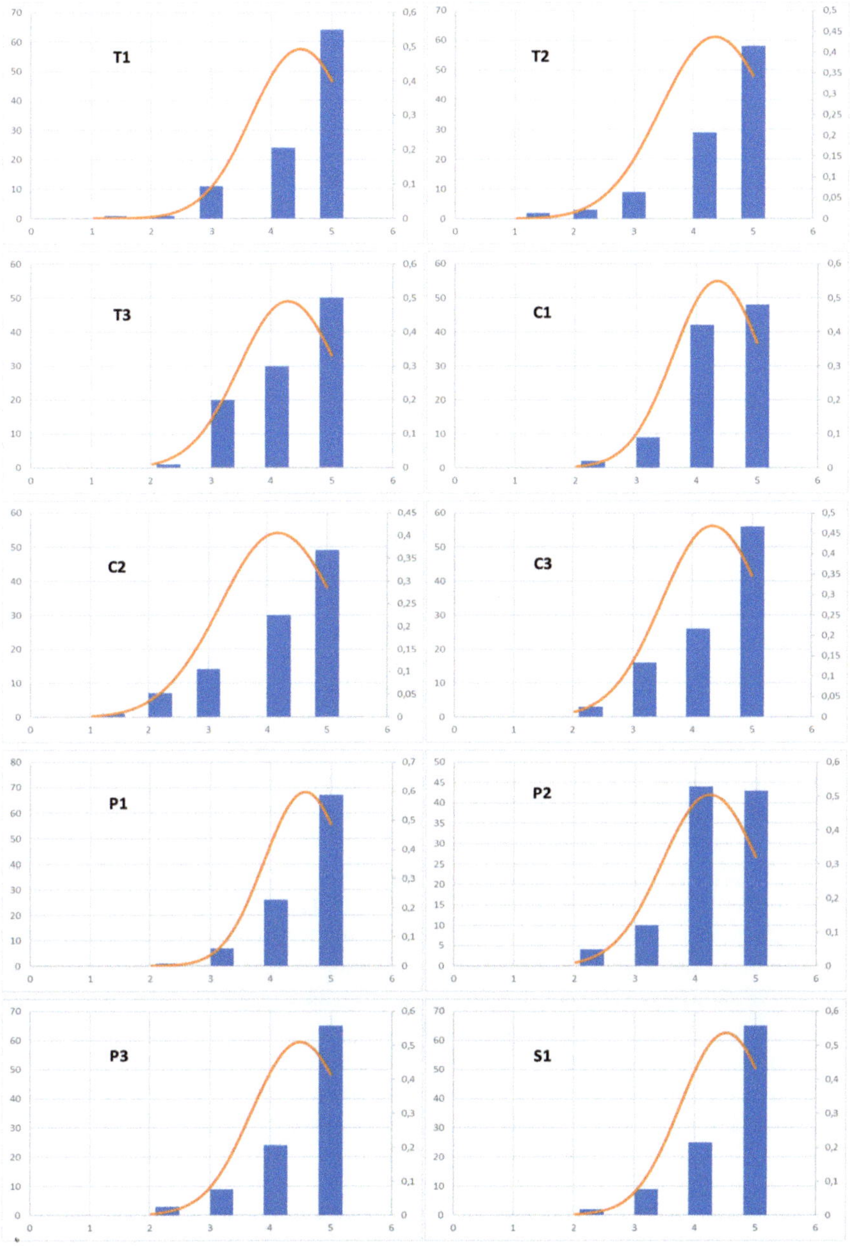

Fig. 7 Distribution of perception data frequencies per question. *Source* own authorship

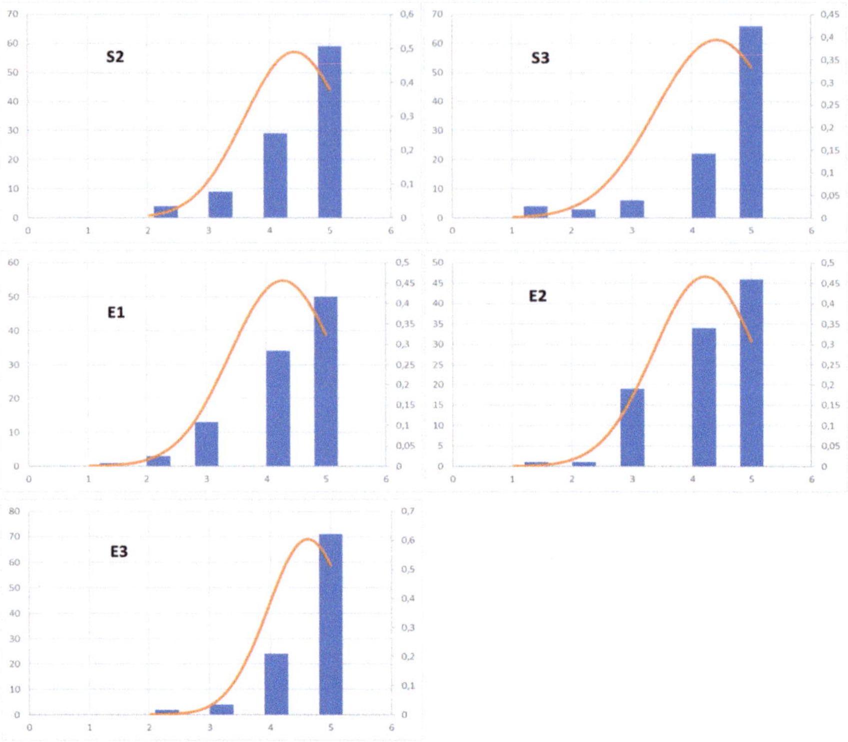

Fig. 7 (continued)

The observation of a negatively skewed distribution in the graphs raises an interesting point in the interpretation of participants' perceptions of the I-GO Assistant tool. The concentration of responses towards the higher end of the Likert scale indicates that most users have a positive view of the tool usage. However, the presence of negative skewness indicates the existence of variability in the responses, despite the overall trend towards higher ratings. This suggests some users may have more critical or negative perceptions warranting a closer examination of this aspect.

This pattern predominantly implies that participants are satisfied with the functionality and usability of the I-GO platform in the context of I5.0 support. However, identifying some lower-scoring responses points to the possibility of areas susceptible to improvements on this tool.

It is important to note that the deviation of data from the normal distribution can be explained by the natural complexities of subjectively evaluating digital technologies and tools. Responses, which are typically positive, tend to cluster around higher values. However, the variety in responses highlights the importance of understanding the results within a broader context. Furthermore, according to Pratiwi and Waluyo (2023) studies focusing on survey scores are susceptible to non-normal data patterns due to diverse cognitive capacities. This context may include factors that

Table 4 Descriptive statistics for the responses regarding the perception of the I-GO Assistant tool usage

	Mean	Standard deviation	Skewness Standard error: 0.240	Kurtosis Standard error: 0.476
T1	4.4752	0.80739	− 1.663	2.944
T2	4.3663	0.91349	− 1.682	2.888
T3	4.2772	0.81386	− 0.667	*− 0.819*
C1	4.3465	0.72713	− 0.960	0.691
C2	4.1782	0.98383	− 1.075	0.410
C3	4.3366	0.85179	− 1.009	*− 0.033*
P1	4.5743	0.66853	− 1.507	1.795
P2	4.2475	0.79254	− 0.967	0.682
P3	4.4950	0.78261	− 1.516	1.632
S1	4.5149	0.74315	− 1.469	1.529
S2	4.4158	0.81569	− 1.358	1.229
S3	4.4158	1.01259	− 2.036	3.768
E1	4.2772	0.87314	− 1.219	1.374
E2	4.2178	0.85562	− 0.926	0.682
E3	4.6238	0.66108	− 1.952	4.004

influence how people perceive the tool usage, such as their previous experiences with technology or their expectations.

In Table 4, we can observe there are negative kurtosis for questions T3 and C3. It is worth examining the coherence of these responses in the context of the specific questions that generated them, as they might provide additional insights or nuances to be explored. These negative kurtosis results indicate the distributions of these responses have a relatively flatter shape than the normal distribution (Hatem et al. 2022). The presence of negative kurtosis may be associated with a more uniform dispersion of data around the mean (Delgado et al. 2022), meaning they indicate a certain degree of uniformity or consistency in the responses, resulting in a distribution shape that does not have long tails or pronounced extremes.

3.3 SERVQUAL Analysis

The SERVQUAL scale commonly uses the difference between perception and expectation of quality to draw conclusions and suggestions (Bentancourt 2015; Oliveira and Ferreira 2008). Negative results indicate that perceptions fall short of expectations, suggesting that users perceive the I-GO Assistant's quality as lower than anticipated. Null results would indicate that perceptions through the use of the tool and expectations were exactly the same, with no gap.

In general, it should be emphasized that the gaps analyzed in this section refer to gap 5, as they pertain to the differences between the expectation scores (before using the tool) and the perception scores (after using the tool. Tables 5, 6, and 7 compile the gap analysis, the tables present the average gaps within each quality dimension and the overall gap result.

Thus, it can be observed that the Engineering classes (Agronomy, Environmental, and Food Engineering) at the Lagoa do Sino campus exhibited larger gaps compared to the Engineering Production classes in Sorocaba city, and with the Administration classes in Lagoa do Sino campus for almost all dimensions of quality, except for the dimensions of promptness and empathy. It can be inferred that the total gap was slightly higher than the gaps for the Engineering Production and Administration classes. This establishes the existence of distinct critical perceptions among students from each course, influenced by different academic background, which led to slight variations among expectations and perceptions after using the tool with the evaluated SMEs. The five questions with more gaps in quality were compiled in Table 8.

In accordance with Parasuraman et al. (1985), the presence of gaps in all evaluated dimensions is acknowledged. Therefore, it is valid to infer there is a discrepancy between what users of the I-GO platform expected to receive and their quality perception after actual usage. Two reliability questions (C1 and C2) exhibited significant gaps, underscoring the need for the I-GO platform organizers to pay more attention to carefully crafting the question formulations to avoid interpretation errors and to ensure that the questions asked truly reflect mechanisms capable of bringing resource efficiency to businesses through the tool's usage.

Table 5 Gap analysis: Production Engineering classes at UFSCar Sorocaba campus

R	T1	T2	T3	C1	C2	C3	P1	P2	P3	S1	S2	S3	E1	E2	E3
Exp	4.92	4.72	4.56	4.92	5.00	4.72	4.68	4.52	4.88	5.00	4.96	4.52	3.92	4.36	4.80
Per	4.60	4.52	4.24	4.24	4.48	4.56	4.60	4.20	4.72	4.72	4.44	4.72	3.84	4.00	4.72
Gap	– 0.32	– 0.20	– 0.32	– 0.68	– 0.52	– 0.16	– 0.08	– 0.32	– 0.16	– 0.28	– 0.52	0.20	– 0.08	– 0.36	– 0.08
Mean	– 0.28			– 0.45			– 0.19			– 0.20			– 0.17		
Mean	– 0.26														

Table 6 Gap analysis: Administration classes at UFSCar Lagoa do Sino campus

R	T1	T2	T3	C1	C2	C3	P1	P2	P3	S1	S2	S3	E1	E2	E3
Exp	4.67	4.80	4.83	4.87	4.83	4.72	4.78	4.74	4.78	4.76	4.76	4.76	4.70	4.61	4.72
Per	4.48	4.46	4.57	4.43	4.24	4.43	4.65	4.41	4.46	4.65	4.50	4.35	4.43	4.46	4.63
Gap	– 0.20	– 0.35	– 0.26	– 0.43	– 0.59	– 0.28	– 0.13	– 0.33	– 0.33	– 0.11	– 0.26	– 0.41	– 0.26	– 0.15	– 0.09
Mean	– 0.27			-0.43			– 0.26			– 0.26			– 0.17		
Mean	– 0.28														

Table 7 Gap analysis: Engineering classes (Agronomy, Environmental, And Food Engineering) at UFSCar Lagoa do Sino campus

R	T1	T2	T3	C1	C2	C3	P1	P2	P3	S1	S2	S3	E1	E2	E3
Exp	4.47	4.47	4.77	4.47	4.50	4.37	4.60	4.47	4.40	4.63	4.63	4.37	4.40	4.23	4.47
Per	4.37	4.10	3.87	4.30	3.83	4.00	4.43	4.03	4.37	4.13	4.27	4.27	4.40	4.03	4.53
Gap	– 0.10	– 0.37	– 0.90	– 0.17	– 0.67	– 0.37	– 0.17	– 0.43	– 0.03	– 0.50	– 0.37	– 0.10	0.00	– 0.20	0.07
Mean	– 0.46			– 0.40			– 0.21			– 0.32			– 0.04		
Mean	– 0.29														

Table 8 Analysis of the highlighted gaps

Quality dimension	Acronym	Answers	Mean gap
Reliability	C2	The explanations provided in each of the questions of the I-GO Assistant tool must/are sufficiently explanatory to avoid interpretation errors	– 0.59
Tangibility	T3	The contact channels provided by the I-GO Assistant tool must/are accessible and easy to use for users	– 0.49
Reliability	C1	The I-GO Assistant tool must/does offer reliability and the ability to meet resource efficiency commitments	– 0.43
Assurance	S2	The I-GO Assistant tool must/does have a section for privacy policies and protection of data provided by users	– 0.38
Responsiveness	P2	The I-GO Assistant tool must/does have an effective problem resolution and internet-based response system	– 0.36
Empathy	E3	The I-GO Assistant tool seems/seems to be concerned about meeting the needs of SMEs	– 0.03

Similarly, the gap in question T3 is noteworthy, as it showed the second largest average gap among the 15 questions. This emphasizes the importance for the I-GO Assistant platform's organizers consider better communication channels to support users of this tool, which could potentially lead to greater quality perception to the tangibility dimension.

On the other hand, it's worth highlighting that question E3 from the empathy dimension exhibited the smallest gap, with a difference of around 0.03. In this context, it's notable that the I-GO Assistant platform seems to have a strong focus on meeting the needs of SMEs in the context of I5.0, confirming the main objective behind creating this platform is being perceived by its users.

Figure 8 compiles average expectations and perceptions for each of the analyzed quality dimensions. Overall, it can be observed that the reliability dimension (C) presents the largest gap, underscoring a greater need for attention from the platform developers regarding the requirements of this dimension to enhance user experience and foster greater user confidence. On the other hand, the tangibility (T), security (S), and responsiveness (P) dimensions exhibited average gaps, indicating a need

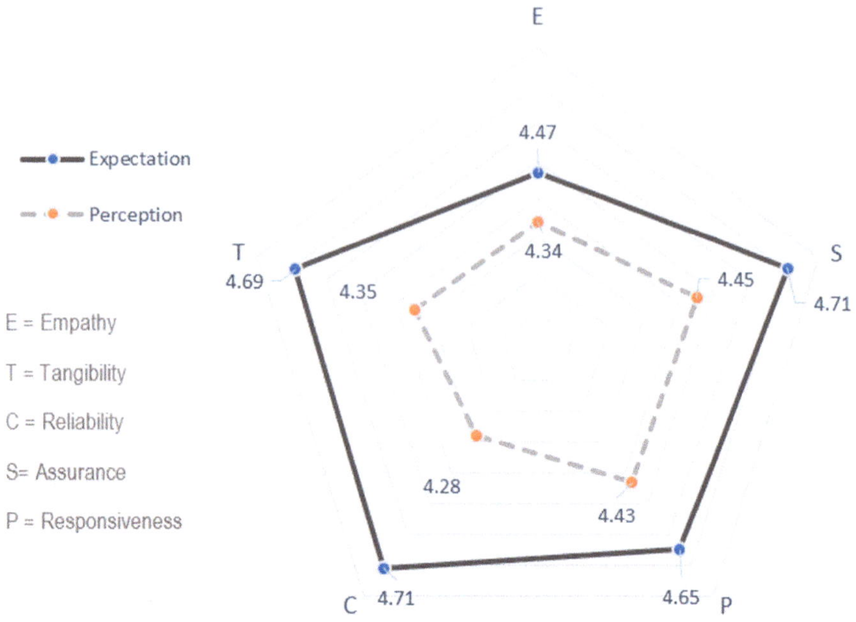

Fig. 8 Expectation and Perception quality per dimension. *Source* own authorship

for attention, albeit at a considerably lower level of urgency and necessity when compared to the requirements of the reliability dimension. The empathy dimension (E), conversely, showed a low average gap, suggesting that there are good practices within the platform that have been recognized by users.

4 Conclusions

This chapter offers actionable suggestions to enhance the training on the use and the performance of the I-GO Assistant tool. Notably, the study highlights knowledge management and strategic trade-off analysis as the most pertinent I5.0 features.

Employing the SERVQUAL model to measure gaps in the use of the I-GO Assistant reveals that users' expectations exceed their perceptions across the five dimensions of service quality (empathy, tangibility, reliability, assurance, responsiveness). Thus, the tool proves effective as a digital resource for training bachelor students on the topic. Although a discernible gap exists, it is relatively low, differing by a margin of 0.2733 (average of values from Tables 5, 6, and 7) or approximately 5.33%.

Future research could investigate the factors contributing to this gap, particularly the variability in results that led to negative perceptions. This analysis will facilitate the development of an action plan aimed at the continuous improvement of the tool. These improvements, once implemented in the I-GO Assistant tool and monitored through the reapplication of the SERVQUAL assessment, will serve as mechanisms for continuous monitoring and internal control.

One of the limitations of this study relates to the potential misinterpretation of questions and the subjective nature of responses in the questionnaire, which reflects each user's individual experience with the usability of digital tools. This is evident in Sect. 3.3, where each group of students, according to their respective courses, displayed different gap indicators influenced by their specific contexts. However, these limitations can serve as opportunities for improvement in the questionnaire design, ensuring that the questions are direct and clear for users.

Finally, the I-GO Assistant proves to be a valuable tool for mapping the current I5.0 features for SMEs. It generates meaningful recommendations based on maturity results in categories such as resource efficiency strategy, energy management, water conservation, material efficiency, sustainable chemicals, and waste management. Advancing quality management in the I5.0 context requires addressing emerging challenges and opportunities, such as incorporating social risks and impacts into the tool's maturity scale analysis.

Acknowledgements The authors are grateful to the "Fundação de Amparo à Pesquisa do Estado de São Paulo" (FAPESP), process number 2022/15134-1.

References

Bentancourt SMP (2015) Servqual como instrumento de gestão da qualidade da informação em ambiente EaD. (Dissertação de mestrado). Universidade Federal de Santa Catarina, Florianópolis

David LO, Nwulu NI, Aigbavboa CO, Adepoju OO (2022) Integrating fourth industrial revolution (4IR) technologies into the water, energy & food nexus for sustainable security: a bibliometric analysis. J Clean Prod 363:132522. https://doi.org/10.1016/j.jclepro.2022.132522

Delgado BC, Farrapo Júnior AC, Mergulhão RC, Selva DAL (2022) Assessment of the quality of digital services provided by an e-learning platform focused on sustainability issues. Production 32:1–20. Fap UNIFESP (SciELO). https://doi.org/10.1590/0103-6513.20210110

European Commission, Directorate-General for Research and Innovation, Industry 5.0: human-centric, sustainable and resilient, Publications Office, 2021. https://data.europa.eu/doi/https://doi.org/10.2777/073781

European Commission. Industry 5.0 Community of Practice Call for Expression of Interest. (2023). https://research-and-innovation.ec.europa.eu/news/all-research-and-innovation-news/industry-50-community-practice-call-expression-interest-2023-06-01_pt

Farrapo Júnior AC (2019) Avaliação da qualidade dos serviços de suporte em Tecnologia da Informação em uma Instituição Federal de Ensino Superior. Dissertação (Mestrado em Engenharia de Produção)—Universidade Federal de São Carlos, Sorocaba. Disponível em: https://repositorio.ufscar.br/handle/ufscar/11774

Gajanan PS (2016) A study of retail service gaps with specific reference to three major color television brands in Pune city. 280 f. Tese (Doutorado)—Course of Philosophy, Savitribai Phule Pune University, Pune

Hatem G, Zeidan J, Goossens M, Moreira C (2022) Normality testing methods and the importance of skewness and kurtosis in statistical analysis. BAU J Sci Technol 3(2):7

Hair JF (2009) Análise Multivariada de dados. Bookman Editora

Hajjar ST (2018) Statistical analysis: internal-consistency reliability and construct validity. Int J Quant Qual Res Meth 6(1):27–38

Hein-Pensel F, Winkler H, Brückner A, Wölke M, Jabs I, Mayan IJ, Kirschenbaum A, Friedrich J, Zinke-Wehlmann C (2023) Maturity assessment for Industry 5.0: a review of existing maturity models. J Manuf Syst 66:200–210, fev. 2023. Elsevier BV

Hussain S, Singh AM, Mohanty P, Gavinolla MR (2023) Next generation employability and career sustainability in the hospitality industry 5.0. Worldw Hosp Tour Themes 15(3):308–321. https://doi.org/10.1108/WHATT-01-2023-0011

Lee M, Yun JJ, Pyka A, Won D, Kodama F, Schiuma G, Park H, Jeon J, Park K, Jung K (2018) How to respond to the fourth industrial revolution, or the second information technology revolution? Dynamic new combinations between technology, market, and society through open innovation. J Open Innov Technol Mark Complex 4(3):21–41. Elsevier BV. https://doi.org/10.3390/joitmc4030021

Nayeri S, Sazvar Z, Heydari J (2023) Towards a responsive supply chain based on the industry 5.0 dimensions: a novel decision-making method. Expert Syst Appl 213:119267. Elsevier BV. https://doi.org/10.1016/j.eswa.2022.119267

Oliveira OJ, Ferreira EC (2008) Adaptação e aplicação da escala SERVQUAL na educação superior. Revista Gestão da Produção, Operações e Sistemas 3(3):133–146. Retrieved in 2021, September 21, from https://revista.feb.unesp.br/index.php/gepros/article/viewFile/473/193

Parasuraman A, Zeithaml VA, Berry LL (1985) A conceptual model of service quality and its implications for future research. J Marketing 49(4):41–50. Retrieved from https://doi.org/10.1177/002224298504900403

Parasuraman A, Zeithaml VA, Berry LL (1988) SERVQUAL: A multiple-item scale for measuring consumer perceptions of service quality. J Retailing 64:12–40

Pratiwi DI, Waluyo B (2023) Autonomous learning and the use of digital technologies in online English classrooms in higher education. Contemp Educ Technol 15(2):ep423

Saniuk S, Grabowska S, Straka M (2022) Identification of social and economic expectations: contextual reasons for the transformation process of industry 4.0 into the industry 5.0 concept. Sustainability 14(3):1391. MDPI AG. https://doi.org/10.3390/su14031391

Stylidis K, Wickman C, Söderberg R (2019) Perceived quality of products: a framework and attributes ranking method. J Eng Des 31(1):37–67. Informa UK Limited. https://doi.org/10.1080/09544828.2019.1669769

Sumi RS, Kabir G (2021) Satisfaction of e-learners with electronic learning service quality using the SERVQUAL model. J Open Innov Technol Mark Complex 7(4):227. Elsevier BV. https://doi.org/10.3390/joitmc7040227

Taber KS (2018) The use of Cronbach's alpha when developing and reporting research instruments in science education. Res Sci Educ 48:1273–1296

The Green Growth Knowledge Partnership (2022) GGKP impact report 2022. Global Green Growth Institute (GGGI), Organisation for Economic Co-operation and Development (OECD), United Nations Environment Programme (UNEP), United Nations Industrial Development Organization (UNIDO) and World Bank, 12 p

Tumsekcali E, Ayyildiz E, Taskin A (2021) Interval valued intuitionistic fuzzy AHP-WASPAS based public transportation service quality evaluation by a new extension of SERVQUAL Model: p-servqual 4.0. Expert Syst Appl 186:115757. Elsevier BV. https://doi.org/10.1016/j.eswa.2021.115757

Higher Education for Sustainable Development in South-East Asia: Problems, Values, Strategies and Practices

Viktor Zinchenko, Yurii Mielkov, Vasyl Levkulych, Mykola Popovych, Iryna Storozhuk, Oksana Bulvinska, and Viktoriia Levkulych

Abstract The chapter is dedicated to the philosophical consideration of the strategies of sustainable development and their implementation in the countries of the South-East Asia region. The authors discuss the major principles and contradictions of sustainable development, the latter being considered as the synergy of ecological, economic, and social vectors of the development of human civilization, and stress the role of universities in achieving such synergy. Special attention is given to sustainable development practices in HEIs in South-East Asia, as it is argued that the noted contradiction reveals itself there in the most clear way by the developing countries trying to achieve the ecologically destructive level of consumption.

Keywords Sustainable development · Higher education · Environmental education · Green university · Campus culture

V. Zinchenko (✉) · Y. Mielkov
Institute of Higher Education of the NAES of Ukraine, Kyiv, Ukraine
e-mail: vvzinchenko@ukr.net

V. Levkulych
Department of Philosophy, Uzhhorod National University, Uzhhorod, Ukraine

M. Popovych
Department of Law, Vocational, Social and Humanitarian Education Higher Educational Institution, «Podillia State University», Kamianets-Podilsky, Ukraine

I. Storozhuk
Interregional Academy of Personnel Management, Kyiv, Ukraine

O. Bulvinska
Boris Grinchenko Kyiv Metropolitan University, Kyiv, Ukraine

V. Levkulych
Department of Accounting and Audit, Uzhhorod National University, Uzhhorod, Ukraine

W. Leal Filho et al. (eds.), *Education for Sustainable Development: The Contribution of Universities*, World Sustainability Series,
https://doi.org/10.1007/978-3-031-86985-3_21

1 Introduction

The phenomenon of sustainable development is one of the most fundamental strategies for building the future of civilization in the twenty-first century, as a realistic response to existing global challenges aimed at ensuring the very preservation of humankind. While being the subject of discussions by scholars and of declarations by politicians for almost forty years already, for now it still presents only one possible direction for the further progress of human civilization out of the many other scenarios—arguably the most desirable one, but sadly not the very probable. The reason for that is a number of problems, difficulties, and contradictions related realizing the idea of sustainable development. In particular, philosophical and sociological analyses of the notion and phenomenon of sustainable development reveals that the major issue is combining social, economic and environmental vectors into one conception of development—without achieving the harmony of the three elements that constitute the idea of sustainability the latter would remain but an empty declaration or just a political slogan.

The region of South-East Asia presents an excellent object of study concerning both the contradictions and the values of sustainable—as well as the strategies for its realization in practices, especially those of higher education. We would argue that higher education serves as a key factor in the implementation of the guidelines of sustainable development, of their embodiment in social values shared and followed by future specialists, both in their professional activities and in everyday life.

One of the first investigations of the contemporary development of the education system in the region of South-East Asia was conducted by Ka-Ho Mok (2006) who has analyzed and outlined major trends and features of such development in China, Hong Kong, Taiwan, South Korea, Singapore, and Japan; while noting the influence of globalization on the social and economic welfare of the noted countries, that work does not focus its attention on sustainable development. On the other hand, one of the most fundamental research on higher education for sustainable development was conducted by Gough and Scott (2007) who considered the relationship between higher education and sustainable development and the ability of universities to promote sustainable development through their teaching, research and self-management; however, their vision and analysis was limited mostly to cases of Europe. At the same time, the book by Ashok Dansana (2013) demonstrates the importance of turning attention to the problems of following the trends of sustainable development for the universities of the countries in the Global South, on the example of the region of Western Odisha in India.

The aim of this chapter could thus be outlined in the following tasks. First, it would be desirable to conduct the general analysis of the problems peculiar to the concept of sustainable development using the methodology of philosophy; second, it is to provide an insight into higher education as the possible factor of realizing its guidelines; and third, to review the practices of such a realization that could be seen on examples of countries in the region of South-East Asia.

2 The Idea of Sustainable Development and Its Major Problems

Sustainable development in its essence is a paradigm or a scenario for the present and future development of human civilization that is made to be consistent, long-lasting, harmonious, and harmless, i.e., the one that does not threaten that very future, the one that is not realized at the expense of younger generations and the one that combines three separate vectors of development—economic, social and environmental (or ecological). Such a conceptualization is directly related to the ecological crisis, which arose as a result of anthropogenic factors, and the presence of which started to be realized by scholars in the 1970s, becoming one of the main global problems of our time.

The most convincing evidence of this realization was the famous first published report of the Club of Rome, an informal international organization founded in—"The Limits to Growth" (Meadows et al. 1972). The authors of this report, having analyzed the current trends in the development of human civilization, concluded that if the then-present growth trends in world population, industrialization, pollution, food production, and resource depletion were to continue unchanged, then the limits to growth on our planet would be reached sometime within the next one hundred years, resulting quite probably in a "sudden and uncontrollable decline in both population and industrial Yet, those trends could be altered by establishing "a condition of ecological and economic stability that is sustainable far into the future" (Meadows et al. 1972, p. 23–24).

It is based on this prediction that both groups of scientists and various intergovernmental commissions have developed the concept of *sustainable development* to designate new strategies for human civilization. Yet, the analysis of the documents actively adopted since the end of the 1980s demonstrates the presence of a certain contradiction inherent in the concept of sustainable development—first of all, it is the lack of fundamental consistency between its "economic" and "ecological" components. Even existing trends of establishing an environment-friendly circular economy are based on the idea of saving resources for the future development of the economy itself (Atstaja et al. 2022). The ecological vector of development is not even elaborated anywhere with the due conceptual clarity: nature itself is presented mostly as a 'resource base' and as a means for realizing the economic development of humankind, with the meaning and content of such a development being considered primarily from the point of view of the Western system of values, which does not always acknowledge the existence of other ethical systems. It should be noted that the authors of the UN declarations about the need to follow the guidelines of sustainable development are aware of the possibility that a significant part of humanity will turn away from Western social ideas in the not-very-unlikely case of their inefficiency in solving environmental problems, seeing their representatives who took part in international summits and conferences "as nothing more than sounding brass or tinkling cymbals" (United Nations 2002).

It is important to emphasize that the ecological dimension of determining strategies for the development of humanity in the twenty-first century is not something external, not a kind of 'limiting factor' imposed over the economic (and/or social) component of its development to outline some limits of its growth—besides, it is difficult to derive values and profound philosophical insights from political declarations. Likewise, we cannot solve humanity's environmental problems without addressing their social aspect. From many examples of attempts to solve this problematic situation in real practice, we can cite just one that relates to the region of South-East Asia. It is the testimony of the American ecologist of Lithuanian origin Barbara Goldoftas, the author of the book "The Green Tiger: The Costs of Ecological Decline in the Philippines" (Goldoftas 2006). After she went to the Philippines to study the forests there, she unwittingly became more and more involved in the problems of social policy and public morality, devoting her final work to studying the lives of people rather than those of trees: "Although trained as a plant ecologist, I had not intended to write about the environment, but I found myself increasingly disturbed by the destruction in the countryside and the disorder it brought to rural communities. As I reported on migration, overseas workers, the widespread prostitution, the spread of HIV, the militarization of the countryside, I realized that they were different parts of the same puzzle and could not be separated from the growing ecological crisis" (Goldoftas 2006, p. ix).

The situation in the Philippines—as well as in many countries of South-East Asia and the Third World in general—provides an excellent perspective on the problems inherent in the very concept of sustainable development. First, the present social and economic development of some parts of the world was made possible at the expense of the exploitation of nature and other, poorer countries. Second, as soon as those poorer countries—mostly, the countries of the said Third World—face globalization and strive to achieve the same level of life quality and especially consumption—the level that is being enjoyed by the more developed countries (and the one they willingly propagate in the media to the rest of the world), they threaten not only their nature (like in the Philippines where it turns out to be the *circulus vitiosus* when trying to provide for social and economic development leads to the destruction of forests, which in turn results in even more poverty), but the Earth itself.

In the 2020s, the perspective of humanity stopping itself from reaching "the limits of growth" looks even more pessimistic than in the 1970s (Zinchenko et al. 2020). The ideas of 'carrying capacity' and 'ecological footprint' present an avid example of how our social and economic development appears to contradict the ecological aspect of sustainability. For example, in 2023, the day of August 2 was designated as "Earth Overshoot Day": that means that humanity has exhausted nature's budget for the year just in seven months, so that "[f]or the rest of the year, we are maintaining our ecological deficit by drawing down local resource stocks and accumulating carbon dioxide in the atmosphere" (Global Footprint Network 2023). That proves that declarations of sustainable development are just not enough, and it could be argued that the current ecological crisis is a crisis of senses, values, and lifestyle no less (or even more so) than it is a crisis of industry and social demography (Mielkov 2020a). And to solve that crisis and to achieve actual sustainable development, more profound measures are needed.

We can mention here the opinion of Japanese historian Syuntaro Ito, who pointed out that the transition to the paradigm of sustainable development should generally constitute a whole global ecological or environmental revolution (Ito 1997). The same idea is being stressed by French philosopher André Gorz: the whole world-view of humans should be changed from growth to *degrowth*: "La décroissance est donc un impératif de survie. Mais elle suppose une autre économie, un autre style de vie, une autre civilisation, d'autres rapports sociaux…" (Gorz 2008, p. 29). Changing the lifestyle is clearly much more difficult than declaring the affinity to sustainable development, but it does present an actual way to achieve a harmonious integration of ecological and social factors that can lead to the formation of the desired future by establishing practices that can realize that future. And that's why from the whole spectrum of possible directions of philosophical analysis of the phenomenon of sustainable development we will turn to those specifically relating to strategies of higher education.

3 Approaches to Realizing the Ideas of Sustainable Development in Higher Education

The guidelines for sustainable development are currently expressed in the most complete way in the UN Resolution "Transforming our world: the 2030 Agenda for Sustainable Development" adopted in September 2015. This document is positioned as "a plan of action for people, planet and prosperity" with the aim of eliminating poverty in all its manifestations and dimensions, strengthening universal peace, healing the planet, and protecting it from destruction. The fourth of the declared seventeen goals is dedicated to education itself: "Ensure inclusive and equitable quality education and promote lifelong learning opportunities for all" (United Nations General Assembly 2015).

However, a more careful and in-depth analysis of these guidelines allows us to conclude that education, in particular higher education, is a factor that can and should ensure the fulfillment of almost all other goals, from achieving social justice and economic well-being for all people to ensuring a sustainable consumption and nature use—as it is education that can lay out the value foundations of appropriate behavior among graduates of higher education institutions, as well as to try to establish the corresponding practices.

In fact, as early as the beginning of the 1990s, Anthony Cortese noted that the role of higher education institutions in the implementation of the goal of sustainable development, although it is most often overlooked, is extremely important. It is a higher education institute that is said to bear a profound, moral responsibility to increase the awareness, knowledge, skills, and values needed to create a just and sustainable future: "Higher education often plays a critical but often overlooked role in making this vision a reality. It prepares most of the professionals who develop, lead, manage, teach, work in, and influence society's institutions" (Cortese 1992, p. 1112).

The problem of translating loud political declarations (and profound philosophical concepts as well, by the way) into the language of educational practices is, of course, quite—and it is a factor that obviously prevents us from paying due attention to the role of higher education institutions in the implementation of sustainable development guidelines. The very term "education for sustainable development" is easily perceived primarily as some attempts to strengthen just the environmental component of education, in the perspective of optimizing methods of managing nature as a "base of resources" for economic development. And that is, with all due respect to ecological education and "green economics", is just not enough to lay out the values of sustainable development in its full glory, where the environmentalist vector of development appears as a value and a goal in itself, not just as a means for achieving other goals.

Of course, we are not talking here about certain specialties only, whether in the field of management, public administration, or biological science, which are directly related to the specified topics so that it would be easy to outline a clear and consistent plan of actions in regard to achieving the practices of "greening education" for them. The problem is that higher education as a whole is to get that aspect and direction: as argued by the Indian researchers of educational innovations A. Kalam and S. Pal Singh, graduates of any specialty and major must have a working knowledge of sustainability (Kalam and Pal Singh 2001).

On the other hand, we should not understand such *knowledge of sustainability* as some external burden encumbering the educational process of various other disciplines with loads of information alien to them. For example, British authors Stephen Gough and William Scott, who devoted the noted fundamental book to considering the perspectives and paradoxes of higher education for sustainable development, demonstrate that ministries of education from different countries around the world are primarily concerned with issues such as literacy and the development of skills and competences, but tend to perceive all other goals as being imposed on education by external agents who try to push them by pursuing their interests, like promoting peace, sports, animal welfare, nation building, health care, etc. Within the scope of the noted approach, the issues of environment and sustainable development naturally fall into this list of wishes for overloading the educational process—and it is far less than obvious that in the eyes of teachers and HEIs administrators, those tasks will be able to take priority over the interests of all other "external agents of influence."

That's why, while reviewing strategies and practices of higher education for sustainable development, we should consider them not as external—either desirable or undesirable—burdens on higher education. Just as sustainable development itself tries to harmoniously combine the ecological vector with the economic and the social ones, in the same way higher education for sustainable development is not some new type of higher education or a separate direction of its reformation, but an expression of its very essence—i.e., in their value dimensions, strategies for the sustainable development of humanity coincide with the strategies of higher education itself. Or, as Stephen Gough and William Scott put it: "Our concern, therefore, is the proper place of sustainable development in what a university does, rather than the role of universities in implementing (any particular conception of) sustainable development" (Gough and Scott 2007, p. 15).

In other words, along with activities in the field of purely ecological (or environmentalist) enlightenment, the mission of universities is to form a developed personality, a person of a high level of culture (Ortega y Gasset 1966), who is able to generate new knowledge and act in a situation of uncertainty and complexity, when any ready-made instructions are missing or not working. That idea is in fact the classical ideal and understanding of the mission of the university, the one we return to at a new level of civilization development: the human personality appears in the center of attention as the main goal, and humanistic dimensions, as well as the human-centered approach (Semenets-Orlova et al. 2022), come to the fore when defining sustainable strategies for the development of higher education.

4 Practices of Higher Education for Sustainable Development in South-East Asia

The region of South-East Asia presents an excellent object of study concerning—as already noted—both the contradictions and the values of sustainable development, as well as their implementation into practices of higher education. Being the most populated region (accounting for more than 60% of the world population), it includes countries with different political systems, radically different levels of welfare, and two-thirds of the world's poor. At the same time, the greatest and most rapid successes in the last century were demonstrated by Hong Kong, Singapore, South Korea, and Taiwan—countries that, lacking minerals and natural resources, were literally forced to build themselves a 'knowledge economy' thanks to the development of higher education as a priority for national politics. And today, according to the conclusions of a team of international analysts, initiatives for sustainable development in this region, while generally coinciding with global trends, nevertheless demonstrate significant success in the aspect of higher education (Ryan et al. 2010).

At the same time, the rapid development of the region in the twentieth century had its costs and drawbacks. Emphasis on massification and professionalization, on engineering and technical specialties led to some difficulties in the twenty-first century, when the guidelines of sustainable development in particular and life under the conditions of a complex and unpredictable world in general started to require a revolutionary transition to human-centered, value-based and ecological education. On the other hand, sustainable development coincides with the trend of decolonization of knowledge, involving the awareness of the fundamental cultural diversity of the world: *ecological revolution* appears in some sense as a way to overcome Eurocentrism and a linear approach to human history, asserting the cultural diversity and humanistic values. It is not surprising then that precisely in the Third World countries there appeared some concepts that are distinguished by a humanistic and value-laden understanding of education and science—from the "pedagogy of the oppressed" by the great Brazilian Paulo Freire to the "humanistic economy" by the Indian Amartya Sen. At the same time, as a result of their colonial past, most of the countries of the region of South-East Asia have experienced more or less significant influence of the European or American education system, and therefore feature higher education that combines local centuries-old traditions with those peculiar to the European universities.

That especially applies to *China*, given its status as the first country in the world both in terms of population and economy. The problem of China is the difficulties that arise while carrying out the reforms in higher education. Since the 1980s, educational reforms in China have been moving somewhat slowly but surely in such directions as decentralization, diversification, and proliferation, which contrast sharply with the hypercentralization of the Mao age (Mielkov 2020b). That trend is aptly denoted by the term "centralized decentralization": on the one hand, HEIs (including the country's first private HEIs, "minban" (民劉), that started to appear since the 1980s) are given more and more autonomy today, and on the other hand, they are still subject to strict state-based regulation to ensure quality and accountability (Mok 2006).

China's global education development strategies are most fully outlined in the national plan titled《中国教育现代化2035》("Modernization of China's Education—2035") that was adopted in 2019. The ideas of the UN "Agenda for Sustainable Development" are implemented here while following the traditions of China's own culture and its approach to the development of world education, including education of ideals and beliefs, moral and aesthetic education, and the development of "smart campuses" (中共中央 2019). Still, as suggested by the results of a comparative study of students' perception of sustainable development in public and private universities in China, the "top-down approach" to address sustainability issues is inefficient as plans issued by the administration do not reflect students' wishes and perception—and opposite democratic approach based on decentralization is more accurate and reliable here (Wang et al. 2020). An example of such a "bottom-up" approach could be presented by Shandong University in Jinan which follows a way to realize the concept of a 'green university' focused on the principles of sustainable development—particularly, by practicing energy efficiency in campuses, from using solar panels up to establishing energy limits for undergraduate and graduate students, as well as for the academic staff (Mu et al. 2015).

According to such trends of efficient decentralization, Chinese experts bring to the fore the concept of *campus culture*, which is perceived as a complex phenomenon of different 'cultures', including material culture, institutional culture, and spiritual culture, thus presenting a sound way to solve the contradictions of sustainable development. As argued by Xi Shen, all those components of campus culture are created jointly by students and university staff, whose efforts are accumulated in the long-term practice of administering and running schools, starting from campus buildings and their environment and up to the ethics of academic integrity (Shen 2012).

In difference from mainland China, HEIs of *Hong Kong* present an avid example of an effective combination of Chinese traditions with the experience gained under British administration. It is no coincidence that the economy of this special administrative region is characterized in such synergistic terms as "social capitalism". Like in those countries of the Global North where social and economic problems are not as acute as in poorer regions, Hong Kong demonstrates quite a successful realization of the environmental component in educational practices, as manifested in particular by the activities of the Hong Kong Sustainable Campus Consortium. That organization founded in 2010 is a collective initiative of eight leading universities of the special administrative region. The declaration signed by their administrators asserts the importance of climate change and sustainable development strategies for the universities' fundamental mission of research, teaching, and the transfer of knowledge, as well as the moral obligation of the staff to provide students, as future leaders of society, with a full understanding of the nature of environmental, social and economic challenges. At the same time, the emphasis is placed on the ability of universities to act as models for their community and on their responsibility for public work in the relevant fields. Hence HEIs are committed to reassessing the lifestyle of their campuses, setting goals for reducing energy consumption, water and greenhouse gas emissions, and waste, and including relevant topics in curricula.

It should be stressed that in this case, the inclination towards sustainable practices is by no means limited to a formal declaration by the management: each of the universities participating in the Consortium provides detailed information on the measures that have been implemented in the corresponding areas, related to both the day-to-day life of the campuses and academic research. The annual report for the year 2021–2022 specifically mentions a reduction in electricity and resources consumption and the wider adoption of renewable energy sources on campuses; in particular, in 2021/22 Greenhouse Gas Emissions were reported to have fallen by 22.6 percent over the comparable 2015/16 figure, which demonstrates the success in pursuing the goal of carbon neutrality by all the member institutions (Hong Kong Sustainable Campus Consortium 2022, p. 12). As a kind of note, it could be added that the issue of introducing the guidelines of sustainable development into the educational process at the level of higher education in Hong Kong is rather not strong enough because the main attention there is paid to this aspect of education in primary and secondary schools, and not at the universities (Zinchenko et al. 2020).

In many ways, the experience of another 'Asian tiger'—*Singapore*—turns out to be similar to that of Hong Kong. A poor island country, the former long-time British colony has developed quite rapidly after gaining independence—albeit under a highly authoritarian rather than democratic government (and without subsequent unification with a powerful socialist state). Despite some differences in the social system, the main guidelines for reforming the education systems of Singapore and Hong Kong are found to be generally the same: market orientation, corporate governance, economic rationalism, and academic capitalism (Lee 2002).

On the one hand, the focus on the market economy has brought significant fruits to Singapore which features one of the best scientific, especially mathematical, higher education in the whole world. On the other hand, such a result was achieved at the expense of the lagging of the humanities in particular and integral education in general: Singapore demonstrates a shortage of humanitarian studies and faces the agenda to make a transition in higher education to one based on creative, critical thinking and holistic learning. Still, it is that practical orientation that has led, in particular, to South-East Asia's first major on sustainability having been launched at Singapore Management University in 2019 as a second major for students planning to receive a Bachelor of Business Management diploma.

It is also worth noting that the practices of Singapore HEIs relate not only to sustainable campuses, as in Hong Kong, but to more global issues like sustainable cities. In particular, the Centre for Sustainable Asian Cities, established at the College of Design and Engineering at the National University of Singapore, is engaged in research in the field of environmental urban design—from industrial ecology and modeling of transport networks for regions with high population density to the study of the impact of climate variability on energy consumption and improvement of the environment to ensure the quality of life of the elderly population and to climatic map studies (Wong et al. 2015).

An interesting and avid example of a similar design could also be found in Malaysia, which is usually not included in the number of the leading countries of South-East Asia and the 'tigers' of this region. Both in terms of the presence of some unsolved environmental problems (the ongoing cutting down of the age-old tropical forest in Sumatra and Kalimantan, along with the ongoing destruction of a large number of endemic animals and plants), and social tensions do not allow us to talk about the successful steps of this state on the way to sustainable development. Thus, only in 2002 was the system of racial quotas in the selection of applicants for higher education abolished—in favor of a more meritocratic approach; at the same time, in particular, the Malay researcher Muhammad Bakri Musa cites examples of condemnation of scientists by government officials for criticizing the government education system or for publishing research on the critical state of the environment in the country (Musa 2007).

However, some practices of individual local HEIs, even under such not very favorable conditions, can be interesting examples. We are talking here about Universiti Sains Malaysia (University of Science Malaysia), one of the oldest and most popular in the country—its main campus, which is located on the island of Penang, as early as 2006 has already demonstrated in its mission statement a strong commitment to the goals of sustainable development, including poverty alleviation and environmental improvement. This last guideline was conceptualized even earlier, in 2001–2002, in the form of the "University in the Garden" concept. This idea is intended to reflect the close relationship between the role and function of the university as an institution of higher education and nature as a component of the global ecological situation—and an invitation to value, preserve, and nurture the campus environment as part of the effort to create and maintain an intellectually conducive environment for the synergy of symbiotic coexistence.

At the time of its foundation in 1969, the Malaysian University of Science used to occupy one small building in the middle of a tropical mountain forest, on the territory of a former British military base. Despite the inevitable expansion that took place over the decades, the various structural units of the HEIs are still placed in mostly two- or three-storey buildings in gardens, which not only harmoniously combine age-old nature and human culture, but are also meant to symbolize knowledge and cultural heritage (including elements of the English garden left over from colonial rule), ethnic and linguistic diversity of the diverse population of Malaysia, as well as the ecological future of the country (Penerbit Universiti Sains Malaysia 2003).

In fact, it is this human-dimensional approach (which is, particularly, based on building and using small houses as opposed to multistoried urban skyscrapers) that well demonstrates the humanistic values of sustainable development. Even if the ecological crisis arose as a result of anthropogenic factors, i.e. due to humans over-reaching their 'limits of growth', it is still humanity alone that can propose different values, an alternative to the destructive consumption that threatens both nature and humanity itself. The way to achieve synergy of social, economic, and ecological vectors of development lies not just in promoting environmentalism, which could be effectively done rather in the countries already enjoying a high level of social and economic well-being, but by ecologically changing human lifestyle, like it was and is practiced in less developed countries.

5 Conclusions

We can thus summarize some ideas expressed in this chapter that was intended to provide a short insight into the major principles and contradictions of sustainable development as well as into some practices of its implementation into higher education in South-East Asia.

First, the sustainable development of society as a reference point for human civilization, despite all its problems and contradictions, is one of the most promising elements of the strategy for the development of human civilization in the twenty-first century, as a relatively realistic response to existing global challenges of "the limits of growth" and as a means of banal preservation of this civilization itself. The main challenge lies in achieving the synergy of the three vectors of development—social, economic, and ecological—so that the last, environmentalist vector of development appears as a value and a goal in itself, not just as a means for achieving the other two goals.

Second, to create and maintain an intellectually conducive environment for the synergetic and symbiotic coexistence of humans and nature, a change in lifestyle is required, made possible by the implementation of the guidelines of sustainable development into the practices of education and especially higher education. As in their value dimension, strategies for the sustainable development of humanity coincide with the strategies of higher education itself, coinciding with the topic of educating a cultural human personality with a broad world outlook, critical and creative thinking, and tolerance for other outlooks.

Third, the said response of sustainable development does sound different in the different regions of the world. The large population and inequalities in welfare mean that social and economic development for most countries in South-East Asia presents a challenge as great as that of environmental development. The paradox is that for the countries of the third world, in difference from the more developed societies, the vectors of social development, economic development—and ecological (environmentally safe) development often present rather alternatives, three opposite ways of life, than something that could be easily combined into one clear set of goals and strategies for their implementation.

It could be argued that while "the first world" faces the challenges of modifying its way of life on the grounds of *degrowth*, poorer countries strive to obtain the unobtainable, that is to achieve the ecologically destructive level of consumption enjoyed by the population of the Global North. Such a contradiction requires not only an 'ecological revolution', but also a revolution in education based on values of sustainable development. Even today, different countries in the region of South-East Asia present different effective practical examples of introducing the guidelines of sustainable development into higher education, and it is important to note that all of them at the same time try to follow those guidelines while still respecting local cultural traditions, which could well serve as an example for the rest of the world. The noted major contradiction could serve as a driving force for the development of strategies for implementing the ideas of sustainable development in higher education not only in South-East Asia but in all other countries based on the experience presented, which could present some prospects for future investigations on the topic.

References

Atstaja D, Koval V, Grasis J, Kalina I, Kryshtal H, Mikhno I (2022) Sharing model in circular economy towards rational use in sustainable production. Energies 15(3):939. https://doi.org/10.3390/en15030939

Cortese AD (1992) Educational for an environmentally sustainable future. Environ Sci Technol 26(6):1108–1114

Dansana A (2013) Higher education and sustainable development: new challenges and opportunities. Regal Publications

Global Footprint Network (2023) Earth overshoot day https://www.footprintnetwork.org/our-work/earth-overshoot-day/

Goldoftas B (2006) The green tiger: the costs of ecological decline in the Philippines. Oxford University Press

Gorz A (2008) Écologica. Galilée

Gough S, Scott W (2007) Higher education and sustainable development: paradox and possibility. Routledge

Hong Kong Sustainable Campus Consortium (2022) 2021–2022 Annual Report https://drive.google.com/file/d/1uA6xLYddqKDJkgBGRM2iVeLAeROdrNk1/view

Ito Sh (1997) A framework for comparative study of civilizations. Comp Civiliz Rev 36(36):3

Kalam APJ, Pal Singh S (2001) Target 3 Billion, PURA: innovative solutions towards sustainable development. Penguin

Lee MH (2002) A tale of two cities: comparing higher education policies and reforms in Hong Kong and Singapore. Aust J Educ 46(3):255–286

Meadows D, Meadows D, Randers J, Behrens W (1972) The limits to growth: a report for the Club of Rome's project on the predicament of mankind. Universe Books
Mielkov Y (2020a) From sustainable development to degrowth: philosophical and educational strategies for sustainability. Philos Educ 26(1):37–53. https://doi.org/10.31874/2309-1606-2020-26-1-2
Mielkov Y (2020b) Implementation of higher education for sustainable development in China. Virtus 45:105–108. http://virtus.conference-ukraine.com.ua/Journal45.pdf
Mok KH (2006) Education reform and education policy in East Asia. Routledge
Mu R, Zhan L, Liu P, Zuo J, Song Y, Ma R, Cao D, Yuan X (2015) Theory and practice of sustainability in higher education—from the perspective of Green University. Asia-Pacific Energy Equipment Engineering Research Conference (AP3ER 2015) https://www.semanticscholar.org/paper/Theory-and-Practice-of-Sustainability-in-Higher-%E2%80%93-Mu-Liu/df5d06560b2e83183e1b7eda3f62b50820ca40ff?p2df
Musa MB (2007) Towards a competitive Malaysia: development challenges in the twenty-first century. Strategic Information and Research Development Centre
Ortega y Gasset J (1966) Misión de la universidad. In: Obras completas. Revista de Occidente, T. IV, pp 143–242
Penerbit Universiti Sains Malaysia (2003) The University in a garden. https://www.usm.my/images/pdf/theuniversityinagarden.pdf
Ryan A, Tilbury D, Corcoran PB, Abe O, Nomura K (2010) Sustainability in higher education in the Asia-Pacific: developments, challenges, and prospects. Int J Sustain High Educ 11(2):106–119
Semenets-Orlova I, Shevchuk R, Plish B, Moshnin A, Chmyr Y, Poliuliakh R (2022) Human-centered approach in new development tendencies of value-oriented public administration: potential of education. Economic Affairs (New Delhi) 67(5):899–906. https://doi.org/10.46852/0424-2513.5.2022.25
Shen X (2012) Academic culture and campus culture of universities. High Educ Stud 2(2):61–65
The Central Committee of the Communist Party of China and the State Council issued "China Education Modernization 2035" (2019) http://www.xinhuanet.com/politics/2019-02/23/c_1124154392.htm
United Nations General Assembly (2015) Transforming our world: the 2030 Agenda for sustainable development. https://sdgs.un.org/2030agenda
United Nations (2002) The Johannesburg declaration on sustainable development. https://www.un.org/esa/sustdev/documents/Johannesburg%20Declaration.doc
Wang J, Yang M, Maresova P (2020) Sustainable development at higher education in china: a comparative study of students' perception in public and private universities. Sustainability 12(6):2158. https://doi.org/10.3390/su12062158
Wong NH, Jusuf SK, Katzchner L, Ng E (2015) Urban climatic map studies in Singapore. Routledge
Zinchenko V, Bazelyuk N, Boychenko M, Gorbunova L, Kurbatov S, Mielkov Yu, Shipko O (2020) Strategies of higher under internationalization for sustainable development of society (In Ukrainian). Institute of Higher Education on NAES of Ukraine https://ihed.org.ua/wp-content/uploads/2021/03/Strategiy_VO_v_umovah_internac_IVO-2020-199p_avtors-kolektiv.pdf

Perceptions of Periti on Sustainable Construction and Development in Malta

James Gabarretta and Mark Mifsud

Abstract The construction and building sectors are among the most significant economic sectors in Malta in terms of environmental, social and economic aspects. As world economies strive to become more sustainable, the importance of Sustainable Construction (SC) and Sustainable Development (SD) in the aforementioned economic sectors is ever more important. This study aims to address three primary research questions in the case of Malta—(i) How do Periti perceive SC and SD? (ii) What initiatives are currently in place to promote SC and SD in Malta's construction industry? (iii) How can Education for Sustainable Development (ESD) enhance sustainable practices within the sector? Using quantitative data from a broader study, this paper provides valuable insights into the prevailing attitudes of Periti and the potential for ESD to advance sustainability in these economic sectors. The findings show a general perception among Periti that Malta's construction sector is not developing sustainably. ESD was recognised as a critical driver for promoting SC and SD practices, with multiple opportunities for ESD initiatives being identified. Additionally, the study underscores the importance of a holistic approach that integrates ESD with broader industry efforts such as regulation, policy development and enforcement. A model proposed in the study's conclusion illustrates the collaborative roles of public, private, and educational institutions in promoting SD in Malta's construction sector. This study provides actionable insights to guide ESD programmes and policies aimed at enhancing sustainability within the industry.

J. Gabarretta (✉) · M. Mifsud
Centre for Environmental Education and Research (CEER), University of Malta, Msida, Malta
e-mail: james.gabarretta.15@um.edu.mt

M. Mifsud
e-mail: mark.c.mifsud@um.edu.mt

W. Leal Filho et al. (eds.), *Education for Sustainable Development: The Contribution of Universities*, World Sustainability Series,
https://doi.org/10.1007/978-3-031-86985-3_22

1 Introduction

Malta's economy, predominantly driven by the services industry, particularly tourism, financial services, and gaming, has experienced a 60% growth between 2013 and 2023 (World Bank 2024). This economic surge has been accompanied by a 25% increase in population during the same period, primarily due to inward migration (Briguglio 2024; NSO 2024)). Meeting the demands of this growing population has been the main driver behind the expansion of the building and construction sectors, which have had a significant impact on Malta's environmental resources, energy consumption, and social aspects. Although the technical dimensions of the Architecture, Engineering, and Construction (AEC) industry—particularly energy-related aspects—have been studied extensively, substantial gaps remain when examining the sector holistically.

Enhancing the quality of the built environment is a crucial aspect of Malta's journey in Sustainable Development (SD). Therefore, understanding the perspectives of the main stakeholders of this important industry, among them architects and civil engineers, known locally as Periti, is essential for embedding sustainable practices within the AEC, building and construction sectors.

The Centre for Environmental Education and Research (CEER) at the University of Malta is committed to advancing research on SD-related issues (Cassar 2022; Cassar and Caruana 2018; Mifsud 2012). As the importance of SD and ESD in the AEC, building and construction sectors becomes increasingly recognised, the CEER may play an important role in shaping a more sustainable version of these industries.

2 Research Aims

This study aims to explore Periti's perceptions of Sustainable Construction (SC) and Sustainable Development (SD) and investigate how Education for Sustainable Development (ESD) can further advance sustainability within the relevant economic sectors. The outcome of this study will serve to inform future efforts to address SD and SC aspects in ESD aimed at the construction industry.

3 Literature Review

3.1 Sustainable Development (SD)

The term "development", is a seemingly straightforward term that can be interpreted differently depending on the context it is used in (Hugé et al. 2013). Sumner and Tribe (2008) note that "development" can, inter alia, imply a long-term societal transformation process or be used as a concept within the dominant discourse of

Western modernity. Meanwhile, development studies have focused a lot of attention on poverty alleviation, particularly in the third world and in relation to education and agriculture (Fischer 2019; Gibson 2020).

The most widely cited definition of SD is that by the 'Our Common Future' report, which defines SD as "development that meets the needs of the present without compromising the ability of future generations to meet their own needs" (Brundtland 1987). On the other hand, the Development Planning Act of Malta (Chapter 552, Article 70) defines development as "the carrying out of building, engineering, quarrying, mining, or other operations for the construction, demolition, or alterations in, on, over, or under any land or the sea, the placing of advertisements or the making of any material change in the use of land, building, or sea" (Development Planning Act 2016).

These starkly different definitions of development have profoundly different implications on how SD is viewed, whether as that related primarily to building and construction or as a wider concept. In discussing SD action and policy, it is therefore crucial to apply a precise understanding to which type of development one is referring to.

Nevertheless, a global commitment towards SD has been shown by governments (European Commission 2020; Government of Malta 2017). The Sustainable Development Goals (SDGs) provide a framework for measuring and working towards achieving SD globally, but most countries, including Malta, are not on track to fulfilling their targets (Malekpour et al. 2023). Identifying what obstacles and barriers lie between the actual uptake of SD practices, is key to devising strategies and plans to overcome these (Lee 2015).

3.2 *Sustainable Construction (SC)*

Precisely defining SC is an equally challenging task. For example, Du Plesis (2002) provided various interpretations of how construction may be interpreted ranging from, inter alia (i). construction as the process of developing a property to (ii). the comprehensive life cycle of the building to; (iii). the economic sector of the construction industry and its related counterparts.

Within its broadest sense, the construction, property development and AEC sectors are responsible for a large share of global energy consumption and greenhouse gas emissions (UNEP 2022). In Malta, the construction sector is a significant economic contributor, yet it remains one of the largest sources of waste and environmental impact (ERA 2021).

Various specific aspects of SC have received attention in recent years. Locally one of the most studied aspects related to the AEC sector is energy efficiency. Gatt and Yousif (2020) discuss at length the various measures that should be aimed at improving the energy sustainability of buildings. Buhagiar (2006) illustrated how buildings can be designed for sustainable development in Malta, while Wismayer et al. (2019) discussed the importance of sustainable regeneration of built heritage through integrating energy needs; demands and the need for improving educational offerings in this area. Nevertheless, several other architectural aspects, particularly those that concern social and environmental principles from a holistic perspective require prioritisation in this field.

The use of sustainability rating systems for buildings, such as the LEED rating system and BREEAM has a limited uptake locally (Borg et al. 2020). This is not surprising considering the type of Maltese development projects which tend to be smaller in nature (KPMG 2023).

Beyond the Sustainable Development Goals (SDGs), which advocate for sustainability across all sectors, several key targets specifically focus on the construction industry. The EU 2030 Climate Targets, European Green Deal, and the New European Bauhaus initiatives are driving progress toward nature protection, decarbonization, and climate neutrality across the EU and its member states (Hainsch et al. 2022). These international initiatives are reshaping the future of sustainable construction and development in Malta and beyond.

3.3 *Education for Sustainable Development (ESD) for Periti*

Periti (collective term for Architects and Civil engineers in Malta) are key stakeholders in the building, construction and AEC sectors (Bonenberg and Kapliński 2018). Their training, perspectives, and professional practices have a direct influence on the adoption of SC practices (Buhagiar 2006; Caruso and Buhagiar 2021; Wismayer et al. 2019).

Among the several areas for improvement in the construction, building, property and AEC sectors of Malta, the Chamber of Architects and Civil Engineers (Kamra tal-Periti/KTP) also highlight the potential of educational offerings for Periti, to improve the quality of the Maltese construction sector (KTP 2007, 2020).

Much of the work on ESD in Malta has focussed on the formal learning environments, particularly within primary and secondary schools (Mifsud et al. 2018). Little to no ESD work has focussed on the educational side of the Periti profession. Defining what educational needs and sustainability aspects of what the profession requires for the future is therefore a core lacuna in literature. There is a clear need, internationally, to modernise education on sustainability within the AEC sector, but doing so requires overcoming several obstacles (Bonenberg and Kapliński 2018; Borg et al. 2020; Caruso and Buhagiar 2021).

4 Methodology

4.1 Research Design

This paper forms part of a broader research project that employed an explanatory sequential mixed-methods approach. The aim of this research is to investigate the perspectives of Maltese Periti on SC and SD in Malta, and employed a questionnaire ($n = 102$) followed by in-depth interviews ($n = 7$) to explore these views comprehensively. The research was guided by three key questions: (i) How do Periti perceive SC and SD in Malta? (ii) What are their views on the current efforts to promote SC and SD within the sector? (iii) How can Education for Sustainable Development (ESD) contribute to advancing SC and SD within the profession? This paper presents the quantitative findings obtained through the questionnaire, which served both as standalone insights and as foundational data for the subsequent qualitative phase of the research.

4.2 Data Collection

The questionnaire was distributed online and was self-administered. In total, 31 questions were presented, split into 4 sections (i). Socio-Demographics (ii). SD (iii). SC (iv). ESD. A mx of Semantic Differential, Likert Scale, Multiple Choice, Tick Box and Open-ended questions were used. Prior to launching to the target audience, the questionnaire was piloted with two subjects to review it and changes were made. The questionnaire was distributed via Social Media, direct communication with the subjects and through personal networks of the researchers.

A target sample size of 288 was calculated using Cochran's Equation and adjusted using the Finite Population Correction Factor (FPC). The calculated sample size was based on a confidence Level of 95%, Proportion of 0.5 and Margin of Error of 5%. Despite all efforts to meet the target sample size, only 102 respondents completed the questionnaire. This resulted in 8.9% of the total population of Periti registered at the time of the study (1143). Provided that this study was carried out in a relatively new terrain for ESD research, this was deemed to be acceptable.

4.3 Data Analysis

The data collected through the questionnaire data was analysed quantitatively in RStudio (version 2023.03.1+446). Descriptive and summary statistics were computed using the psych package while graphs were plotted using Tidyverse (GGPlot2). The data was tested for normality using the Shapiro-Wilkinson test, whereby it resulted that none of the 81 variables of the questionnaire (that arose from the different sets of answer options) were above the threshold p-value of 0.05, thereby rejecting the hypothesis that the data is normally distributed. As a result, non-parametric tests were used for all possible combinations of the variables (n = 3240) using the Pearson's Chi-Square test, revealing 258 statistically significant correlations. Most of these combinations did not convey any useful findings. However, combinations between variables which were deemed to carry insightful findings but were not statistically significant were equally considered in the questionnaire data analysis.

5 Results

5.1 Socio-demographics

Out of the 102 questionnaire respondents, 70.6% were male and 29.4% were female. The most common age bracket was 1981–1990 (n = 40), followed by 1971–1980 (n = 23) then 1991–2000 (n = 22). The most common specialisations of Periti were (i). Architecture (n = 74) (ii). Structural engineering (n = 45) and (iii). Building Conservation (n = 34). Most respondents work on a Full-Time basis (56.9%); followed by Full-Time self-employed (35.5%).

5.2 Sustainable Construction (SC)

The three most common chosen understandings of SC were (i). Renewable energy and energy efficiency in buildings (n = 95) (ii). Building designs that minimise construction waste (n = 91); (iii). Buildings that respect nature and biodiversity (n = 80). The most common level was 4 (n = 48) when Periti were asked to rate themselves on a scale of 1 (Not Knowledgeable) to 5 (Highly Knowledgeable), indicating a significant level of knowledge. Periti also leaned towards placing a high importance on SC with a mean score of 3.9 and median of 4 when asked to rate between 1 (Not important) and 5 (Very Important) the importance of SC in their work ethos. Periti marked a significant level of difficulty in promoting SC principles with their clients with a mean of 3.95 and median of 4 on a scale of 1 (Very easy) to 5 (Very difficult). Periti also remarked that SC projects incur higher upfront costs but save costs in the long run with a mean score of 4.22 and median of 4 when asked to rate from 1 (Completely Disagree) and 5 (Completely Agree). The role of the Energy Performance Certification (EPC) process in advancing SC practices was ranked relatively low with the mean being 2.45 and median 2 on the scale of 1 (No Role) to 5 (Extremely important Role). The stakeholder which most Periti chose as having a deciding role as to whether a sustainable approach is adopted in a construction project or not was assigned to the Client/Developer with 69.6%

of respondents choosing this option followed by the Government authority selected only by 13.7%.

Periti were asked to rank the current level of sustainability in the construction industry with 2 (Almost unsustainable) being the most chosen for the Economic and Social aspect, while the Environmental aspect received 1 (Completely Unsustainable) as the most popular choice. Respondents showed low scores for the question "Is enough being done in the industry to promote SC", the worst performer was seen as the government with a mean score of 1.7, followed by the private sector (mean = 1.9) and finally Civil Society (mean 2.7). The three most chosen perceived barriers to SC practices were "Low interest in SC projects by clients" (n = 72), "Cost of SC projects" (n = 55) and "Lack of education on SC for the general public" (n = 64). The most popular ways of achieving SC were "Financial incentives for SC projects" (n = 73) followed by "Standards for sustainability in construction projects" (n = 69) and "Training for developers on SC" (n = 43).

5.3 *Sustainable Development (SD)*

Respondents were also queried about the concept of SD whereby they were queried about various aspects of the topic. The first question dealt with their level of agreement with the Brundtland (WCED) report definition of SD and one can see a high level of agreement with this definition with the a mean score of 4.31 and medium score of 4 on a scale of 1 (Does not represent my understanding) to 5 (Fully represents my understanding). When asked whether SD and SC mean the same thing to them a more neutral and mixed set of results are observed with a mean score of 2.69 and median score of 3 (1—Completely Disagree; 5—Completely Agree). Asked whether respondents often refer to the SDGs for guidance in their work the results are fairly spread out however the most popular choice is the 1 (Completely Disagree). The mean score of 2.54 and median of 2.5 are observed for this question. Asked to rate to what degree respondents agree with the statement that "Malta is developing sustainably as a country", the results lean towards 1(Completely disagree), with this being the most selected option, and a mean score of 1.51 and median of 1.79.

5.4 *Education for Sustainable Development (ESD)*

The final part of the questionnaire dealt with ESD. When asked about whether they think that the mandatory University of Malta courses prepare Periti for real-world sustainability challenges, a low score can be observed with the mean being 2.31 and median 2 on a scale of 1—Do not prepare at all to 5- Fully prepare. Most respondents (59.8%) said that they did not take any courses related to SC after they became warranted, while 40.2% did. Periti were also asked to rate the extent of availability of local training courses on SC and SD and a clear bias towards a lower score is observed with the mean score being 2.15 and the median 2 (1—Inexistent to 5—Saturated). Respondents rated a relatively high perceived potential positive effect in offering training on SC to developers (mean = 4.13; median 4) (1—No improvement; 5—Complete improvement). This contrasted with the significantly lower perceived potential positive effect of offering training on SC to Periti (mean = 3.42; median = 3). The three most frequently selected area of improvement that Periti chose when asked which areas they would welcome more training on were—"Using innovative building materials with smaller environmental footprints" (n = 89); "Reducing energy and water demand of my designs" (n = 68) and "Improving the well-being of people using and affected by my designs" (n = 47).

Question	1	5	Mean	SD	Median	Min	Max
SC knowledge	Not knowledgeable	Highly knowledgeable	3.49	0.75	4	2	5
SC importance	Not important at all	Very important	3.87	0.92	4	1	5
Difficulty promoting SC	Very easy	Very difficult	3.95	0.89	4	2	5
SC costs more but saves in the long run	Completely disagree	Completely agree	4.23	0.70	4	2	5
Role of EPC in promoting SC	No role	Extremely important role	2.45	1.20	2	1	5
Social sustainability	Completely unsustainable	Completely sustainable	1.89	0.80	2	1	4
Environmental sustainability	Completely unsustainable	Completely sustainable	1.73	0.78	2	1	4
Economic sustainability	Completely unsustainable	Completely sustainable	2.46	0.99	2	1	5
Civil society—enough to promote SC	Completely disagree	Completely agree	2.76	1.17	3	1	5
Government—enough to promote SC	Completely disagree	Completely agree	1.72	0.91	1.5	1	5
Private—enough to promote SC	Completely disagree	Completely agree	1.95	0.93	2	1	5
Definition SD (Brundtland)	Does not represent my understanding	Fully represents my understanding	4.31	0.77	4	2	5
SD/SC mean same thing?	Completely disagree	Completely agree	2.7	1.13	3	1	5
Refer to SDGs in work?	Completely disagree	Completely agree	2.55	1.30	2.5	1	5
Malta developing sustainably?	Completely disagree	Completely agree	1.52	0.79	1	1	5

(continued)

(continued)

Question	1	5	Mean	SD	Median	Min	Max
Education prepares Periti for real-world challenges?	Do not prepare at all	Fully prepare	2.31	1.07	2	1	5
Availability of education on SC locally?	Inexistent	Saturated	2.16	0.79	2	1	5
Impact of education (Periti)	No improvement at all	Complete improvement	3.42	1.02	3	1	5
Impact of education (Developers)	No improvement at all	Complete improvement	4.14	0.80	4	2	5

6 Discussion

The findings of this study reveal several significantly important perceptions of Periti regarding SC, SD and ESD. The results also suggest a gap between the theoretical understanding of the concepts and their practical implementation locally, but also globally (Lee 2015). The transition from knowledge to practice may be hindered by several barriers that this study identified.

6.1 Barriers to Uptake of SD and SC Practices

6.1.1 Financial Constraints

The high upfront costs associated with sustainable building materials and technologies and SC practices are seen as a significant obstacle by Periti and their clients. Although sustainable buildings may offer long-term savings, the results show that this is the top barrier to SC uptake. On the other hand, KPMG (2023) estimated that around half of property buyers are willing to pay more for sustainable properties (e.g. those that include solar panels). Notwithstanding this, the past years have seen the ballooning of property prices and therefore additional initial investment costs that increase the sustainability of certain aspects of buildings (e.g. energy use) may reduce the marketability of sustainable incentives.

6.1.2 Regulatory and Political Challenges

Many respondents highlighted the lack of stringent regulatory enforcement that infuse sustainability aspects in construction projects. Current policies, while present, are perceived as being poorly implemented. The EPC process was seen as being very ineffective in promoting SC practices, and has also been studied previously (Mallia and Prizeman 2018). Similarly the lack of local building standards, especially for the core sustainability aspects leaves the door relatively open to developers to get by with the bare minimum. Introducing mandatory standards, also called for by the KTP (2020), is likely to increase the quality of development projects while, if properly designed and enforced.

6.1.3 Educational Gaps

Many Periti expressed that their initial training and continuous professional development do not adequately cover the practical aspects of sustainability, limiting their ability to advocate for and implement SC principles effectively. The questionnaire results also show a very large potential for educating developers on aspects related to sustainability in the AEC and property market. The style of education would need to follow good practices, and be specifically geared towards the specific stakeholders that the efforts are targeting.

6.2 Opportunities for Improvement

While several initiatives are already in place to promote SD and SC, such the Energy Performance for Building Regulation which mandates the use of Energy Performance Certificates (EPCs) for all new properties and those which are being sold, the results of the questionnaire show that Periti do not perceive these as being sufficient. Periti expressed dissatisfaction with the current state of regulation and policies. The construction sector's reliance on traditional methods and the resistance to change further complicate efforts to adopt sustainable practices. While the questionnaire shows various barriers, it also suggests that there are several opportunities to advance SD and SC in Malta.

6.2.1 Financial Incentives

Introducing tax breaks, grants, or subsidies for sustainable building projects could encourage more clients to invest in SC. This has been shown as an effective solution for renewable energy and energy efficiency in buildings (Gatt and Yousif 2020); but could be mainstreamed into several other fields of SC.

6.2.2 Regulatory Strengthening

Introducing stricter sustainability codes and standards which can complement existing efforts, will play a crucial part in the broader adoption of SC practices. KTP (2020) had made similar recommendations in their report. The enforcement of existing and new regulations will pose a significant challenge with vast anecdotal evidence suggesting that more effort is needed in this area.

6.3 The Role of Education for Sustainable Development

ESD is recognised as a core component in bridging the gap between knowledge and uptake of sustainability and SD principles (Mifsud 2012). The results of the questionnaire show that most Periti believe their formal education does not adequately prepare Periti for real world sustainability challenges. The results also show a perceived lack of Continuous Professional Development (CPD) training opportunities that focus on sustainability aspects. The initiatives mentioned previously (i.e. financial incentives and regulatory strengthening) would work best if coupled with efforts for ESD.

ESD for SC and SD in the building and construction sectors may be developed accordingly for various stakeholders, such as Periti, developers and contractors. The recent introduction of the Periti Act mandates that Periti undergo regular CPD training, which presents a promising opportunity to improve the profession's capacity in implementing sustainable decisions (Buhagiar 2006; Caruso and Buhagiar 2021; Wismayer et al. 2019).

7 Way Forward

The model shown below summarises the findings of this research, by providing a holistic way forward to improve the existing situation vis-à-vis sustainability within the construction, property, building and AEC sector. Action needs to be synchronised across multiple fronts to meet the demands of a growing population, the need to develop more sustainably and our various local and international commitments.

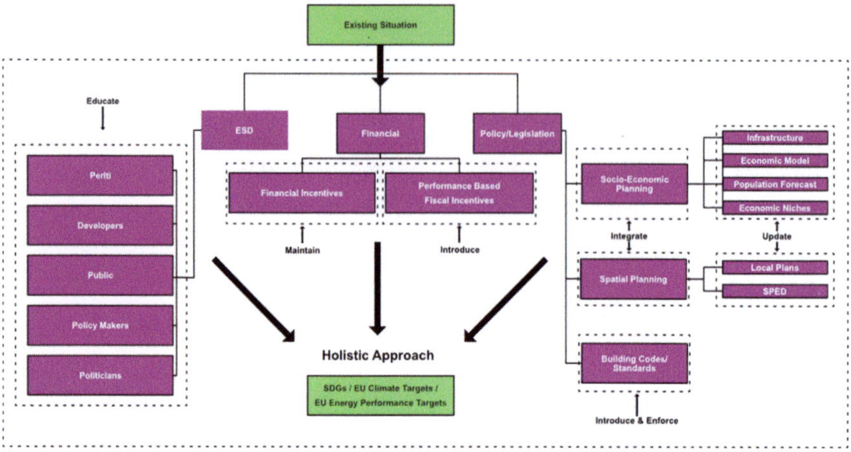

8 Limitations and Future Prospects

This study focused primarily on SC and SD in building projects, with less attention given to areas like civil engineering (e.g. transport infrastructure). As the researcher was an outsider to the profession and sector, the researcher faced challenges, including extensive background research and establishing rapport with professionals. The complex and politically sensitive nature of the research, along with limited literature, posed interpretive challenges, but efforts were made to maintain objectivity and minimize bias. Significant gaps in research were identified in the Periti profession, and a virtual absence was found on the property developers side. In studying the building and construction sectors, a more comprehensive perspective should be taken by taking into account other key stakeholders such as property developers. Further research on the educational offerings within Malta's building, construction and AEC sector is recommended, focusing on formal programs and CPD offered by institutions like the University of Malta, Building and Construction Authority and The Malta College of Arts, Science and Technology. Future research should also assess the impact of introducing building standards and codes, fiscal incentives, and spatial planning policies.

9 Conclusion

This study highlights the importance of aligning the perceptions of Periti with practical measures to promote SC and SD within the AEC and property sectors of Malta in Malta. While awareness of SD and SC is relatively high, the actual uptake of sustainable practices in the building and construction sector remains limited due to financial, regulatory, and educational barriers. By adopting a more integrated approach that involves policy setting, educational improvements and further research, Malta can make significant progress toward achieving its sustainability goals in the building and construction sectors.

References

Bonenberg W, Kapliński O (2018) The architect and the paradigms of sustainable development: a review of dilemmas. Sustainability 10(1):100. https://doi.org/10.3390/su10010100

Borg R, Dalli Gonzi R, Borg S (2020) Building sustainably: a pilot study on the project manager's contribution in delivering sustainable construction projects—a Maltese and international perspective. Sustainability 12(23):10162. https://doi.org/10.3390/su122310162

Briguglio L (2024) The sustainable development implications of major economic changes in Malta between 2012 and 2021. Xjenza Online 12(Special Issue):19–30. https://doi.org/10.7423/XJENZA.2024.1.02

Brundtland GH (1987) Our common future: report of the world commission on environment and development. Geneva, UN-Dokument A/42/427. http://www.un-documents.net/ocf-ov.htm

Buhagiar V (2006) Sustainable development and building design in Malta. University of Malta MSD, Faculty of Architecture and Civil Engineering, p 10

Caruso L, Buhagiar V (2021) Educating and training for a new European Bauhaus: a proposed renovation wave for students and professional AEC curricula

Cassar C (2022) Designing curriculum about governance and sustainability in higher education: a case study. In: Grima S, Özen E, Romānova I (eds) Contemporary studies in economic and financial analysis. Emerald Publishing Limited, pp 1–35. https://doi.org/10.1108/S1569-37592022000108A030

Cassar C, Caruana V (2018) Empowering youths as lifelong learners through education for sustainable development: a critique from the perspective of Maltese youths. In: Leal Filho W, Mifsud M, Pace P (eds) Handbook of lifelong learning for sustainable development. Springer International Publishing, pp 87–107. https://doi.org/10.1007/978-3-319-63534-7_7

Development Planning Act, Pub. L. No. Development Planning Act, 552 CAP (2016)

Du Plessis C (2002) Agenda 21 for sustainable construction in developing countries, published for CIB and UNEP by CSIR building and construction technology, Pretoria

ERA (2021) Construction and demolition waste strategy for Malta (No. 9789918952212)

European Commission (2020) Delivering on the UN's sustainable development goals—a comprehensive approach 53(9):1689–1699

Fischer AM (2019) Bringing development back into development studies. Dev Chang 50(2):426–444. https://doi.org/10.1111/dech.12490

Gatt D, Yousif C (2020) Policy measures addressing nearly zero-energy buildings in the small island state of Malta. In: Byron J, Moncada S, Veenendaal W (eds) Handbook of governance in small states, 1st edn. Routledge, pp 117–140. https://doi.org/10.4324/9780429061356-9

Gibson M (ed) (2020) Chapter 6—The beginnings of modern development theory. In: Food and society. Academic Press, pp 153–163. https://doi.org/10.1016/B978-0-12-811808-5.00006-4

Government of Malta (2017) Malta's sustainable development vision for 2050, pp 23–23

Hainsch K, Löffler K, Burandt T, Auer H, Crespo Del Granado P, Pisciella P, Zwickl-Bernhard S (2022) Energy transition scenarios: what policies, societal attitudes, and technology developments will realize the EU Green Deal? Energy 239:122067. https://doi.org/10.1016/j.energy.2021.122067

Hugé J, Waas T, Dahdouh-Guebas F, Koedam N, Block T (2013) A discourse-analytical perspective on sustainability assessment: interpreting sustainable development in practice. Sustain Sci 8(2):187–198. https://doi.org/10.1007/s11625-012-0184-2

KPMG (2023) KPMG—MDA construction industry and property market report 2023 (Construction industry and property market report). MDA. KPMG—MDA Construction industry and property market report 2023

KTP (2007) The urban challenge

KTP (2020) A modern building and construction framework for Malta. https://kamratalperiti.org/wp-content/uploads/A-MODERN-BUILDING-CONSTRUCTION-FRAMEWORK-FOR-MALTA.pdf

Lee K-H (2015) Drivers and barriers to energy efficiency management for sustainable development: organizational drivers and barriers for energy efficiency. Sustain Dev 23(1):16–25. https://doi.org/10.1002/sd.1567

Malekpour S, Allen C, Sagar A, Scholz I, Persson Å, Miranda JJ, Bennich T, Dube OP, Kanie N, Madise N, Shackell N, Montoya JC, Pan J, Hathie I, Bobylev SN, Agard J, Al-Ghanim K (2023) What scientists need to do to accelerate progress on the SDGs. Nature 621(7978):250–254. https://doi.org/10.1038/d41586-023-02808-x

Mallia MR, Prizeman O (2018) Energy performance certification is the software currently used in Malta suitable for the energy assessment of its historic buildings? Energy Effic Hist Build

Mifsud M (2012) Environmental education development in Malta: a contextual study of the events that have shaped the development of environmental education in Malta. J Teach Educ Sustain 14(1):52–66. https://doi.org/10.2478/v10099-012-0004-x

Mifsud M, Cassar C, Cordina L, Grixti B, Spiteri L, Zahra A (2018) Engaging families in sustainable development through sustainability walks. In: Leal Filho W, Mifsud M, Pace P (eds) Handbook of lifelong learning for sustainable development. Springer International Publishing, pp 307–320. https://doi.org/10.1007/978-3-319-63534-7_21

NSO (2024) World population day: 11 July 2024. National Statistics Office

Sumner A, Tribe M (2008) What could development studies be? Develop Pract 18(6):755–766. JSTOR

UNEP (2022) 2022 global status report for buildings and construction: towards a zero-emission, efficient and resilient buildings and construction sector. (Nos. 978-92-807-3984-8). www.globalabc.org

Wismayer A, Hayles CS, McCullen N (2019) The role of education in the sustainable regeneration of built heritage: a case study of Malta. Sustainability 11(9):2563. https://doi.org/10.3390/su11092563

World Bank (2024) GDP (current LCU)—Malta (World Bank National Accounts Data, and OECD National Accounts Data Files.) [Data]. https://data.worldbank.org/indicator/NY.GDP.MKTP.CN?end=2023&locations=MT&start=2013

Enacting Sustainability at the University—Staffs' Hope and Motivation in Relation to Their Actions

Ilona Södervik, Heidi Hyytinen, Rami Ratvio, Tuija Veintie, Janne Salovaara, Janna Pietikäinen, Milja Heikkinen, and Tarja Tuononen

Abstract The aim of this study is to investigate university staff's sustainability-related hope and motivation, their participation in sustainability actions, and their perceptions of sustainability debates, exploring how these factors are interrelated. Data were collected in Spring 2023 through an online survey of 413 University of Helsinki staff members. The survey incorporated the Hope Concerning Climate

---

I. Södervik (✉) · R. Ratvio
Faculty of Educational Sciences, Centre for University Teaching and Learning and Helsinki Institute of Sustainability Science, University of Helsinki, Helsinki, Finland
e-mail: ilona.sodervik@helsinki.fi

R. Ratvio
e-mail: rami.ratvio@helsinki.fi

H. Hyytinen
Philosophical Faculty, School of Educational Sciences and Psychology, University of Eastern Finland, Joensuu, Finland
e-mail: heidi.hyytinen@uef.fi

T. Veintie · J. Pietikäinen · M. Heikkinen
Faculty of Agriculture and Forestry, Department of Forest Sciences, Institute of Sustainability Science, University of Helsinki, Helsinki, Finland
e-mail: tuija.veintie@helsinki.fi

J. Pietikäinen
e-mail: janna.pietikainen@helsinki.fi

M. Heikkinen
e-mail: milja.e.heikkinen@helsinki.fi

J. Salovaara
Institute for Atmospheric and Earth Systems Research and Helsinki Institute of Sustainability Science, University of Helsinki, Helsinki, Finland
e-mail: janne.salovaara@helsinki.fi

T. Tuononen
Faculty of Educational Sciences, Centre for University Teaching and Learning, University of Helsinki, Helsinki, Finland
e-mail: tarja.tuononen@helsinki.fi

W. Leal Filho et al. (eds.), *Education for Sustainable Development: The Contribution of Universities*, World Sustainability Series,
https://doi.org/10.1007/978-3-031-86985-3_23

Change Scale, and the Multidimensional Work Motivation Scale adapted for the sustainability context. Participants were also asked about their involvement in sustainability actions and their views on the adequacy of sustainability debates at the university. The results revealed positive correlations between hopefulness and intrinsic motivation, while denial was related to the dimensions of external motivation. Notably, denial was more prevalent among those with lower participation in sustainability actions. A significant portion of participants (63%) felt that sustainability issues are insufficiently discussed, and this perception was linked to their level of hopefulness. Those who answered that there is too much talk about sustainability had a higher score in denial than the others. Despite the university's strategic emphasis on sustainability, there is considerable variation among staff in terms of their sustainability-related hope, motivation, and actions. The possible reasons together with consequences and implications of the results are discussed.

Keywords Sustainability hope · Sustainability motivation · Sustainability action · University staff · Higher education

1 Introduction

Universities are engines of social change and train the experts of the future (Filho et al. 2024). This role places a moral obligation on universities to promote and model sustainable working practices in their own activities (Shephard 2015). Globally, Higher Education Institutions (HEIs) are mandated to integrate sustainability across their fundamental operations and daily functions. In Europe, the European Union's European Strategy for Universities underscores the pivotal role of universities in the dual transition towards sustainability (European Commission 2022).

Despite the well-established scientific consensus on anthropogenic (human-induced) sustainability threats like climate change, and the fact that many people are worried about such threats, there remains a troubling paradox, namely that necessary advancements in concrete mitigation and adaptation acts are still inadequate, and the human impact is sometimes altogether questioned (Johnson et al. 2017; Pidgeon 2012). In this article, sustainability refers to the wellbeing and existence of human communities and other species within the Earth's biosphere capacity and includes ecological, social, cultural and economic sustainability (Folke et al. 2016). Previous research addresses the effects of sustainability-related hope and hopelessness on young people's motivation and attitudes towards their pro-sustainable behaviour, focusing on climate change (Ojala 2012). However, studies focusing on university staff are lacking. It could be hypothesized, based on previous research in other fields, that university staff members' sustainability-related hope and motivation have an impact on their pro-sustainable behaviour or vice versa. For example, the conceptions on teaching and learning that academics have influence their pedagogical decisions (cf. Hyytinen and Löfström 2017; Tuononen et al. 2023).

Finnish universities occupy a unique organizational position, particularly because their employees, especially academic staff, have a considerable amount of autonomy in shaping their work practices (Universities Act 558/2009). Such autonomy can be deduced to significantly influence the nature and quality of research, teaching and other activities conducted at universities. Therefore, it is important to understand factors that influence all university staff members' sustainability engagement and participation in concrete actions to create more sustainable working practices.

Thus, the aims of this article are to inspect: first, the sustainability-related hope and motivation among university staff members and explore how the factors are interrelated. Second, to evaluate the staff's participation in sustainability actions and examine the relationship between their hope, motivation, and level of participation and lastly, to analyse the staff's perceptions of sustainability debates and explore how these perceptions are connected to their hope and motivation.

2 Theoretical Framework

2.1 Sustainability-Related Hope

In the field of sustainability education, many researchers define hope based on Snyder's hope theory (Snyder 2000). This theory emphasizes the will- and waypower to achieve a desired outcome or avoid negative consequences, and agency thinking, which involves the effort to implement these methods and actively pursue goals (Snyder 2000; Li and Monroe 2018). Therefore, hope emerges when a positive objective is perceived as attainable (Snyder et al. 2002). Hope goes hand in hand with critical consciousness, an urge to transform the world through action based on critical reflection on the current reality and the existing socio-environmental problems (Freire 1968/2005). When the goal is a change to more sustainable behaviour, knowledge and skills alone are not enough, and a more holistic approach is needed (Cantell et al. 2019). Climate education, particularly, has emphasized climate-related emotions, such as how to face climate anxiety, and particularly, the creation of hope (Ojala 2012, 2015). Without hope, people can become incapable of action and stop acting altogether. Maintaining hope is crucial for both the success of sustainability efforts, and for people's emotional well-being. Education has drawn on Ojala's (2012) model of dividing perceptions of hope into non-constructive and constructive hope. Constructive hope arises from the combination of action and emotional processing (Pihkala 2017). Whereas, mere wishful thinking, for example, that someone else will solve the problems, is not constructive.

Ojala created the "Hope Concerning Climate Change" scale to examine the relatedness between hope and environmental engagement among Swedish high-school students (both junior and senior) and young adults (Ojala 2012, 2015). The 2015 version of the scale measured hope with 12 items of different sources of hope. Items were classified into the following subcategories: positive re-appraisal, trust-others, trust-self and denial. *Positive re-appraisal* means strategies where concerns about environmental issues are viewed from a perspective that fosters hope. *Trust-others* encompasses the views where hope is directed towards various social actors, such as researchers and environmental organizations, whereas *trust-self* means trust in one's ability to positively influence environmental problems (Ojala 2012). The last category *denial* includes views where, for example, the negative impacts of climate change or climate change as a phenomenon are entirely denied. The first three subcategories belonged to the "constructive hope" category, while denial was the sole member of the "hope based on denial" category (Ojala 2015). Studies show that constructive hope correlates with pro-environmental behaviour and vice versa when it comes to hope based on denial (Ojala 2012, 2015).

2.2 Sustainability-Related Motivation

Self-determination theory (SDT) identifies three basic psychological needs that explain human motivation, namely autonomy (i.e., sense of freedom and ownership in one's actions), competence (i.e., the experience and feeling of mastery), and relatedness (i.e., the need to feel connected and belongingness with others (Deci and Ryan 2000, 2020). There are individual and social contextual differences in how these three basic needs are optimally supported (Deci and Ryan 2000). SDT distinguished qualitatively different types of motivation (Trépanier et al. 2023). First, *amotivation* refers to lack of motivation towards a task or activity. Second, *intrinsic motivation* refers to doing something because it is interesting and enjoyable. It also includes the subtype of *internal effort*, which means a willingness to put effort into something. Third, *extrinsic motivation* relates to doing something for external reasons, such as receiving rewards and approval, avoiding criticism, enhancing one's self-esteem, or achieving a personally valued goal (Gagné et al. 2015). Furthermore, two subtypes of extrinsic motivation are identified as *external regulation* and *introjected regulation.* External regulation refers to doing something to obtain rewards or avoid punishments. Rewards and punishments can be material (i.e., money) and social (i.e., to get approval from others). Introjected regulation, in turn, involves regulating behaviour due to internally pressuring forces, such as ego-involvement, shame, and guilt. SDT-based work motivation scales have been used and validated in different domains and contexts (Ferner et al. 2008; Gagné et al. 2015; Tremblay et al. 2009), and The Multidimensional Work Motivation Scale (MWMS) (Gagné et al. 2015) has become a widely used instrument to measure work motivation (Trépanier et al. 2023).

Previous research has shown that intrinsic motivation and the autonomous forms of extrinsic motivation are related to engagement and positive outcomes in varied educational and cultural contexts (Ryan and Deci 2020). Additionally, both types of motivation can be facilitated by supporting basic psychological needs. Taken together, prior research has shown that SDT plays a crucial role in behaviour (Ryan and Deci 2020; Nota et al. 2011).

2.3 Sustainability-Related Hope and Motivation in Relation to Sustainability Actions

Understanding the factors influencing people's behaviour is a complex question. Previous studies of reasons have proved that, for example, an increased amount of knowledge is not typically enough to make peoples' daily practices more sustainable (Kollmuss and Agyeman 2002). In certain studies, hope has been considered crucial for fostering environmental engagement (Lueck 2007). Empirical studies from the health field have shown that individuals with high levels of hope tend to absorb more information about how to protect themselves from illnesses and use this information more proactively compared to those with lower levels of hope (Snyder et al. 2002). Conflicting results do exist, however, as hope alone, without a feeling of worry, was found to relate negatively to pro-sustainable behaviour (Ojala 2007, see also Vandaele and Stålhammar 2022). Instead, hope was positively related to behaviour, if one was also highly worried of sustainability issues (Ojala 2007). Thus, it is essential to develop a more detailed understanding of how hope influences action. In addition, the way one assesses their own performance plays a key role in guiding behaviour. SDT theory emphasizes the idea that one's belief in one's ability to influence the status quo is necessary to induce action (Deci and Ryan 2000, 2020). If a person believes that they do not have the capability to complete the task, they are unlikely to even attempt to perform it (Usher and Schunk 2018).

Thus, it is presumed that factors beyond the knowledge level, such as motivation, constructive hope, together with worry—a sense that the issue demands more attention—play a significant role in influencing pro-sustainable actions among university staff, but this relation has not been studied before. Our research questions, based on above-described premises, are presented in the following section.

2.4 Aims

1. What are the university staff's levels of sustainability-related hope and motivation and how are these interrelated?

1.1 What kind of differences are there between academic and administrative staff?

2. How actively do staff participate in sustainability actions at university and how are their hope and motivation related to this?
3. What are the staff's perceptions on adequacy of sustainability debates at university and how are these perceptions related to their hope and motivation?

3 Methods and Materials

3.1 Participants

Altogether 413 University of Helsinki staff members participated in this study by filling in an online questionnaire. 54% of them were academic staff, meaning teaching and research staff, and 46% were administrative and other staff (e.g. staff providing support for students and academics). The median age of the participants was 45, ranging from 20 to 76 years. The majority (71%) of the participants self-identified as woman, 24% as man, 1% as other, and 4% did not want to indicate their gender.

The participants represented about 5% of the total staff of more than 8,000 employees at the University of Helsinki. The distribution of the study participants by staff category and age was very similar to the distribution of the whole University's staff, among which 58% were teaching and research staff and the average age was 41 years.

The study was performed according to the ethical instructions of The Finnish National Board on Research Integrity (TENK). Each participant gave their information-based consent before answering the questionnaire.

3.2 Data Collection

The study applied an online survey methodology to collect data from the staff members of University of Helsinki. Completing the questionnaire took about 15 min. We invited staff members to participate through diverse channels, including the university intra-net, on-campus information screens, as well as e-mails, to reach staff in different sectors and faculties. The data were collected during a four-week period between March and April 2023.

3.3 Instruments

To answer the research questions, the following research instruments were used.

The survey used the *Hope Concerning Climate Change Scale* (12 items; Ojala 2012, 2015), which aimed to measure four dimensions of relatedness between hope and environmental engagement: (1) positive re-appraisal, meaning strategies where concerns about environmental issues are viewed from a perspective that fosters hope, (2) trust-others, which encompassed the views where hope is directed towards various social actors, such as researchers and environmental organizations and (3) trust-self, meaning trust in one's ability to positively influence environmental problems and (4) denial, which included views where, for example, the negative impacts of climate change or climate change as a whole are denied (Ojala 2012). The first three dimensions belonged to the "constructive hope" category, while denial was the sole member of the "hope based on denial" category (Ojala 2015). This survey instrument has been tested and reported in prior studies (Ojala 2012, 2015; Li and Monroe 2018). A five-point scale (1 = not at all, 2 = little, 3 = moderately, 4 = quite a lot, 5 = very much) was used to measure the items.

The Multidimensional Work Motivation Scale (19 items; Gagné et al. 2015) was used to measure motivation to invest in sustainability issues at work. The original instrument has been tested and validated in prior studies (Gagné et al. 2015; Trépanier et al. 2023). It includes scales of autonomous motivation, introjected and external regulation, and amotivation (Gagné et al. 2015). In the present study, the items of Multidimensional Work Motivation Scale were modified to the sustainability context. Employees were asked: How well do the following statements describe the reasons why you invest in or would like to invest in sustainability issues in your work at the University of Helsinki? A five-point scale (1 = not at all, 2 = little, 3 = somewhat, 4 = a lot, 5 = very much) was used to measure the items.

In addition, staff were asked about their participation in sustainability actions and the adequacy of sustainability debates at Helsinki University. The staff's sustainability actions at the university were asked in ten yes/no questions (*In the last 6 months, which of the following have you done in relation to your job or workplace?*). In addition, to investigate the staff's opinion whether the issue of sustainability demands more attention, the adequacy of sustainability debates was asked with this question: *How much do you think sustainability is discussed in the university community?* It was measured on a five-point scale (1 = completely too little, 2 = slightly too little, 3 = appropriate, 4 = slightly too much, 5 = completely too much).

All survey items can be found in Appendix 1.

3.4 Data Analysis

To explore how the measures worked in the new context, the explorative factor analysis (EFA) was conducted. Two sets of items (12 items measuring sustainability-related hope, 19 items measuring sustainability-related motivation) were separately subjected to an EFA. Principal Axis Factoring was used for extraction and Varimax for rotation. The Kaiser-Meyer Olkin measure of sampling adequacy suggested that the data were considered acceptable to conduct factor analysis: sustainability-related Hope (KMO = 0.829; Bartlett's test < 0.001) and sustainability-related motivation (KMO = 0.853; Bartlett's test < 0.001). A three-factor solution was identified for sustainability-related Hope. However, two items were deleted from the subsequent analysis because of low communalities (i.e., item 8.10 0.274 and item 8.11 0.247). In addition, one factor was found to be problematic in our study: only one item (8.13) loaded onto this factor. Because the use of a one-item scale has been reported as problematic (Allen et al. 2022), this scale was removed from the analysis. Finally, only two factors for sustainability-related Hope were used and labelled as: hopeful (six items), denial (three items). Cronbach's Alphas were all above 0.70, which can be considered acceptable (Tavakol and Dennick 2011). For the sustainability-related Motivation, EFA yielded five factors solution, namely Mot_F1: Internal effort, Mot_F2: External regulation, social, Mot_F3: Introjected motivation, Mot_F4: Intrinsic motivation, and Mot_F5: External motivation, material. Next, we examined communalities of the items and factor leadings. Cronbach's Alphas were all above 0.70, which can be considered acceptable (Tavakol and Dennick 2011). The scales and items used in the present study, their factor loadings from EFA and their Cronbach's alphas are presented in Appendix 2.

The following statistical comparisons were conducted. First, sustainability-related hope and motivation between the staff groups (academics and administrative staff) were compared using independent t-tests. Pearson's correlations between sustainability hope and motivation were inspected. To answer the second research question, the staff's sustainability actions at the university were asked by means of ten yes/no questions, from which a sum was calculated for each participant. Staff were divided into three groups as follows: (1) low activity (0–3 sustainability action types), (2) medium activity (4–6 sustainability action types), (3) high activity (7–10 sustainability action types). These three categories were used in comparing staff's sustainability-related hope and motivation using One-way ANOVA. Tukey's test was used to conduct the post-hoc comparisons. To answer the third research question, we created three groups by combining answers 1 and 2 (too little), 3 (appropriately) and 4 and 5 (too much). We then compared these three groups in sustainability-related hope and motivation by using one-way ANOVA.

4 Results

4.1 Sustainability-Related Hope and Motivation of University Staff

Our first aim addressed sustainability-related hope and motivation among university staff. The results show that, in general, university staff were relatively hopeful ($M = 3.73$; $SD = 0.76$) and only very little denial existed ($M = 1.25$; $SD = 0.47$) (Table 1). Additionally, concerning sustainability-related motivation, internal effort received the highest mean score, whereas the dimension of external motivation, focusing on material, received the lowest score.

We also examined whether there are differences between academics and administrative staff (Table 2). The results showed that the groups differed only in one dimension, namely External motivation (material) Mot 5 ($t(411) = 3.347$, $p < 0.001$) academics having more external motivation compared to administrative staff.

Table 1 Sustainability-related hope and motivation of university staff

	N	M	SD	Min	Max
Hopeful	413	3.73	0.76	1.00	5.00
Denial	413	1.25	0.47	1.00	3.67
Internal effort	413	4.55	0.63	1.00	5.00
External regulation, social	413	2.03	0.89	1.00	5.00
Introjected motivation	413	2.93	1.14	1.00	5.00
Intrinsic motivation	413	3.58	1.03	1.00	5.00
External motivation, material	413	1.65	0.83	1.00	5.00

Positive correlations were found between sustainability hopefulness and internal effort, intrinsic motivation and external regulation (social). Correspondingly, denial correlated nega

tively with internal effort and intrinsic motivation (Table 3). Denial correlated positively with external regulation (social) and external motivation (material).

4.2 *Participation in Sustainability Actions at University*

Our second aim was to investigate staff's participation in sustainability actions at university and how their sustainability-related hope and motivation differ depending on their activity level. In general, the staff participated on average in 4.62/10.00 sustainability actions (SD: 2.08; Min: 1.00; Max: 10.00) during the last six months.

Significant differences regarding sustainability-related hope and motivation were found between the groups (see Table 4). Denial was higher among the group of low activity ($F(2) = 6.957$; $p < 0.001$; $\eta^2 = 0.033$). The dimensions of internal effort, intrinsic motivation and external motivation (material) were higher among the of high activity group ($F(2) = 25.325$; $p < 0.001$, $\eta^2 = 0.112$; $F(2) = 22.676$; $p < 0.001$; $\eta^2 = 0.101$; $F(2) = 8.395$, $p < 0.001$; $\eta^2 = 0.040$).

4.3 *Staff's Perceptions on Adequacy of Sustainability Debates at University and Their Relations to Sustainability-Related Hope and Motivation*

Our third aim was to explore participants' perceptions of how much sustainability matters are discussed in the university community. Most of the participants (63%, $n = 262$) perceived that sustainability matters are discussed too little, 31% ($n = 128$) appropriately, and 6% ($n = 23$) reported that there is too much discussion on sustainability. We then explored how these different groups of participants

Table 2 Comparison of sustainability-related hope and motivation between teaching and research staff and other staff

	N		M		SD		Min		Max	
	Academic	Administrative	Academic	Administrative	Academic	Administrative	Academic	Administrative	Academic	Administrative
Hopeful	223	190	3.67	3.80	0.78	0.74	1.00	1.00	5.00	5.00
Denial	223	190	1.23	1.28	0.44	0.50	1.00	1.00	3.33	3.67
Internal effort	223	190	4.57	4.53	0.66	0.60	1.00	1.00	5.00	5.00
External regulation, social	223	190	2.01	2.05	0.90	0.88	1.00	1.00	5.00	4.50
Introjected motivation	223	190	2.96	2.90	1.17	1.10	1.00	1.00	5.00	5.00
Intrinsic motivation	223	190	3.55	3.62	1.08	0.97	1.00	1.00	5.00	5.00
External motivation, material	223	190	1.78	*1.51	0.83	0.80	1.00	1.00	5.00	4.67

* $t(411) = 3.347, p < 0.001$

Table 3 Correlations between sustainability hope and motivation

	Denial	Internal effort	External regulation, social	Introjected motivation	Intrinsic motivation	External motivation, material
Hopeful	– 0.058	0.275**	0.197**	0.070	0.455**	0.072
Denial		– 0.330**	0.150**	– 0.95	– 0.117*	0.127**
Internal effort			0.026	0.378**	0.588**	– 0.037
External regulation, social				0.365**	0.230**	0.502**
Introjected motivation					0.387**	0.190**
Intrinsic motivation						0.195**
External motivation, material						

perceived sustainability-related hope and motivation. The results showed that the groups differed in both dimensions of hope (see Table 5). Post hoc tests revealed that those who perceived that there is too much discussion of sustainability scored lower on hopeful than two other groups. Regarding the denial factor, the third group (Too much discussion) scored the highest and differed from the other groups. In addition, the first group (Too little discussion) scored the lowest denial factor, and they also differed from the second group (Appropriately discussion). In other words, participants' perceptions of how much sustainability matters are discussed was related to how hopeful they are regarding sustainability.

The results revealed that these groups also differed in three of the motivation factors, namely, internal motivation, introjected motivation and intrinsic motivation. Group 3 (Too much discussion) differed in internal motivation from groups 1 and 2 and group 1 from group 2. In other words, those participants who perceived that there is too little discussion about sustainability at university scored higher on internal motivation than those who perceived that there is too much or appropriate discussion. Regarding the introjected motivation and intrinsic motivation, group 3 differed from group 1. Thus, perceptions of too little discussion were related to higher scores on both motivation types (i.e., introjected and intrinsic motivation).

Table 4 Sustainability-related hope and motivation in relation to sustainability actions

	N			M			SD			Min			Max		
Activity level scale	Low	Medium	High	Low	Medium	High	Low	Medium	High	Low	Medium	High	Low	Medium	High
Hopeful	135	191	79	3.63	3.76	3.87	0.77	0.76	0.74	1.00	1.00	1.83	5.00	5.00	5.00
Denial	135	191	79	1.38	1.20	1.18	0.57	0.43	0.36	1.00	1.00	1.00	3.33	3.67	2.33
Internal effort	135	191	79	4.28	4.66	4.81	0.82	0.44	0.35	1.00	2.67	3.33	5.00	5.00	5.00
External regulation, social	135	191	79	2.02	1.98	2.14	0.92	0.84	0.91	1.00	1.00	1.00	5.00	4.75	4.50
Introjected motivation	135	191	79	2.86	2.93	3.10	1.16	1.07	1.23	1.00	1.00	1.00	5.00	5.00	5.00
Intrinsic motivation	135	191	79	3.19	3.69	4.08	1.11	0.90	0.90	1.00	1.00	1.00	5.00	5.00	5.00
External motivation material	135	191	79	1.64	1.53	1.98	0.89	0.69	0.95	1.00	1.00	1.00	5.00	5.00	4.67

Table 5 Sustainability-related hope and motivation differences between the groups of adequacies of sustainable debates

Sustainability-related hope and motivation scales	Group 1: Too little discussion (n = 262)		Group 2: Appropriately (n = 128)		Group 3: Too much discussion (n = 23)		F	p	Eta-squared
	Mean	SD	Mean	SD	Mean	SD			
Hopeful	3.73	0.78	3.80	0.67	3.28	0.92	4.72	$p < 0.05$	$\eta^2 = 0.02$
Denial	1.17	0.36	1.35	0.56	1.65	0.73	16.05	$p < 0.001$	$\eta^2 = 0.07$
Internal effort	4.68	0.41	4.39	0.74	3.90	1.23	24.29	$p < 0.001$	$\eta^2 = 0.11$
External regulation, social	2.00	0.90	2.08	0.85	1.99	1.01	0.38	$p = 0.684$	$\eta^2 = 0.002$
Introjected motivation	3.13	1.10	2.61	1.09	2.38	1.30	12.55	$p < 0.001$	$\eta^2 = 0.06$
Intrinsic motivation	3.70	0.97	3.45	1.03	2.94	1.37	7.45	$p < 0.001$	$\eta^2 = 0.04$
External motivation, material	1.62	0.81	1.68	0.81	1.87	1.08	1.05	$p = 0.351$	$\eta^2 = 0.005$

5 Discussion and Conclusions

The purpose of this study was to investigate: (1) university staff members' sustainability-related hope and motivation and explore how the factors are interrelated (2) to evaluate the staff's participation in sustainability actions and examine the relationship between their hope, motivation, and level of participation and (3) to analyse the staff's perceptions of sustainability debates and explore how these perceptions are connected to their hope and motivation.

The results showed that in general, the hopefulness of staff members was higher compared to denial. Furthermore, the results also showed positive correlations between hopefulness and internal effort and intrinsic motivation. Surprisingly, positive correlation was also found between hopefulness and external regulation, which might indicate that the collective efforts of the social community may be seen as important factors in maintaining hopefulness. On the contrary, denial was negatively correlated with internal effort and intrinsic motivation, while it showed a positive correlation with two dimensions of external motivation.

In comparing academic and administrative staff, the major difference was found in academics, who exhibited a higher level of external motivation. This could be attributed, for example, to the fact that many current and highly competitive funding calls require explicit guidelines on incorporating sustainability into research plans—exemplifying the typologies of Deci and Ryan (2000, 2020). While there is a clear benefit in asking academics to consider sustainability in their teaching and research, it can also result in a somewhat superficial, merely external motivation, to address it.

Differences in sustainability actions were also found, and denial was higher in the lower level of participation group. This may mean that those who are skeptical about sustainability issues are not willing to engage in sustainability actions because they do not consider them to be necessary. They may also doubt the willingness of others to commit to sustainability efforts. The motivation of people from privileged groups to engage with social justice issues tends to be related to empathy. This helps to connect with and care for others, with moral and spiritual values, as well as with matters concerning self-interest or experiencing a sense that social justice is also personally beneficial (Goodman 2000). Acting sustainably sometimes requires giving up some of the customary practices without knowing if others are willing to make similar sacrifices, and this is thought to affect people's willingness to act (Jagers et al. 2019). Furthermore, pro-sustainable behaviour often requires deviating from the norm, even rebellion (Racimo et al. 2022). This can be challenging in a large and rather traditional organization like a university (Galperin 2012). Thus, examples of practices that engage and provide positive examples that inspire and build confidence in the actions of other (peer) individuals should be highlighted.

However, there is also reason to believe that sustainability-related hope and motivation can be an effect from participating in different activities. Engaging in collective sustainability action may combat feelings of anxiety and helplessness and foster feelings of hope (Schwartz et al. 2023). As elaborated by, for example, Pihkala (2017), constructive hope also arises from action—not only the other way around of having hope which leads to action. Collective action may also strengthen community connection and social support that supports well-being (Lorimer et al. 2020). As suggested by, for example, Cantell et al. (2019), creating more sustainable behaviour requires a holistic approach that scopes beyond knowledge and skills to the dimensions of hope, motivation, participation and such addressed here.

Altogether, participants' perceptions on how much sustainability matters are addressed by and at the university showed a correlation to the hopefulness they have concerning sustainability. A clear majority of the staff felt that sustainability issues were not addressed at the university at a sufficiently profound level—a level that would reflect the importance of the issues and commitment to addressing them. Thus, the level of attention given to the topics can be perceived to suggest a lack of support from the university, perhaps assessed as universities disregarding their role and responsibility to engage with sustainability issues (Shepard 2015; Pidgeon 2012). Although also contrary to the majority, the minority group that perceived there to be too much discussion about sustainability scored lowest on hope and highest on denial. This could exhibit, for example, general frustration among the staff members on the advancement of sustainability issues, which can at times seem rhetorical rather than concrete.

5.1 Limitations and Future Research

In the data collection, we had a limited number of respondents, which is a typical challenge particularly for web and mail surveys and is linked to a trend where the general willingness to respond to surveys has declined, men being even less likely to respond than women (Becker 2017). We assume that personnel who are motivated and active concerning sustainability issues are more likely to respond to the survey than those with less interest, and this may introduce some bias into the dataset. This limits the generalizability of findings. Additionally, the response rates varied between faculties and sectors within the multi-faculty university, and this limited our opportunities to compare responses between different groups of participants, for example, based on the faculty in which they work. Thus, comparisons were made between two broad groups, academics and administrative staff, as the participants include a good number of people representing both groups in the same proportion as each is represented among the entire personnel of the university. In future, it would be important to explore disciplinary differences in more detail. Another limitation is that the measures we used did not load on the factors as expected; for example, the Hope questionnaire only formed two, rather than four, factors in our sample. Nevertheless, according to Ojala (2012, 2015), two broader dimensions can be distinguished in the measure: constructive hope and denial. Similarly, in the MWMS measure, which was adapted to a sustainability context for this study purpose, two factors from the original instrument were combined. As the hope measure focuses on climate change, the measures somewhat lean towards the ecological dimension of sustainability. The

MWMS measure, in its adapted form, relates to sustainability in general, which in practice means the respondents have answered based on different sustainability perceptions, some focusing only on ecological aspects, while others have considered it to be a holistic term covering several aspects, or something in between (Veintie et al. 2024). New measures that capture the holistic nature of the phenomenon are needed in the future.

5.2 Conclusions

This study is to our knowledge the first to examine the sustainability-related hope and motivation of university staff in relation to their engagement in sustainability actions and perceptions of adequacy discussions at university. Data was collected in one university in Finland. Although the modest number of respondents limits the generalizability of the findings, our results clearly highlight the fact that despite the university's strategic emphasis on sustainability, there is considerable variation among staff in terms of their sustainability-related hope, motivation, and actions. We argue that fostering pro-sustainable behaviour and sustainability action at universities necessitates a holistic approach that extends beyond mere knowledge and skills to include dimensions such as hope, motivation, and participation. To advance sustainability efforts at the university, it is also essential to establish structures (such as HR policies) that incentivize and mandate sustainable actions, fostering collective action at a concrete level. Additionally, supporting a culture of action and providing collective opportunities will encourage participation, even from those who might not act independently. Universities have the potential to act as hubs for interdisciplinary collaboration, promoting a comprehensive grasp of sustainability challenges. We call on universities to encourage and develop constructive hope at different levels, such as in education and working practices (see also Vandaele and Stålhammar 2022). Overall, universities have an exceptional opportunity to contribute to a more sustainable future for society, and we should make the most of this opportunity.

Appendix 1: Reported Survey Items

1. **How well do the following statements describe the reasons why you put effort or would like to put effort into sustainability issues in your work at the University of Helsinki? (1 = not at all, 2 = a little, 3 = somewhat, 4 = a lot, 5 = very much)**

    1.1 Because I have to prove to myself that I can.
    1.2 Because things related to sustainability are interesting in my work.
    1.3 Because others will respect me more (e.g. supervisor, colleagues, students, family, customers…).

1.4 Because it gives me pleasure to learn about sustainability in my work.
1.5 To avoid being criticised by others (e.g. supervisor, colleagues, students, family, customers...).
1.6 I don't put effort into sustainability in my work because it feels like I'm wasting my time.
1.7 Because putting effort into sustainability aligns with my personal values.
1.8 Because I can get research funding/project funding/salary increase only if I put effort into sustainability in my work.
1.9 To get the approval of others (e.g. supervisor, colleagues, students, family, customers...)
1.10 Because putting effort into sustainability has personal significance to me.
1.11 I do little because I think things related to sustainability are not worth putting effort into in my work.
1.12 Because I am afraid that I will lose my job if I don't put effort into sustainability.
1.13 Because otherwise I will feel ashamed of myself.
1.14 Because I personally consider it important to put effort into sustainability in my work.
1.15 Because others offer me greater job security if I put effort into sustainability in my work (e.g. employer, supervisor, research or project funder...).
1.16 Because I have fun putting effort into sustainability at work.
1.17 Because otherwise I will feel bad about myself.
1.18 I don't know why I should put effort into sustainability in my work, because it's pointless.
1.19 Because it makes me feel proud of myself.

2. **How much do you think sustainability is discussed in the university community?** 1 = completely too little, 2 = slightly too little, 3 = appropriate, 4 = slightly too much, 5 = completely too much
3. **In the last 6 months, have you done the following things related to your work or workplace?** You can choose as many as you want. You don't have to choose anything.

    1. Participated in a discussion, public lecture, workshop or similar event on the theme of sustainability and responsibility.
    2. Participated in a discussion on sustainability or responsibility with your workmates, in the workplace or on the university's social media forum.
    3. Chosen a climate-friendly lunch or fair trade products at the lunch and coffee table.
    4. Sent a message or initiative on sustainability to the management of the university, faculty or educational programme, or collected names under an initiative, appeal or address on the theme of sustainability.
    5. Participated in a working group, committee or network on sustainability, responsibility, environmental issues, justice or equality.

6. Considered sustainability, circular economy, or social responsibility in procurement.
7. Acted to promote recycling or to reduce consumption (e.g. of electricity or water) in the workplace.
8. Chosen an environmentally friendly way to move between work and home
9. Chosen, for environmental reasons, to travel by land or sea instead of flying to an event organised far away.
10. Done researched or teaching on topics related to sustainability.

4. **To what extent do the statements below reflect your thinking on what makes you feel hopeful about climate change? (1 = Not at all, 2 = Little, 3 = Moderately, 4 = Quite a lot, 5 = Very much) I am hopeful about climate change …**

   4.1 Because I believe that research and technological solutions will help reduce problems related to climate change.
   4.2 Because I don't believe that climate change is as big a problem as some researchers claim.
   4.3 Because I think more people will be willing to take action to help solve the problems caused by climate change.
   4.4 Because we as individuals can change our behaviour, and together we can make a positive impact on climate change.
   4.5 Because I believe that climate change is natural and I have doubts that climate change is caused by emissions made by people.
   4.6 Because awareness of the problem of climate change has increased significantly in recent years.
   4.7 Because I am willing to take action to help solve the problems caused by climate change.
   4.8 Because politicians in more and more countries begin to take climate change seriously.
   4.9 Because sooner or later we will have to take climate change issues seriously and start taking responsibility.
   4.10 Because I feel doubtful that there is any change in the climate.
   4.11 Because as long as there are people active in environmental organisations, there is always a possibility that the climate issue will be solved.
   4.12 Because I know there are several things I can do to help reduce the problems caused by climate change.
   4.13 Because I try to focus on positive news about climate change in the media.
   4.14 Because I think it is good that summers in Finland are getting warmer as a result of climate change.
   4.15 Climate change is beyond my control, so I will not even try to solve the problems caused by it.

Appendix 2: Scales, Items, Factor Loadings and Cronbach's Alphas of Sustainability-Related Hope and Motivation Measures

	Factor loadings
H1 Hopeful α = 0.871	
8.1. Because I believe that research and technological solutions will help reduce problems related to climate change	0.651
8.4. Because we as individuals can change our behaviour, and together we can make a positive impact on climate change	0.688
8.6. Because awareness of the problem of climate change has increased significantly in recent years	0.702
8.8. Because politicians in more and more countries begin to take climate change seriously	0.529
8.9. Because sooner or later we will have to take climate change issues seriously and start taking responsibility	0.684
8.12. Because I know there are several things I can do to help reduce the problems caused by climate change	0.603
H2 Denial α = 0.734	
8.2. Because I don't believe that climate change is as big a problem as some researchers claim	0.706
8.5. Because I believe that climate change is natural and I have doubts that climate change is caused by emissions made by people	0.783
8.14. Because I think it is good that summers in Finland are getting warmer as a result of climate change	0.543
M1 Internal effort α = 0.869	
3.6. I don't put effort into sustainability in my work because it feels like I'm wasting my time. (R)	0.762
3.7. Because putting effort into sustainability aligns with my personal values	0.600
3.10. Because putting effort into sustainability has personal significance to me	0.597
3.11. I do little because I think things related to sustainability are not worth putting effort into in my work. (R)	0.696
3.14. Because I personally consider it important to put effort into sustainability in my work	0.657
3.18. I don't know why I should put effort into sustainability in my work, because it's pointless. (R)	0.723
M2 External regulation, social α = 0.818	
3.1. Because I have to prove to myself that I can	0.514
3.3 Because others will respect me more (e.g. supervisor, colleagues, students, family, customers…)	0.833
3.5. To avoid being criticised by others (e.g. supervisor, colleagues, students, family, customers…)	0.677

(continued)

(continued)

	Factor loadings
3.9. To get the approval of others (e.g. supervisor, colleagues, students, family, customers…)	0.719
M3 Introjected motivation α = 0.817	
3.13. Because otherwise I will feel ashamed of myself	0.805
3.17. Because otherwise I will feel bad about myself	0.817
3.19. Because it makes me feel proud of myself	0.507
M4 Intrinsic motivation α = 0.800	
3.2. Because things related to sustainability are interesting in my work	0.565
3.4. Because it gives me pleasure to learn about sustainability in my work	0.735
3.16. Because I have fun putting effort into sustainability at work	0.696
M5 External motivation, material α = 0.746	
3.8. Because I can get research funding/project funding/salary increase only if I put effort into sustainability in my work	0.677
3.12. Because I am afraid that I will lose my job if I don't put effort into sustainability	0.523
3.15. Because others offer me greater job security if I put effort into sustainability in my work (e.g. employer, supervisor, research or project funder…)	0.803

References

Allen MS, Iliescu D, Greiff S (2022) Single item measures in psychological science. Eur J Psychol Assess

Becker R (2017) Gender and survey participation: an event history analysis of the gender effects of survey participation in a probability-based multi-wave panel study with a sequential mixed-mode design. Methods Data 29:1–29. https://doi.org/10.12758/MDA.2021.08

Cantell H, Tolppanen S, Aarnio-Linnanvuori E, Lehtonen A (2019) Bicycle model on climate change education: presenting and evaluating a model. Environ Educ Res 25(5):717–731. https://doi.org/10.1080/13504622.2019.1570487

Deci EL, Ryan RM (2000) The "what" and "why" of goal pursuits: human needs and the self-determination of behavior. Psychol Inq 11(4):227–268. https://doi.org/10.1207/S15327965PLI1104_01

European Commission (2022) Communication from the Commission to the European Parliament, the Council, the European Economic and Social Committee and the Committee of the Regions on a European strategy for universities. https://eur-lex.europa.eu/legal-content/EN/TXT/?uri=COM%3A2022%3A16%3AFIN

Fernet C, Senécal C, Guay F, Marsh H, Dowson M (2008) The work tasks motivation scale for teachers (WTMST). J Career Assess 16(2):256–279. https://doi.org/10.1177/1069072707305764

Filho LW, Sierra J, Price E, Eustachio JHPP, Novikau A, Kirrane M, Dinis MAP, Salvia AL (2024) The role of universities in accelerating the sustainable development goals in Europe. Sci Rep 14(1), Article 15464. https://doi.org/10.1038/s41598-024-65820-9[1]

Folke C, Biggs R, Norström AV, Reyers B, Rockström J (2016) Social-ecological resilience and biosphere-based sustainability science. Ecol Soc 21(3):41. https://doi.org/10.5751/ES-08748-210341

Freire P (2005) Pedagogy of the oppressed. The Continuum International Publishing Group. (Original work published 1968), New York

Gagné M, Forest J, Vansteenkiste M, Crevier-Braud L, van den Broeck A, Aspeli AK, Bellerose J, Benabou C, Chemolli E, Güntert ST, Halvari H, Indiyastuti DL, Johnson PA, Hauan Molstad M, Naudin M, Ndao A, Hagen Olafsen A, Roussel P, Wang Z, Westbye C (2015) The multidimensional work motivation scale: validation evidence in seven languages and nine countries. Eur J Work Organ Psy 24(2):178–196. https://doi.org/10.1080/1359432X.2013.877892

Galperin BL (2012) Exploring the nomological network of workplace deviance: developing and validating a measure of constructive deviance. J Appl Soc Psychol 42(12):2988–3025. https://doi.org/10.1111/j.1559-1816.2012.00971.x

Goodman DJ (2000) Motivating people from privileged groups to support social justice. Teach Coll Rec 102(6):1061–1085. https://doi.org/10.1111/0161-4681.00092

Hyytinen H, Löfström E (2017) The role of assessment in promoting academic integrity. J Acad Ethics 15(1):5–28. https://doi.org/10.1080/03075079.2019.1582015

Jagers SC, Harring N, Löfgren Å, Sjöstedt M, Alpizar F, Brülde B, Langlet D, Nilsson A, Carney AB, Dupont S, Steffen W (2019) On the preconditions for large-scale collective action. Ambio 1–15. https://doi.org/10.1007/s13280-019-01284-w

Johnson CN, Balmford A, Brook BW, Buettel J, Galetti M, Lei G, Wilmshurst JM (2017) Biodiversity losses and conservation responses in the Anthropocene. Science 356:270–275. https://doi.org/10.17863/CAM.10996

Kollmuss A, Agyeman J (2002) Mind the gap: why do people act environmentally and what are the barriers to pro-environmental behavior? Environ Educ Res 8(3):239–260. https://doi.org/10.1080/13504620220145401

Li C, Monroe MC (2018) Development and validation of the climate change hope scale for high school students. Environ Behav 50(4):454–479. https://doi.org/10.1177/0013916517708325

Lorimer K, Knight R, Shoveller J (2020) Improving the health and social wellbeing of young people: exploring the potential of and for collective agency. Crit Public Health 32(2):145–152. https://doi.org/10.1080/09581596.2020.1786501

Lueck MAM (2007) Hope for a cause as cause for hope: the need for hope in environmental sociology. Am Sociol 38:250–261. https://doi.org/10.1007/s12108-007-9017-7

Nota L, Soresi S, Ferrari L, Wehmeyer ML (2011) A multivariate analysis of the self-determination of adolescents. J Happiness Stud 12(2):245–266. https://doi.org/10.1007/s10902-010-9191-0

Ojala M (2007) Confronting macrosocial worries: worry about environmental problems and proactive coping among a group of young volunteers. Futures 39(6):729–745. https://doi.org/10.1016/j.futures.2006.11.007

Ojala M (2012) Hope and climate change: the importance of hope for environmental engagement among young people. Environ Educ Res 18(5):625–642. https://doi.org/10.1080/13504622.2011.637157

Ojala M (2015) Hope in the face of climate change: associations with environmental engagement and student perceptions of teachers' emotion communication style and future orientation. J Environ Educ 46(3):133–148. https://doi.org/10.1080/00958964.2015.1021662

Pidgeon N (2012) Public understanding of, and attitudes to, climate change: UK and international perspectives and policy. Climate Policy 12(sup01):85–106. https://doi.org/10.1080/14693062.2012.702982

Pihkala P (2017) Environmental education after sustainability: hope in the midst of tragedy. Global Discourse 7(1):109–127. https://doi.org/10.1080/23269995.2017.1300412

Racimo F, Valentini E, Rijo De León G, Santos TL, Norberg A, Atmore LM, Murray M, Hakala SM, Olsen FA, Gardner CJ, Halder JB (2022) Point of view: the biospheric emergency calls for scientists to change tactics. eLife 11:e83292. https://doi.org/10.7554/eLife.83292

Ryan RM, Deci EL (2020) Intrinsic and extrinsic motivation from a self-determination theory perspective: definitions, theory, practices, and future directions. Contemp Educ Psychol 61, Article 101860. https://doi.org/10.1016/j.cedpsych.2020.101860

Schwartz SEO, Benoit L, Clayton S, Parnes MF, Swenson L, Lowe SR (2023) Climate change anxiety and mental health: environmental activism as buffer. Curr Psychol 42:16708–16721. https://doi.org/10.1007/s12144-022-02735-6

Shephard K (2015) Higher education for sustainable development. Palgrave Pivot

Snyder CR (2000) Genesis: the birth and growth of hope. In: Snyder CR (ed) Handbook of hope: theory, measures, and applications. Academic Press, pp 25–38. https://doi.org/10.1016/B978-012654050-5/50004-X

Snyder CR, Rand KL, Sigmon DR (2002) Hope theory: a member of the positive psychology family. In: Snyder CR, Lopez SJ (eds) Handbook of positive psychology. Oxford University Press, pp 257–276. https://doi.org/10.1093/oxfordhb/9780199399314.013.3

Tavakol M, Dennick R (2011) Making sense of Cronbach's alpha. Int J Med Educ 2:53–55. https://doi.org/10.5116/ijme.4dfb.8dfd

Tremblay MA, Blanchard CM, Taylor S, Pelletier LG, Villeneuve M (2009) Work extrinsic and intrinsic motivation scale: its value for organizational psychology research. Can J Behav Sci 41(4):213

Trépanier S-G, Peterson C, Gagné M, Fernet C, Levesque-Côté J, Howard JL (2023) Revisiting the multidimensional work motivation scale (MWMS). Eur J Work Organ Psy 32(2):157–172. https://doi.org/10.1080/1359432X.2022.2116315

Tuononen T, Hyytinen H, Kleemola K, Hailikari T, Toom A (2023) Generic skills in higher education—teachers' conceptions, pedagogical practices and pedagogical training. Teach Higher Educ. https://doi.org/10.1080/13562517.2023.2248003 Universities Act 558/2009. https://www.finlex.fi/en/laki/kaannokset/2009/en20090558_20160644.pdf

Usher EL, Schunk DH (2018) Social cognitive theoretical perspective of self-regulation. In: Schunk DH, Greene JA (eds) Handbook of self-regulation of learning and performance, 2nd edn. Educational Psychology Handbook Series. Routledge, Taylor & Francis Group, New York, NY, pp 17–35

Vandaele M, Stålhammar S (2022) Hope dies, action begins?" The role of hope for proactive sustainability engagement among university students. Int J Sustain High Educ 23(8):272–289. https://doi.org/10.1108/IJSHE-11-2021-0463

Veintie T, Heikkinen M, Kettunen P, Salovaara J, Södervik I, Ratvio R, Tuononen T, Pietikäinen J (2024) From sustainability perception to action in higher education: a study of faculty and staff perspectives. [Manuscript submitted for publication.]

Sustainable Development in Higher Education: A Review of Effective Pedagogical Tools

Katalin Ásványi and Eszter Gedeon

Abstract In the twenty-first century, global society faces significant social and environmental challenges, highlighting the necessity of integrating education for sustainable development (ESD) within university curricula. Higher education institutions are crucial in cultivating environmentally and socially responsible individuals, aligning with Sustainable Development Goal (SDG) 4.7. The paper seeks to systematically review pedagogical tools used in higher education to teach sustainable development, aiming to increase students' sustainability awareness. Through analyzing over 300 articles, pedagogical tools were based on effectiveness in students' engagement, levels of digitalization, and real-world applicability. By developing a structured framework, it guides university professors in selecting tools to foster sustainable development education, supporting educators in creating curricula that promote both foundational knowledge and real-world application.

Keywords Sustainability · Education · Pedagogical tools · Effective

1 Introduction

In the twenty-first century, we are increasingly faced with social and environmental challenges, which makes education for sustainable development (ESD) a priority across all university curricula. Higher education can and should help students to become environmentally and socially responsible individuals, as consumers, as employees or entrepreneurs and as citizens. University education plays a crucial role in society, as it is responsible for training professionals, conveying the latest

K. Ásványi (✉)
Institute of Sustainable Development, Corvinus University of Budapest, Budapest, Hungary
e-mail: katalin.asvanyi@uni-corvinus.hu

E. Gedeon
Laboratoire d'Economie et de Gestion de l'Ouest, Universite de Bretagne Sud, Lorient, France
e-mail: eszter.gedeon@univ-ubs.fr

W. Leal Filho et al. (eds.), *Education for Sustainable Development: The Contribution of Universities*, World Sustainability Series,
https://doi.org/10.1007/978-3-031-86985-3_24

knowledge, and shaping perspectives, which serve as the driving force for societal development. The Talloires Declaration, in 1990, was the first to highlight the importance of sustainability in higher education (Hoover and Harder, 2015). Several solutions can be found for its implementation, ranging from the integration of sustainability principles and topics into curricula (Lozano et al., 2015) to applying these in research and/or university operations (Leal Filho and Manolas 2011).

The field of education for sustainable development (ESD) has increasingly embraced diverse pedagogical approaches and tools, each aimed at fostering sustainability-related knowledge, skills, and values among students. However, challenges remain in implementing ESD: time constraints, limited resources, and the need for cross-disciplinary collaboration often hinder the successful integration of these methodologies (Park et al., 2022). Addressing these challenges is crucial, as pedagogical approaches that overcome these limitations are more likely to inspire lasting engagement and behavioral change among students, supporting the overarching goals of ESD.

While the literature on education for sustainable development (ESD) has explored various aspects of teaching methodologies, there remains a notable gap in the comprehensive categorization of pedagogical tools specifically tailored for ESD. Existing studies often focus on isolated elements, such as the integration of technology or the development of specific competencies, but lack a cohesive framework that can guide educators in selecting tools across different ESD contexts (Thürer et al., 2018; Oliveira et al., 2019). This fragmented approach limits educators' ability to implement ESD effectively, as the absence of a structured tool categorization hinders the development of curricula that fully address the multifaceted nature of sustainability education. Addressing this gap, the present study aims to create an organized framework of pedagogical tools that aligns with ESD goals, thereby providing educators with practical guidance for enhancing sustainability education.

The primary goal is to provide a comprehensive framework for evaluating and selecting the most effective pedagogical tools for sustainability education at the university level. The research does not aim to present research findings of the different analyzed articles. The choice of topic was fundamentally influenced by the following factors: (a) the integration of sustainability into higher education curricula is increasingly pertinent, and (b) it is filling a gap in the sense that, narrowed down specifically to this topic, there is not yet a comprehensive literature review on the topic of sustainability pedagogical tools.

The study contributes to the literature by focusing specifically on a systematic categorization and analysis of pedagogical tools to enhance educators' ability to teach sustainability effectively.

This research would provide benefits for various stakeholders: for students to experience innovative and various pedagogical tools for ESD, for lecturers to increase their teaching effectiveness, and for researchers to work closely with colleagues from another country both in research and teaching fields. Based on research findings, we will make recommendations for lecturers wishing to raise their students' awareness of sustainable development. Appropriate pedagogical tools can contribute to the success of education for sustainable development, which can be seen, for example, in a greater awareness of sustainable development among graduates or an increase in the number of students changing their minds towards sustainability.

2 Education for Sustainability

Learning about sustainability not only yields collective benefits through investigating the necessary societal transformation but also offers significant individual advantages for students. Among the key benefits for students are the development of problem-solving skills, honing critical thinking abilities, and fostering collaboration skills (Coppens et al., 2020; Kinoshita et al., 2019; Howlett et al., 2016). Additionally, it equips them with a critical understanding of intricate real-world issues. However, teaching sustainability entails several challenges (Park et al., 2022), including the allocation of requisite time and effort (for course restructuring and reorganization), limited awareness of sustainability among both lecturers and students, hurdles in community collaboration, cross-disciplinary miscommunication, misconceptions regarding sustainability, as well as inadequate funding and financial constraints.

3 Methodology

The aim of a systematic literature review is to summarize the chosen topic for a given period, present the main research findings, or highlight changes (Fenyvesi and Vágány, 2019). The method provides new results regarding the given topic through meta-analysis (Mack, 2016). By reviewing the literature, research gaps can be identified, which may form the basis for new empirical research (Webster and Watson, 2002) and may assist in the development of new research questions (Rowley and Slack, 2004).

The primary objective of this study is to develop a structured framework for categorizing pedagogical tools in education for sustainable development (ESD) at the university level. To achieve this, the study employs a systematic literature review (SLR) methodology, based on the PRISMA (Preferred Reporting Items for Systematic Reviews and Meta-Analyses) protocol (Moher et al., 2009). This rigorous approach ensures comprehensive data collection, accurate filtering, and consistent analysis of the relevant literature on ESD pedagogical tools. As a first step, we conducted a preliminary study on which topics previously appeared in systematic literature reviews related to the pedagogical tools of sustainability education, which serves to substantiate the relevance of our research. Using a Google Scholar search, we found articles that analyse specific aspects of sustainability education, such as transformative learning (Aboytes and Barth 2019), the exploration of the relationship between competencies and pedagogical approach (Lozano et al., 2017), or the integration of sustainability into the curriculum (Thürer et al., 2018), and some others which analysed pedagogical tools, such as technologies as pedagogical tools for teaching and learning science (Oliveira et al., 2019) or innovative pedagogical tools (Bizami et al., 2023). However, no similar study has yet been conducted specifically on the exploration of pedagogical tools for sustainability education. Based on the gaps identified, the authors conducted a detailed search in the Scopus database, focusing on publications that examined ESD pedagogical tools across various university settings. During the title-abstract and keyword search, the following keyword combinations were used along five keywords: 'pedagogical method AND sustainability AND higher education OR university' and 'educational method AND sustainability AND higher education OR university'. Only English-language publications were included in the database. Initially, based on the two search combinations, a total of 390 articles were found. All data was downloaded and duplicate articles were removed from the database, leaving 348 articles, which were imported into an Excel file. Within the topic, we precisely defined which topics to exclude from the examined articles, thus leaving only articles in the sample that are relevant to our research question. Participants in the research group independently selected relevant articles by reviewing the titles and abstracts. In cases where an article was selected by only one researcher, we discussed it together and, based on a consensus decision, either included or excluded the article from the final sample.

Despite the applied filters, several articles were found not to focus on pedagogical tools for teaching sustainability. These were excluded from the final database. During the review process, three key dilemmas emerged:

1. Should articles discussing the education of non-management courses, such as architecture or medicine, be included?
2. Should all types of education—online, offline, and hybrid—be considered?
3. Should teaching methods used for purposes other than ESD be included?

In addressing these dilemmas, articles related to non-management disciplines were included only if the pedagogical tools could be reasonably applied to management education. All types of education—online, offline, and hybrid—were included, recognizing the evolving landscape of educational delivery. However, only articles that specifically focused on pedagogical methods for ESD were retained, as tools designed for other purposes were deemed outside the scope of this study.

The remaining 173 articles were downloaded, and after reading the full texts, the researchers consulted again to select the final sample. Here, we further excluded 14 articles that, upon closer examination, did not address the topic defined by our research question, which was not yet clear when only reading the abstract; an additional 16 studies, which were scientific publications but not published in scientific journals; and 14 more articles that were not available in full text. Thus, 143 relevant articles remained in the end. The researchers independently organized the information found in the studies according to predefined criteria, which helped eliminate errors and establish common interpretations. The process along the exclusion criteria is illustrated in Fig. 1.

The articles dealing with the pedagogical tools of sustainability education was carried out along the following research questions:

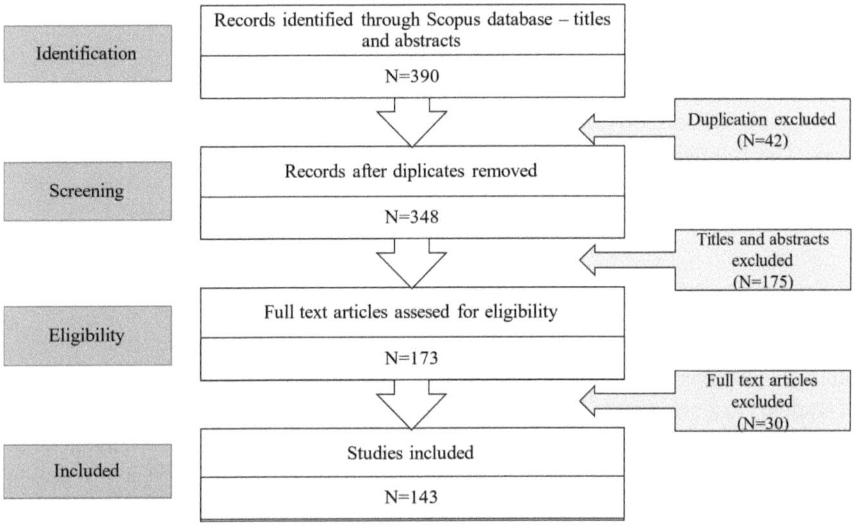

Fig. 1 Steps of PRISMA protocol. *Source* Own editing based on the PRISMA protocol methodology (Moher et al. 2009)

- Based on which effectiveness elements can the pedagogical tools of sustainability education be interpreted?
- Along which categories can the pedagogical tools of sustainability education be grouped?

The following information was extracted from the articles in table form in Excel:

- Basic data: author's name, year of publication
- Data related to the research: pedagogical tool, definition of the pedagogical tool, effectiveness elements

To answer the research questions, the pedagogical tools were then categorized by various frameworks, including student engagement level, digitalization, real-world engagement, and elements of effectiveness. This systematic categorization enabled the development of a structured framework that supports educators in selecting ESD tools suitable for diverse instructional goals and contexts.

In the following, the studies based on their main characteristics are presented, a detailed overview of the pedagogical tools and their effectiveness elements are provided (Pautasso 2013). The aim is not to draw general conclusions but rather to gain a deeper understanding of the specific pedagogical tools of sustainability education and, based on these, to answer the research questions.

4 The Findings of the Literature Review

At the end of this procedure a total of 105 tools were found in the articles. The tools can be categorized based on various factors, depending on the teaching context and objectives. The following frameworks can help professors to decide which pedagogical tools to prioritize in ESD.

4.1 By Student Engagement Level

Students' engagement (Table 1) refers to the level of necessary investments on behalf of students in learning activities (Mandernach, 2015). It comprises three main dimensions: cognitive engagement, affective engagement and behavioral engagement. Cognitive engagement indexes the mental effort students direct towards learning, understanding and mastering the knowledge (ex. thinking about dilemmas, solutions related to the course). Affective engagement can be the emotional intensity experienced by students during learning but also their reactions of students to learning (ex. enjoyment or interest). Behavioral engagement refers to students' attendance to class as well as to the level of participation in learning activities (ex. actively participating in students-teacher discussions) (Kahu, 2013; Mandernach, 2015).

There is a consensus about the several positive effects of students' engagement on learning outcomes (Mandernach, 2015). It fosters students' learning, enhances knowledge retention and quality (Rincón et al., 2020) and also strengthens students' persistence in learning. Among the immediate effects we can also find satisfaction and learning achievement, and the long-term effects comprise work success and personal growth (Kahu, 2013).

4.2 By Application of Technology

Digitalization in higher education has permeated numerous dimensions of academic institutions. Within the teaching domain specifically, the importance of digitalization manifests through distinct elements such as digital platforms and content, innovative pedagogical methodologies, digital literacy and teaching administrative process (referring to flexibility in learning, course delivery and quality) (Benavides et al., 2020). The integration of mobile devices, Big Data and various AI-powered tools reshape teaching methodologies, driving educators to adapt to these evolving technologies (Rudolph et al., 2023).

Digital technology also redefines the roles of both teachers and students. It creates new, motivating and flexible learning environments, that require more autonomy and collaboration on behalf of students while necessitating pedagogical and technological guidance from teachers (Benavides et al., 2020).

Table 1 Categorization of pedagogical tools based on students' engagement level

Level of engagement	Examples of tools	Definition	Mentioned by
High	Gamification	Gamification is an educational strategy that uses elements of games for educational purposes	Rincón et al. (2020), Rincón et al. (2019)
	Serious games	Entertaining games with non-entertainment goals that educate, train and inform	Ameerbakhsh et al. (2019), Riedel and Hauge (2011)
	Role-play	A game which consists of simulating a real-life situation	Martinez et al. (2021), Moreno-Guerrero et al. (2020)
	Challenge-based learning	Challenge based learning: participation in university challenges	Buerkle et al. (2023), Fantozzi et al. (2024)
	Art-based methodology	Embedding a crafting project and reflection about it in teaching	Molderez et al. (2021), Vasileva et al. (2021)
Moderate	Group discussions	Free verbal exchange of ideas between group members or teachers and students	Martinez et al. (2021)
	Debates	The introduction of a debate topic while teaching to foster discussion among students	Martinez et al. (2021), Zare and Othman (2015)
	Case-based teaching	Case methods rely on the presentation of concrete management situations, in which students must actively engage in discussions to find valuable solutions	Buerkle et al. (2023), Schimperna et al. (2022), Haselbach and Langfitt (2016)
Low	Lecture	Classic university lecture	Delaney and Wei (2023)
	Visioning multimedia presentations (videos)	Involving visual elements, mainly photos and pre-recorded videos in teaching	Tomasella et al. (2024)

Source Authors' construction

Digitalization is crucial to adapt to the new learning needs of students and satisfy the need of employers to a workforce that is equipped with digital skills. Digitalization can reduce the cost of delivering education, provide personalized learning experiences adapting to diverse student profiles. By improving quality and accessibility of education, digitalization broadens educational reach and has the potential to make high-quality learning more widely available (Abad-Segura et al., 2020). Table 2 presents pedagogical tools based on their level of digitalization.

Table 2 Classification of pedagogical tools by digitalisation levels

Level of digitalization	Examples of tools	Definition	Mentioned by
Fully digital	Virtual reality	The use of Virtual Reality, an interactive, immersive digital environment in education	Shanti and Al-Tarazi (2023)
	Metaverse	The integration of Metaverse, a three-dimensional virtual world into education	Salloum et al. (2023)
	MOOC	Massive Open Online Course: online learning courses that a student accesses through the internet	Pedersen et al. (2022)
	Online service learning	Combines learning and community service processes where the participants learn while working on real needs of the environment, the service or/and instruction occur online	Aramburuzabala et al. (2024)
Blended or hybrid	Flipped classroom	The key element is that the traditional way of teaching and learning is inverted	Peters et al. (2024), Santos and Serpa (2020)
	Problem visualization	Visual representations of concepts and problems	Lukman and Virtic (2018), Marcone (2022)
	Real world data analysis	Students use real world data to analyze problems	Anastasiadis et al. (2021), Yoshida (2022)
Non-digital	Field trips	Education outside of the classroom	Park et al. (2024)
	Internships	Working in an organization while learning to gain experience for a qualification	Park et al. (2024)
	Community involvement	Interact and/or engage local population in learning	Shephard (2008), Bello (2017)

Source Authors' construction

4.3 By Real-World Engagement

Real-world engagement means connecting academic concepts to practical, real-life contexts. The connection can be done through integrating practical, industry-relevant experiences into academic curricula, aligning education with students' career objectives and market needs.

This approach aims to make learning more relevant by showing students how theories and ideas apply to professional scenarios. Applied to ESD, it connects students with real-world sustainability challenges and applications organizations face (O'Neill and Short, 2023).

Table 3 Pedagogical tools based on real-world engagement

Level of real-world engagement	Examples of tools	Definition	Mentioned by
Direct real-world engagement	Internships	Working in an organization while learning to gain experience for a qualification	Park et al. (2024)
	Community-based learning	Community-based learning is an academic course-based type of experiential learning in which the student provides some meaningful work for and with disadvantaged individuals or groups	Shephard (2008)
	Live projects	Projects involving business people, SMEs, larger companies as well as Non Governmental Organisations (NGOs)	De Eyto et al. (2008)
	Field trips	Education outside of the classroom	Park et al. (2024)
Simulated real-world engagement	Virtual reality	The use of Virtual Reality, an interactive, immersive digital environment in education	Shanti and Al-Tarazi (2023)
	Metaverse	The integration of Metaverse, a three-dimensional virtual world into education	Salloum et al. (2023)
	Business simulations	A dynamic, streamlined and practical representation of a business scenario	Sheikh et al. (2023)
	Serious games	Entertaining games with non-entertainment goals that educate, train and inform	Ameerbakhsh et al. (2019), Riedel and Hauge (2011)

Source Authors' construction

This approach is crucial because it addresses the gap between traditional education and the evolving demands of the labour market, enhancing student employability and lifelong learning skills (Morley and Jamil, 2021).

Table 3 presents tools that work through direct real-world engagement and tools that simulate real-world engagement in teaching.

4.4 By Elements of Effectiveness

To enhance sustainability education, teaching tools can be categorized by their primary elements of effectiveness: Knowledge Acquisition, Behavioral Change and Values, Collaboration and Teamwork, and Critical Thinking and Reflection. Tools focused on Knowledge Acquisition deliver essential theories and foundational knowledge, forming the basis for understanding complex sustainability concepts. Those aimed at Behavioral Change and Values are designed to influence students' attitudes and behaviors, encouraging values that align with sustainable practices. Tools that enhance Collaboration and Teamwork foster communication and cooperative problem-solving skills, crucial for addressing sustainability issues that often require interdisciplinary and collective approaches. Finally, tools promoting Critical Thinking and Reflection encourage students to deeply analyze and reflect on sustainability challenges, helping them develop the ability to critically evaluate and propose solutions to real-world issues. By combining these elements, educators can cultivate a holistic and impactful learning experience that not only informs but also empowers students to act sustainably (Table 4).

5 Conclusion

The integration of education for sustainable development (ESD) into higher education has become increasingly critical. Universities, as influential institutions for developing future professionals, are uniquely positioned to equip students with the knowledge, skills, and values needed to address these complex issues. However, despite widespread recognition of the importance of ESD, the educational landscape lacks a cohesive framework to guide educators in selecting and applying pedagogical tools that effectively promote sustainability education. This study responds to this gap by presenting a structured categorization of ESD tools, enabling educators to foster a more informed, engaged, and sustainable mindset among students.

Through a systematic literature review, this study categorizes over 100 pedagogical tools based on dimensions such as engagement level, digitalization, real-world applicability. These findings provide a contextualized, evidence-based framework that aligns with the Sustainable Development Goals (SDG) framework, particularly SDG 4.7, which emphasizes education as a foundation for sustainable development. By offering a comprehensive guide to these pedagogical tools, the study supports educators in designing curricula that go beyond theoretical understanding, fostering practical skills and ethical commitments necessary for real-world sustainability challenges.

This study highlights a variety of pedagogical tools that can enhance students' awareness of sustainability by promoting engagement. High-engagement tools, such as role-play, challenge-based learning, and serious games, were identified as particularly effective for fostering immersive, active learning experiences. These tools encourage students to apply sustainability concepts practically, promoting not only knowledge acquisition but also shifts in behavior and values. The findings underline that while traditional lecture-based methods provide foundational knowledge, interactive approaches are crucial for cultivating a deeper understanding and long-term commitment to sustainable practices among students. For instance, tools requiring high engagement, such as community-based learning, are ideal for courses that emphasize experiential learning, while digital tools like virtual reality (VR) and

Table 4 Pedagogical tools based on their effectiveness

Elements of effectiveness	Examples of tools	Definition	Mentioned by
Knowledge acquisition	Guest lecture	A guest lecture is a presentation or talk given by an external expert, to share specialized knowledge, insights, or professional experience with student	Park et al. (2024), Tomasella et al. (2024)
	Flipped classroom	The key element is that the traditional way of teaching and learning is inverted	Foster and Stagl (2018)
	Sustainability workshops	Sustainability workshops provide students with knowledge about sustainable practices and their importance	Olawale (2019)
	Inquiry-based learning	The inquiry-based learning methodology stimulates learning by posing questions or problems that help the student to build and acquire knowledge through discovery	Zarco-Periñan and Zarco-Soto, (2022), Kennelly and Taylor (2007)
Behavioral change and values	Challenged based learning	Challenge based learning: participation in university challenges	Vatovec and Ferrer (2019), Andrási et al. (2022), Fantozzi et al. (2024)
	Deliberate practice	Deliberate practice in teaching is defined as deliberate activities aimed at self-improvement and the promotion of students' learning and development	Sihua et al. (2023)
	Partnership with employers and industry	Introduced industry partners to support and provide inspiration to students, for example, setting live briefs to simulate real life sustainability issues	Delaney and Wei (2023), Tomasella et al. (2024)
Collaboration and teamwork	Cooperative games	Interactive activities where students work together toward shared goals	De Abreu et al. (2022)
	Role play	A play that consists of simulating a real-life situation	Martinez et al. (2021), Moreno-Guerrero et al. (2020)

(continued)

Table 4 (continued)

Elements of effectiveness	Examples of tools	Definition	Mentioned by
	Project-based learning	This method enables students to learn through the completion of complex projects that require them to acquire and use knowledge independently. It involves supervised, small collaborative groups working on specific projects	Cohen (2009), Buerkle et al. (2023), Assali (2017)
	Problem-based learning	PBL relies on the creation of a real problem and the involvement of students to find a valuable solution, exchanging information with their colleagues to acquire knowledge and critically reflect on the underlying issue	Schimperna et al. (2022), Peters et al. (2024)
Critical thinking and problem solving	Ecological Footprint calculator/ analysis	It allows individuals to observe and realize the impact of their personal actions on the planet through a simple message and visually appealing results	Gündüz and Alsagher (2018), Moreno et al. (2022)
	Service learning	Students engage in community service, reflect critically on this experience, and learn from it personally, socially and academically	Aramburuzabala et al. (2024), Schimperna et al. (2022)
	Reflexive journals	Students maintain journals to reflect on their learning process	Holden et al. (2011)
	Problem-based learning	PBL relies on the creation of a real problem and the involvement of students to find a valuable solution, exchanging information with their colleagues to acquire knowledge and critically reflect on the underlying issue	Schimperna et al. (2022), Peters et al. (2024)
	Project-based learning	This method enables students to learn through the completion of complex projects that require them to acquire and use knowledge independently. It involves supervised, small collaborative groups working on specific projects	Cohen (2009), Buerkle et al. (2023), Assali (2017)

Source Authors' construction

online service learning are suitable for hybrid or fully online teaching environments. By offering this categorization, the study aids educators in tailoring sustainability education to various learning contexts, enhancing its relevance and effectiveness.

The systematic literature review conducted in this research revealed several gaps in the current academic discourse on sustainability education. Notably, there is limited empirical data on the long-term impact of certain pedagogical tools in fostering sustained behavioral change post-graduation. Additionally, while the literature addresses various individual tools, there is a scarcity of studies exploring the integrated use of multiple pedagogical tools to reinforce sustainability principles across diverse educational settings. Addressing these gaps would provide a more comprehensive understanding of the optimal approaches to embedding sustainability in higher education.

For educators seeking to integrate sustainability into their courses, this study provides actionable insights and practical guidance on choosing effective pedagogical tools. By using the categorized tools and effectiveness elements as a reference, educators can design curricula that not only communicate core sustainability concepts but also engage students on a deeper, more impactful level. Tools that encourage collaboration, problem-solving, and real-world application are particularly recommended to prepare students to address complex sustainability issues as future professionals. For institutions, adopting these categorized tools offers a pathway to aligning educational programs with Sustainable Development Goal (SDG) 4.7, which emphasizes quality education and lifelong learning for sustainable development.

This study, while providing a comprehensive categorization of pedagogical tools for education in sustainable development (ESD), faces several limitations that should be considered when interpreting the findings. First, the study is limited by its reliance on published literature in English, which may exclude relevant insights from non-English publications and regional studies that could offer culturally specific perspectives on ESD pedagogical tools. Additionally, the study's focus on peer-reviewed articles means that potentially valuable case studies, reports, or unpublished data were not included, which may limit the framework's applicability across diverse educational settings.

Another constraint is the scope of articles reviewed, as the rapidly evolving landscape of sustainability education and digital pedagogical tools may lead to emerging practices that are not yet fully represented in the current literature. Consequently, the framework may require periodic updates to reflect new tools and methods that align with technological advancements and shifting educational paradigms. Furthermore, the study's reliance on systematic literature review methods inherently limits the findings to documented research, without empirical testing or validation of the tools in real classroom settings. Future research could benefit from empirical studies that test and refine the framework across various educational contexts. This study points to several directions for future research, such as examining the influence of cultural and institutional contexts on the adoption and effectiveness of sustainability teaching methods. Future studies could also explore how pedagogical tools for sustainability education can be adapted to fields beyond management sciences, such as engineering or medicine, and assess the effectiveness of interdisciplinary approaches. Longitudinal studies assessing the impact of sustainability education on graduates' professional behaviors and values would also contribute valuable insights to the field. Additionally, cross-cultural studies incorporating non-English literature could expand the framework's applicability and highlight unique pedagogical tools and practices in diverse cultural and institutional contexts.

This research contributes significantly to the discourse on Education for Sustainable Development (ESD) by providing a structured approach to embedding sustainability in higher education through targeted pedagogical tools. The categorization and analysis of these tools not only support educators in curriculum design but also align with broader ESD goals to create environmentally and socially responsible individuals. By enhancing the educational strategies available for teaching sustainability, this study offers potential for broader educational impact and societal benefit, advancing the role of higher education in fostering sustainable development.

Acknowledgements This research was supported by the Hubert Curien Program of Campus France and Tempus for the 2024-2026 funding period.

References

Abad-Segura E, González-Zamar MD, Infante-Moro JC, Ruipérez Garcia G (2020) Sustainable management of digital transformation in higher education: global research trends. Sustainability 12:2107

Aboytes RJG, Barth M (2020) Transformative learning in the field of sustainability: a systematic literature review (1999–2019). Int J Sustain High Educ 21(5):993–1013

Ameerbakhsh O, Maharaj S, Amir H, McAdam B (2019) A comparison of two methods of using a serious game for teaching marine ecology in a university setting. Int J Hum Comput Stud 127:181–189

Anastasiadis S, Perkiss S, Dean BA, Bayerlein L, Gonzalez-Perez MA, Wersun A, Acosta P, Jun H, Gibbons B (2021) Teaching sustainability: complexity and compromises. J Appl Res High Educ 13(1):272–286

Andrási G, Körtvési D, Szegedi K (2022) Developing ethical future managers embracing sustainability: case study of Budapest Business School. Int J Web-Based Learn Teach Technol 17:1–19

Aramburuzabala P, Culcasi I, Cerrillo R (2024) Service-learning and digital empowerment: the potential for the digital education transition in higher education. Sustainability 16:2448

Assali IM (2017) The contribution of interior design studio in promoting sustainable design in Bahrain. International Journal of Mechanical Engineering and Technology 8:10

Bello J (2017) Cooking with the sun: teaching and capacitating about solar energy. Renewable Energy and Power Quality Journal 1:808–812

Benavides LMC, Arias TAJ, Serna ADM, Bedoya BWJ, Burgos D (2020) Digital transformation in higher education institutions: a systematic literature review. Sensors 20:3291

Bizami NA, Tasir Z, Kew SN (2023) Innovative pedagogical principles and technological tools capabilities for immersive blended learning: a systematic literature review. Educ Inf Technol 28(2):1373–1425

Buerkle A, O'Dell A, Matharu H, Buerkle L, Ferreira P (2023) Recommendations to align higher education teaching with the UN sustainability goals: a scoping survey. International Journal of Educational Research Open 5

Cohen N (2009) Designing the sustainable foodshed: a cross-disciplinary undergraduate environmental studies course. Innov High Educ 35:51–60

Coppens T, Pineda AV, Henao K, Rybels S, Samoilovich D, De Jonghe N, Camacho H (2020) Innovating education for sustainable urban development through problem-based learning in Latin America. Journal of Problem Based Learning in Higher Education 8:1–18

De Abreu S, Vogt K, Schreier A, Fawcett P, Marchand M, Vogt D, Tulee M, Wendell G, Lacey M, Matsumoto K (2022) Teaching holistic environmental thought: a classroom approach. Thinking Skills and Creativity 46:101141

De Eyto A, Mahon M, Hadfield M, Hutchings M (2008) Strategies for developing sustainable design practice for students and SME professionals. Eur J Eng Educ 33:331–342

Delaney E, Liu W (2023) Postgraduate design education and sustainability—an investigation into the current state of higher education and the challenges of educating for sustainability. Frontiers in Sustainability 4. https://doi.org/10.3389/frsus.2023.1148685

Fantozzi I, Luozzo S, Schiraldi M (2024) The impact of university challenges on students' attitudes and career paths in industrial engineering: a comparative study. Sustainability 16:1600

Fenyvesi É, Vágány J (2020) A rejtett gazdaság néhány területének szisztematikus szakirodalmi áttekintése. Közgazdasági Szemle 67(5):512–532

Foster G, Stagl S (2018) Design, implementation, and evaluation of an inverted (flipped) classroom model economics for sustainable education course. Journal of Cleaner Production 183

Gündüz S, Abdusalam EA (2018) Consciousness levels of Libyan higher education students on ecological footprint and sustainable life. Quality & Quantity: International Journal of Methodology 52(1):67–78

Haselbach L, Langfitt Q (2016) Incorporating prerecorded environmental lifecycle assessment modules in a classroom setting. Journal of Professional Issues in Engineering Education and Practice 143

Holden R, Griggs V (2011) Not more learning logs! A research based perspective on teaching reflective learning within HR professional education. Hum Resour Dev Int 14(4):483–491

Hoover E, Harder MK (2015) What lies beneath the surface? The hidden complexities of organizational change for sustainability in higher education. J Clean Prod 106:175–188

Howlett C, Ferreira J-A, Blomfield J (2016) Teaching sustainable development in higher education: building critical, reflective thinkers through an interdisciplinary approach. Int J Sustain High Educ 17:305–321

Kahu ER (2013) Framing student engagement in higher education. Stud High Educ 38(5):758–773

Kennelly J, Taylor N (2007) Education for sustainability for the K-6 curriculum: a unit of work for pre-service primary teachers in NSW. Australian Journal of Environmental Education 23

Kinoshita A, Mori K, Rustiadi E, Muramatsu S, Kato H (2019) Effectiveness of incorporating the concept of city sustainability into sustainability education programs. Sustainability 11:4736

Leal Filho W, Manolas E (2011) Making sustainable development in higher education a reality: lessons learned from leading institutions. GUNI Higher Education in the World 4: Higher Education's Commitment to Sustainability from Understanding Action. Palgrave Macmillan:28–31

Lozano R, Ceulemans K, Scarff Seatter C (2015) Teaching organizational change management for sustainability: designing and delivering a course at the University of Leeds to better prepare future sustainability change agents. J Clean Prod 106:205–215

Lozano R, Merrill MY, Sammalisto K, Ceulemans K, Lozano FJ (2017) Connecting competences and pedagogical approaches for sustainable development in higher education: A literature review and framework proposal. Sustainability 9(10):1889.

Lukman R, Virtic P (2018) Developing energy concept maps – an innovative educational tool for energy planning. Journal of Sustainable Development of Energy, Water and Environment Systems 6:742–754

Mack C (2016) How to write a good scientific paper: review articles. J Micro/nanolithogr MEMS MOEMS 15(2):020101

Mandernach J (2015) Assessment of student engagement in higher education: a synthesis of literature and assessment tools. International Journal of Learning, Teaching and Educational Research 12(2):1–14

Marcone G (2022) Humanities and social sciences in relation to sustainable development goals and STEM education. Sustainability 14(6):3279

Martinez C, Ruiz MN, Buil-Fabregá M (2021) Higher education: the best practices for fostering competences for sustainable development through the use of active learning methodologies. International Journal of Sustainability in Higher Education ahead-of-print

Moher D, Liberati A, Tetzlaff J, Altman DG (2009) The PRISMA Group Preferred Reporting Items for Systematic Reviews and Meta Analyses: The PRISMA statement

Molderez I, Baraniuk D, Lambrechts W (2021) The role of poetry in promoting a sustainability mindset: Walter Benjamin as a guide toward a slow journey. Frontiers in Sustainability 2.

Moreno PS, Mapar M, Nicolau M, Patrizi N, Malandrakis G, Pulselli F, Bacelar-Nicolau P, Caeiro S, Theodossiou N, Mancini M, Galli A (2022) Teaching sustainability within the context of everyday life: steps toward achieving the Sustainable Development Goals through the EUSTEPs module. Frontiers in Education 7. https://doi.org/10.3389/feduc.2022.639793

Moreno-Guerrero AJ, Rodríguez-Jiménez C, Gómez-García G, Ramos Navas-Parejo M (2020) Educational innovation in higher education: use of role playing and educational video in future teachers' training. Sustainability 12(6):2558

Morley DA, Jamil MG (2021) Introduction: real world learning—recalibrating the higher education response towards application to lifelong learning and diverse career paths. In: Morley DA, Jamil MG (eds) Applied pedagogies for higher education. Palgrave Macmillan, Cham

O'Neill G, Short A (2023) Relevant, practical and connected to the real world: what higher education students say engages them in the curriculum. Irish Educational Studies:1–18

Olawale F (2019) Sustainability orientation and sustainable entrepreneurial intentions of university students in South Africa. Entrepreneurship and Sustainability Issues 7(2):990–999

Oliveira A, Feyzi Behnagh R, Ni L, Mohsinah AA, Burgess KJ, Guo L (2019) Emerging technologies as pedagogical tools for teaching and learning science: a literature review. Human Behavior and Emerging Technologies 1(2):149–160

Park HY, Licon CV, Sleipness OR (2022) Teaching sustainability in planning and design education: a systematic review of pedagogical approaches. Sustainability 14:9485

Pautasso M (2013) Ten simple rules for writing a literature review. PLoS Computational Biology 9(7)

Pedersen ERG, Andersen KR, Diaz Schiavon AL (2022) Diffusing corporate sustainability knowledge with online education: experiences from a massive open online course (MOOC) on sustainable fashion. J Int Educ Bus 15(1):89–105

Peters A-K, Capilla R, Constantin V, Coroamă RH, Lago P, Leifler O, Moreira A, Fernandes J, Penzenstadler B, Porras J, Venters C (2024) Sustainability in computing education: a systematic literature review. ACM Transactions on Computing Education 24(1):13

Riedel JCKH, Hauge JB (2011) State of the art of serious games for business and industry. Proceedings of the 2011 17th International Conference on Concurrent Enterprising (ICE 2011)

Rincón F, Elvira G, Mena J, Ramírez-Montoya M, María-Soledad, Ramirez-Velarde R (2019) The use of gamification in xMOOCs about energy: effects and predictive models for participants' learning. Australasian Journal of Educational Technology:43–59

Rincón F, Elvira G, Mena J, Ramírez-Montoya M, María-Soledad (2020) Gamification: a new key for enhancing engagement in MOOCs on energy? International Journal on Interactive Design and Manufacturing (IJIDeM) 14:10

Rowley J, Slack F (2004) Conducting a literature review. Manag Res News 27(6):31–39

Rudolph J, Tan S, Tan S (2023) ChatGPT: bullshit spewer or the end of traditional assessment in higher education? Journal of Applied Learning & Teaching 6(1)

Salloum S, Al Marzouqi A, Alderbashi KY, Shwedeh F, Aburayya A, Al Saidat MR, Al-Maroof RS (2023) Sustainability model for the continuous intention to use metaverse technology in higher education: a case study from Oman. Sustainability 15(6):5257

Santos A, Serpa S (2020) Flipped classroom for an active learning. Journal of Education and e-Learning Research 7:167–173

Schimperna F, Nappo F, Collaretti F (2022) Universities and CSR teaching: new challenges and trends. Administrative Sciences 12(2):55

Shanti Z, Al-Tarazi D (2023) Virtual reality technology in architectural theory learning: an experiment on the module of history of architecture. Sustainability 15(23):16394

Sheikh R, Abdalkrim G, Shehawy Y (2023) Assessing the impact of business simulation as a teaching method for developing 21st century future skills. Journal of International Education in Business 16

Shephard K (2008) Higher education for sustainability: seeking affective learning outcomes. International Journal of Sustainability in Higher Education 9

Sihua L, Ying L, Haohan L (2023) Research on sustainable teaching models of new business—take Chinese university business school as an example. Sustainability 15:8037

Thürer M, Tomašević I, Stevenson M, Qu T, Huisingh D (2018) A systematic review of the literature on integrating sustainability into engineering curricula. J Clean Prod 181:608–617

Tomasella B, Akbar B, Lawson A, Howarth R, Bedford R (2024) Embedding the Sustainable Development Goals into higher education institutions' marketing curriculum. Journal of Marketing Education 0(0)

Vasileva P, Golubev V, Ibragimov I (2021) Trash to treasure: integrating environmental awareness into university curriculum. Journal of Teaching English for Specific and Academic Purposes 205:https://doi.org/10.22190/JTESAP2102205V

Vatovec C, Ferrer H (2019) Sustainable well-being challenge: a student-centered pedagogical tool linking human well-being to ecological flourishing. Sustainability 11(24):7178

Webster J, Watson RT (2002) Analyzing the paths to prepare for the future: writing a literature review. MIS Q 26(2):13–23

Yoshida M (2022) Effects of using world indicators for online ES learning. Sustainability 14(21):13919

Zarco-Periñan PJ, Zarco-Soto F (2022) Towards the achievement of the Sustainable Development Goals through engineering training for labor market reintegration of older workers. Appl Sci 12:8020

Zare P, Othman M (2015) Students' perceptions toward using classroom debate to develop critical thinking and oral communication ability. Asian Social Science 11(9)

Enhancing Sustainable Development Literacy in Higher Education: Insights from Accounting and Pedagogy Courses

Sara Moggi and Gabriella Calvano

Abstract Universities have an essential role to play in sustainable development processes. The 2030 Agenda recognises the centrality of education for sustainable development, including at the university level. These processes go beyond disciplinary curricula. They require processes that can develop university students into active, future-oriented citizens. Considering the urgent need for pedagogical development, the present research presents the results of an explorative study that assesses how a common framework for teaching sustainable development can affect students' literacy on sustainable development in different disciplines, such as accounting and pedagogy. Through a survey, authors developed data collection to examine higher education students' literacy level and how their perception of sustainable development issues can change by embedding sustainability content during these courses. Data were collected at the beginning of two courses proposing a 6-h dedicated sustainability section. The same survey was proposed at the end of these courses to determine the change in the attending students' sustainable development literacy level. The results underline how the same approach can obtain different results on students' literacy levels in two subjects (accounting and pedagogy). The present study suggests the importance of a tailor-made education according to various subjects and proposes some approaches for embedding this perspective in the educational design.

S. Moggi (✉)
Department of Management, Università Degli Studi Di Verona, Verona, Italy
e-mail: sara.moggi@univr.it

G. Calvano
Department of Research and Innovation in the Humanities, Università Degli Studi Di Bari, Bari, Italy
e-mail: gabriella.calvano@uniba.it

W. Leal Filho et al. (eds.), *Education for Sustainable Development: The Contribution of Universities*, World Sustainability Series,
https://doi.org/10.1007/978-3-031-86985-3_25

1 Introduction

Universities are key actors in global competitiveness and sustainable development (Hessen & Schmelkes, 2022; Loorbach & Wittmayer, 2024; Leal Filho, 2011). They are called upon to adapt and respond to stakeholder needs innovatively and responsibly to address rapid changes and complex social, environmental, and economic challenges (Disterheft et al., 2015). These challenges require universities to engage on several fronts: a) making campuses more sustainable (Leal Filho et al., 2020); b) reviewing and updating curricula (Dlouhá et al., 2019); and c) enhancing public and social engagement activities (Trencher et al., 2014; Leal et al., 2021). This latter point is mainly related to addressing the needs of local and urban communities (Leal et al., 2021). The crucial role of universities has been further highlighted in the post-pandemic era, where the demand for sustainable and resilient solutions has become even more apparent (Leal Filho et al., 2020).

The proximity of universities to their territorial communities has generated numerous highly relevant practices and processes of collaborative learning and participation (Trencher et al., 2014; Shiel et al., 2016). These initiatives go beyond traditional knowledge transfer, creating dialogue, co-creation, and social innovation spaces. In this context, sustainability invites universities to become sites of critical resistance, where nothing is exempt from being questioned (Derrida, 2001) and fosters innovative thinking to address global challenges.

As the 2030 Agenda for Sustainable Development (UN, 2015) highlights, building a sustainable world requires a profound cultural shift in how we live, think and act toward 2030 Agenda goals (Unesco, 2017). In this process, education for sustainable development plays a key role in forming active and responsible citizens, actual citizens of sustainability (Unesco, 2020). In the context of sustainable development, universities should play an active role as agents of social change, demonstrating a commitment to rediscovering their civic role (Goddard et al., 2020), reclaiming their predisposition to educate for citizenship, creating spaces for discussion and the co-construction of solutions (Ávila et al., 2019), and promoting sustainable innovation. As catalysts for change, universities must ensure that students become active citizens with a clear understanding of future challenges and how to address them with several tools (Jacobi et al., 2016). This context demands an educational approach beyond mere transmission of knowledge, incorporating practical skills, critical reflection, and active involvement with real-world challenges (Unesco, 2017; Leal Filho et al., 2018). Therefore, educating about sustainability at the university level is essential, as well as promoting educational transformative processes that look to the future and are built with an approach rooted in global citizenship (Unesco, 2020; 2017).

Considering the pedagogical need for sustainable development literacy, the present study presents the results of explorative research exploring how a common framework for teaching sustainable development can (similarly) affect students from different disciplines, such as accounting and pedagogy. Data was collected through a survey to analyse the evolution of the students' literacy and their perception of sustainable development issues (before and after the courses). Specifically, the first survey, exploring the level of sustainable development literacy, collected data at the beginning of two courses (accounting and pedagogy). The same survey was proposed at the end of these courses to determine the change in the sustainable development literacy level of the attending students.

The remainder of this paper is as follows: Sects. 2 and 3 present an overview of the literature review on education for sustainable development and the 2030 Agenda. Section 4 describes the methodology and how data collection was carried out. Section 5 presents the present study's findings; Sect. 6 presents the discussion, conclusions and suggestions for future research development.

2 Education for Sustainable Development in Higher Education: Perspectives and Challenges

Higher education faces a crucial and increasingly urgent challenge in today's educational landscape: effectively integrating Education for Sustainable Development (ESD) into academic curricula, institutional practices, governance, and leadership dynamics (Leal Filho et al., 2019a). This challenge arises from a growing awareness of the impact of sustainability issues on global society and the need to train professionals who can address our time's complex environmental, social, and economic challenges.

Sustainability is far from a static concept; it is a dynamic and evolving process (Wals & Jickling, 2002). Bologna (2018) highlighted that sustainability involves continuously learning new, equitable, and planet-respecting behaviours and lifestyles. This perspective positions sustainability as an educational process characterised by constant change and systemic co-evolution. This evolving process profoundly impacts how universities structure their programs and teaching practices (Leal Filho et al., 2019b; 2024), which must be adaptive, historically and geographically determined, and tailored to the context in which they operate while also encouraging reflection on the global impact of local actions.

Higher education institutions are called to become proper sustainability laboratories, where theory and practice merge into a holistic approach that Morin (2015) would describe as complex. Leal Filho et al. (2019a) emphasise that universities must embody sustainability principles in every aspect of their operations, from governance and internal policies to the everyday actions of the academic community. This transformation requires a radical rethinking of curricula, organisational structures, and relationships with the surrounding community, focusing on social responsibility and the third mission.

To make these processes of change and transformation practical and tangible, it is essential to implement educational processes oriented towards sustainable development. In higher education, ESD should be wider than transmitting specific knowledge, although this is necessary. As Lambrechts and Van Petegem (2016) highlighted, ESD must foster critical, complex thinking that connects disciplines and generates social and technological innovation. This approach requires a fundamental revision of teaching content and pedagogical methods (Tilbury, 2011), shifting from a traditional education model based on information transmission to one that promotes the active construction of knowledge, skills, and competencies (Gaffney & O'Neil, 2019). Through ESD pathways, students are no longer merely 'consumers of knowledge' but become producers, collaborating with other university community members to explore existing problems and identify potential solutions to sustainability issues (Calvano et al., 2019).

Promoting the transition from 'education about/to sustainable development' to 'education for sustainable development' and eventually to 'sustainable education' (Sterling, 2006) is essential. This latter dimension represents the most advanced and transformative level of education, where the entire educational process is re-envisioned through the lens of sustainability. This concept aligns with the idea of the transformative university proposed by Guzmán-Valenzuela (2016), which views the academic institution as an agent of social change, promoting democratic values and justice. In this context, universities prepare students for the workforce and shape them into active, aware citizens capable of contributing to a more sustainable society.

The challenges of ESD require a rethinking of pedagogical practices. Mulà et al. (2017) emphasise the importance of developing pedagogical approaches that equip students to respond to the complexities and uncertainties of the future. This challenge involves active learning experiences, problem-based learning, and collaborative projects that tackle real-world sustainability challenges. Such approaches enable students to develop transversal skills, including systems thinking, the ability to anticipate future scenarios, and decision-making competence under uncertainty.

Integrating ESD into university curricula is challenging (Leal Filho et al., 2019b; 2024). Lozano et al. (2019) identify common barriers, including resistance to change, a lack of specific skills among faculty, and the difficulty of integrating interdisciplinary topics into traditionally compartmentalised academic structures. However, the same authors note an increasing number of good practices and innovative initiatives in this area, such as creating interdisciplinary courses, introducing sustainability modules into existing programs, and developing action research projects involving students and local communities. Another important aspect is the need to contextualise ESD according to the different cultural and socio-economic realities in which educational processes are implemented. As Stephens et al. (2008) point out, universities must adapt their ESD approaches to local contexts while maintaining a global perspective. This requires developing international partnerships and exchanging best practices among institutions worldwide.

Finally, ESD's ethical and political dimension is crucial to higher education. As Makrakis (2017) argues, it cannot overlook critical reflection on social and environmental justice issues. This approach involves training professionals who are technically competent, ethically aware, and socially responsible. Students should be encouraged to reflect on the ethical implications of their future professional decisions and their role in promoting a sustainable future.

3 Sustainable Futures: Developing Competencies at University Level

In the context of liquid modernity, as theorised by Bauman (2000), we face an ever-changing reality in which it is difficult to find solid reference points. The effectiveness of knowledge and skills developed through traditional teaching methods, including those in higher education, is often questioned (Barnett, 2000). Consequently, universities need to redefine the role of preparing individuals capable of navigating the uncertainty and complexity of the twenty-first century (Nussbaum, 2010).

Higher education stands at a critical crossroads: the choice between educating highly specialised professionals who may be civically disengaged or educating active, reflective and participatory citizens (Baldacci, 2019). This dichotomy reflects the tension between labour market demands and the need to prepare students to address increasingly complex and interconnected global challenges. However, growing awareness of environmental, economic and social crises has underlined the urgency of educating individuals who can excel in their professions and actively contribute to building a sustainable future.

Rieckmann (2012) identifies two possible paths for universities committed to sustainable development: a) sustainable universities ensure that students develop the transversal competencies needed to tackle complex problems and contribute to building a fairer and more just future; b) universities commit to building concrete sustainable development through their work. Universities can commit to sustainable development training future professionals who will then design and implement sustainability policies. For this reason, universities should create new professional profiles (green and sustainable) and design their educational pathways to train both prepared and flexible professionals and responsible citizens. In this way, universities can also reclaim their educational role (Calvano, 2021), as they do not limit themselves to guaranteeing the acquisition of new knowledge.

Sustainability competencies, understood as a set of interrelated skills, attributes and knowledge that enable individuals to address and solve complex problems related to sustainable development, have been at the centre of the international research debate on education for sustainable development for several years (Wiek et al., 2011; Fedeeva & Mochizuki, 2010). According to Unesco (2017), sustainability competencies are essential for achieving the Sustainable Development Goals (UN, 2015), particularly SDG 4, which aims to ensure inclusive, equitable, and quality education and promote lifelong learning opportunities for all. Educational pathways that foster the development of these competencies are the university's response to the need to educate about uncertainty and complexity (Fedeeva & Mochizuki, 2010).

Many frameworks for sustainability competencies have been developed. Wiek et al. (2011) describe a framework of seven key and transversal competencies crucial in sustainability: problem-solving, strategic, social, interpersonal, systemic thinking, normative and anticipatory. Unesco (2017) adds to this list by including collaborative competence, critical thinking and self-awareness, highlighting the need for a more integrated educational approach that combines cognitive, socio-emotional and behavioural aspects. It is important to note that these competencies are not isolated but interrelated and mutually reinforcing. For example, systems thinking is fundamental to integrated problem-solving, while self-awareness supports the development of interpersonal skills.

Furthermore, these competencies need to be developed contextually, considering the cultural, social and environmental specificities of the context in which each student lives and works. The decision to prioritise these two frameworks and not consider others (e.g., the significant ones by Barth et al., 2007; Lozano et al., 2017; Rieckmann et al., 2018; Brundiers et al., 2021; European Commission, 2022) was driven by their relevance to the research described in this article. These frameworks were deemed more suitable for achieving the study's objectives.

These transversal competencies do not replace specific disciplinary knowledge but integrate it, providing students with the tools to tackle complex challenges in the real world (Lozano et al., 2017). Integrating these competencies into university curricula requires significantly rethinking traditional educational practices. Unesco (2020), through its Education for Sustainable Development for 2030 framework, has further emphasised the importance of these competencies in promoting action for sustainable development. This approach recognises that increasing basic sustainability literacy, as taught in most countries, will not move societies towards sustainable development (Id., 2020). Therefore, a more transformative approach to education is needed to foster the sustainability competencies necessary for all students to become change agents in their communities and beyond (Rieckmann et al., 2018; Unesco, 2020).

Moreover, the shift towards interdisciplinarity and transdisciplinarity is emerging as a key challenge for education for sustainable development (Balsiger et al., 2017). This approach requires universities to promote learning processes that transcend traditional disciplinary boundaries and facilitate dialogue with surrounding realities, businesses and territorial institutions. Achieving this vision will require a significant change in university teaching practices, which must be participatory, active and experiential.

By integrating sustainability competencies into curricula, adopting interdisciplinary approaches and promoting transformative learning experiences, higher education institutions can shape a new generation of leaders: competent professionals and global citizens ready to lead the transition to a fairer and more sustainable world (Rieckmann et al., 2018; Unesco, 2020).

4 Methodology

The two authors collected data from September 2021 to January 2022 at the University of Bari and the University of Verona. Data collection involves two separate rounds of survey submission for each university. The survey aims to show how students' sustainable development literacy levels can change due to introducing sustainability items into the attended course (e.g. accounting and pedagogy). The survey and its questions were designed considering the pedagogical need for sustainable development and the main literature on education for sustainable development. Queries were divided into two separate parts. The first part collected data on the personal features of the respondent (e.g., age, studies). The second part explores the level of students' literacy on sustainable development. A 5-point Likert scale was employed, in which 1 was the lower lever rather than 5, which is the maximum literacy rate or importance of a specific issue (e.g., 2030 Agenda, sustainability).

In summary, since the aim was to determine the influence that teaching sustainable development had on the level of literacy in the attending students, the same survey was submitted at the beginning and end of two different courses. These two courses focus of our analysis were carried out during the first academic semester of the Italian academic year 2021/2022 and are on accounting and pedagogy with an average of 150 students per class. This approach permitted the collection of information on the students' evolution in terms of literacy level and students' perception of sustainable development concerns.

The authors employed the same teaching approach in both courses to improve students' literacy on sustainable development: general information was provided throughout the course, with a dedicated section focusing specifically on sustainable development. In particular, frontal lessons were dedicated to presenting the concept, history, and main issues of sustainable development. Then, students were involved in a workshop where they were required to show a case study on how the SDGs impact a selected case (e.g., a university, company, or non-profit organisation). The case analysis permitted students to reflect on the sustainability of their university.

Finally, divided into groups of up to eight, the students identified unsustainable situations in the studied contexts and collaboratively developed potential solutions to address these issues. Each group was also asked to design information, awareness, and education campaigns on sustainable development involving participants from universities, companies, and non-profit organisations.

These several trigger actions during the 6-h lesson dedicated to sustainable development enhanced the active learning process. Students were involved in a learning process based on direct experience. In particular, university teachers asked them to reflect on sustainable solutions that improve common contexts (e.g., universities, firms, cities) and concretely solve problems that directly or indirectly affect their daily lives. These requests have enabled them to cultivate a stronger sense of belonging and increased motivation to participate.

5 Results

This section presents the first attempt at data analysis of 370 valid answers from the two surveys. Data are presented by comparing the results, considering the answers at the beginning and end of the two courses. The first subsection proposes the sample description, and the second one presents the main results on students' perception and literacy on sustainable development.

Table 1 Sample description

	Ex Ante	Ex post facto
Gender	166 Female; 23 Male; 1 I prefer not to answer	152 Female; 27 Male; 1 I prefer not to answer
Age	Mean 22; Min 20; Max 31	Mean 21,76; Min 20; Max 31
Courses	Accounting 87; Pedagogy 103	Accounting 40; Pedagogy 140
Level of degree	Bachelor's degree 126; master's degree 64	Bachelor's degree 147; master's degree 33

Data are presented through a selection of the main results of the two surveys. The soft-colour graphs present the ex-ante data collection, and the stronger-colour graph the ex post facto data collection.

5.1 *Sample Description*

The data collection ensures 370 valid answers: 190 from the ex-ante and 180 from the ex-post facto (Table 1).

5.2 *Level of Sustainable Development Literacy*

The results from the core part of the questionnaire highlight an increasing general awareness of "Sustainability" (Fig. 1), the "2030 Agenda" (Fig. 2) and the concept of "sustainable development" (Fig. 3) in both types of courses.

Similar results have been raised on their knowledge of "circular economy" (Fig. 4) and "inclusion and equal opportunities" (Fig. 5), topics that were embedded into the program and directly discussed during both courses.

The survey also explores how has been changed their perception in terms of the importance of the 2030 Agenda (Fig. 6) in different aspects of the student's life, such

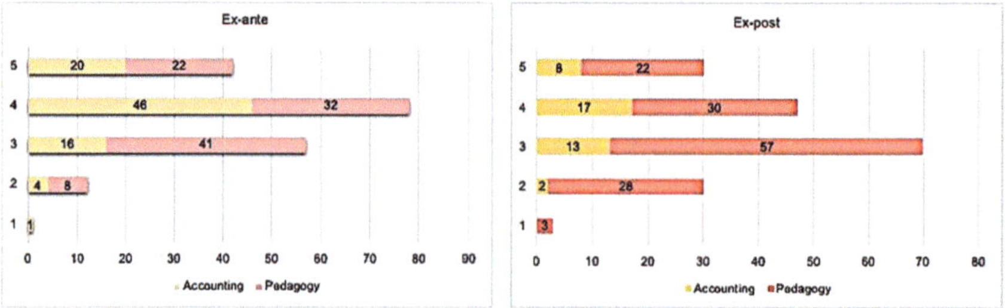

Fig. 1 How much have you heard about "Sustainability"?

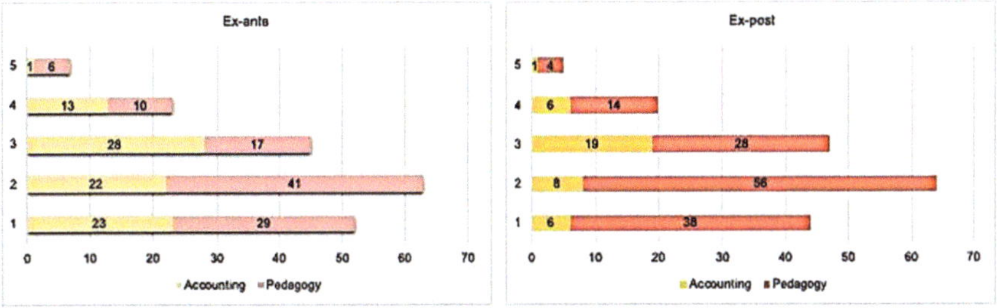

Fig. 2 How much have you heard about "2030 Agenda"?

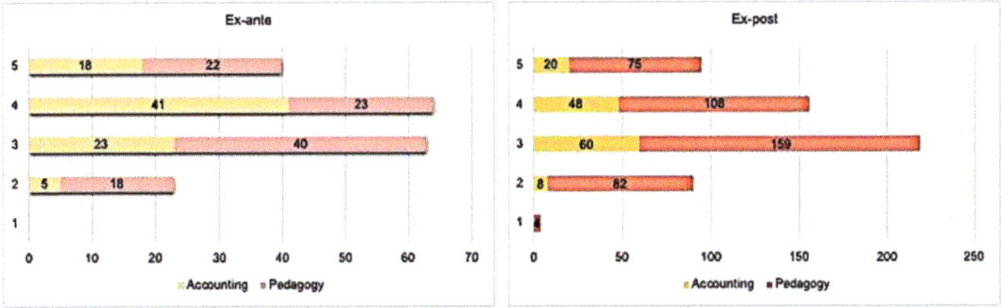

Fig. 3 How much have you heard about "Sustainable Development"?

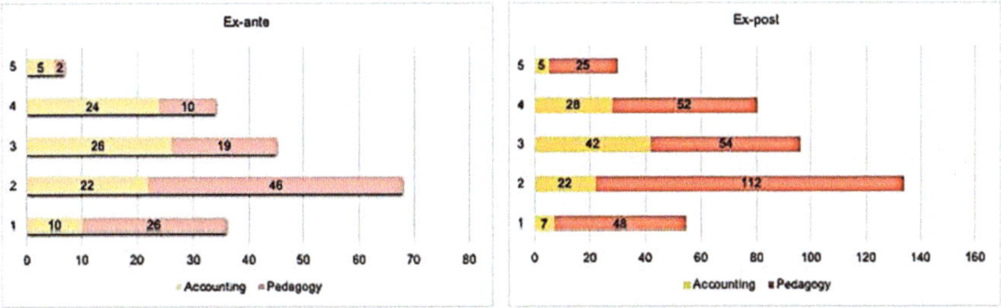

Fig. 4 How much have you heard about "Circular economy"?

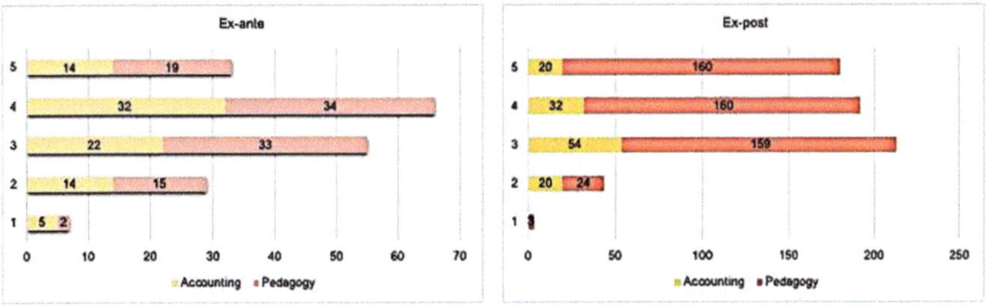

Fig. 5 How much have you heard about "inclusion and equal opportunities"?

as personal (Fig. 7) and professional growth (Fig. 8). In addition, we also explore their thought on the importance that other students (not attending) can benefit from participation in the course (Fig. 9) as well as the need for other professors to learn more about the 2030 Agenda (Fig. 10).

During the courses, the students were involved in project works, which required them to critically analyse concrete cases in which SDGs were employed as attitudes, strategies, and actions toward sustainability. In doing so, students learn by considering organisations from different fields, such as NGOs, the public sector, and for-profit entities. Considering these aspects, the survey also requests opinions on the importance of the 2030 Agenda for different kinds of organisations, such as

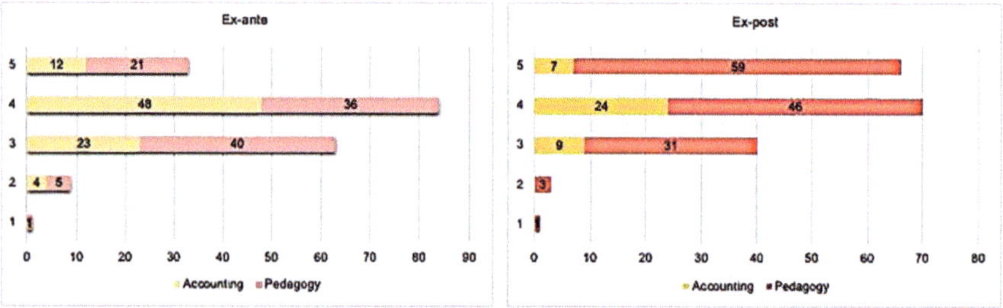

Fig. 6 How much do you think is important to learn about the 2030 Agenda and the 17 SDGs?

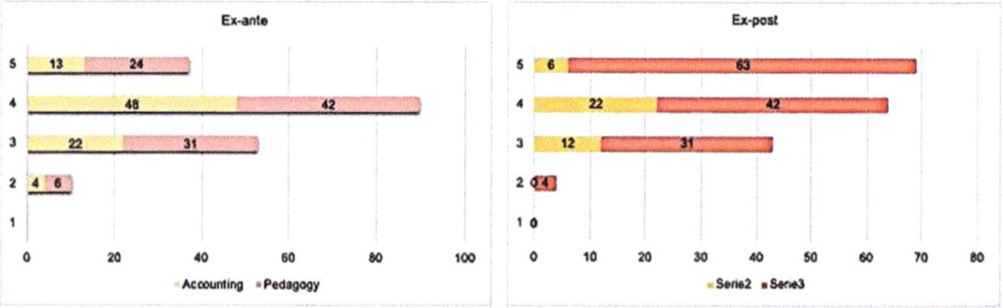

Fig. 7 How much is important to learn about the 2030 Agenda for your personal growth?

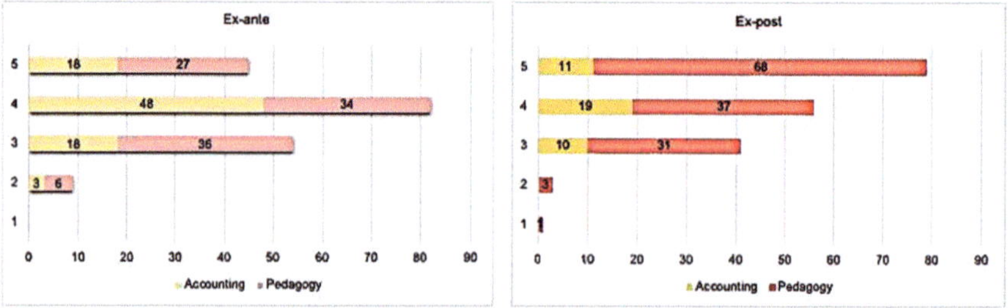

Fig. 8 How important is learning the 2030 Agenda for your professional growth?

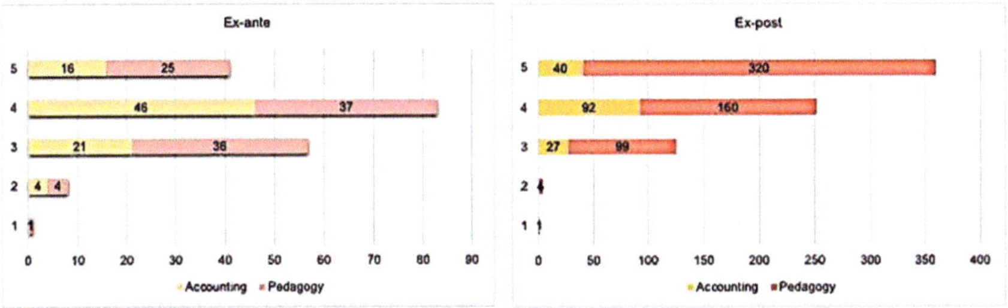

Fig. 9 How much do other students need to learn about the 2030 Agenda?

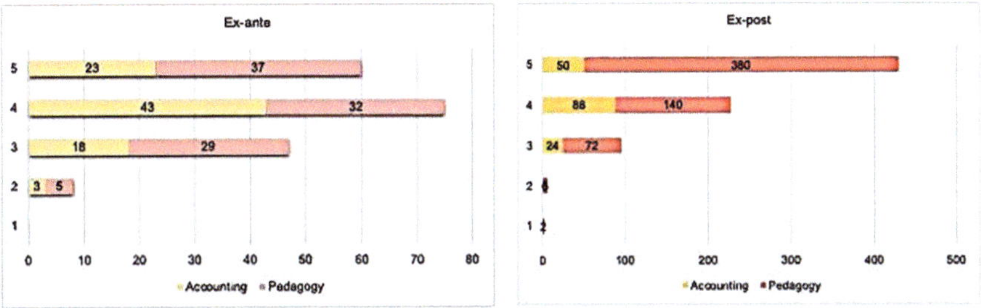

Fig. 10 How much do other professors need to learn about the 2030 Agenda?

businesses, universities, non-profit organisations, and civil society (Figs. 11, 12, 13, 14 and 15).

Finally, the students were asked how to improve their learning during the related course, and many of them requested more practice and a concrete project to apply to steer the change toward a more sustainable development.

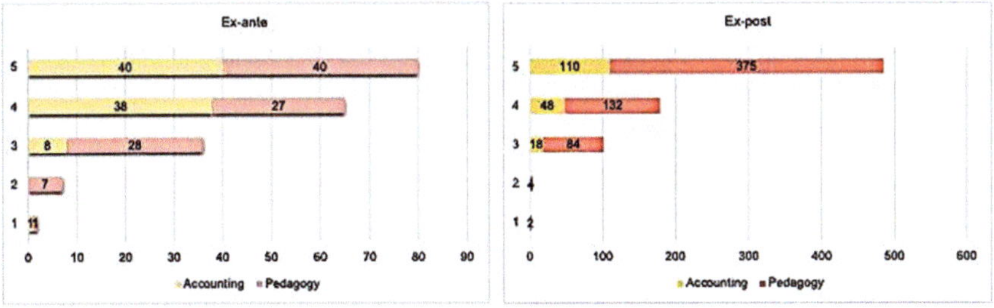

Fig. 11 How important do you think the pursuit of the 2030 Agenda Goals can be for businesses?

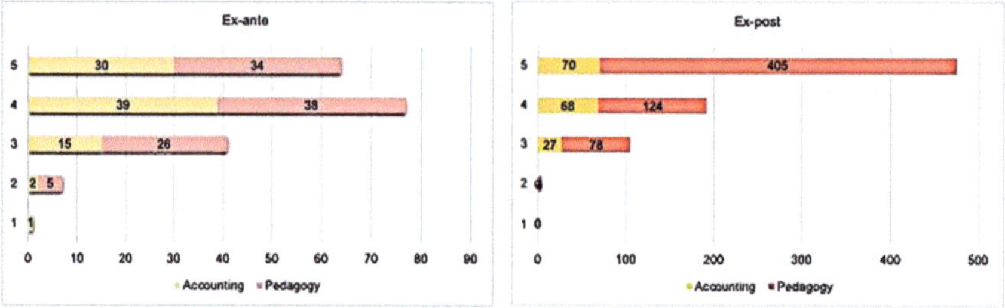

Fig. 12 How important do you think pursuing the 2030 Agenda Goals can be for universities?

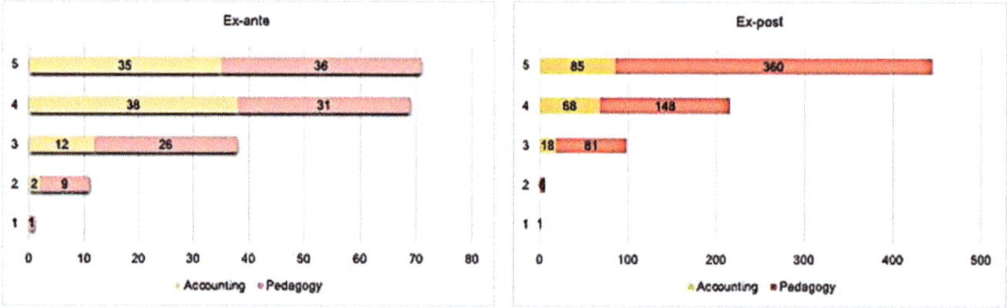

Fig. 13 How important do you think pursuing the 2030 Agenda Goals can be for the public sector?

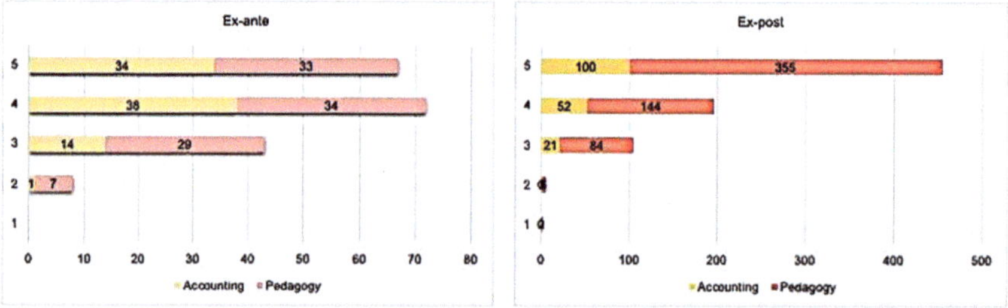

Fig. 14 How important do you think pursuing the 2030 Agenda Goals can be for the NPOs?

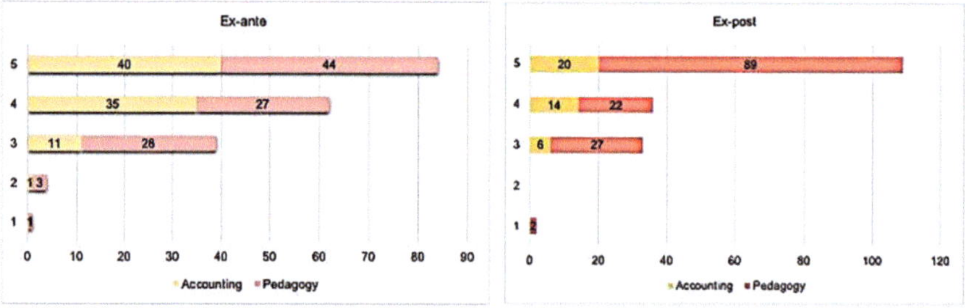

Fig. 15 How important do you think pursuing the 2030 Agenda Goals can be for civil society?

6 Conclusions

This study aimed to investigate the impact of a common framework for teaching sustainable development on students from different disciplines, accounting and pedagogy. By analysing the evolution of students' sustainable development literacy, we sought to understand the potential of such a framework to foster a more holistic understanding of sustainability issues.

The challenges posed by ESD necessitate fundamentally rethinking traditional teaching and training methods. Scholars underscore the importance of developing pedagogical approaches that prepare students to steer the complexities and uncertainties of the future (e.g., climate change). This challenge involves adopting active learning strategies, problem-based learning, and collaborative projects that address real-world sustainability issues. Such approaches foster the development of transversal skills, including systems thinking, the ability to anticipate future scenarios, and decision-making in uncertain contexts.

To address the complexities of ESD, educational approaches must be rethinking. Mulà et al. (2017) emphasise the need for pedagogical strategies that empower students to steer the uncertainties of the future toward sustainable development. This need demands active learning, problem-based learning, and collaborative projects that tackle real-world sustainability challenges. Such approaches promote cross-skills among fields of study. By integrating sustainability competencies into curricula, adopting interdisciplinary approaches, and promoting transformative learning experiences, higher education institutions can cultivate a new generation of leaders: competent professionals and global citizens equipped to steer the transition towards a more equitable and sustainable future (Rieckmann et al., 2018; Unesco, 2020). In this vein, interdisciplinary and transdisciplinary approaches are becoming increasingly crucial for ESD (Unesco, 2017). Universities must foster learning processes that surpass traditional disciplinary boundaries and engage with external stakeholders (e.g., entrepreneurs, local entities, and citizens). This requires a fundamental shift towards participatory, active, and experiential learning (Leal et al., 2024).

The study aimed to assess the impact of a specific course on students' understanding and perception of sustainable development in courses on accounting and pedagogy. The results indicate a significant increase in awareness of key sustainability concepts such as the 2030 Agenda, sustainable development, and circular economy in both courses. Students also demonstrated an amplified awareness of the importance of the 2030 Agenda in various aspects of their lives, including personal and professional growth, and they recognised the need for other students and professors to be more knowledgeable about the 2030 Agenda. This issue doubts the knowledge of scholars working in higher education about sustainability and requests urgent and appropriate training for teachers at all levels of the Italian education system.

Furthermore, the study revealed a positive perception of the role of different organisations in achieving the 2030 Agenda goals, highlighting the importance of collaboration between businesses, universities, public sector entities, and civil society, incorporating more practical exercises and real-world projects. This feedback underscores the need for experiential learning opportunities to reinforce the third mission also in this sense (Gaffney & O'Neil, 2019).

The present study proposes a first overview of a study that involved the Universities of Bari and Verona in Italy. The primary purpose is to explore how the same teaching approach toward the 2030 Agenda and SDGs can differently (or not) affect students' literacy levels from different courses (accounting and pedagogy). The collected data presents some suggestions, as the reactions from students are similar in both courses. This suggests that a common approach can be applied.

At a policy level, this implies the definition of a common national framework to implement sustainable development literacy processes in higher education. In this research field's future venues, further studies should apply a common framework on pilot educational cases to provide more insights on how this framework can work or not) at different levels (e.g., bachelor's, Master's degree, PhD).

Since its explorative nature, the present research has some limitations. First, it stemmed from a limited number of observations and one specific academic year. Second. It refers to two specific courses. Further research should enlarge the scale of the study to other courses rooted in STEM subjects. Additionally, an experimental approach could identify differences between the adaptation of the approach and 2030 topics according to the area of study.

In conclusion, this study provides valuable insights into the potential of a common framework for teaching sustainable development. By fostering a deeper understanding of sustainability issues and promoting active citizenship, universities can play a pivotal role in addressing global challenges and building a more sustainable future.

References

Ávila LV, Beuron TA, Brandli LL, Damke LI, Pereira RS, Klein LL (2019) Barriers to innovation and sustainability in universities: An international comparison. Int J Sustain High Educ 20(5):805–821

Baldacci M (2019) La scuola al bivio. Mercato o democrazia. Franco Angeli, Milano

Balsiger J, Förster R, Mader C, Nagel U, Sironi H, Wilhelm S, Zimmermann AB (2017) Transformative learning and education for sustainable development. GAIA-Ecological Perspectives for Science and Society 26(4):357–359

Barnett R (2000) University knowledge in an age of supercomplexity. High Educ 40:409–422

Barth M, Godemann J, Rieckmann M, Stoltenberg U (2007) Developing key competencies for sustainable development in higher education. Int J Sustain High Educ 8(4):416–430

Bauman Z (2000) Liquid modernity. Polity Press, Cambridge

Bologna, G. (2018). È possibile lo sviluppo sostenibile nell'Antropocene? In Von Weizsacker, E.U., Wijkman, A., *Come On! Come fermare la distruzione del pianeta.* Firenze: Giunti, 19–48.

Brundiers K, Barth M, Cebrián G, Cohen M, Diaz L, Doucette-Remington S, Zint M (2021) Key competencies in sustainability in higher education an agreed-upon reference framework. Sustain Sci 16(1):13–29

Calvano G (2017) Educare per lo sviluppo sostenibile. L'impegno degli Atenei italiani: esperienze in corso e buone pratiche. Aracne, Roma

Calvano G (2021) The third mission that looks to the future is for a sustainable, transformative, civic university. Formazione Lavoro Persona 33:75–89
Calvano G, Paletta A, Bonoli A (2019) How the structures of a Green Campus promote the development of sustainability competences. The experience of the University of Bologna. In: Leal Filho W, Bardi U (eds) Sustainability on University Campuses: Learning, Skills Building and Best Practices. Springer, Cham, pp 31–44
Derrida, J. (2002). *L'Università senza condizione.* Milano: Raffaello Cortina.
Disterheft A, Caeiro S, Azeteiro UM, Leal Filho W (2015) Sustainable universities -a study of critical success factors for participatory approaches. J Clean Prod 106:11–21
Dlouhá J, Henderson L, Kapitulčinová D, Mader C (2018) Sustainability-oriented higher education networks: Characteristics and achievements in the context of the UN DESD. J Clean Prod 172:4263–4276
European Commission (2022). *GreenComp, the European sustainability competence framework.* Luxemburg: Publications Office of the European Union
Fedeeva Z, Mochizuki Y (2010) Higher education for today and tomorrow: university appraisal for diversity, innovation and change towards sustainable development. Sustain Sci 5:249–256
Gaffney, J.L., O'Neil, J.K. (2019). Experiential Learning and Sustainable Development. In: Leal Filho, W. (eds) Encyclopedia of Sustainability in Higher Education. Springer, Cham. https://doi.org/10.1007/978-3-030-11352-0_348
Goddard J, Vallance P (2013) The University and the City. Routledge, London
Guzman-Valenzuela C (2016) Unfolding the meaning of public(s) in universities: toward the transformative university. High Educ 71:667–679
Hessen, D. and Schmelkes, S. (2022). *Higher Education and the SDGs. A synthesis based of the Report of the Global Independent Expert Group on the Universities and the 2030 Agenda (EGU2030).* Paper commissioned for the World Higher Education Conference 18–20 May 2022. Paris: Unesco. https://unesdoc.unesco.org/ark:/48223/pf0000389861
Jacobi, P. R., Toledo, R. F., & Grandisoli, E. (2016). Education, sustainability and social learning. *Brazilian Journal of Science and Technology, 3*(3).
Lambrechts W, Van Petegem P (2016) The interrelations between competences for sustainable development and research competences. Int J Sustain High Educ 17(6):776–795
Leal Filho W (2011) About the Role of Universities and Their Contribution to Sustainable Development. High Educ Pol 24:427–438
Leal Filho W, Raath S, Lazzarini B, Vargas VR, de Souza L, Anholon R, Quelhas OLG, Haddad R, Klavins M, Orlovic VL (2018) The role of transformation in learning and education for sustainability. J Clean Prod 199:286–295
Leal Filho W, Azeiteiro U, Alves F, Pace P, Mifsud M, Brandli L, Disterheft A (2019) Reinvigorating the sustainable development research agenda: the role of the sustainable development goals (SDG). Int J Sust Dev World 26(2):176–188
Leal Filho W, Shiel C, Paço A, Mifsud M, Veiga Ávila L, Londero Brandli L, Molthan-Hill P, Pace P, Azeiteiro UM, Ruiz Vargas V, Caeiro S (2019) Sustainable Development Goals and sustainability teaching at universities: Falling behind or getting ahead of the pack? J Clean Prod 232:285–294
Leal Filho, W., Salvia, A. L., Pretorius, R. W., Brandli, L. L., Manolas, E., Alves, F., ... & Vargas, V. R. (2020). *Universities as Living Labs for Sustainable Development: Supporting the Implementation of the Sustainable Development Goals.* Cham: Springer Nature.
Leal Filho, W., Salvia, A. L., Brandli, L., Azeteiro U. M., Pretorius, R. (Eds.) (2020). *Universities, Sustainability and Society: Supporting the Implementation of the Sustainable Development Goals.* Cham: Springer Nature.
Leal Filho, W., Dibbern, T., de Maya, S.R., Alarcón-del-Amo, M., Marin Rives, L. (Eds.) (2024). *The Contribution of Universities Towards Education for Sustainable Development.* Cham: Springer Nature.
Loorbach DA, Wittmayer J (2024) Transforming universities. Sustain Sci 19:19–33. https://doi.org/10.1007/s11625-023-01335-y

Lozano R, Merrill MY, Sammalisto K, Ceulemans K, Lozano FJ (2017) Connecting competences and pedagogical approaches for sustainable development in higher education: A literature review and framework proposal. Sustainability 9(10):1889

Makrakis, V. (2017). Developing and validating a sustainability justice instrument to transform curriculum, learning and teaching. *Proceeding 9th International Conference in Open & Distance Learning*, November, Athens.

Morin E (2015) Insegnare a vivere. Manifesto per cambiare l'educazione. Raffaello Cortina Editore, Milano

Mulà, I., Tilbury, D., Ryan, A., Mader, M., Dlouha, J., Mader, C., Benayas, J., Dlouhy, J., Alba, D. (2017). Catalysing Change in Higher Education for Sustainable Development. A review of professional development initiatives for university educators. *International Journal of Sustainability in Higher Education, 18*(5), 798–820.

Nussbaum MC (2010) Not for profit: Why democracy needs the humanities. Princeton University Press, Pricenton

Rieckmann M (2012) Future-oriented higher education: Which key competencies should be fostered through university teaching and learning? Futures 44(2):127–135

Rieckmann M (2018) Learning to transform the world: Key competencies in Education for Sustainable Development. In: Leicht A, Heiss J, Byun WJ (eds) Issues and trends in Education for Sustainable Development. Unesco Publishing, Parigi, pp 39–59

Shiel, C., Leal Filho, W., do Paço, A., & Brandli, L. (2016). Evaluating the engagement of universities in capacity building for sustainable development in local communities. Evaluation and Program Planning 54:123–134

Stephens JC, Romàn M, Graham AC, Scholz RW (2008) Higher education as a change agent for sustainability in different cultures and contexts. Int J Sustain High Educ 9:317–338

Sterling, S. (2006). *Educazione sostenibile.* Cesena: Anima Mundi.

Tilbury D (2011) Education for Sustainable Development: an Expert Review of Processes and Learning. Unesco, Paris

Trencher G, Yarime M, McCormick KB, Doll CN, Kraines SB (2014) Beyond the third mission: exploring the emerging university function of co-creation for sustainability. Science and Public Policy 41:151–179

Unesco (2017). Education for sustainable development goals. Learning objectives. Unesco, Paris

Unesco (2020) Education for sustainable development: a roadmap. Unesco, Paris

United Nations (2015). *Transforming our World. The 2030 Agenda for sustainable development.* In https://sdgs.un.org/2030agenda.

Wals AEJ, Jickling B (2002) "Sustainability" in higher education: from doublethink and newspeak to critical thinking and meaningful learning. Int J Sustain High Educ 3:221–232

Wiek A, Withycombe L, Redman CL (2011) Key competencies in sustainability: A reference framework for academic program development. Sustain Sci 6:203–213

Sustainability in University Curricula. Comparison of the Environmental Competences of Cross-Disciplinary Subjects in a University in the Metropolitan Area of Buenos Aires, Argentina

Norma Gabriela Faitani

Abstract Universities around the world are showing an increasing tendency to respond positively to calls for sustainability on their campuses. As generators of knowledge through research, teaching and outreach, universities have a responsibility to create and maintain a strong environmental education base for all their students and users. In this context, this paper investigates the curricula of subjects that cut across more than one degree programme in a Higher Education Institution (HEI) in the Buenos Aires Metropolitan Area. The main objective is to identify subjects common to more than one degree or undergraduate course and to analyse how they are aligned with the key competences for sustainability. To this end, a survey was sent to the teachers responsible for teaching these subjects, which included criteria on subjects taught, time/unit coverage, specific competences, techniques and forms of assessment. The information gathered was used to compile the Higher Education Sustainability Indicator (HESI), assessing the competences in the dimensions of knowing, doing and being of the student and future graduate. Among the most relevant results, it is worth highlighting that the competences promoted by teachers focus more on reflecting (knowing and doing) than on acting (doing and being). Furthermore, it was identified that some graduates do not have competences for sustainability in their curricula due to the lack of compulsory or optional subjects with environmental themes. This forces students to seek extracurricular environmental training, which is limited at HEIs.

Keywords Education for sustainability · Higher education · Competences · Curriculum analysis · Sustainability indicator

---

N. G. Faitani (✉)
Universidad Nacional de General Sarmiento-Becaria Post Doctoral-CONICET, Los Polvorines, Argentina
e-mail: nfaitani@campus.edu.ungs.ar

W. Leal Filho et al. (eds.), *Education for Sustainable Development: The Contribution of Universities*, World Sustainability Series,
https://doi.org/10.1007/978-3-031-86985-3_26

1 Introduction

The incorporation of environmental issues in university curricula is increasingly essential for the incorporation of the environmental perspective and environmental sustainability in the training of any professional graduating from a higher education institution. Environmental education (EE) must move away from being just a line of intervention in primary and secondary schools and become a transversal axis from initial education to higher education and continue at the highest levels of training. Sustainability in higher education has gained prominence in the context of AE following the adoption of the 2030 Agenda for Sustainable Development. It is now considered one of the key indicators of educational quality (Anghel and Neculau 2022).

EE articulates current socio-environmental issues by combining theory with approach and action. Current issues vary according to context, but the most global are those related to climate change, natural resource depletion and associated environmental degradation, declining environmental quality in urban and rural areas, loss of biodiversity, migration and loss of indigenous peoples, frequent zoonoses and pandemics, among others.

According to Mora and Guerrero (2022), tackling these issues and their causes has to do with educational processes in which students learn to participate in successful activities, in relation to relevant and strategic policies of society. In the same way, they must act critically in the face of these issues, take part, and deliberate not only about the solution of the problem but also about understanding the focus or source of the issue. In this framework, the authors argue that it is not enough for higher education institutions to transform themselves into green or sustainable spaces in management, i.e. in the installation of differentiated baskets, saving energy or encouraging the use of bicycles, but it must be a more profound change and teach them to learn, understand and adopt the problem and its effects as their own and make decisions about it.

The term 'competence' is used in various fields, professional and academic fields (Mulder 2007). Commonly, an architect, a teacher or any professional is competent when he/she performs his/her task well or recreates the work or job he/she is asked to do. However, due to its everyday and ambiguous use, this term requires some clarification. Weinert (2001) and other authors have analysed the elements that characterise competences. According to Weinert, competences are defined as the cognitive abilities and skills that individuals possess or can learn in order to solve certain problems. In addition, they include the motivation, willingness and skills needed to successfully and responsibly use available solutions and tools in variable situations.

Competences involve the integration of knowledge, procedures and attitudes, oriented towards action. Being competent implies knowing how to apply what one knows and is to solve problems or issues efficiently (Cano 2008; Martínez et al. 2012). In university teaching, the definition of competence implies integrating knowledge, skills, attitudes and values oriented towards reflective action. Therefore, it is necessary to design methodological scenarios that allow these factors to be developed and guarantee the acquisition of competences.

From the review of the literature on competences, the characteristics that Pulido Trullén (2008) and other authors such as Blanco Blanco (2008), Bisquerra Alzina and Pérez Escoda (2007), Cano (2008) and Martínez et al. (2012) consider essential for competences were identified. The basic elements of a competence are the integration of knowledge, skills, attitudes and values. Competences are developed continuously, must apply to real problems and cover personal, professional and social aspects, assessing effectiveness versus efficiency. In addition, competences are defined and updated through action, are linked to a specific context and enable problem situations to be solved, facilitating effective solutions to known and new problems.

In higher education, the curriculum stands for the entire planned educational experience. This encompasses not only the content taught, but also the methods used and the purposes behind the teaching. Incorporating sustainability into the curriculum implies a comprehensive approach that integrates sustainable principles, values and practices into all these aspects of education (Lund University 2024). By working through competences, people can solve problems and perform activities, enabling them to construct knowledge and transform reality. This involves integrating knowing how to be, knowing how to know and knowing how to do, taking into account the requirements of the environment and needs through critical awareness and self-knowledge. Competency-based education promotes autonomous learning, personal growth, socio-economic development and curricular organisation, transcending the subject-based curriculum. In the field of sustainability, competences must respect three specific areas: knowing, being and doing. Being competent implies knowing how to apply what one knows and what one is, to solve problems or issues efficiently (Godínez López 2013; Albareda-Tiana and Gonzalvo-Cirac 2013; Martínez et al. 2012).

This paper investigates the curricula of subjects that cut across more than one degree programme in a Higher Education Institution (HEI) in the Metropolitan Area of Buenos Aires, Argentina. Firstly, subjects common to more than one degree or undergraduate course were explored. Subsequently, a comparison was made by analysing the key competences for sustainability. This was done by a survey sent to teachers responsible for teaching these subjects, including criteria such as topics taught, time/unit coverage, specific and technical competences, and forms of assessment. The information gathered was used to form the Higher Education Sustainability Indicator (HEIS), assessing competences in the dimensions of knowing, doing and being of the student and future graduate.

2 Key Competences in Sustainability (KCS)

Lozano et al. (2017), report that there is little research on the connection between the way courses are taught (pedagogical approaches) and how they can affect sustainability competences, however, the literature highlights the need to integrate sustainability in higher education. Lozano et al. (2021), developed a framework for connecting competences with pedagogical approaches in higher education for sustainability. This framework aims to help educators create and update their courses to provide a more comprehensive, holistic, and systemic education for future leaders, decision-makers, educators and change agents. Key competences mentioned include systems thinking, interdisciplinary work, anticipatory thinking, justice, responsibility and ethics, and communication and use of media. According to Barth et al. (2007), competences for sustainability include systems thinking, long-term foresight and cross-sectoral collaboration. Brundiers et al. (2010) emphasise the importance of competences for sustainable action, while Wiek et al. stress the development of problem-solving skills and creativity.

Key competences for sustainability are understood as a set of functionally linked knowledge, skills and attitudes that enable them to successfully perform a task and solve a problem in their personal and professional lives (Baartman et al. 2007). In terms of environmental education and in particular in higher education, key competences are a fundamental reference point for developing the knowledge and skills profile that is intended for future graduates to be critical, to propose solutions and to be managers of the transition towards more sustainable actions. Education for sustainability should empower students to analyse and address sustainability challenges, prepare for and mitigate future issues, and create and capitalize on opportunities to achieve short-term sustainability goals. Gago Cortés mentions that the most developed competences in higher education at a university in Spain were personal commitment, justice, responsibility and ethics, strategic action, assessment and evaluation, and tolerance of ambiguity and uncertainty. Jiménez-Munguía, after conducting a study on key competences for sustainability at a university in Mexico, found that the student community developed anticipatory thinking, assessment and evaluation, and interpersonal relationships and peer collaboration for problem solving.

Specifically, to achieve these key competences in sustainability, some authors suggest the use of the Socratic method, group discussion, role play, personal or group diaries, brainstorming, peer assessment, internships, solving real community problems, outdoor education, literature research, problem analysis, values clarification, case studies and concept maps (Lambrechts et al. 2013). (Lambrechts et al. 2013). Pietikäinen and D'Amato (2021) identify project- or problem-based learning, the case study approach and lecture teaching (Favi 2021) in higher education as pedagogical approaches. Cotton and Winter (2010) suggested the use of role-plays and simulations, group discussions, class starter activities such as watching a mind-opening video, brainstorming and fieldwork, among others. Lozano et al., considers that progress has been made in incorporating competences for sustainable development and pedagogical approaches. However, he argues that in order to achieve sustainability, it is necessary to combine traditional pedagogical approaches and approaches designed for sustainability. In the first group we can find lectures and case studies, which need to be rethought and redesigned to better develop competences and ultimately education for sustainability. Capacities must also be built to connect the pedagogical approaches with the greatest potential to right competences.

On the other hand, Redman and Wiek (2021) carried out a study and survey of the competences related to sustainability in higher education by analysing publications in specialised education journals from 2005 to 2020. The competences they were able to identify are systems thinking competence, anticipatory competence, normative competence, strategic competence, collaborative competence, critical thinking competence, self-awareness competence, integrated problem-solving competence. In addition, Brundiers et al. (2021) and Redman and Wiek (2021) also propose adding the following competences: intrapersonal competence (self-awareness competence), integration competence and application competence.

All the competences, according to the descriptions provided by the authors, were classified into the dimensions of Knowing, Doing and Being. The same is represented in Table 1.

Although key competences are widely applied in sustainability courses and programmes worldwide, their implementation is often insufficient (Trencher et al. 2018). Wiek and Redman (2022), mention the main drawbacks in the adoption of the key competence's framework in educational practice for sustainability competences, in parallel they list several effective strategies to overcome them. Both are presented in Table 2:

3 Materials and Methods

Case Study

This paper takes as a case study the Universidad Nacional de General Sarmiento (UNGS), located in the district of Malvinas Argentina's, in the Metropolitan Area of Buenos Aires, Argentina. Figure 1 shows the location of the HEI selected for this study.

The method of this work is divided into four steps designed to identify and evaluate cross-cutting subjects related to sustainability in more than one degree programme. The steps are described in detail below:

- Step 1. Diagnosis

The aim of this step is to recognise, by analysing the curricula, the cross-cutting subjects in more than one degree or undergraduate programme at UNGS. Cross-cutting subjects are those that address sustainability issues from a holistic and multidisciplinary perspective, allowing for a broader vision applicable to different academic and professional contexts. A comprehensive review of the curricula was carried out to identify these subjects.

- Step 2. Survey

In this step, the lecturers in charge of the subjects identified in Step 1 were contacted. They were sent a survey designed to inquire about the topics covered, the coverage in time and units, the specific competences and techniques taught, and the forms of assessment used in each subject. The survey aimed to gather detailed information on how the key competences for sustainability are taught.

Table 1 Dimensions of knowing, doing and being and associated competencies

Dimension	Competence	Description
Knowing	• *Cognitive competences* • *Systems thinking competence* • *Integration competences*	Represents a critical understanding of the problem at all levels (global, national, local). Uses concepts underlying socio-environmental issues from various disciplines, mastering theories on environmental development, social and economic policies, and production systems. Integrates the different environmental dimensions (social, cultural, economic, political, aesthetic, physical, biological). Develop critical thinking and the ability to communicate coherently and clearly, formulate relevant judgements, and evaluate values. Apply modelling skills and approach complex systems. Consider the three aspects of sustainable development (society, environment, and economy). Apply collective problem solving at different scales. Sustainable action plans
Doing	• *Methodological competences* • *Anticipatory thinking competence* • *Strategic thinking competences* • *Implementation competences*	Abilities, techniques, strategies and skills in methodological procedures applied in specific areas (organising, planning, designing, carrying out, manipulating). It involves making timely decisions, intervening methodologically, and carrying out actions related to sustainability and the rational use of resources. It fosters initiative-taking attitudes, creative thinking and problem solving. Build possible and dynamic future states through simulations and scenarios. Building a future vision to implement action, intervention, adaptation, transfer and efficiency plans
Being	• *Attitudinal competences* • *Competences to think values* • *Intra- and interpersonal competences*	Formation of attitudes and values necessary for the exercise of knowledge: responsibility, initiative, justice, equity, autonomy, respect, among others. Promotes new attitudes and values consistent with sustainability, fostering commitments to harmonious interaction with the environment and values that define rights in human relations. Establish links, generate negotiations, apply principles of sustainability. Interdisciplinary collaboration. Increase personal resilience, self-regulation and individual awareness

Source Constructed based on Brundiers et al. (2021), Redman and Wiek (2021), Godínez López (2013), Aznar and Ull (2009) and Blanco Blanco (2008)

Table 2 Challenges and effective strategies around the adoption of competences for sustainability

Inconveniences in the adoption of competences for sustainability	Effective strategies for the adoption of competences for sustainability
Insufficient coverage and integration of competences	Covering and integrating competences
Insufficient operationalisation of competences	Operationalisation of competences
Insufficient alignment between competences and pedagogies	Matching pedagogies to competences
Insufficient assessment of competence development	Evaluating the development of competences
Insufficient assessment of professional success	Evaluating professional success

Source Constructed based on Wiek and Redman (2022)

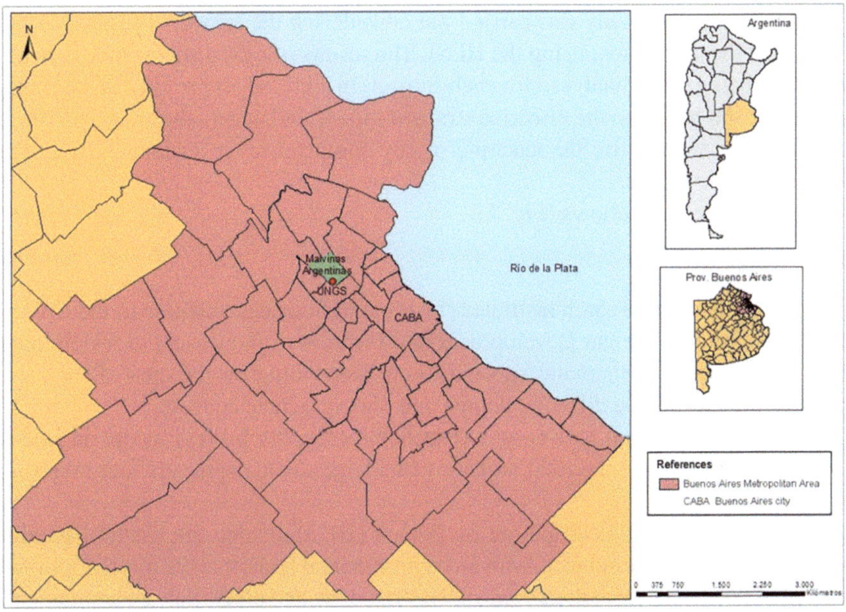

Fig. 1 Location of the National University of General Sarmiento. *Source* Geographic Information System Laboratory of the National University of General Sarmiento

- Step 3. Indicator

With the answers obtained from the survey, Higher Education Sustainability Indicator in (HESI) was assessed. This indicator assesses the competences in the dimensions of 'knowing', 'doing' and 'being' of the student and future graduate. The information collected was used to assess how the subjects contribute to the development of sustainability competences in each of these dimensions.

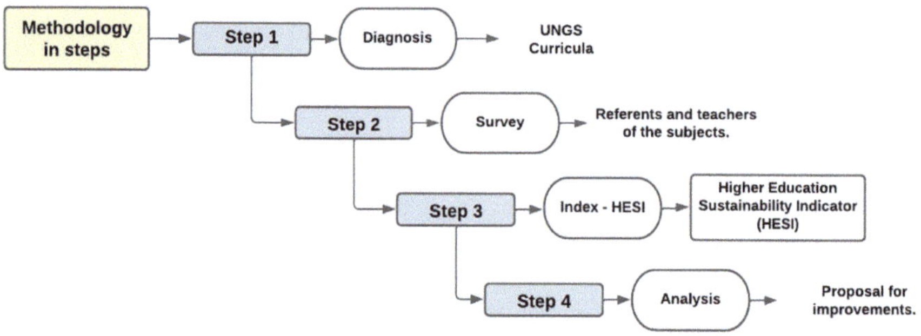

Fig. 2 Methodological scheme

- Step 4. Analysis

The analysis of the results was carried out considering the answers obtained in the survey and the assessment using the HESI. The results of each dimension (knowing, doing and being) were analysed for each subject. In addition, improvement measures were proposed based on the effective strategies identified during the analysis. These measures aim to optimise the teaching of key competences in sustainability in the cross-curricular subjects.

The steps were as follows (Fig. 2).

Step 1-Diagnosis

The UNGS has four research institutes (Institute of Sciences, Institute of the Conurbation, Institute of Human Development and Institute of Industries) in which there are different areas of thematic specialisation. Depending on the area of thematic specificity, they oversee different degree and undergraduate courses.

The UNGS has a total of twenty-two bachelor's degrees, professorships and engineering degrees. While the total number of undergraduate degrees is four (Website UNGS n.d).

According to a study carried out by Faitani (2023), among the UNGS bachelor degrees, there are some subjects with an environmental focus that are mostly optional and can be taken and accredited by any UNGS student, and others that are compulsory and must be taken to complete the study plan.

At the optional level, the optional course 'Bases of life and its evolution', although it has a marked tendency towards biology, has an environmental imprint, as research is carried out on environmental preservation strategies. This subject is shared by two of the teaching programmes offered by the UNGS. Another optional subject is 'Interdisciplinary Laboratory: Environmental Diagnosis'. This is an optional subject for all the teaching programmes.

Among the bachelor's degrees, something similar occurs as in the case of the professorships, except that there is one degree course with an entirely environmental focus, the Bachelor's Degree in Ecology. The Bachelor's Degree in Urban Planning has four subjects, three compulsory and one optional with an environmental focus ('City and Environment', 'Urban Sustainability', 'Urban Environmental Law' and 'Interdisciplinary Laboratory: Environmental Diagnosis', respectively). The other bachelor's degrees have only one optional subject, 'Interdisciplinary Laboratory: Environmental Diagnosis'.

Engineering degrees do not have the option of taking 'Interdisciplinary Laboratory: Environmental Diagnosis' but among their compulsory subjects they have 'Industrial Hygiene and Safety and the Environment.' Chemical Engineering also has two other optional subjects: 'Environmental Chemistry' and 'Environmental Technology and Management'.

In short, most of the subjects with an environmental focus are optional, which does not guarantee that all graduates will have knowledge of environmental care in their professional practice (see Fig. 3). On the other hand, the cross-cutting subject among degree courses, especially in engineering, is 'Industrial Hygiene and Safety and the Environment'. Among the teaching and degree courses, the cross-cutting subject is 'Interdisciplinary Laboratory: Environmental Diagnosis'. Although both are transversal, the difference lies in the fact that the former is compulsory for engineering degrees and the latter is optional for other degrees.

Step 2-Survey

Once the two subjects to be studied, 'Industrial Hygiene and Safety and the Environment' and 'Interdisciplinary Laboratory: Environmental Diagnosis', had been identified, a survey was sent to the heads of these subjects.

The survey has three sections. The first section asks about the teacher and their contact, another about the degree or undergraduate courses they belong to, and the minimum content plan. These first two sections are specific to the university where the study is being carried out, as all degree programmes offered by the HEI under study must be listed. The third section asks about the subjects, the coverage, the key competences in sustainability and the assessment mechanisms used.

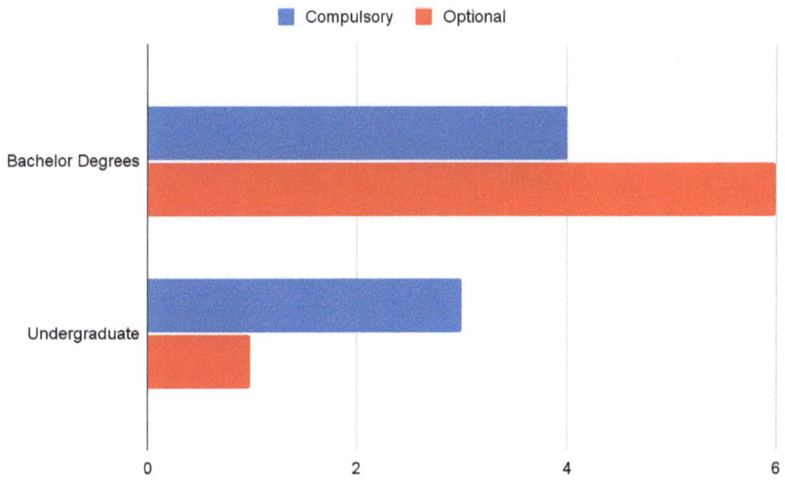

Fig. 3 Number of subjects with an environmental focus. Optional and compulsory subjects

The questions that make up the third section are as follows, only the presence or absence of the issues, coverage and competencies are probed:

- **Topics covered in the subject**: Sustainability, Sustainable development, Climate change, Energy efficiency, Circular economy, Safe water, Environmental management, Waste management, other topics.
- **Number of units in which these topics are found**: Whole subject, Unit 1, Unit 2, Unit 3, Unit 4, Other.
- **What are the key competences in sustainability that students are expected to acquire from their subject**: Understanding of concepts and principles applicable to environmental issues; Projecting environmental values from an ecocentric perspective; Implementing actions to improve environmental quality, social well-being and economic stability; Integration of ideas and knowledge in order to improve environmental quality; Ethical awareness and commitment. Understanding social and environmental responsibility in decision making; Empathy for the situation being addressed through participatory or remote research; Critical and systemic thinking, analysing complex problems and considering long-term impacts; Collaboration between members of different disciplines and actors; Using effective communication, conveying ideas and knowledge in order to improve environmental quality; Collaboration between members of different disciplines and actors. Use effective communication, convey ideas and values related to sustainability; Employ strategic foresight to create practical solutions to current conflicts; Strategic planning according to the socio-environmental and economic context; Promote sustainable changes and solutions from their professional perspectives; Adaptability and resilience. Facing challenges and changes in a dynamic environment; Other.
- **Technical competences (linked to the execution of tasks and procedures)**: Data analysis and modelling. Use tools to assess impacts and simulate scenarios; Sustainable design. Create products and services that minimize environmental impacts; Clean technologies. Apply renewable energies and efficient technologies according to contexts; Waste management. Create integrated waste management plans applying specific regulations; Integrated environmental management. Create environmental management plans with reference to regulations or international standards; Other.
- **Specific competences (linked to problem solving and challenge solving)**: Knowledge about the 2030 agenda. Understand the seventeen (17) goals and their interconnectedness; Life cycle analysis. Assess environmental and social impacts of products and services; Circular economy. Design closed and efficient systems; Natural resource management. Manage renewable and non-renewable resources sustainably; Climate change and mitigation. Understand causes, effects, and strategies to reduce emissions; Biodiversity and conservation. Protect ecosystems and species; Health and well-being. Promote healthy and just environments; Sustainable innovation. Develop innovative and sustainable solutions; Others.
- **Tools used to achieve the competences**: Theoretical exposition of specific topics; Video/film viewing; Analysis of academic/journalistic articles; Problem-based

learning—Analysis of real/fictional cases in order to solve problems; Debate; Class in workshop format; Research on case studies; Collaborative/collective mapping; Field trips and surveys; Technical visits; Surveys or interviews with key informants; Others.
- **Type of evaluation of the contents seen in class—especially those related to environmental issues**: Group oral presentation; Group report presentation; Individual written exam; Individual oral exam; Individual oral presentation; Individual oral presentation; Individual oral presentation; Others.

Step 3-Index HESI

The HESI index is composed of three dimensions: Knowing, Doing and Being (K, D, B). Each dimension has the Key Competences in Sustainability (KCS), which are mentioned in Table 1. This indicator assesses the presence or absence of this competence according to the answers of the survey of the subject leaders. In this sense, if the competence does not exist, it is assigned zero (zero points), and if it is present, one point.

Once each indicator is scored, all the scores assigned are added together. The total score that the HESI can reach is 10 and the lowest score is 0.

The following Eq. 1 is the HESI, composed of the six indicators that represent the KCS by KDB dimensions, mentioned in Table 1:

$$HESI = K(Knowing) + D(Doing) + B(Being) \quad (1)$$

Step 4-Analysis

The analysis was carried out based on the results of the surveys of subject leaders. On the one hand, we analysed the topics addressed in the subjects, how much time is devoted to them in the subject, and on the other hand, the KCS of each dimension according to the HESI and the type of assessment and tools used.

4 Results and Analysis

The results of the survey are presented below, and the two subjects are compared.

First, it is worth mentioning that both subjects are four-term subjects of four hours per week. The subject Environmental Diagnosis (ED) devotes the entire subject to environmental issues, compared to the subject Hygiene and Safety (HS), which only devotes the last unit of its subject. Figure 4 shows that ED focuses on four specific topics in the entire course, while HS addresses seven topics in a single term.

The KCS were analysed according to the dimensions of Knowing, Doing and Being. A comparative table is presented below (see Table 3), which assigns a value of 1 when the competence is present in the subject and 0 when it is not. Both subjects

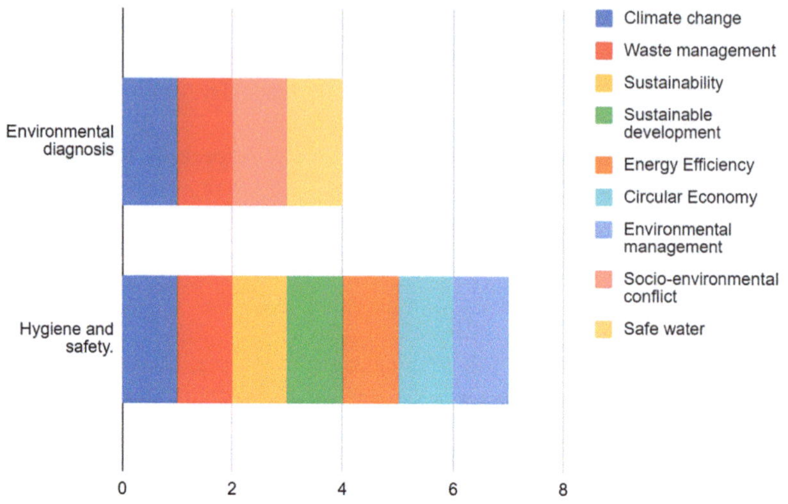

Fig. 4 Topics covered in the subject "environmental diagnosis" and "hygiene and safety"

have all the competences in the knowing dimension. However, in the doing dimension, neither subject includes the competence of anticipatory thinking or the creation of actions or activities with strategic foresight. However, the HS subject does have the competence of strategic thinking with a focus on current affairs, a competence that is not present in the ED subject. In the dimension of being, the ED subject covers the full range of competences because its longer field activities ease a close association with local people, fostering a greater sense of belonging. On the other hand, the HS subject does not promote the competence of creating environmental values, which could be attributed to its more engineering focus and the limited dedication of only one unit during the four-month period. This may be due to different pedagogical approaches and training aims. The ED subject prioritises more extensive field activities and a community focus, which fosters a greater sense of belonging and the development of attitudinal competences. In contrast, the HS subject, with its engineering focus, devotes less time to these aspects and more to technical skills, limiting the development of environmental values. It is a question of balance between theory, practice and values according to the goals of each programme.

Figure 5 presents the overall results graphically. It shows that both subjects score 8/10 in HESI, showing that they address issues and promote key competences for sustainability.

Regarding the tools used to assess competencies (see Fig. 6), the ED subject presents a wider range of assessment methods and tools, while HS limits itself to using only three of them. Technical visits are deeply rooted in engineering education. In contrast, collaborative work, which encompasses participatory mapping, key informant surveys and case studies, is explored in greater depth in diagnostic-oriented subjects. Finally, neither the analysis of newspaper articles nor the Problem Based Learning method is employed by any of the subjects. The difference in the

Table 3 Presence or absence of KCS in the subjects studied by KDB dimension

Dimensions	Key competences in sustainability	Environmental diagnosis	Hygiene and safety
Knowing	Cognitive competences	1	1
	Systems thinking competence	1	1
	Integration competences	1	1
Doing	Methodological competencies	1	1
	Anticipatory thinking competence	0	0
	Strategic thinking competencies	0	1
	Implementation competences	1	1
Being	Attitudinal competencies	1	1
	Competences to think values	1	0
	Intra- and interpersonal competencies	1	1
	Total score	8	8

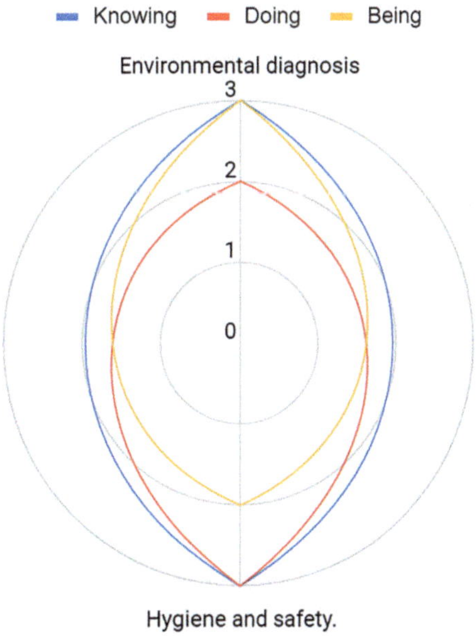

Fig. 5 Key competences in sustainability. Comparison of subjects

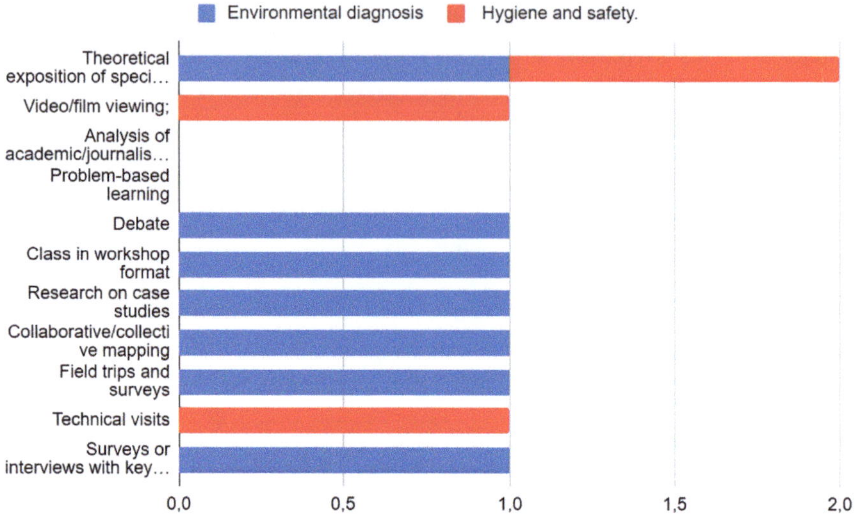

Fig. 6 Tools used to achieve the competences

tools used to assess competencies is due to the nature and aims of each subject. ED, having a broader and more integrative approach, looks to assess competencies from different perspectives, incorporating methods such as collaborative work and participatory mapping, which are fundamental for the diagnosis and comprehensive understanding of the problems. In addition, it addresses a specific case over four months, while HS, only in four weeks. On the other hand, HS, with a more technical approach, focuses on technical visits, which are essential for engineering education.

Finally, both subjects use the individual written examination method to assess the competences acquired. However, the ED subject also incorporates group presentations and group reports, as it encourages more frequent group work dynamics, while individual work is done occasionally. In contrast, the HS subject is characterised by an intensive focus on individual work, with group activities taking place sporadically (see Fig. 7).

5 Conclusions

The analysis reveals that both subjects share a four-hour-per-week four-term structure but differ significantly in their thematic focus and depth. Environmental Diagnosis (ED) is devoted exclusively to environmental issues throughout the course, while Hygiene and Safety (HS) addresses these issues only in its last unit. This allows ED to explore the topics in greater depth, while HS covers a greater number of topics, albeit in a more superficial manner.

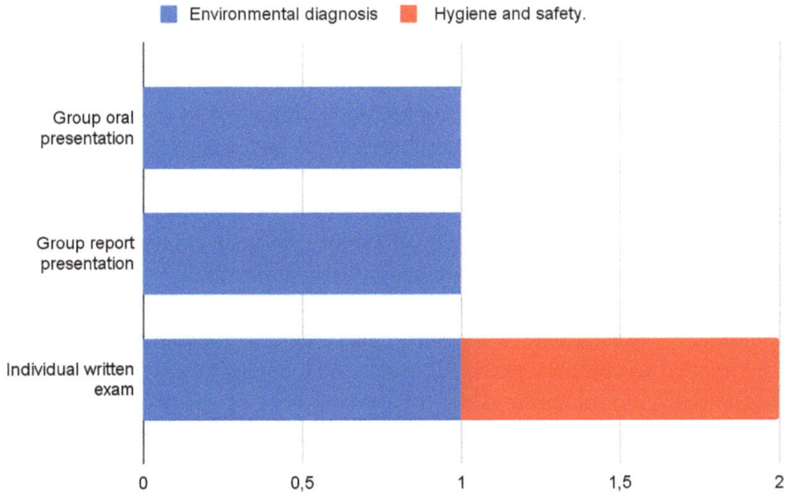

Fig. 7 Type of evaluation of the contents

In terms of competences (KCS), both subjects meet the competences in the Knowing dimension. However, in the dimension of Doing, neither includes competences of anticipatory thinking nor creating actions with strategic foresight, although HS incorporates strategic thinking focused on current affairs, which is absent in ED. In the dimension of Being, ED stands out for covering all competences, thanks to its extended field activities that foster a greater sense of belonging. In contrast, HS does not promote the creation of environmental values, probably due to its engineering focus and limited dedication to these issues.

The assessment tools also show differences. While ED uses a wider variety of methods, including collaborative work and participatory mapping, while HS is limited to technical visits. Neither uses newspaper article analysis or Problem Based Learning methodology.

Finally, ED incorporates group presentations and reports, promoting teamwork dynamics, while HS focuses more on individual work. These differences reflect the different pedagogical approaches and training objectives: ED prioritises a holistic and community-based approach, while HS focuses on engineering-specific technical competences. These choices are aligned with the goals and contexts of each subject.

6 Recommendations for Achieving the KSCs

In order to improve the Higher Education Sustainability Indicator (HESI) in the competences not addressed, the following improvements are proposed for each subject:

For Health and Safety (HS), anticipatory thinking should be incorporated and modules that develop the ability to foresee and plan for future environmental scenarios should be integrated. This may include simulation exercises and prospective case studies. Environmental values should also be promoted by devoting more units and activities to building environmental awareness and values. Practical activities such as sustainability projects in the community can be highly effective. Collaborative work can allow inter and intra competences to be further developed by introducing more group dynamics, such as collaborative projects, participatory mapping and case studies, to develop teamwork and collaboration skills.

For Environmental Diagnosis (ED), strategic thinking should be encouraged, incorporating exercises and workshops that emphasise planning and strategic thinking applied to current realities. This may include policy analysis and strategic decision-making in real cases. Also, the development of specific technical skills, extending the use of more engineering-specific technical tools, such as technical visits and the implementation of specific technologies in the field. Finally, article analysis and problem-based learning can create an enriching learning space, as it introduces critical analysis of journalistic and academic articles related to sustainability from other parts of the world and broadens the global outlook.

To build on the strengths of both subjects and mutually enrich each other's competences, a collaborative and interdisciplinary approach focusing on integrated projects is recommended. The implementation of joint projects will allow students of ED and HS subjects to work on real sustainability problems, thus combining their competences. For example, a project integrating environmental diagnosis and health and safety improvements in a local community will allow ED students to apply their field knowledge and community sensitivity, while HS students will contribute their technical and safety skills.

In addition, the organisation of joint seminars and workshops can address sustainability issues from multiple perspectives, benefiting students from both subjects by learning strategic thinking skills, anticipatory thinking, and environmental values. Sharing assessment and teaching methodologies is also crucial. Integrating effective methods from one subject into the other, such as including more collaborative work and participatory mapping activities in HS and implementing more technical visits and analysis of contemporary cases in ED, will enrich learning.

Facilitating joint reflection sessions, where students analyse and discuss the learning and competences acquired in the interdisciplinary projects, will enable a holistic and practical understanding of sustainability. These strategies will not only strengthen subject-specific competences, but also promote a holistic view of sustainability, benefiting all students and enhancing the university's programmes.

7 Limitations

The internal validity of this research might be limited by certain factors related to the survey used. Firstly, the survey may not have provided a complete picture of the sustainability competences taught in Higher Education Institutions (HEIs). Although the framework used, HESI, aims to provide sufficient indicators to cover sustainability and to be manageable, problems of interpretation of the criteria, reliability derived from self-assessment and understanding of the terms from different perspectives or academic backgrounds may have arisen.

The small number of respondents (two) does not allow for a complete generalisation of sustainability teaching in all HEIs and KCS. However, the choice of only two subjects is justified by the characteristics of the subjects, as they are crosscutting to several degree programmes and address sustainability-related issues. However, the selection of cross-cutting subjects may not capture all the sustainability competences that might be present in specific subjects of each degree programme. This may limit the full understanding of how sustainability is taught at the National University of General Sarmiento (UNGS). It is also worth considering the bias introduced by non-response, due to some individuals refusing to participate or to complete the survey. In addition, the research focused on a specific academic period, which may not reflect changes or improvements in sustainability teaching that may occur over time.

In terms of restrictions, obtaining accurate and complete syllabus data and detailed responses from teachers may be affected by limitations in access to information and the collaboration of the parties involved; syllabuses for each subject are not public, and must be requested from each teacher teaching the subject. The institutional context and administrative structure of UNGS may influence the applicability of the findings to other institutions, due to differences in administrative structures, educational policies and pedagogical approaches.

On the other hand, the HESI index was evaluated exclusively at UNGS. Therefore, it is necessary to test its effectiveness in other Higher Education Institutions (HEIs) that offer cross-cutting subjects with sustainability themes.

8 Future Perspectives

Firstly, it is essential to extend the study to other universities and educational institutions in the Buenos Aires Metropolitan Area and other regions to validate the findings and improve the generalisability of the results. Furthermore, it is necessary to increase the number of subjects and respondents in order to obtain a more representative and complete picture of how sustainability competences are integrated into curricula.

Longitudinal monitoring is also recommended to assess how sustainability competences evolve in students and graduates over time and how they impact on their professional performance. Furthermore, it is essential to incorporate more qualitative methods, such as in-depth interviews and focus groups, in order to complement the data obtained through surveys and gain a richer and more nuanced understanding.

Finally, it is crucial to create and validate new educational tools and resources to strengthen the teaching of sustainability in different disciplines and educational levels.

Acknowledgements We would like to thank the National Scientific and Technical Research Council (Consejo Nacional de Investigaciones Científicas y Técnicas-CONICET) of Argentina for funding this research. I would also like to thank the teachers who responded to the survey and provided valuable information.

References

Albareda-Tiana T, Gonzalvo-Cirac M (2013) Competencias genéricas en sostenibilidad en la educación superior. Revisión y compilación. Revista de Comunicación de la SEECI. Año XV (32):141–159. ISSN 1576-3420

Anghel GA, Neculau BC (2022) Quality education as a predictor of education for sustainability. J Educ Soc Multicult 3:132–141. https://doi.org/10.2478/jesm-2022-0008

Aznar P, Ull M (2009) La Formación de Competencias Básicas para el Desarrollo Sostenible: El papel de la universidad. Revista de Educación, número extraordinario 200. Universidad de Valencia, España, pp 21–237

Baartman LKJ, Bastiaens TJ, Kirschner PA, Van der Vleuten CPM (2007) Evaluation assessment quality in competence-based education: a qualitative comparison of two frameworks. Educ Res Rev 2:114–129

Barth M, Godemann J, Rieckmann M, Stoltenberg U (2007) (2007) Developing key competencies for sustainable development in higher education. Int J Sustain High Educ 8:416–430

Bisquerra Alzina R, Pérez Escoda N (2007) Las competencias emocionales. Educación XXI: revista de la Facultad Educ (10):61–82

Blanco Blanco A (2008) Competence-based university education. Book, Learning-centred university teaching: useful strategies for teachers. Editorial: Octaedro, pp 31–59. ISBN: 978-84-8063-924-8

Brundiers K, Barth M, Cebrián G, Cohen M, Diaz L et al (2021) Key competencies in sustainability in higher education—toward an agreed-upon reference framework. Sustain Sci 16:13–29

Brundiers K, Wiek A, Redman CL (2010) Real-world learning opportunities in sustainability: from the real world. Int J Sustain High Educ 11(4):308–324. https://doi.org/10.1108/14676371011077540

Cano E (2008) La evaluación por competencias en la educación superior. Profesorado: revista de currículum y formación del profesorado 12(3):11

Cotton D, Winter J (2010) 'It's not just bits of paper and light bulbs': a review of sustainability pedagogies and their potential for use in higher education. In: Jones P, Selby D, Sterling S (eds) Sustainability Education: Perspectives and Practice Across Higher Education. Earthscan

Faitani N (2023) Environmental sustainability of a higher education institution in the Conurbano Bonaerense. Ph.D. thesis work in Science and Technology at the National University of General Sarmiento. http://repositorio.ungs.edu.ar:8080/xmlui/handle/UNGS/1453

Favi C (2021) Sustainability competences and pedagogical approaches at the University of Parma. In: Lozano R, Barreiro-Gen M (eds) Developing sustainability competences through pedagogical approaches. Strategies for Sustainability. Springer, Cham. https://doi.org/10.1007/978-3-030-64965-4_5

Godínez López A (2013) Development of Environmental Competences in the academic training of secondary school students. Master's Thesis in Education with Orientation in Environment. University of San Carlos de Guatemala. Available at: Tesis de posgrado. Godinez Lopez
Jiménez-Munguía MT, Flores-Cervantes DX, Martins R, Palou E, López-Malo A, Ramírez-Corona N (2021) Sustainability competences and pedagogical approaches at the Universidad de las Américas Puebla. In: Lozano R, Barreiro-Gen M (eds) Developing sustainability competences through pedagogical approaches. Strategies for Sustainability. Springer, Cham. https://doi.org/10.1007/978-3-030-64965-4_7
Lambrechts W, Mulà I, Ceulemans K, Molderez I, Gaeremynck V (2013) The integration of competences for sustainable development in higher education: an analysis of bachelor programs in management. J Clean Prod 48:65–73. https://doi.org/10.1016/j.jclepro.2011.12.034
Lozano R, Merrill MY, Sammalisto K, Ceulemans K, Lozano FJ (2017) Connecting competences and pedagogical approaches for sustainable development in higher education: a literature review and framework proposal. Sustainability 9(11):1889. https://doi.org/10.3390/su9101889
Lozano R, Barreiro-Gen M, Temel M (2021) Literature review and methods. In: Lozano R, Barreiro-Gen M (eds) Developing sustainability competences through pedagogical approaches. Strategies for Sustainability. Springer, Cham. https://doi.org/10.1007/978-3-030-64965-4_2
Lund University (2024) Sustainability in curriculum: a pathway to a sustainable future? Obtenido de. https://www.ahu.lu.se/en/educational-development/seminars/sustainable-teaching-learning-in-higher-education-principles-practices/sustainability-in-curriculum/
Martínez AM, Sánchez JAR, Navarro JGC (2012) Aprendizaje basado en competencias: Una propuesta para la autoevaluación del docente. Profesorado. Rev Currículum Formación Profesorado 16(2):325–338
Mora P, Guerrero G (2022) Las competencias ambientales clave en las actividades docentes del profesorado de ciencias. Tecné, Episteme y Didaxis TED (51):299–316
Mulder M (2007) Competence: the essence and use of the concept in initial and continuing vocational initial and continuing vocational training. Eur J Vocat Educ Train 40(1):5–24
Pietikäinen J, D'Amato D (2021) Sustainability competences and pedagogical approaches at the University of Helsinki. In: Lozano R, Barreiro-Gen M (eds) Developing sustainability competences through pedagogical approaches. Strategies for Sustainability. Springer, Cham. https://doi.org/10.1007/978-3-030-64965-4_4
Pulido Trullén JI (2008) Generic competences—what are they? In: Generic and transversal competences of university graduates. ICE de la Universidad de Zaragoza, pp 35–41
Redman A, Wiek A (2021) Competencies for advancing transformations towards sustainability. Front Educ
Trencher G, Vincent S, Bahr K, Kudo S, Markham K, Yamanaka Y (2018) Evaluating core competencies development in sustainability and environmental master's programs—an empirical analysis. J Clean Prod 181:829–841
Weinert FE (2001) Concept of competence: a conceptual clarification. In: Rychen DS, Salganik LH (eds) Defining and selecting key competencies. Hogrefe, Göttingen
Wiek A, Redman A (2022) What do key competencies in sustainability offer and how to use them. In: Vare P, Lausselet N, Rieckmann M (eds) Competences in education for sustainable development. Sustainable Development Goals Series. Springer, Cham. https://doi.org/10.1007/978-3-030-91055-6_4

The TRAIN Framework for AI-Assisted Development of Novice Accounting Instructors

Semen Son-Turan

Abstract Novice accounting instructors in higher education, and particularly in specialized fields like accounting, face significant challenges, including the need to frequently update curricula, balance theoretical and practical instruction, master new technologies, and integrate sustainability principles into their teaching. These demands are compounded by limited access to professional development resources and growing pressures for educational sustainability. This study evaluates the potential of AI chatbots to address these challenges by enhancing the capacity of junior instructors to create and deliver effective, sustainability-aligned course materials. It introduces the TRAIN framework (Technology-enhanced Responsible AI Integration for Novice Instructors), a structured model designed to align AI capabilities with Sustainable Development Goals (SDGs), particularly SDG 4 (Quality Education) and SDG 4.c (Teacher Training). Through a systematic literature review (2019–2024), the research identifies key themes, including chatbot integration in education, AI's capabilities and limitations, and its broader implications for accounting education. The findings demonstrate that AI chatbots can streamline content creation, provide personalized instructional support, and foster sustainable teaching practices. By bridging critical gaps in AI literacy and sustainability integration, the TRAIN framework offers a roadmap for empowering educators to meet evolving educational demands while advancing environmentally responsible teaching methodologies.

Keywords AI · Accounting education · Sustainable development · Technology · Chatbot

S. Son-Turan (✉)
Faculty of Economics, Administrative and Social Sciences, MEF University, Istanbul, Türkiye
e-mail: sons@mef.edu.tr

W. Leal Filho et al. (eds.), *Education for Sustainable Development: The Contribution of Universities*, World Sustainability Series,
https://doi.org/10.1007/978-3-031-86985-3_27

1 Introduction

The teaching profession faces unprecedented challenges globally, particularly in higher education and specialized fields like accounting. Recent studies reveal that nearly a third of campus communities, including faculty, experience symptoms of depression, anxiety, and stress (Meeks et al. 2023). These pressures are intensified by increasing demands for teaching excellence, increasing administrative workloads, rigorous research obligations, and the necessity to integrate sustainability principles into curricula.

Junior accounting instructors encounter multifaceted challenges in times of rapid educational shifts, especially within the frameworks of Era 5.0 and competency-based education. Key challenges identified in the literature are:

1. **Curriculum Development and Adaptation**: With the accelerated integration of artificial intelligence (AI) and technology, accounting curricula require frequent updates to remain relevant and harmonious with industry practices. This task can be burdensome, especially given the pace of technological advancements (Tavares et al. 2023).
2. **Balancing Theory with Practical Skills**: Competency-based education prioritizes developing practical skills next to theoretical knowledge. For junior instructors, creating assessments and activities that effectively measure both knowledge and competencies can be overwhelming (Cheng et al. 2024).
3. **Technological Proficiency**: Effective incorporation of AI tools and digital platforms into teaching demands technological expertise. Acquiring proficiency in these tools can be time-consuming for novice instructors, often necessitating additional training (Abeysekera 2024).
4. **Student Engagement and Performance**: Junior instructors must engage students with diverse backgrounds in accounting knowledge and technological proficiency. This diversity requires strategies to address diverse learning needs while managing issues like accounting-related anxiety, which can impact performance (Khalid and Abdul Rauf 2023).
5. **Professional Development and Support**: Access to mentorship and professional development is essential, however, often limited. Without adequate resources, junior instructors may struggle to meet evolving educational demands (Samkin and Stainbank 2016).

Contemporary literature highlights promising developments in addressing these challenges through AI integration. Studies show that chatbots and AI tools are becoming increasingly essential in educational settings, with proven effectiveness in managing complex tasks and delivering personalized assistance and further research indicates that while AI tools like ChatGPT show proficiency in solving accounting cases and supporting assessments, they require careful instructor oversight and intervention (Cheng et al. 2024; Abeysekera 2024).

Broader implications of AI integration in educational systems underscore both opportunities and challenges. Only about 25% of secondary school teachers report feeling competent with ICT tools (Baytar et al. 2023), emphasizing the need for comprehensive training initiatives. In accounting education specifically, emerging technologies like AI and Web 3.0 are reshaping curricula and teaching methods (Li and Zhao 2022), while creating new demands for AI literacy among educators (Wilton et al. 2022).

However, significant gaps remain in current research and practice:

- Limited empirical research exists on integrating chatbots into instructor training programs
- Current AI literacy frameworks for educators need expansion to cover both AI and data competencies
- Long-term impacts of AI integration on student outcomes and ethical considerations require further investigation
- More research is needed on interactive, case-based learning applications of chatbots in accounting education.

On top of these challenges come teacher shortages and high attrition rates. The European Commission identified 13 countries facing significant teacher shortages (Carlo et al., cited in Viac and Fraser 2020). In Australia, 30–50% of teachers leave the profession within five years, while in the UK, over 20% leave within two years (den Brok et al. 2017). Additionally, the post-COVID educational landscape has heightened demands for hybrid teaching modes, adding pressure on instructors already managing post-traumatic stress issues (Jelińska and Paradowski 2021).

AI, and particularly AI chatbots, offer promising solutions to these challenges. The intersection of technology and education can be framed through various theoretical lenses such as Technology Acceptance Models (TAM and UTAUT) (Davis 1989; Venkatesh et al. 2003), Technological Pedagogical Content Knowledge (TPACK) (Mishra and Koehler 2006), Constructivist Learning Theory and Cognitive Load Theory (Vygotsky 1978; Sweller 1988), and Ecological Modernization Theory and the Triple Bottom Line Framework (Mol and Spaargaren 2000; Jeurissen and Elkington 2000).

To address the multifaceted challenges faced by novice accounting instructors, this study introduces the TRAIN framework (Technology-enhanced Responsible AI Integration for Novice Instructors). This model provides a structured and sustainable approach to leveraging AI tools, particularly chatbots, to enhance instructors' ability to develop and deliver effective, sustainability-aligned accounting course materials. The TRAIN framework aligns with the UNESCO AI Competency Framework for Teachers (UNESCO 2024), which emphasizes a human-centered approach to AI in education by prioritizing accessibility, ethical considerations, and inclusivity.

The primary objective of this research is to systematically review literature from the Web of Science and Google Scholar databases (2019–2024) to explore the integration of AI chatbots in accounting education. This review focuses on supporting the professional growth of novice instructors by addressing pedagogical challenges through a framework grounded in adaptive learning theory (Lee and Park 2015) and sustainable innovation frameworks (Boons and Lüdeke-Freund 2013). The study hypothesizes that AI tools, when effectively integrated into educational practices and aligned with sustainability and professional development goals, can significantly improve teaching quality. These improvements include automating content creation, providing personalized instructional support, and facilitating the incorporation of Sustainable Development Goal (SDG)-aligned principles in course design. Additionally, the study highlights the potential for AI to contribute to instructors' mental well-being by reducing workload-related stress, which in turn fosters improved teaching efficacy and better student outcomes.

The research makes two key contributions: it addresses the gap in AI tools for supporting junior accounting instructors and introduces the TRAIN framework for sustainable AI integration in accounting education, providing strategies for leveraging AI to develop junior instructors while advancing sustainability goals. These practitioners will play a pivotal role in advancing an ecosystem and society where AI and human intelligence coalesce to foster effective, inclusive, and sustainable learning environments aligned with the Sustainable Development Goals.

2 Theoretical Background

This research is particularly grounded in adaptive learning theory (Lee and Park 2015) and sustainable innovation frameworks (Boons and Lüdeke-Freund 2013) to investigate how AI chatbots can assist junior instructors in creating and delivering sustainability-aligned accounting course materials. Adaptive learning theory emphasizes that personalized, responsive support tailored to individual needs can enhance learning outcomes. In this study, AI chatbots serve as adaptive tools that address junior instructors' unique needs, providing real-time support and streamlining course material preparation and delivery through automation. This aligns with SDGs, which focus on embedding sustainable practices within organizational processes; in this case, integrating sustainability into academic teaching.

The adoption of AI in education is further illuminated by the Technology Acceptance Model (TAM) (Davis 1989) and the Unified Theory of Acceptance and Use of Technology (UTAUT) (Venkatesh et al. 2003), which explain how educators adopt new technologies based on ease of use and perceived benefits. For junior instructors, AI chatbot adoption depends on how user-friendly and beneficial these tools are in enhancing teaching practice while supporting sustainability objectives.

Mishra and Koehler's (2006) Technological Pedagogical Content Knowledge (TPACK) framework underscores the interaction between content, pedagogy, and technology in effective teaching. Junior accounting instructors often face challenges across all three domains: mastering technical content (e.g., IFRS and US GAAP), developing strong pedagogical approaches, and incorporating sustainability concepts. AI chatbots can help alleviate these challenges by providing support that enhances both the technological and pedagogical dimensions of their teaching, thereby easing the integration of sustainability-aligned content into lessons.

This study hypothesizes that when aligned with sustainability and professional development goals, AI tools can significantly transform teaching quality. AI tools automate routine tasks, improve instructional design, and promote the integration of sustainable educational frameworks like the Sustainable Development Goals (SDGs). The theoretical foundation also draws from constructivist learning theory (Vygotsky 1978) and cognitive load theory (Sweller 1988), where AI chatbots act as scaffolding tools, assisting instructors in navigating the "zone of proximal development" (Vygotsky 1978). Moreover, by automating routine tasks, AI tools help manage cognitive load, allowing instructors to focus on developing interactive learning materials. This technological background supports SDG 4, particularly Target 4.c, which emphasizes the importance of teacher capacity building and quality education delivery (United Nations 2015).

The sustainability dimension of AI integration is further understood through ecological modernization theory (Mol and Spaargaren 2000), which proposes that technological advancement can complement and even promote environmental sustainability. However, AI integration also poses sustainability challenges, especially concerning the energy demands of AI systems like large language models (LLMs). Addressing the balance between AI's benefits and environmental costs requires a nuanced approach, guided by the triple bottom line framework (Jeurissen and Elkington 2000), which accounts for educational, environmental, and economic impacts.

3 Methodology: Systematic Literature Review with Dual Independent Search

This study employs a systematic literature review (SLR) methodology to critically evaluate and synthesize existing research on the integration of AI tools, specifically chatbots, in accounting education. The goal is to identify themes, challenges, and opportunities for supporting novice accounting instructors while advancing sustainability-aligned teaching practices. The review provides the foundation for developing the TRAIN framework, aligning with the study's overarching objective of improving teaching quality and supporting Sustainable Development Goals (SDGs).

3.1 Research Framework and Approach

The methodology is guided by Snyder's (2019) four-phase systematic review approach, structured as follows:

1. **Design**: Define the scope and objectives of the review, focusing on AI integration in accounting education and its potential to address instructional and sustainability challenges.
2. **Conduct**: Perform comprehensive searches across scholarly databases using well-defined inclusion and exclusion criteria.

3. **Analysis**: Thematically synthesize the findings to identify patterns and gaps in the literature.
4. **Structuring and Writing**: Organize the insights into a coherent framework that informs the TRAIN model.

A dual independent search process was adopted to enhance the rigor and comprehensiveness of the review. Each researcher conducted independent searches, screened studies, and analyzed findings before reconciling discrepancies through collaborative discussion. This method is widely recognized for its ability to reduce bias and increase reliability (Waffenschmidt et al. 2019).

3.2 Research Questions and Focus

The review aimed to answer the following key research questions:

1. How have AI tools, particularly chatbots, been integrated into accounting education?
2. What challenges and opportunities do novice instructors face in adopting AI-driven teaching tools?
3. How can AI tools align with sustainability goals to enhance instructional quality and professional development?

3.3 Search Strategy and Keywords

The review focused on peer-reviewed articles published between January 2019 and November 2024 to capture the latest developments in AI integration within education. Searches were conducted across **Web of Science** and **Google Scholar** using the following search string:

> ("artificial intelligence" OR "chatbot" OR "AI") AND ("accounting education" OR "business education") AND ("instructor develop" OR "faculty develop*" OR "teacher train*") AND ("sustainab*" OR "SDG*").

3.4 Selection and Screening Process

1. **Initial Screening**: Titles and abstracts of 891 articles were reviewed for relevance. Studies unrelated to higher education, lacking a focus on AI in accounting education, or addressing only technical aspects were excluded. After duplicate removal, 379 articles remained.
2. **Exclusion Criteria**: Studies were excluded if they:
    - Did not address AI integration clearly.

- Lacked a focus on social sciences or education.
- Had no explicit link to sustainability.
- Were overly technical without educational application.

3. **Full-Text Review**: A detailed review of 161 articles evaluated their methodological rigor, relevance to the research questions, and applicability to novice instructor development. Articles related to K-12 were included if their findings were generalizable to higher education contexts.

3.5 Quality Assessment

Each study was assessed for:

- **Clarity of focus**: Whether the study explicitly addressed AI in educational contexts.
- **Methodological robustness**: The reliability and validity of the findings.
- **Relevance**: Alignment with AI-driven educational development, sustainability integration, or instructor challenges.

A total of 32 high-quality studies were retained for thematic synthesis.

3.6 Thematic Synthesis

The selected studies were categorized into four thematic areas to align with the study objectives:

1. **AI Integration in Education**: Examining how chatbots enhance teaching practices.
2. **Instructor Development**: Addressing challenges and support mechanisms for novice instructors.
3. **Sustainability Integration**: Exploring the role of AI in fostering sustainability in education.
4. **Challenges and Benefits of AI in Teaching**: Highlighting barriers and opportunities for effective adoption.

3.7 Study Objectives Linked to Methodology

This systematic review provides the foundation for the TRAIN framework by identifying evidence-based strategies and gaps in existing research. It supports the development of an actionable model that aligns AI tools with professional development goals and sustainability principles, offering practical solutions to enhance instructional practices and improve educational outcomes.

Table 1 presents a systematic summary of individual studies, outlining key research questions and findings related to AI's role in supporting educational goals, such as curriculum adaptation, skill development, and ethical considerations in AI literacy. Each article's insights help identify the primary areas where AI, particularly chatbots, can assist junior instructors in accounting and broader educational settings. The data reveals that 34.38% of the studies focus on AI in university accounting education, with an additional 12.5% exploring AI's role in teacher training within this context. The spread across categories illustrates AI's versatility in different educational applications, including university and non-university settings, with a clear trend toward supporting junior instructors' teaching capacities in diverse environments.

Graph 1 summarizes descriptive statistics on the main themes, challenges, and benefits of AI chatbot use in education. Notably, instructor support (34.4%) and curriculum development (28.1%) emerge as primary themes, with key challenges like technological proficiency (25%) and the need for professional development (28.1%). Benefits such as enhanced student engagement (31.3%) and skill development (25%) align well with the study's goal to assist junior instructors in overcoming instructional challenges through AI-supported teaching.

These findings effectively illustrate how AI chatbots can support junior instructors, enhance curriculum development, and improve engagement against existing barriers such as technology training and professional support.

Table 1 Summary of research on AI chatbot ıntegration in education

Category	Related articles (author and year)	Total number	Percentage (%)
University accounting education	Cheng et al. (2024), Abeysekera (2024), Al Ghatrifi et al. (2023), Tavares et al. (2023), Li and Zhao (2022), Cai (2022), Holmes and Douglass (2022), Qasim et al. (2022), Hu (2022), Grabińska et al. (2021)	11	34.38
University accounting education and AI-assisted teacher training	Tavares et al. (2023), Cai (2022), Hu (2022)	4	12.5
University education and AI-assisted teacher training	Bustamante Bula and Camacho Bonilla (2024), Duan et al. (2024), Sperling et al. (2024), Wilton et al. (2022), Çelik et al. (2022)	7	21.88
Non-university education but AI-assisted teacher training	Yue et al. (2024), Folmeg et al. (2024), Okolo (2024), Shin and Lee (2024), Chai et al. (2023), Jung et al. (2022), Baytar et al. (2023)	8	25
Uncategorized articles (not specifically related to the four categories above)	Okulich-Kazarin et al. (2024), Folmeg et al. (2024)	2	6.25

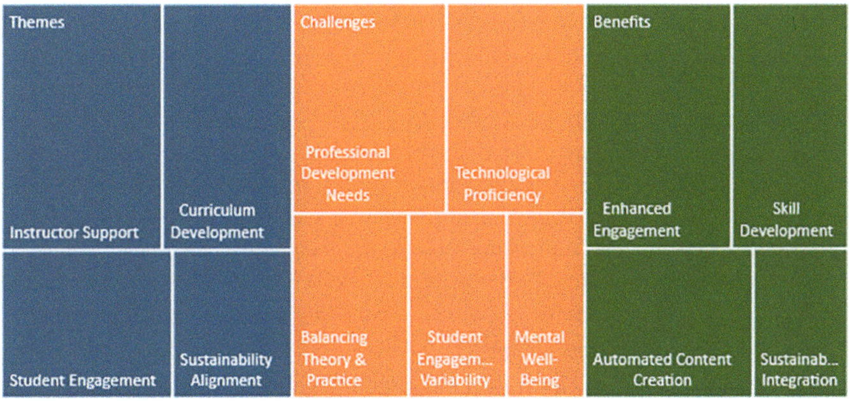

Graph 1 Key themes, challenges, and benefits of AI chatbot ıntegration in accounting education

4 Discussion of Findings

The reviewed literature can be grouped into four main themes: (1) Integration of Chatbots in Education, (2) Capabilities and Limitations of AI Tools, (3) Broader Implications of AI and Technology in Educational Systems, and (4) AI and Emerging Technologies in Accounting Education.

Theme 1: Integration of Chatbots in Education

The use of chatbots in educational settings is a recurring theme in the literature (Abeysekera 2024; Tavares et al. 2023; Jung et al. 2022; Shin and Lee 2024; Duan et al. 2024). Jayasuriya and Dow develop a chatbot for accounting and finance, which supports both, instructors and students by managing complex tasks. Wutzler highlighted how chatbots improve classroom learning by providing personalized assistance. Abeysekera (2024) focuses on ChatGPT's role in assessments, demonstrating its effectiveness as a teaching and assessment support tool. Tavares et al. (2023) underscores the utility of chatbots in accounting education while emphasizing the need for human oversight. Jung et al. (2022) and Shin and Lee (2024) illustrate the versatility of chatbots across diverse educational settings, including language learning. Duan et al. (2024) show how AI-based prediction systems, designed with instructor input, enhance trust and transparency. These studies highlight the potential of chatbots to improve instructional support and administrative efficiency, but they also call for ongoing development to ensure reliability and alignment with varied educational needs.

Theme 2: Capabilities and Limitations of AI Tools

Several articles address the strengths and limitations of AI tools in education (Cheng et al. 2024; Bustamante Bula and Camacho Bonilla 2024; Holmes and Douglass 2022; Gabriel et al. 2022; Olari and Romeike 2021; Karatas et al. 2024; Yue et al.

2024). Cheng et al. (2024) examine ChatGPT's effectiveness in solving accounting cases, noting that while it performs well, its limitations require instructor intervention. Bustamante Bula and Camacho Bonilla (2024) analyze AI in schools, highlighting its transformative potential and ethical challenges. Holmes and Douglass (2022) explore how AI reshapes accounting education, stressing the need for educators to adapt. Gabriel et al. (2022) discuss AI integration in education globally, emphasizing data-driven decision-making and digital maturity. Olari and Romeike (2021) point to gaps in AI literacy for educators, suggesting more comprehensive frameworks covering AI and data skills. Karatas et al. (2024) highlight AI's role in curriculum adaptation, and Yue et al. (2024) note gaps in K-12 teachers' technological pedagogical knowledge, recommending targeted professional development.

Theme 3: Broader Implications of AI and Technology in Educational Systems

The broader impacts of AI and technology in education are discussed in several studies (Al Ghatrifi et al. 2023; Weuffen et al. 2023; Baytar et al. 2023; Çelik et al. 2022; Wilton et al. 2022; Sperling et al. 2024). Al Ghatrifi et al. (2023) examine the integration of emerging technologies like AI and Big Data to modernize accounting education, voicing challenges and benefits. Weuffen et al. (2023) emphasize inclusive, sustainable educational practices to support diversity and equity. Baytar et al. (2023) determine that only a quarter of secondary school teachers felt competent using ICT, highlighting a need for training. Çelik et al. (2022) identify barriers to AI integration, including skill gaps. Wilton et al. (2022) stress the importance of AI literacy among educators, recommending professional development initiatives. Williams et al. introduce Doodlebot, an educational robot, showing that hands-on interaction enhances AI literacy and creativity. Sperling et al. (2024) identify gaps in AI literacy for teachers, especially in practical and ethical aspects.

Theme 4: AI and Emerging Technologies in Accounting Education

These studies focus on AI and other emerging technologies in accounting education (Cai 2022; Qasim et al. 2022; Hu 2022; Li and Zhao 2022; Grabińska et al. 2021; Lin et al. 2022; Tavares et al. 2023). Cai (2022) discusses training modes for accounting talents, highlighting AI's role in fostering effective learning. Qasim et al. (2022) analyze AI integration in UAE accounting education, identifying challenges and opportunities. Hu (2022) proposes a collaborative AI-assisted model to make learning more interactive, while Li and Zhao (2022) explore the influence of AI and Web 3.0 in enhancing accounting curricula. Grabińska et al. (2021) underscore the value of AI skills in finance and accounting and Lin et al. (2022) explore teachers' perspectives on sustainable AI education, recommending design frameworks for effective curriculum delivery.

As a result, several gaps are ıdentified, and grouped in the categories below:

1. **Limited Research on Instructor Training via Chatbots**: Although several studies discuss chatbots for instructor support (Tavares et al. 2023; Jung et al. 2022), empirical research on their integration into instructor training, particularly outside accounting, is limited.

2. **Insufficient Focus on AI Literacy for Educators**: Olari and Romeike (2021) highlight gaps in AI literacy frameworks, suggesting the need for initiatives that cover both AI and data competencies, particularly given the rapid adoption of AI tools.
3. **Ethical and Long-Term Implications of AI in Education**: Studies by Bustamante Bula and Camacho Bonilla (2024) and Gabriel et al. (2022) emphasize ethical considerations but highlight that limited research on AI's long-term impacts on student outcomes and ethical challenges exists.
4. **Expanded Role of Chatbots in Accounting Education**: While studies largely focus on assessment and instructional support (Abeysekera 2024), there is a gap in research on using chatbots for more interactive, case-based learning in accounting.

These insights underscore AI's transformative potential in educational contexts and highlight areas for future research.

5 The TRAIN Framework: A Sustainable Approach to AI-Enhanced Instructor Development

Based on the systematic literature review's findings and the gaps identified in the previous section, the TRAIN framework (Technology-enhanced Responsible AI Integration for Novice instructors) emerges as a structured, sustainable approach to AI-enhanced instructor development, aligning with key Sustainable Development Goals (SDGs). The framework emphasizes eco-friendly AI practices (SDG 13, SDG 12) and regulatory support for industry-aligned skills (SDG 9), while at the same time integrating tools like Web 3.0 for real-time curriculum updates. AI-enhanced pedagogical tools promote AI literacy and effective lesson planning (SDG 4), while instructor well-being initiatives support professional development and inclusivity (SDG 3). Through guided phases involving foundation building, skill development, advanced integration, and sustainability & growth, the framework aids in transitioning instructors from novice to expert and facilitates best practices while fostering a collaborative environment (SDG 17). This approach ensures that instructors are prepared to leverage AI responsibly and in this way meet current educational needs in a sustainable manner (Table 2).

6 Conclusion

Recent literature demonstrates that artificial intelligence and chatbots hold significant promise for improving teaching and learning outcomes, as evidenced by multiple studies (Abeysekera 2024; Tavares et al. 2023). The TRAIN framework has been

Table 2 The TRAIN framework

TRAIN framework component	Phase of implementation	Theme	Targeted SDGs	Citations
Technical foundation & environmental responsibility	Phase 1: Foundation building	Green AI and sustainable practices	SDG 13 (climate action)	Cheng et al. (2024), Weuffen et al. (2023), Lin et al. (2022)
	Phase 2: Skill development	Resource optimization and carbon footprint monitoring	SDG 12 (responsible consumption and production)	Gabriel et al. (2022), Qasim et al. (2022), Wilton et al. (2022)
	Phase 3: Advanced integration	Selection of energy-efficient AI tools	SDG 9 (industry innovation and infrastructure)	Li and Zhao (2022), Cai (2022)
Regulatory & curriculum integration support	Phase 1: Foundation building	AI for standards tracking and ESG reporting	SDG 9, SDG 12	Hu (2022), Grabińska et al. (2021), Gabriel et al. (2022)
	Phase 2: Skill development	Industry-aligned skills and Web 3.0 integration	SDG 4 (quality education), SDG 9	Al Ghatrifi et al. (2023), Li and Zhao (2022)
	Phase 3: Advanced integration	Data-driven decision-making and interactive learning	SDG 4, SDG 9	Bustamante Bula and Camacho Bonilla (2024), Cai (2022)
AI-Enhanced pedagogical development	Phase 2: Skill development	Personalized AI literacy and assessment design	SDG 4, SDG 4.c (equitable teacher training and capacity building)	Abeysekera (2024), Karatas et al. (2024), Olari and Romeike (2021)
	Phase 3: advanced integration	Curriculum adaptation and content optimization	SDG 4; SDG 4.c	Duan et al. (2024), Yue et al. (2024), Çelik et al. (2022)
	Phase 4: Sustainability & growth	Co-design for trustworthy AI and TPACK integration	SDG 4; SDG 17 (partnerships)	Duan et al. (2024), Yue et al. (2024), Çelik et al. (2022)
Instructor well-being & growth	Phase 2: Skill development	Targeted professional development	SDG 4.c; SDG 3 (health and well-being)	Baytar et al. (2023), Wilton et al. (2022), Bustamante Bula and Camacho Bonilla (2024)

(continued)

Table 2 (continued)

TRAIN framework component	Phase of implementation	Theme	Targeted SDGs	Citations
	Phase 3: Advanced integration	Reducing technical barriers and ethical oversight	SDG 3; SDG 10 (reduced inequalities)	Weuffen et al. (2023), Çelik et al. (2022), Tavares et al. (2023)
	Phase 4: Sustainability & growth	Inclusive practices and mental well-being	SDG 3; SDG 4	Gabriel et al. (2022), Baytar et al. (2023)
Novice-to-expert transition support	Phase 3: Advanced integration	Practical and ethical skills for modern accounting	SDG 4; SDG 9	Sperling et al. (2024), Cai (2022), Al Ghatrifi et al. (2023)
	Phase 4: Sustainability & growth	Transparent AI and system maturity	SDG 9; SDG 12	Duan et al. (2024), Gabriel et al. (2022)
	Phase 4: Sustainability & growth	Curriculum model adaptation and talent training	SDG 4; SDG 9	Qasim et al. (2022), Cai (2022), Holmes and Douglass (2022)

developed in response to these opportunities, addressing key challenges from technological development to instructor training, ethical considerations, and human oversight (Bustamante Bula and Camacho Bonilla 2024; Gabriel et al. 2022; Yue et al. 2024).

The research was guided by an essential question: how AI tools, particularly chatbots, can be leveraged to support junior accounting instructors' professional development while promoting environmental sustainability and broader Sustainable Development Goals. Through systematic review, it has been discovered that AI applications can be utilized to enhance both instructional support and administrative efficiency (Duan et al. 2024; Tavares et al. 2023). It has been demonstrated that AI-driven decision-making facilitates a more nuanced, personalized approach to education (Cheng et al. 2024; Holmes and Douglass 2022).

A significant finding reveals how modern technologies can be integrated with sustainable practices to construct more resilient and innovative educational models. The work of Son-Turan (2022) has been particularly influential in this regard. It has been observed that chatbots and similar AI tools can be employed to revolutionize administrative tasks while enhancing the overall educational experience (Duan et al. 2024).

Several limitations of this study must be acknowledged. The systematic review was restricted to recent research (2019–2024), potentially excluding valuable insights from earlier foundational work. The framework's applicability beyond accounting education remains to be validated, as the research was primarily contextualized within this discipline. The reliance on theoretical insights, without empirical validation of the TRAIN framework, has constrained its immediate practical application. Additionally, the environmental impact of AI tools, particularly their energy demands, continues to be identified as a significant challenge that must be addressed.

Future research directions have been identified across several crucial areas. Empirical validation of the TRAIN framework must be conducted across diverse educational settings and regions, as outlined in SDG 4c. Pilot studies should be implemented to measure both teaching quality and student outcomes. The framework's scope should be expanded beyond accounting into priority fields such as STEM and healthcare education. Further investigation into other AI tools, including adaptive learning platforms and augmented reality, should be undertaken to complement existing chatbot functionalities. Ethical challenges such as data privacy and algorithmic bias must be thoroughly examined, while more energy-efficient AI models should be developed and tested to align with the framework's emphasis on sustainable practices (Baytar et al. 2023; Al Ghatrifi et al. 2023; Abeysekera 2024).

Stakeholder engagement, particularly among educators, students, and policymakers, has been identified as essential for maintaining transparency and trust in these systems (Duan et al. 2024). The framework's advocacy for Small Language Models (SLMs) as environmentally conscious alternatives to larger models has been highlighted. SLMs, including Mistral's 7B, Microsoft's Phi-2, and Google's Gemma, have been found to be particularly well-suited for resource-constrained environments (Arthur Team 2024), aligning with the framework's sustainability objectives.

For effective implementation, responsible innovation must be embraced by educational institutions, and educators must be adequately prepared to utilize AI within the TRAIN framework's structured guidance (Çelik et al. 2022; Weuffen et al. 2023). Comprehensive AI literacy and professional development programs have been emphasized to address skill gaps (Olari and Romeike 2021; Wilton et al. 2022), while ethical and environmental impacts have been carefully considered (Bustamante Bula and Camacho Bonilla 2024; Gabriel et al. 2022).

The TRAIN framework can be viewed as a comprehensive roadmap for developing educators capable of integrating AI and human intelligence in collaborative, technology-rich environments. While the study's limitations are acknowledged, these constraints can be understood as indicators of valuable future research directions. It is anticipated that educators will play a crucial role in advancing both educational ecosystems and society at large, fostering learning environments that are effective, inclusive, and sustainable. The framework's potential to address instructor shortages and retention in education has been identified, though this potential must be validated through empirical research.

References

Abeysekera I (2024) ChatGPT and academia on accounting assessments. J Open Innov: Technol, Market, Complexity 10(1):100213

Al Ghatrifi MOM, Al Amairi JSS, Thottoli MM (2023) Surfing the technology wave: an international perspective on enhancing teaching and learning in accounting. Comput Educ: Artif Intell 4:100144

Arthur Team (2024) The beginner's guide to small language models. Arthur. https://www.arthur.ai/blog/the-beginners-guide-to-small-language-models

Baytar EM, Ettourouri A, Saqri N, Ouchaouka L (2023) Moroccan teachers' perceptions and concerns about ICT integration. IAFOR J Educ11(2):79–96

Boons F, Lüdeke-Freund F (2013) Business models for sustainable innovation: state-of-the-art and steps towards a research agenda. J Clean Prod 45:9–19

Bustamante Bula R, Camacho Bonilla A (2024) Inteligencia artificial (IA) en las escuelas: una revisión sistemática (2019–2023). Enunciación 29(1):62–82

Cai C (2022) Training mode of innovative accounting talents in colleges using artificial intelligence. Mob Inf Syst

Çelik I, Dindar M, Järvelä S (2022) The promises and challenges of artificial intelligence for teachers: a systematic review of research. TechTrends 66(4):616–630

Chai CS, Liang SY, Wang XW (2023) A survey study of Chinese teachers' continuous intentions to teach artificial intelligence. Educ Inf Technol, Early Access

Cheng X, Dunn R, Holt T, Inger K, Jenkins JG, Jones J, Wood DA (2024) Artificial intelligence's capabilities, limitations, and impact on accounting education: investigating ChatGPT's performance on educational accounting cases. Issues Account Educ 39(2):23–47

Davis FD (1989) Perceived usefulness, perceived ease of use, and user acceptance of information technology. MIS Q 13(3):319–340

den Brok P, Wubbels T, van Tartwijk J (2017) Exploring beginning teachers' attrition in the Netherlands. Teachers Teach 23(8):881–895

Duan XJ, Pei B, Wang CL (2024) Towards transparent and trustworthy prediction of student learning achievement by including instructors as co-designers: a case study. Educ Inf Technol 29(3):3075–3096

Folmeg M, Fekete I, Koris R (2024). Towards identifying the components of students' AI literacy: An exploratory study based on Hungarian higher education students' perceptions. J Univ Teach Learn Pract 21(6)

Gabriel F, Marrone R, Van Sebille Y, Kovanovic V, de Laat M (2022) Digital education strategies around the world: practices and policies. Irish Educ Stud 41(1):85–106

Grabińska B, Andrzejewski M, Grabiński K (2021) The students' and graduates' perception of the potential usefulness of artificial intelligence (AI) in the academic curricula of finance and accounting courses. E-Mentor 2021(5):16–25

Holmes AF, Douglass A (2022) Artificial intelligence: Reshaping the accounting profession and the disruption to accounting education. J Emerg Technol Account 19(1):53–68

Hu JN (2022) Partial differential equation-assisted accounting professional education and training: artificial intelligence collaborative course system construction. Sci Program 2022

Jelińska M, Paradowski MB (2021) The impact of demographics, life and work circumstances on college and university instructors' well-being during quaranteaching. Front Psychol 12:643229

Jeurissen R, Elkington J (2000) Cannibals with forks: The triple bottom line of 21st century business. J Bus Ethics 23(2):229–231

Jung J, Lee J, Kim H (2022) A study on the development of AI chatbot for Korean language education using chatbot builder. J Int Netw Korean Language Cult 19(3):405–441

Karatas F, Eriçok B, Tanrikulu L (2024) Reshaping curriculum adaptation in the age of artificial intelligence: mapping teachers' AI-driven curriculum adaptation patterns. Br Educ Res J, Early Access

Khalid FM, Abdul Rauf FH (2023) Factors affecting accounting students' academic performance. Eur Proc Educ Sci 3:95–104
Lee WK, Park JH (2015) The factors influencing information technology non-acceptance: approach by grounded theory. J Inf Syst 24(4):83–97
Li CZ, Zhao X (2022) Research on the influence of artificial intelligence technology with Web 3.0 on accounting education and its countermeasures. ACM Trans Asian Low-Resource Language Inf Process 21(6)
Lin XF, Chen L, Hu QT (2022) Teachers' perceptions of teaching sustainable artificial intelligence: a design frame perspective. Sustainability 14(13)
Meeks K, Peak AS, Dreihaus A (2023) Depression, anxiety, and stress among students, faculty, and staff. J Am Coll Health 71:348–354
Mishra P, Koehler MJ (2006) Technological pedagogical content knowledge: a framework for teacher knowledge. Teach Coll Rec 108(6):1017–1054
Mol AP, Spaargaren G (2000) Ecological modernisation theory in debate: a review. Environ Polit 9(1):17–49
Okolo, C. T. (2024). Beyond AI hype: A hands-on workshop series for enhancing AI literacy in middle and high school students. In: Proceedings of the conference for research on equitable and sustained participation in engineering, computing, and technology, RESPECT 2024, pp 86–93
Okulich-Kazarin V, Artyukhov A, Wolowiec T (2024) Will AI become a threat to higher education sustainability? A study of students' views Sustainability 16(11)
Olari V, Romeike R (2021) Addressing AI and data literacy in teacher education: a review of existing educational frameworks. In: Proceedings of the 16th workshop in primary and secondary computing education. ACM, pp 1–2
Qasim A, El Refae GA, Eletter S (2022) Embracing emerging technologies and artificial intelligence into the undergraduate accounting curriculum: reflections from the UAE. J Emerg Technol Account 19(2):155–169
Samkin G, Stainbank L (2016) Teaching and learning: current and future challenges facing accounting academics, and the development of an agenda for future research. Meditari Accountancy Res 24(3):294–317
Shin D, Lee JH (2024) Moving from off-the-shelf chatbots to a user-designed bespoke L2 chatbot. Lang Learn Technol 28(1)
Snyder H (2019) Literature review as a research methodology: an overview and guidelines. J Bus Res 104:333–339
Son-Turan S (2022) Fostering equality in education: the blockchain business model for higher education (BBM-HE). Sustainability 14(5):2955
Sperling K, Stenberg CJ, Stenliden L (2024) In search of artificial intelligence (AI) literacy in teacher education: a scoping review. Comput Educ Open 6
Sweller J (1988) Cognitive load during problem solving: effects on learning. Cogn Sci 12(2):257–285
Tavares MC, Azevedo G, Marques RP, Bastos MA (2023) Challenges of education in the accounting profession in Era 5.0: a systematic review. Cogent Bus Manag 10(2):2220198
UNESCO (2024) AI competency framework for teachers. https://unesdoc.unesco.org/ark:/48223/pf0000391104
Venkatesh V, Morris MG, Davis GB, Davis FD (2003) User acceptance of information technology: toward a unified view. MIS Q 27(3):425–478
Viac, C., & Fraser, P. (2020). Teachers' well-being: A framework for data collection and analysis. OECD Education Working Papers, No. 213, OECD Publishing.
Vygotsky LS (1978) Mind in society: the development of higher psychological processes. Harvard University Press
Waffenschmidt S, Knelangen M, Sieben W, Bühn S, Pieper D (2019) Single screening versus conventional double screening for study selection in systematic reviews: a methodological systematic review. BMC Med Res Methodol 19:1–9

Weuffen S, Burke J, Goriss-Hunter A, Plunkett M, Emmett S (2023) Working towards a sustainable, responsive, inclusive, and diverse global education future. In: Weuffen S, Burke J, Plunkett M, Goriss-Hunter A, Emmett S (eds) Inclusion, equity, diversity, and social justice in education: a critical exploration of the sustainable development goals. Springer, pp 277–286

Wilton L, Ip S, Fan F (2022) Where is the AI? AI literacy for educators. In: 23rd International conference on artificial ıntelligence in education (AIED)

Yue M, Jong MSY, Ng DTK (2024) Understanding K-12 teachers' technological pedagogical content knowledge readiness and attitudes toward AI education. Educ Inf Technol Early Access

A Critical Analysis of Sustainability Indicators for Education and Curricula in Higher Education Institutions and Their Adaptation to the Portuguese Context

Maria Constança Simões Rigueiro, Sandra Caeiro, Patrícia Câmara, Ana Fonseca, Teresa Godinho, Mahsa Mapar, Anabela Mendes Moreira, Maria Alzira Pimenta Dinis, Carla Sá Farinha, João Simão, and Manuela Vieira da Silva

Abstract The sustainability assessment within Higher Education Institutions (HEI) serves as a critical learning and capacity-building tool, fostering reflection on sustainability initiatives and advancing alignment with the Sustainable Development Goals (SDG). Current sustainability assessment tools include various indicators across different dimensions of sustainability implementation within HEI. However, there is a significant gap in categorising specific indicators relevant to learning and curricula.

M. C. S. Rigueiro
ISISE, Instituto Politécnico de Castelo Branco, Castelo Branco, Portugal
e-mail: constanca@ipcb.pt

S. Caeiro · M. Mapar · J. Simão (✉)
Center for Global Studies, Universidade Aberta, Lisbon, Portugal
e-mail: joao.simao@uab.pt

S. Caeiro
e-mail: scaeiro@uab.pt

M. Mapar
e-mail: mahsa.mapar@uab.pt

P. Câmara
Escola Superior de Enfermagem de São José de Cluny (ESESJCluny), Madeira, Portugal
e-mail: pcamara@esesjcluny.pt

A. Fonseca · M. A. P. Dinis
Fernando Pessoa Research, Innovation and Development Institute (FP-I3ID), Porto, Portugal
e-mail: afonseca@ufp.edu.pt

M. A. P. Dinis
e-mail: madinis@ufp.edu.pt

T. Godinho
Setúbal Polythecnic University, Setúbal, Portugal
e-mail: teresa.godinho@esce.ips.pt

W. Leal Filho et al. (eds.), *Education for Sustainable Development: The Contribution of Universities*, World Sustainability Series,
https://doi.org/10.1007/978-3-031-86985-3_28

In line with a recent initiative by the Sustainable Campus Network (RCS) in Portugal, this study aims to analyse how existing sustainability assessment tools address indicators within the domains of education and curricula. The research aims to identify and categorise indicators relevant to the education and curricula dimension, offering a preliminary list complemented by implementation examples. This analysis draws on a variety of available sustainability assessment tools for HEI. This goal is part of a broader research context in future development of sustainability indicators to systematically assess and monitor sustainability practices within Portuguese HEI. A systematic literature review was done to extract the education and curricula indicators from two key sources (i) existing sustainability assessment tools used globally, and (ii) existing sustainability reports of Portuguese HEI. As a result, 216 "education and curricula" indicators were collected, laying a solid foundation for further analysis and application in the Portuguese HEI context. The indicators were grouped into four different categories, *Teaching*, *Learning*, *Outreach*, and *Administrative and Organisational Framework*. Within the Portuguese context, only 13 sustainability reports were analysed (out of 97 HEI), in which six institutions addressed the 'education' category, primarily focusing on the number of courses, subjects, research initiatives, and projects aligned with the SDG. Although the present analysis was limited to publicly available sustainability reports, which may not fully reflect the breadth of initiatives undertaken by HEI, the findings highlight a significant gap in sustainability reporting among Portuguese HEI. The results of this study are expected to guide HEI towards more effective integration and promotion of a global culture of sustainability within the domains of curricula and education.

Keywords Higher education institutions (HEI) · Sustainability indicators · Sustainable development goals (SDG) · SDG 4 · Sustainable campus network (RCS) · Portugal · Education and curricula

A. M. Moreira
TECHN&ART, Technology, Restoration and Arts Enhancement Center, Polytechnic Institute of Tomar, Tomar, Portugal
e-mail: anamoreira@ipt.pt

M. A. P. Dinis
Marine and Environmental Sciences Centre (MARE), University of Coimbra, Coimbra, Portugal

C. S. Farinha
CENSE, Center for Environmental and Sustainability Research, NOVA School of Science and Technology, Lisboa, Portugal
e-mail: carla.farinha@ine.pt

J. Simão
CAPP—Centre for Public Administration and Public Policies, Universidade de Lisboa, Lisboa, Portugal

M. V. da Silva
REQUIMTE/LAQV, ESS, Polytechnic of Porto, Porto, Portugal
e-mail: mvsilva@ess.ipp.pt

1 Introduction

Higher education institutions have been addressing sustainability challenges, broadly encompassing their institutional advancements in three dimensions: social, environmental, and economic. However, given the breadth and diversity of their activities, the best approach to assessing their performance and progress towards becoming sustainable universities remains under development. It is widely recognised that several elements of sustainability should be emphasised among HEI, which include education, research, campus operation, community engagement, and reporting (Lozano et al. 2015; Husaini et al. 2018). However, some difficulties persist, when developing an integrative model of HEI strategies that encompasses all sustainability measures for use in decision-making by top management (Lozano et al. 2015).

Sustainability Assessment Tools (SAT) have become fundamental in HEI, enabling strategic planning and development towards sustainable universities (Berzosa et al. 2017; Bashir et al. 2023). Numerous SATs and approaches have emerged, with many focusing strongly on campus-based assessments. These tools employ qualitative, quantitative, or mixed methodologies. Examples include the Campus Sustainability Assessment Framework (CSAF), Auditing Instrument for Sustainability in Higher Education (AISHE), Sustainability Access Questionnaire (SAQ), Unit-based Sustainability Assessment Tool (USAT), Alternative University Appraisal Model (AUA), Conference of Rectors of Spanish Universities (CRUE), and Sustainability Tracking, Assessment & Rating System (STARS) (Fischer et al. 2015; Berzosa et al. 2017). Despite their variety, many of these tools place limited emphasis on curricula and education. The Sustainable Tool for Auditing Universities' Curricula in Higher Education (STAUNCH) addresses this gap by specifically focusing on education and curricula (Lozano and Barreiro-Gen 2019). Developing a consistent set of sustainability indicators is vital to support the monitoring of progress towards sustainability in HEI. Existing sustainability assessment tools are complex, encompassing various indicators and criteria related to education and curricula. A careful analysis is needed to categorise the indicators most relevant to this area. The selection of adequate assessment indicators must consider that these should be simple, measurable, achievable, and time-bound to ensure comparability, consistency, and usability (Bauler 2012).

In Portugal, higher education plays a significant role and is organised into a binary system of university and polytechnic education, provided by both public and private institutions. The *Direção-Geral de Ensino Superior* (DGES), which operates under the Portuguese Ministry of Education, is responsible for managing higher education in Portugal and plays a crucial role in coordinating and overseeing HEI. According to DGES, there are 97 higher education institutions in Portugal, comprising 61 private and 36 public institutions, with 61 polytechnic and 36 university institutions. The total number of students enrolled in higher education stands at 446,028 (DGESa n.d.). Participating in cooperation networks and adopting shared governance procedures can also enhance the development of sustainability integration strategies in HEI (Vargas et al. 2019; Farinha et al. 2020). In this scope, RCS is a collaborative initiative involving individuals from national HEI. Its mission is to promote the adoption of sustainable development principles and practices, addressing environmental, social, and economic dimensions across campuses.

The present study aims to analyse international sustainability assessment tools for HEI and sustainability reports of Portuguese HEI to identify and categorise indicators relevant to the dimension of education and curricula. This study forms part of broader research within the scope of the Portuguese RCS, focusing on defining tools and indicators to systematically assess and monitor sustainability practices in Portuguese HEI.

This study's innovation lies in the pre-selection and categorisation of a set of indicators related to teaching and curricula, which will form part of a broader system of sustainability indicators in Portuguese HEI, within the context of the future National Observatory of HEI in Portugal Although it does not complete the final identification of the indicators to be included in this Observatory, it represents an important step, as no prior progress has been made in this area within the Portuguese context.

The chapter is organised as follows. Section 2 reviews the integration of sustainability in HEI and evaluates sustainability in teaching and curricula, with a particular focus on Portugal. Section 3 details the research method, which comprises four phases. Section 4 examines the results of the analysis. Finally, Sect. 5 provides the conclusion of the study.

2 Literature Review

2.1 Sustainability Integration in Higher Education Institutions

Sustainability education has become a fundamental aspect of higher education, driven by the urgent need to address global environmental challenges and promote sustainable development across all sectors of society. In recent years, sustainability education in HEI has attracted significant research attention, with notable progress in integrating sustainability principles into curricula and institutional practices (Machado and Davim 2023; Basheer et al. 2024). HEI recognise their crucial role in fostering sustainability education, helping students develop responsible and participatory citizenship aligned with sustainability principles and equipping them to address complex environmental and social challenges. For sustainability frameworks to be effective within HEI, they must support both incremental and transformative changes that drive changes in existing structures and practices. These frameworks should not only encourage students and staff to engage with sustainability issues critically and innovatively but also align with the unique contexts and operational realities of each institution (Caeiro et al. 2020; Mair and Druckman 2023). This alignment enables HEI to integrate sustainability into their mission, equipping students to contribute to sustainable development and preparing future leaders to implement solutions across diverse fields. Through this commitment, HEI become hubs of sustainable transformation, setting benchmarks for education and action that extend beyond campus boundaries and contribute considerably to local, national, and global sustainability goals. Varouchas et al. (2018) argue that integrating quality into curricula is both a pedagogical concern and a strategic approach, emphasising that quality should drive HEI transformation towards academic excellence, emphasising that quality should be regarded as the central driver for transforming HEI. Similarly, a strategic approach is essential for embedding sustainability within HEI, as these institutions play a crucial role in driving societal transformation toward sustainability. This is achieved particularly by incorporating sustainability into curricula and research practices (Caeiro et al. 2020; Mair and Druckman 2023; Nikolaou et al. 2023), a process that requires both incremental and transformative shifts in educational content and institutional practices. HEI are encouraged to adopt a holistic approach that addresses social and economic dimensions, moving beyond a focus on environmental issues. Buerkle et al. (2023) highlight the importance of aligning higher education curricula with the United Nations' SDG, arguing that sustainability should be integrated into both teaching methods and content. This integration ensures that students acquire the skills and knowledge necessary to address global sustainability challenges. The same authors emphasise that embedding sustainability within pedagogical practices, rather than simply adding it as supplementary content, is essential for equipping future generations with the tools to meaningfully contribute to the SDG. Supporting this view, Leal Filho et al. (2023) underscore the transformative potential of education in fostering societal change toward sustainability, arguing that curricula should not merely treat sustainability as an additional topic but rather integrate it as a core element. This integration helps students develop critical thinking and problem-solving skills, vital for addressing global challenges. To fully engage with sustainable development, universities must embed sustainability throughout their academic, research, and operational activities. Accountability is also a key issue in relation to sustainability management in HEI. Moggi (2023) emphasises that HEI should communicate their sustainability strategies and achievements through sustainability reporting, particularly by adopting frameworks such as the Global Reporting Initiative (GRI). The same author argues that such frameworks should be adapted to align with universities' specific missions, which prioritise teaching and research. This adaptation ensures that sustainability reporting remains consistent with their core functions while maintaining transparency and accountability. In the broader context of sustainability reporting aligned with global standards such as the SDG, Nikolaou et al. (2023) highlight the need for universities to enhance their sustainability reporting practices, aligning with global standards such as the SDG to provide a more comprehensive view of their efforts. Greening the curriculum' is considered a critical strategy for advancing sustainability integration, as aligning academic programmes with the SDG enables HEI to contribute more effectively to global sustainability goals.

2.2 Assessing Sustainability in Higher Education Institutions: Impacts on Education and Curricula

Assessing education for sustainability within HEI is essential for understanding its effectiveness and ensuring continuous improvement. With HEI increasingly integrating sustainability into curricula, research, and campus operations, evaluating the outcomes of these efforts has become increasingly urgent. Effective assessment allows HEI to evaluate how well they prepare students to tackle complex sustainability challenges while promoting accountability in sustainable development. According to Lozano et al. (2015), assessment and reporting are critical dimensions of sustainability integration in HEI. As noted earlier, several tools have been developed to assess how sustainability concerns are integrated into HEI, measuring performance across various dimensions. These tools offer a framework for evaluating the incorporation of sustainability principles and guide strategic planning and decision-making (Bullock and Wilder 2016; Caeiro et al. 2020). By using sustainability assessment tools and frameworks, HEI can benchmark their performance against peers, set improvement goals, and publicly report on progress. This practice enhances transparency and fosters a culture of continuous improvement in sustainability. Sustainability assessment tools with ranking systems attract university administrators' attention, making them valuable for improving sustainability integration processes (Veidemane 2022; Leal Filho et al. 2024a). However, most sustainability assessment tools for HEI primarily rely on self-assessments and often fail to adequately incorporate all stakeholders' perspectives (Disterheft et al. 2016; Caeiro et al. 2020). A participatory approach involving both internal and external stakeholders is essential for effectively implementing sustainability assessment tools (Mapar et al. 2022).

Despite the critical importance of assessment and the abundance of sustainability assessment tools available for HEI, their implementation is still in its early stages (Findler et al. 2019; Caeiro et al. 2020). Nevertheless, stakeholders within HEI are increasingly interested in information on the adoption of education for sustainability practices, and institutions are committed to monitoring relevant indicators (Veidemane 2022). To facilitate this, systematic and standardised data collection methods are needed, creating opportunities for innovative technologies to monitor sustainability indicators in HEI (Basheer et al. 2024). Most sustainability assessment tools available for HEI are indicator-based, aiming to provide measurable, consistent, comparable, and transparent results (Findler et al. 2019; Ramos 2019; Caeiro et al. 2020). HEI commonly use sustainability assessment indicators to: (i) monitor performance by tracking progress towards established goals; (ii) support decision-making with strategic and operational insights; and (iii) identify opportunities for improvement to meet targets. These indicator systems, identified as crucial for achieving sustainability in institutions (e.g., Mendoza et al. 2019), are most effective when aligned with strategic objectives and integrated into a management system for assessing performance. Developing comprehensive sustainability assessment indicators is essential for measuring an HEI's holistic impact on students and academic programmes, ensuring continuous improvement across all aspects of institutional practices (Varouchas et al. 2018). However, sustainability assessment indicators for HEI should be redesigned through transdisciplinary collaborative research to align these tools with new paradigms and sustainability principles (Ramos 2019).

Regarding education and curricula, sustainability assessment indicators should provide a framework for evaluating how effectively sustainability is integrated into educational programmes, ensuring that HEI equip students with the knowledge and skills needed to tackle global sustainability challenges. This has a direct impact on the quality of education and the readiness of graduates to contribute to a sustainable future. Bashir et al. (2023) emphasise the importance of selecting relevant indicators for sustainability assessment and call for clear frameworks and tools to help HEI evaluate their efforts across various domains. Using comprehensive and consistent evaluation metrics allows institutions to track progress more effectively and make informed decisions to improve sustainability strategies and campus practices. Madeira (2008) highlights the importance of robust data systems for supporting comprehensive sustainability assessments in HEI. This study developed 110 sustainability assessment indicators across five key areas: academic community, teaching, research, operations, and external community impact. A case study applying these indicators at the Faculty of Engineering, University of Porto (Portugal), highlighted challenges in addressing all proposed indicators due to limited data availability.

Adding another perspective, Gutiérrez-Mijares et al. (2023) provide an extensive review of methods used to assess sustainability performance at HEI, stressing the importance of having clear frameworks and assessment tools that account for the diverse manifestations of sustainability in higher education. The same authors underscore that comprehensive assessment methodologies are critical for ensuring that sustainability principles are thoroughly integrated into both teaching and institutional practices. By developing adaptable and effective frameworks for assessing sustainability, HEI can ensure that sustainability principles are embedded not only in teaching but also in broader academic and operational strategies. Methods such as Social Network Analysis (SNA), a method for visualising and analysing the relationships among players by creating a network of nodes and connections, and Cross-Impact Matrix Multiplication (MICMAC), a technique used to assess the driving and dependence powers of variables, categorising them into autonomous, dependent, linkage, and independent groups based on their direct and indirect relationships) can assist institutions in visualising and analysing the relationships between assessment indicators, leading to more effective sustainability strategies and continuous improvement in sustainable campus practices (Bashir et al. 2023).

2.3 *Sustainability Assessment Tools in Portuguese Higher Education Institutions*

Sustainability assessment tools should be adjusted to specific contexts and regions (Mapar et al. 2022; Basheer et al. 2024). In Portugal, strategies to promote sustainability education in HEI are becoming more common, driven primarily by institutional proactivity rather than governmental incentives (Farinha et al. 2020). Similar to many other HEI worldwide, Portuguese institutions initially focused their sustainability integration efforts on improving environmental performance, including waste management and reducing energy, water, and paper consumption (Amaral et al. 2015; Leal Filho et al. 2021). The main challenges faced by Portuguese HEI in integrating sustainability strategies mirror those encountered by institutions worldwide, including a lack of funding, insufficiently trained personnel, and the absence of national policy guidelines (Aleixo et al. 2018; Farinha et al. 2020). In a study covering 13 Portuguese public universities, Farinha et al. (2020) concluded that assessment and reporting were the least developed dimension of sustainability integration. This finding is consistent with a global study on sustainable development practices in HEI (Lozano et al. 2015). To address these challenges, Rodrigues (2023) developed a comprehensive framework for assessing sustainability in Portuguese HEI. This framework introduces 36 sustainability assessment indicators spanning five dimensions: environmental, economic, social, academic, and institutional. These indicators provide a straightforward "yes/no" evaluation method, facilitating internal assessments and comparable sustainability reporting, thereby driving institutional improvements. Recent studies (Duarte et al. 2023; Farinha et al. 2020; Leal et al. 2024a, b) highlight the need for Portuguese HEI to adopt holistic sustainability strategies rather than relying solely on isolated measures. Although some practices have emerged in integrating sustainability into education and curricula, significant improvements are needed to align these efforts with HEI strategic planning (Duarte et al. 2023; Leal et al. 2024a, b).

3 Methods

To achieve the objective of selecting and categorising indicators related to education and curricula, a four-stage process was developed. The process involved analysing sustainability assessment tools (phase one) and sustainability reports (phase two), followed by identifying indicators (phase three) and categorising them (phase four). SAT and sustainability reports, as dynamic instruments for evaluating and communicating sustainability, serve as the most reliable sources for selecting indicators currently used by HEI.

The approach used for the preliminary proposal of the indicators involved four distinct phases (Fig. 1). In *phase one*, indicators and their associated categories and subcategories were analysed and extracted from sustainability assessment tools in

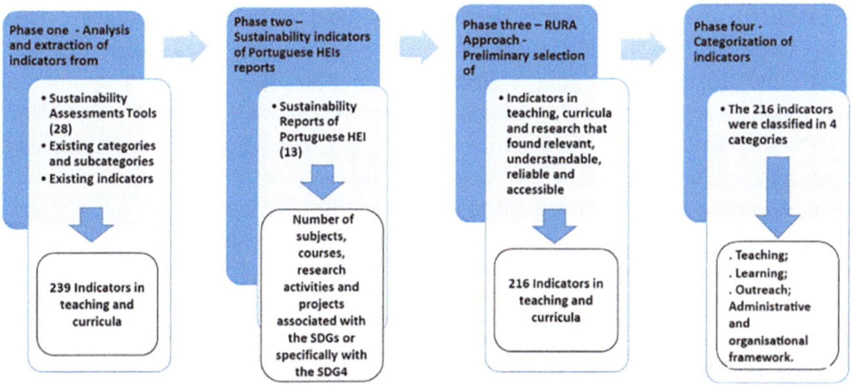

Fig. 1 Phased process for selecting sustainability indicators within the domains of education and curricula

HEI, updating the earlier works of Caeiro et al. (2020), Mapar et al. (2022) and Gutiérrez-Mijares et al. (2023). For this purpose, from June to August 2024, a search was conducted on Google Scholar using the terms "sustainability assessment", "higher education", "university", "campus", "indicator", and "evaluation". These tools were then selected based on the following criteria, previously used by Caeiro et al. (2020) and Mapar et al. (2022):

(1) Tools developed specifically for assessing the performance of sustainability implementation in HEI.
(2) Tools covering at least two key dimensions of sustainability implementation in HEI: (i) facilities or campus operations; (ii) teaching and curricula; (iii) organisational management; (iv) external community; (v) research; (vi) assessment and communication.
(3) Tools covering at least two of the sustainability pillars, i.e., environmental, social, and economic, to ensure that they adopt a holistic, whole-institution approach, as most tools focus on only two of these pillars.

While this work focuses solely on the dimension of teaching and curricula, exploring tools that conduct integrative sustainability assessments at decision-making and strategic levels was considered important. This aligns with the broader research objective of developing sustainability indicators for HEI sustainability implementation. A total of 28 assessment tools were found that met these criteria (see Annex). In *phase one*, only indicators related to 'teaching and curricula' from the 28 tools were extracted. Some tools classified 'education and research' indicators together, but only those related to education were included in this study. In total, 239 indicators related to teaching and curricula were selected during this phase.

In *phase two*, the focus was on identifying the sustainability assessment indicators Portuguese HEI already use for teaching and curricula. To achieve this, the websites of institutions listed by DGES were consulted for information on sustainability management practices, particularly sustainability reports. The research method involved a systematic and targeted search on the institutional websites of Portuguese HEI, using terms such as 'sustainability', 'sustainability report', 'sustainable development', and 'sustainability plan' to locate and analyse the presence or absence of sustainability reports. The documents found were analysed to identify indicators used by Portuguese HEI for education and curricula, to supplement or compare with those selected in *phase one*.

In *phase three*, the research team analysed all indicators from the previous phases, resulting in a final selection of 216 indicators. Indicators should convey complex information and support decision-making in a simplified and comprehensive format. Therefore, only indicators considered Relevant (providing important and necessary information), Understandable (easily comprehensible), Reliable (ensuring trust in the system), and based on Accessible data (ensuring applicability) were selected. This approach, designated as RURA, was adopted to achieve the ultimate goal of this study.

Finally, in *phase four*, the 216 indicators obtained in *phase three* were classified into four categories, based on the United Nations Environment Programme sustainability university framework (UNEP 2021) and Lozano et al. (2015):

- *Teaching*. Indicators related to the learning approaches employed by teachers.
- *Learning*. Indicators focusing on learners, including formal, informal, and non-formal students or trainees.
- *Outreach*. Indicators addressing the external community beyond the HEI.
- *Administrative and organisational framework*. Indicators focusing on organisational strategies and top management.

A focus group of three experts conducted the classification, assigning each category a scale of 0 to 1 (target/not target). While some indicators could fall into multiple categories, only their primary target was considered. The focus group conducted a two-round process to discuss and resolve disagreements, reaching consensus and improving agreement levels, as outlined by Sampieri et al. (2006).

This study seeks to contribute to the evolving field of sustainability assessment in higher education by focusing specifically on the underexplored dimension of teaching and curricula. While previous studies have proposed frameworks for holistic sustainability assessments, this research distinguishes itself by isolating and categorising education-specific indicators, a critical yet often overlooked domain. Additionally, the RURA approach adds methodological rigour and practical relevance, ensuring the indicators are actionable and aligned with HEI decision-making processes. This focused yet integrative approach provides a refined understanding of how sustainability can be effectively embedded into education and curricula, addressing a critical gap in existing tools and frameworks within the Portuguese context.

4 Results and Discussion

Phase one resulted in a list of 28 assessment tools (see Annex), while *phase two* examined 97 Portuguese HEI. In *phase two*, which focused on identifying sustainability indicators in Portuguese HEI, the analysis revealed that only 13 out of 97 institutions had published sustainability reports on their websites. Among these 13 institutions, three had adopted the GRI framework, adapting it to the context of higher education. This adaptation was necessary because the GRI framework lacks sector-specific standards for HEI. These sustainability reports varied in detail but offered valuable insights into each institution's contributions to the SDG outlined in the 2030 Agenda. These reports were essential for transparency and accountability, allowing stakeholders—such as students, staff, and the wider community—to assess progress in sustainability initiatives. Typically, they addressed diverse aspects of sustainability, including environmental management, social responsibility, and economic viability. By aligning their reporting with GRI standards, these institutions demonstrated a commitment to best practices, enabling better comparisons and evaluations across HEI. However, the findings highlighted the need for broader adoption of sustainability reporting practices to enhance accountability and advance sustainability in higher education. Of the sustainability reports analysed, only six institutions included the 'education' category, using the SDG and the Principles for Responsible Management Education (PRME), a United Nations-supported initiative, as monitoring criteria. These institutions systematically mapped their educational offerings, research activities, and scientific outputs against the SDGs. They specifically focused on indicators related to the number of courses, subjects, research initiatives, and projects aligned with SDG 4, which emphasises quality education. Notably, only two indicators, namely 'number of courses' and 'number of formal programmes', were consistently reported across these sustainability reports. These indicators, already recognized as effective in various international tools, were included in the list from the study's *phase one*, highlighting their relevance in assessing educational contributions to sustainability. However, this limited focus on education within sustainability reports exposed a significant gap in evaluating how HEI integrate sustainability into their curricula and teaching practices. While the inclusion of SDG-related indicators was a positive step toward greater accountability and transparency, it also underscored the need for a broader range of indicators capable of comprehensively assessing educational contributions to sustainability. In *phase three*, the initial combined list of indicators from the 28 assessment tools included 239 indicators. Among them, 24 indicators were not considered due to their lack of clarity or because they did not meet the criteria for being considered valid indicators, as outlined by Ramos (2019). The remaining indicators were evaluated using the RURA approach, thus ensuring that they were Relevant, Understandable, Reliable, and based on Accessible data. This process led to the refinement of the list, resulting in a comprehensive set of 216 indicators that align with the study's goals. Finally, in *phase four*, the 216 indicators were classified into the four defined categories, i.e., *Teaching*, *Learning*, *Outreach* and *Administrative and organisational framework*, by determining the proportion

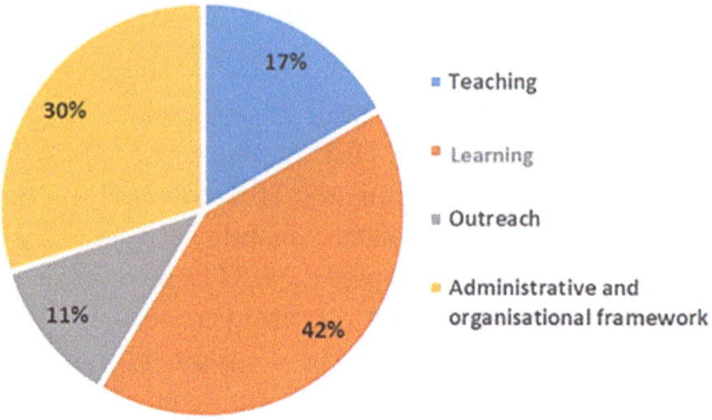

Fig. 2 Proportion of Education and Curricula Indicators in different categories obtained in the ***phase four***

of each category using descriptive data analysis (Fig. 2). The extensive number of tools and indicators in the four categories revealed recurring patterns, both across and within categories. This repetition stems from differences in writing style and semantics, despite the indicators' underlying content being essentially the same.

The findings from the four categories, namely *Teaching*, *Learning*, *Outreach*, and *Administrative and organisational framework*, are detailed below, based on insights from the selected 216 indicators. The results are presented by contribution percentages for each category, arranged in descending order. The analysis not only provides an overview of the contributions but also includes illustrative examples that showcase the role and impact of each category. Ranking the categories emphasises the comparative influence and significance of the indicators, helping to clarify how each category performs and supports the overall objectives. *Teaching* highlights methods, curriculum design, or educator impact; *Learning* focuses on outcomes, engagement, or skill development; *Outreach* illustrates community and external collaborative efforts; and *Administrative and Organisational Framework* evaluates governance, operational efficiency, and support systems.

(i) The *Learning* category is the one with the highest number of indicators (42%). Some examples of indicators belonging to this category are:

- Sustainability-related final projects (final year projects or master's theses related to sustainability)
- Improvement of students' Sustainable Development and Social Responsibility skills
- Undergraduate programs with sustainability-focused learning requirements
- Applied learning for sustainability program
- Students' environmental and sustainability awareness programmes
- Competences for sustainability acquisition

Notably, the two indicators ('number of courses' and 'number of formal programmes') identified in *phase two* sustainability reports are already included in the comprehensive list derived from sustainability assessment tools. The above examples illustrate diverse approaches to integrating sustainability into educational contexts, ranging from awareness-raising programmes to formal academic courses

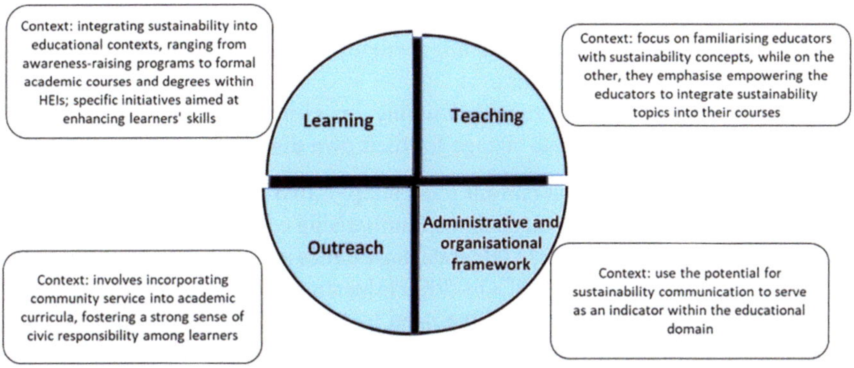

Fig. 3 *Teaching*, *learning*, *outreach*, and *administrative and organisational framework* considered and related context

and degrees within HEI. The scope of these indicators also extends beyond mere knowledge development, and it also includes specific initiatives aimed at enhancing learners' skills (Fig. 3). Indicators This, in turn, contributes to fostering groups of sustainability-literate learners, as emphasised by Décamps et al. (2017).

(ii) The Administrative and Organisational Framework stands as the second level, with 30% of the indicators. Examples of indicators into this category are:

- Internal university awards for innovation and good practices
- Existence of a university-run sustainability website
- Existence of published sustainability report
- The university has developed or uses a framework or strategy for Education for Sustainable Development (ESD)

An interesting aspect among these examples is the potential for sustainability communication to serve as an indicator within the educational domain (Fig. 3). Indicators such as sustainability reports and institutional websites are used as media to represent and disseminate educational efforts, promoting sustainability awareness among internal and external audiences. This aligns with the study of Amey et al. (2020), which highlights the role of HEI sustainability websites as a key resource for engaging the campus community in sustainability initiatives. Reports and websites enhance interaction, networking (Leal Filho et al. 2024b), and communication among students, staff, and stakeholders, contributing to greater awareness and deeper knowledge of sustainability practices within and beyond the institution.

(iii) The *Teaching* approaches category is linked with 17% of the indicators used to achieve sustainability education in the curricula. Examples vary:

- Faculty Training, with training activities being conducted for faculties to enhance the inclusion of sustainability topics in their courses
- Collaborating in teaching sustainability programmes at the faculty level

- The curricula of teacher education programs include courses or specific content on how to adapt teaching methods for students with disabilities and special educational needs
- The HEI's lecturers receive training on climate change, including the implications of climate change for their own disciplines

These indicators show that current tools adopt a two-fold approach to teaching sustainability (Fig. 3). They focus both on familiarising educators with sustainability concepts and empowering them to integrate these topics into their courses. This aligns with the third principle of UNESCO's 2020 Education for Sustainability Roadmap (UNESCO 2020), which highlights the need to equip educators with the skills to embed sustainability within educational programs effectively.

(iv) *Outreach* activities is the last category, targeting the community outside the HEI, with 11% of the indicators. Some examples of the applied indicators within these categories are:

- Number of scholarly events related to environment and sustainability
- Local education programmes or campaigns on climate change risks; impacts, mitigation, impact reduction and early warning
- Community activity and learning service
- The HEI offers service-learning opportunities within its programs in order to tackle local challenges, particularly of disadvantaged groups.

Community activity and service learning are important components of the SDG holistic approach (Fig. 3). It incorporates community service into academic curricula, fostering a strong sense of civic responsibility among learners. This approach enhances their ability to reflect critically on their experiences, making it an effective tool in education for sustainable development, as highlighted by Aramburuzabala and Cerrillo (2023). It involves incorporating community service into academic curricula, fostering a strong sense of civic responsibility among learners while enhancing their ability to reflect critically on their experiences, making it an effective tool in education for sustainable development, as highlighted by Aramburuzabala and Cerrillo (2023).

This research identifies necessary future steps, such as selecting sustainability indicators for a future National Observatory of HEI in Portugal. This study demonstrates that assessing HEI sustainability is a dynamic, multifaceted process encompassing diverse topics, categories, and subcategories. Each sustainability assessment tool uses a distinct nomenclature and framework for its system of indicators, which complicates comparisons between tools. This diversity underscores the complexity of achieving standardisation or alignment across various systems. Addressing these challenges will require future analysis of existing indicators, evaluating them against their specific objectives and contexts. This process should aim to identify commonalities and overlaps while ensuring that the selected indicators align with the strategic goals of sustainability at the national and institutional levels. This approach would facilitate more effective benchmarking and comparison of HEI sustainability practices, supporting alignment with broader frameworks such as the SDG. Additionally, refining the selection process for indicators will enhance the clarity, relevance, and usability of the Portuguese National Observatory's assessment tools.

5 Conclusions

Although SATs are fundamental in HEI, their complexity includes various indicators and criteria related to education and curricula, requiring careful analysis to select and categorise relevant indicators. Objective and consistent sustainability indicators are essential for monitoring progress towards sustainability in HEI. In Portugal, HEI need to adopt holistic sustainability integration strategies, rather than relying solely on isolated measures to improve sustainability performance, remaining considerable room for improvement in aligning these practices with their strategic planning.

This research aimed to identify and categorise indicators focused on education and curricula, for which purpose it analysed existing international sustainability assessment tools and sustainability reports of Portuguese HEI.

The methodology included a comprehensive review and selection of indicators categorised into four main areas: *Teaching*, *Learning*, *Outreach*, and *Administrative and Organisational Framework*. The analysis of the 216 selected indicators across these four categories highlights their distinct contributions to HEI sustainability. Results reveal that:

- The *Learning* category, with the largest share (42%), emphasises sustainability-focused education through initiatives such as sustainability-related final projects, programs enhancing skills, and awareness campaigns;
- *Administrative and Organisational Framework* (following with 30% share), spotlighting tools like sustainability reports and institutional websites that communicate and operationalize sustainability goals;
- *Teaching* (accounting for 17%), integrates faculty training and curriculum adaptation to include sustainability topics, aligning with UNESCO's focus on educator empowerment;
- *Outreach* (representing 11%) underscores community engagement through service learning and educational programs addressing local challenges.

These categories illustrate a comprehensive approach to embedding sustainability in HEI, fostering institutional and societal transformation. Through identifying informative and commonly used sustainability implementation indicators in each category, this study provides a foundational list intended for future use within a Portuguese National Observatory under the auspices of the RCS.

The study also identified a significant gap in sustainability reporting among Portuguese HEI. Of the 97 institutions analysed, only 13 published sustainability reports, and among those, only six incorporated the educational dimension as a category. This limited focus on education in sustainability reporting highlights opportunities for development and improvement. The reports that did address education primarily focused on a narrow set of indicators, notably the "number of courses" and "number of formal programs" aligned with the SDG, particularly SDG 4, emphasising quality education. This restricted focus underscores the need for broader and more diverse indicators that reflect the comprehensive role of education in advancing sustainability initiatives.

The implications of these findings are multifaceted. Firstly, there is a clear need for increased awareness and integration of educational sustainability within institutional practices and reporting frameworks; by adopting a more extensive range of indicators, HEI can better demonstrate their contributions to sustainability and accountability in education. Enhancing the educational dimension of sustainability reporting can help institutions align curricula and teaching with global sustainability goals, fostering a culture of sustainability among students and staff.

However, this study also has limitations. The study was limited to publicly available sustainability reports, which may not fully capture the breadth of initiatives undertaken by HEIs. The preliminary nature of the proposed indicators requires further validation and refinement to ensure their real-world applicability and effectiveness.

Looking ahead, future research could explore the development of additional indicators that encompass various aspects of sustainability education, facilitating a more holistic evaluation of HEI' contributions. Moreover, longitudinal studies could track changes in sustainability reporting practices over time, assessing the impact of implemented indicators on institutional sustainability efforts. Collaboration among HEI, policymakers, and stakeholders within the RCS is essential for advancing sustainability and achieving a more sustainable educational landscape. By addressing the identified gaps, HEI can take significant steps toward fulfilling their roles as leaders in sustainability education, ultimately contributing to the broader goals of the 2030 Agenda for Sustainable Development.

The expected impact of this work extends beyond national borders, promoting a global culture of sustainability in education and contributing significantly to achieving SDG 4, target 4.7.

Annex

HEIS' sustainability assessment tools, searched in this study (adapted and updated from Mapar et al. 2022)

Number	Abbreviation	Full name	Developer	Country	Year	Definition	Link or Reference of the Tool
1	AISHE 2.0	Assessment Instrument for Sustainability in Higher Education	By a researcher in Europe	Netherland/ International	2000	A tool for the development of a policy towards sustainable development	https://www.researchgate.net/publication/327551379_AISHE_20_Manual_-_English_Edition
2	AMAS	Adaptable Model for Assessing, Sustainability in Higher Education	Pontificia Universidad Católica de Chile/Chile (by a researcher)	Chile	2014	Assessing Sustainability in HEIs among different implementation stages	Gómez, F., Sáez-Navarrete, C., Lioi, S., Marzuca, V., Adaptable Model for Assessing Sustainability in Higher Education. Journal of Cleaner Production 2015, 107, 475–485. 10.1016/j.jclepro.2014.07.047
3	ASSC	Assessment System for Sustainable Campus	Hokkaido University in within CAS-NET JAPAN (Campus Sustainability Network in Japan)	Japan	2013	Strengths and weaknesses of implementing sustainability in HEIs and helps them decide future strategies	https://www.osc.hokudai.ac.jp/en/action/assc
4	AUSP (Crue)	Evaluación de las políticas universitarias de sostenibilidad como facilitadoras para el desarrollo de los campus de excelencia internacional	De La Crue Sectoral Commission for Environmental Quality, Sustainable Development and Risk Prevention of CRUE—Conference of Rectors of Spanish Universities	Spain	2007	To improve performance and policies in terms of social responsibility, environment (including public procurement), economy and implementation of the SDGs	https://www.crue.org/wp-content/uploads/2020/02/22.INFORME_EVALUACION_COMPLETO.pdf
5	BIQ-AUA	Alternative University Appraisal Model	Asia–Pacific academic alliance named ProSPER.Net (promotion of sustainability in postgraduate education and research network)	Asian-Pacific	2009	ESD at HEIs (a set of quantitative questions about the overall maturity of the university's efforts towards sustainability)	https://prospernet.ias.unu.edu/projects/past-projects/alternative-university-appraisal-aua

(continued)

(continued)

Number	Abbreviation	Full name	Developer	Country	Year	Definition	Link or Reference of the Tool
6	CITE AMB	Red de Ciencia, Tecnologia, Innovacion y Educación Ambienal en Iberoamérica	Network of Science, Technology, Innovation and Environmental Education in Iberoamerica	Colombia	2014	To make a diagnosis of the current situation of the institutionalisation of environmental commitment in universities, **mainly on top level of universities**	CITE-AMB (Red de Ciencia, Tecnología, Innovación y Educación Ambiental en Iberoamérica) et al. (2014) 'Encuesta Para El Diagnóstico De La Institucionalización Del Compromiso Ambiental En Las Universidades Colombianas [Inquiry for Diagnosing the Institutionalisation of Environmental Committment of Colombian Universities].' Bogotá
7	DUK	German Commission for UNESCO AG HS (2011)	German Commission for UNESCO	Germany	2011	Sustainability self-assessment concept for HEIs	Findler, F.; Schönherr, N.; Lozano, R.; Stacherl, B. Assessing the Impacts of Higher Education Institutions on Sustainable Development—An Analysis of Tools and Indicators. Sustainability 2018, 11, 59. doi: 10.3390/su11010059
8	ESDGC	Education for Sustainable Development and Global Citizenship	Outlined by the Government of Wales	Wales / United Kingdom	2012	Assessment of implementation of ESD in Universities (specifically for HEIs in Wales, UK)	Glover, A.; Jones, Y.; Claricoates, J.; Morgan, J.; Peters, C., Developing and Piloting a Baselining Tool for Education for Sustainable Development and Global Citizenship (ESDGC) in Welsh Higher Education. Innovative Higher Education 2012, 38. 10.1007/s10755-012-9225-0
9	GASU	Graphical Assessment of Sustainability in Universities tool	A researcher in Europe and marketed through a company	United Kingdom	2006	Comparison and benchmarking of universities' sustainability efforts and achievement by graphical overview	Lozano, R. (2006). A tool for a Graphical Assessment of Sustainability in Universities (GASU). Journal of Cleaner Production, 14(9–11), 963–972. http://doi.org/10.1016/j.jclepro.2005.11.041
10	GC	Good Company's Sustainable Pathways Toolkit	US company (Good company) and without support to key experts / actors	USA/ International	2004	Evaluates the **social and environmental** impacts of HEIs	Good Company. Sustainability Assessment of the University of Oregon Based on Good Company's Sustainable Pathways Toolkit. Available online: https://cpfm.uoregon.edu/sites/cpfm2.uoregon.edu/files/sust_assessment_-_sust_pathways_toolkit_0.pd
11	GM	UI Green Metrics University Ranking	Universitas Indonesia	Indonesia	2010	A world university ranking system for universities to assess and compare campus efforts towards sustainability	http://greenmetric.ui.ac.id/

(continued)

(continued)

Number	Abbreviation	Full name	Developer	Country	Year	Definition	Link or Reference of the Tool
12	GMID	Graz Model for Integrative Development	Clemens Mader	Australia	2012	Evaluates the transformative potentials of sustainability processes on ESD, and thus focuses on the **interrelations between an HEI and regional stakeholders**	Mader, C. Sustainability Process Assessment on Transformative Potentials: The Graz Model for Integrative Development. J. Clean. Prod. 2013, 49, 54–63
13	GP	Green Plan and the Label DD&RS or Plan Vert	Conférence des Grandes Ecoles, Conference of University Presidents, French Government and Non-Governmental Organisations within the Grenelle Environment Round TableRoundtable	France	2010	Assist HEIs in drawing up their own sustainability plans	Alghamdi, N., den Heijer, A., & de Jonge, H. (2017). Assessment tools' indicators for sustainability in universities: an analytical overview. International Journal of Sustainability in Higher Education, 18(1), 84–115. http://doi.org/10.1108/IJSHE-04–2015-0071
14	HE 21 HEPS	Higher Education 21's sustainability Indicators or HEPS Higher Education Partnership for Sustainability	Forum for the Future	United kingdom	2001 (Latest version in 2003)	To establish a pioneering partnership group of HEIs that are seen to be achieving strategic objectives through positive engagement with the sustainable development agenda	Buckland, H., Brookes, F., Seddon, D., Johnston, A., Parkin, S., The UK Higher Education Partnership for Sustainability (HEPS) The Declaration 2001, 5 (1). Available online: https://ulsf.org/the-uk-higher-education-partnership-for-sustainability-heps/
15	PSIR	Penn State Indicator Report	Penn State Green Destiny Council to be applied at US universities,	USA, Pennsylvania State	1998 (last version 2000)	Evaluate the performance at Pennsylvania State University through the lens of sustainability	Penn State Green Destiny Council, Penn State Indicators Report 2000: Steps Toward A Sustainable University; Penn State Green Destiny Council: 2000. Available at: http://www.willamette.edu/~nboyce/assessment/PennState.pdf
16	P & P	People & Planet University League	Developed by a network of UK students	United Kingdom	2007 (Updated each year)	Ranking of United Kingdom universities by environmental and ethical performance	People & Planet University League. How Sustainable Is Your University? Available online: https://peopleandplanet.org/university-league
17	SAQ	Sustainability Assessment Questionnaire	The secretariat of the signatories of the Talloires Declaration—Association of University Leaders for a Sustainable Future (ULSF)	US	2001	A qualitative questionnaire designed to help to assess how sustainable a university's teaching, research, operations, and outreach	http://ulsf.org/sustainability-assessment-questionnaire/

(continued)

(continued)

Number	Abbreviation	Full name	Developer	Country	Year	Definition	Link or Reference of the Tool
18	SRC	College Sustainability Report Card	A North American Non-Governmental Institution—Sustainable Endowments Institute	USA/ Canada	2010 **(Suspended in 2012)**	Examines colleges and universities, as institutions, through the lens of sustainability	Sustainable Endowments Institute. The College Sustainability Report Card. Available online: http://www.greenreportcard.org/index.html
19	STARS	Sustainability Tracking, Assessment & Rating System	North American Non-Governmental Institution-Association for the Advancement of Sustainability in Higher Education (AASHE)	USA	2010 **(Last version: STARS 3-2024)**	Self-reporting framework for colleges and universities to measure their sustainability performance	https://stars.aashe.org/resources-support/technical-manual/
20	SUM	Sustainable University Model	By a researcher Luis Velazques	Mexico	2006	A structured framework for visualising and achieving a sustainable university system	Velazquez, L.; Munguia, N.; Platt, A.; Taddei, J., Sustainable university: What can be the matter? Journal of Cleaner Production 2006, 14, 810–819. 10.1016/j.jclepro.2005.12.008
21	SLS	Sustainability Leadership Scorecard	Non-Governmental Association of the United Kingdom and Ireland—The Alliance for Sustainability Leadership in Education	United Kingdom / Ireland	2016	Self-assessment developed specifically for colleges and universities to improve social responsibility and environmental performance through a whole-school approach	EUAC. Sustainability Leadership Scorecard—Launching a Transformational Tool for the Education Sector —The Sustainability Leadership Scorecard. Available online: http://www.eauc.org.uk/sustainability_leadership_scorecard
22	PSAT	Program Sustainability Assessment Tool	North American university—Washington University	USA	**Revised version: 2023 (V3)**	SELF-Assessment tool to to evaluate the sustainability capacity of a program (a small set of organisational and contextual domains that can help build the capacity for maintaining a program)	https://sustaintool.org/psat/
23	THE Impact Ranking	Times Higher Education Impact University Ranking	THE Impact Ranking team	International	2019 **Last version: 2024**	Assess universities against the United Nations' Sustainable Development Goals (SDGs)	https://www.timeshighereducation.com/how-participate-times-higher-education-rankings
24	TUR	Three Dimensional University Ranking	Developed by researchers in Europe	International	2009	Comparison between universities regarding their research, educational and environmental performances	Lukman, R.; Krajnc, D.; Glavič, P., University ranking using research, educational and environmental indicators. Journal of Cleaner Production 2010, 18, 619–628. https://doi.org/10.1016/j.jclepro.2009.09.015

(continued)

(continued)

Number	Abbreviation	Full name	Developer	Country	Year	Definition	Link or Reference of the Tool
25	UEMS	University Environmental Management System	By Alshuwaikhat and Abubakar (researchers in Saudi Arabia)	Saudi Arabia	2008	To achieve campus sustainability through overcoming the lack of environmental management practices for HEIs	Alshuwaikhat, H. M.; Abubakar, I., An integrated approach to achieving campus sustainability: assessment of the current campus environmental management practices. Journal of Cleaner Production 2008, 16 (16), 1777–1785. https://doi.org/10.1016/j.jclepro.2007.12.002
26	USAT	Unit-Based Sustainability Assessment tool	By a Ph.D. research at Rhodes University, South Africa (supported by the United Nations Environmental Programme (UNEP) and Mainstreaming Environment and Sustainability into African Universities)	Africa	2009	Not only to be a guide for educating and aiding universities towards sustainability but also to be a flexible tool used at the departmental, faculty and unit level	Togo, M., & Lotz-Sisitka, H. Unit Based Sustainability Assessment Tool. A resource book to complement the UNEP Mainstreaming Environment and Sustainability in African Universities Partnership; Howick, 2009. https://www.ru.ac.za/media/rhodesuniversity/content/elrc/documents/USAT_Tool.pdf
27	uD-SiM model	Uncertainty-based quantitative assessment of sustainability for HEIs	Developed by researchers	Canada	2011	Sustainability assessment tool using a driving force-pressure-state-exposure-effect-action (DPSEEA) framework to achieve a causality-based impact assessment	Waheed, B.; Khan, F.; Veitch, B.; Hawboldt, K., Uncertainty-based quantitative assessment of sustainability for higher education institutions. Journal of Cleaner Production 2011, 19, 720–732. 10.1016/j.jclepro.2010.12.013
28	SET4HEI	Sustainability Evaluation Tool for Higher Education Institutions	UNESCO International Institute for Higher Education in collaboration with other UN Agencies and higher education experts across the world	International	2024	The goal of this platform is to provide higher education institutions (HEIs) across the world with a self-assessment tool allowing them to map their contributions to the Sustainable Development Goals (SDGs) across time and identify areas for improvement	https://set4hei.org/

References

Aleixo AM, Leal S, Azeiteiro UM (2018) Conceptualization of sustainable higher education institutions, roles, barriers, and challenges for sustainability: an exploratory study in Portugal. J Clean Prod 172:1664–1673. https://doi.org/10.1016/j.jclepro.2016.11.010

Amaral LP, Martins N, Gouveia JB (2015) Quest for a sustainable university: a review. Int J Sustain High Educ 16(2):155–172. https://doi.org/10.1108/IJSHE-02-2013-0017

Amey L, Plummer R, Pickering G (2020) Website communications for campus sustainability: an analysis of Canadian universities. Int J Sustain High Educ 21(3):531–556. https://doi.org/10.1108/IJSHE-04-2019-0137

Aramburuzabala P, Cerrillo R (2023) Service-learning as an approach to educating for sustainable development. Sustainability 15(14):11231. https://doi.org/10.3390/su151411231

Basheer N, Ahmed V, Bahroun Z, Anane C (2024) Exploring sustainability assessment practices in higher education: a comprehensive review through content and bibliometric analyses. Sustainability 16(13). https://doi.org/10.3390/su16135799

Bashir H, Araci ZC, Obaideen K, Alsyouf I (2023) An approach for analyzing and visualizing the relationships among key performance indicators for creating sustainable campuses in higher education institutions. Environ Sustain Indic 19:100267. https://doi.org/10.1016/j.indic.2023.100267

Bauler T (2012) An analytical framework to discuss the usability of (environmental) indicators for policy. Ecol Ind 17:38–45. https://doi.org/10.1016/j.ecolind.2011.05.013

Berzosa A, Bernaldo MO, Fernández-Sanchez G (2017) Sustainability assessment tools for higher education: an empirical comparative analysis. J Clean Prod 161:812–820. https://doi.org/10.1016/j.jclepro.2017.05.194

Buerkle A, O'Dell A, Matharu H, Buerkle L, Ferreira P (2023) Recommendations to align higher education teaching with the UN sustainability goals—a scoping survey. Int J Educ Res Open 5:100280. https://doi.org/10.1016/j.ijedro.2023.100280

Bullock G, Wilder N (2016) The comprehensiveness of competing higher education sustainability assessments. Int J Sustain High Educ 17(3):282–304. https://doi.org/10.1108/IJSHE-05-2014-0078

Caeiro S, Angélica L, Hamón S, Martins R, Elizabeth C, Aldaz B (2020) Sustainability assessment and benchmarking in higher education institutions—a critical reflection. Sustainability 12(2):543. https://doi.org/10.3390/su12020543

Décamps A, Barbat G, Carteron J-C, Hands V, Parkes C (2017) Sulitest: a collaborative initiative to support and assess sustainability literacy in higher education. Int J Manag Educ 15(2):138–152. https://doi.org/10.1016/j.ijme.2017.02.006

Direção-Geral do Ensino Superior [General directorate of higher education] (DGESa, n.d.) Ensino Superior em números [Higher education in numbers]. Available at: https://www.dges.gov.pt/pt/pagina/ensino-superior-em-numeros. Consulted on September of 2024

Direção-Geral do Ensino Superior [General directorate of higher education] (DGESb n.d.) Acesso ao Ensino Superior - Índices de Cursos (por distrito e instituição) [Access to higher education—course indexes (by district and institution)]. Available at: https://www.dges.gov.pt/guias/inddist.asp. Consulted on September of 2024

Disterheft A, Caeiro SS, Leal Filho W, Azeiteiro UM (2016) The INDICARE-model—measuring and caring about participation in higher education's sustainability assessment. Ecol Ind 63:172–186. https://doi.org/10.1016/j.ecolind.2015.11.057

Duarte M, Caeiro SS, Farinha CS, Moreira A, Santos-Reis M, Rigueiro C, Simão J (2023) Integration of sustainability in the curricula of public higher education institutions in Portugal: do strategic plans and self-report align? Int J Sustain High Educ 24(9):299–317. https://doi.org/10.1108/IJSHE-01-2023-0001

Farinha CS, Caeiro SS, Azeiteiro U (2020) Universities speak up regarding the implementation of sustainable development challenges. Int J Sustain High Educ 21(3):465–506. https://doi.org/10.1108/IJSHE-08-2019-0250

Findler F, Schönherr N, Lozano R, Stacherl B (2019) Assessing the impacts of higher education institutions on sustainable development—an analysis of tools and indicators. Sustainability 11(1). https://doi.org/10.3390/su11010059

Fischer D, Jenssen S, Tappeser V (2015) Getting an empirical hold of the sustainable university: a comparative analysis of evaluation frameworks across contemporary sustainability assessment tools. Assess Eval High Educ 40(6):785–800. https://doi.org/10.1080/02602938.2015.1043234

Gutiérrez-Mijares ME, Josa I, Casanovas-Rubio MM, Aguado A (2023) Methods for assessing sustainability performance at higher education institutions: a review. Stud High Educ 48(8):1137–1158. https://doi.org/10.1080/03075079.2023.2185774

Husaini MZ, Jusoh A, Kassim MS (2018) Analysis of sustainability assessment tools (SATs) for higher education institutions (HEIs). Int J Acad Res Bus Soc Sci 8(8):558–571. https://doi.org/10.6007/IJARBSS/v8-i8/4612

Leal S, Azeiteiro UM, Aleixo AM (2024) Sustainable development in Portuguese higher education institutions from the faculty perspective. J Clean Prod 434:139863. https://doi.org/10.1016/j.jclepro.2023.139863

Leal Filho W, Amaro N, Avila LV, Brandli L, Damke LI, Vasconcelos CRP, Hernandez-Diaz PM, Frankenberger F, Fritzen B, Velazquez L, Salvia A (2021) Mapping sustainability initiatives in higher education institutions in Latin America. J Clean Prod 315:128093. https://doi.org/10.1016/j.jclepro.2021.128093

Leal Filho W, Brandli LL, Dinis MAP, Vidal DG, Paço A, Levesque V, Salvia AL, Kozlova V, Ávila LV, Fritzen B, Abubakar IR, Pace P (2023) International trends on transformative learning for urban sustainability. Discov Sustain 4(1):31. https://doi.org/10.1007/s43621-023-00145-7

Leal Filho W, Trevisan LV, Wahaj Z, Deda D, Vasconcelos CRP, Dibbern TA, Vargas VR, Iyer-Raniga U, Anholon R, Novikau A, Abubakar IR, Dinis MAP (2024a) University rankings and sustainable development: the state of the art. Int J Sustain High Educ. ahead-of-print. https://doi.org/10.1108/IJSHE-11-2023-0530

Leal Filho W, Fritzen B, Salvia AL, Dinis MAP, Vasconcelos CRP (2024b) The transformative power of networking in the implementation of the sustainable development goals. Discov Sustain 5(1):1–14, Article 380. https://doi.org/10.1007/s43621-024-00591-x

Lozano R, Ceulemans K, Alonso-Almeida M, Huisingh D, Lozano FJ, Waas T, Lambrechts W, Lukman R, Hugé J (2015) A review of commitment and implementation of sustainable development in higher education: results from a worldwide survey. J Clean Prod 108:1–18. https://doi.org/10.1016/j.jclepro.2014.09.048

Lozano R, Barreiro-Gen M (2019) Analysing the factors affecting the incorporation of sustainable development into European Higher Education Institutions' curricula. Sustain Dev 27(6):965–975. https://doi.org/10.1002/sd.1987

Machado CF, Davim JP (2023) Sustainability in the modernization of higher education: curricular transformation and sustainable campus—a literature review. Sustainability 15(11). https://doi.org/10.3390/su15118615

Madeira ACFD (2008) Indicadores de Sustentabilidade para Instituições de Ensino Superior [Mestrado, UNIVERSIDADE DO PORTO]. https://hdl.handle.net/10216/12228

Mair S, Druckman A (2023) Assessing the suitability of sustainability frameworks for embedding sustainability in higher education curricula: pragmatism versus transformation. Int J Sustain High Educ 24(9):318–334. https://doi.org/10.1108/IJSHE-08-2020-0315

Mapar M, Bacelar-Nicolau P, Caeiro S (2022) Sustainability assessment tools in higher education institutions. In: KAA Gamage N Gunawardhana (eds) The Wiley Handbook of sustainability in higher education learning and teaching, pp 153–186. https://doi.org/10.1002/9781119852858.ch8

Mendoza JMF, Gallego-Schmid A, Azapagic A (2019) A methodological framework for the implementation of circular economy thinking in higher education institutions: towards sustainable campus management. J Clean Prod 226:831–844. https://doi.org/10.1016/j.jclepro.2019.04.060

Moggi S (2023) Sustainability reporting, universities and global reporting initiative applicability: a still open issue. Sustain Account, Manag Policy J 14(4):699–742. https://doi.org/10.1108/SAMPJ-05-2022-0257

Nikolaou II, Tsalis TA, Trevlopoulos NS, Mathea A, Avlogiaris G, Vatalis KI (2023) Exploring the sustainable reporting practices of universities in relation to the United Nations' 2030 Agenda for sustainable development. Discov Sustain. https://doi.org/10.1007/s43621-023-00167-1

Ramos TB (2019) Sustainability assessment: exploring the frontiers and paradigms of indicator approaches. Sustainability 11(3). https://doi.org/10.3390/su11030824

Rodrigues SAM (2023) Indicadores-Chave de Sustentabilidade e das Instituições de Ensino Superior em Portugal - referencial Comum para a avaliação do desempenho [Mestrado, Coimbra]. https://hdl.handle.net/10316/109583

Sampieri RH, Lucio MPB, Collado CF (2006) Metodologia de Pesquisa [Research Methodology]. McGraw-Hill, São Paulo

UNEP (2021) The UNEP sustainability university framework. United Nations Environmental Program. Available at https://www.unep.org/resources/toolkits-manuals-and-guides/uneps-sustainable-university-framework

UNESCO (2020) Education for sustainable development—a roadmap. UNESCO Paris. Available at: https://unesdoc.unesco.org/ark:/48223/pf0000374802

Vargas VR, Lawthom R, Prowse A, Randles S, Tzoulas K (2019) Sustainable development stakeholder networks for organisational change in higher education institutions: a case study from the UK. J Cleaner Prod 208:470–478. https://doi.org/10.1016/j.jclepro.2018.10.078

Varouchas E, Sicilia M-Á, Sánchez-Alonso S (2018) Academics' perceptions on quality in higher education shaping key performance indicators. Sustainability 10(12). https://doi.org/10.3390/su10124752

Veidemane A (2022) Education for sustainable development in higher education rankings: challenges and opportunities for developing internationally comparable indicators. Sustainability 14(9). https://doi.org/10.3390/su14095102

MIX
Papier aus verantwortungsvollen Quellen
Paper from responsible sources
FSC® C105338

If you have any concerns about our products,
you can contact us on
ProductSafety@springernature.com

In case Publisher is established outside the EU,
the EU authorized representative is:
Springer Nature Customer Service Center GmbH
Europaplatz 3, 69115 Heidelberg, Germany

Printed by Libri Plureos GmbH
in Hamburg, Germany